Advanced Machining Science

As machining processes become more advanced, so does the science behind them. This book emphasizes these scientific developments, in addition to the more widely covered technological aspects, providing a full understanding of how machining has adapted to material constraints and moved beyond conventional methods in recent years.

Numerous machining processes have been developed to allow the use of increasingly tough, corrosion-resistant, and temperature-resistant materials for different applications. The advanced machining processes covered in this book range from mechanical, thermoelectric, and electrochemical, including abrasive water jet machining, electric discharge machining and micromachining, ion beam machining, and hybrid processes. It also addresses the sustainability issues raised by these processes. The underlying science of machining is centered throughout, as none of these processes can reach their full potential without both technical expertise and scientific understanding.

Advanced Machining Science and its scientific approach will be of particular interest to students, researchers, and shop floor engineers.

Advanced Machining Science

Edited by
V. K. Jain

CRC Press
Taylor & Francis Group
Boca Raton London New York

CRC Press is an imprint of the
Taylor & Francis Group, an **informa** business

First edition published 2023
by CRC Press
6000 Broken Sound Parkway NW, Suite 300, Boca Raton, FL 33487-2742

and by CRC Press
4 Park Square, Milton Park, Abingdon, Oxon, OX14 4RN

CRC Press is an imprint of Taylor & Francis Group, LLC

ISBN: 978-1-482-21109-2 (hbk)
ISBN: 978-1-032-32824-9 (pbk)
ISBN: 978-0-429-16001-1 (ebk)

DOI: 10.1201/9780429160011

Typeset in Times
by Deanta Global Publishing Services, Chennai, India

Dedicated to those who believe

- *Failure is a step towards success. The Lazarenko brothers developed the electrospark machining (nowadays, known as EDM) on their 42nd attempt.*

- *To fulfill your ambitious goals: Be honest, sincere, hardworking and have the Junoon (passion).*

- *There is no substitute for hard work.*

Contents

Preface

On the shop floor, conventional/traditional machining operations (say, milling, turning, drilling, etc.) are usually performed to shape e and size e different kinds of materials. With the advancements in material science, very hard, corrosion-resistant, very tough, and high temperature-resistant materials are being developed; it is very difficult to machine these materials with conventional machining methods due to constraints in the tool material's properties. Hence, *Advanced Machining Processes (AMPs)* have been developed where energy in its direct form is used to remove material. For example, laser beam machining, electrochemical machining, abrasive jet machining, etc.

There are many books on *AMPs* (also called non-traditional machining processes, Modern Machining Processes, Advanced Machining Methods, and New Technology), but the majority of them are mainly technology centered. In my opinion, readers (students, teachers, researchers) need to know sufficient mathematical/scientific analysis of these different types of AMPs. Without a scientific basis, process performance cannot be properly understood and it cannot be improved beyond a limit. Hence, teachers, students, researchers, and shop floor engineers would like to adopt this book to enhance their understanding of AMPs. Therefore, the chapters in this book start with the necessary description of the process and the technology involved, and then they take the reader to the science involved in the material removal mechanism of each process. After discussing the mathematical/scientific analysis of these processes, it details the parametric analysis, applications, and in some cases, the areas in which further research can be carried out.

This book deals with the commonly used *AMPs* in general, with specific emphasis on the science behind each of these processes. These processes can be classified as *Mechanical Advanced Machining Processes* (USM, AWJM, etc.), *Thermoelectric Advanced Machining Processes* (EDM, LBM, PAM, etc.), and *Electrochemical and Chemical Advanced Machining Processes (ECM, ChM)* including *Biochemical Machining*. Some of these processes have been discussed along with miscellaneous chapters, namely, molecular dynamic simulation (MDS) and sustainability issues of AMPs. Some chapters have a section related to the process applications to micro/nanomachining. It also deals with how to scale down the processes for micromachining by modifying the input parameters.

I would like to thank the authors of different chapters who have really worked hard in writing their chapters during the very tough pandemic conditions throughout the world. Thanks are definitely due to Prof. A. Y. C. Nee who has written the Foreword to this book. I would also like to thank the CRC press team in general, and Cindy in particular, who has helped me at different stages of editing this book.

V. K. Jain (Retired Professor)
IIT Kanpur (India)

Foreword

Prof. Jain and his colleagues are to be congratulated for producing this excellent comprehensive book focusing on the most commonly used unconventional machining processes.

Unconventional machining processes have a very important role to play in the handling of hard, difficult-to-machine components requiring high precision and good finish. This continues to be a dominant technology despite the rise of additive manufacturing processes which generally lack speed, surface finish, and dimensional accuracy.

The chapters cover electrical, mechanical, thermal, chemical, and a hybrid of these processes in removing materials to achieve the final specifications desired by the product designers.

Simulation of advanced machining processes would provide the necessary insight into the physics and mechanism, as well as the basic understanding of these processes. Sustainability issues address the environmental concerns and curb the effects of climate change. The last two chapters complement the other chapters well.

This book will be invaluable to undergraduates, researchers, and industrial organizations, for learning, consulting, and applying some of the details deliberated by a team of outstanding researchers.

It is my great pleasure to write this Foreword and renew my long association with Prof. Jain.

Andrew Y. C. Nee
Editor-in-Chief, International Journal of
Advanced Manufacturing Technology (Springer)
Executive Editor-in-Chief Advances in Manufacturing (Springer)
January 5, 2022

About the Editor

Vijay K. Jain, PhD, earned his BE from Vikram University, Ujjain (MACT, Bhopal), and ME and PhD from University of Roorkee (now, IIT Roorkee). He has 47 years of teaching and research experience. He has served as a Visiting Professor at the University of California at Berkeley (USA) and University of Nebraska at Lincoln (USA). He retired as a Professor from the Indian Institute of Technology Kanpur (India) in 2016. He has also served as a faculty member at other Indian institutions, namely, MREC Jaipur, BITS Pilani, MNREC Allahabad, and MACT Bhopal.

Dr. Jain has won three gold medals, two silver medals, and one best paper award in recognition of his research work. The Institution of Engineers (India), Khosla Research Awards committee, and AIMTDR conference-organizing committee have given this honor to him. He has been awarded a Lifetime Achievement Award by the AIMTDR (NAC). He has written more than nine books (including edited books) and 16+ chapters for different books published by different publishers.

Dr. Jain has been appointed as Editor-in-Chief for three international journals (IJPTech, JAMS, JMF) and Associate Editor of two international journals (JEM and IJMST). He has also served as a Guest Editor for 20+ special issues of different international journals and as a member of the editorial board of 12+ international journals. Dr. Jain has guided 20 PhD, 96+ MTech and ME theses, and many BTech/BE Projects. He has around 400+ research publications to his credit. He has 13+ Indian patents and 1 U.S. patent to his credit. Dr. Jain has organized 24+ summer/winter/short-term schools on various topics such as Micromanufacturing, Micromachining, Advanced Machining, Precision Engineering, CIM Systems, Design of Machine Tools, NC Machine Tools, Advanced Manufacturing Technology, Tools and Die Making, etc.

Contributors

Julfekar Arab
Indian Institute of Technology Bombay
Mumbai, India

Gokhan Aydin
Karadeniz Technical University
Trabzon, Turkey

Palivela Bhargav Chandan
Indian Institute of Technology Tirupati
Tirupati, India

Manas Das
Indian Institute of Technology
Guwahati
Guwahati, India

Shivansh Dhaka
Indian Institute of Information
Technology Kancheepuram
Chennai, India

Pradeep Dixit
Indian Institute of Technology Bombay
Mumbai, India

Jian Gao
University of Strathclyde
Glasgow, UK

Saurav Goel
London South Bank University
London, UK

Xiaoguang Guo
Dalian University of Technology
Dalian, P. R. China

Vijay Kumar Jain (Retired)
Indian Institute of Technology Kanpur
Kanpur, India

Bhaveshkumar Kamaliya
Indian Institute of Technology Bombay
Mumbai, India

Izzet Karakurt
Karadeniz Technical University
Trabzon, Turkey

Manjesh Kumar
Indian Institute of Technology
Guwahati
Guwahati, India

Kishor Kumar Gajrani
Indian Institute of Information
Technology Kancheepuram
Chennai, India

Dileep Kumar Mishra
Indian Institute of Technology Bombay
Mumbai, India

Ranajit Mahanti
Indian Institute of Technology
Guwahati
Guwahati, India

Aluri Manoj
Rajiv Gandhi University of Knowledge
Technologies
Basar, India

Deepak Marla
Indian Institute of Technology Bombay
Mumbai, India

Rakesh G. Mote
Indian Institute of Technology Bombay
Mumbai, India

Satish Mullya
Annasaheb Dange College of
  Engineering and Technology
Ashta, India

Hari Narayan Singh Yadav
Indian Institute of Technology
  Guwahati
Guwahati, India

Jinming Lu
TechnicalUniversityofMunich
Garching, Germany

Xichun Luo
University of Strathclyde
Glasgow, UK

Divyansh Patel
Birla Institute of Technology and
  Science
Pilani, India.

J. Ramkumar
Indian Institute of Technology Kanpur
Kanpur, India

Mamilla Ravi Sankar
Indian Institute of Technology Tirupati
Tirupati, India

Tribeni Roy
Birla Institute of Technology and Science
Pilani, India

and

London South Bank University
London, UK

Vyom Sharma
Indian Institute of Technology Kanpur
Kanpur, India

Mahavir Singh
Indian Institute of Technology Kanpur
Kanpur, India

EwaldA.Werner
TechnicalUniversityofMunich
Garching, Germany

Saeed Zare Chavoshi
Imperial College London
London, UK

1 Advanced Machining Processes and Operations

V. K. Jain

CONTENTS

1.1 INTRODUCTION

Machining plays an important role in manufacturing. Even in cases of precision casting, forming, and molding processes, machining operation as post processing often becomes inevitable. Machining processes can be basically classified into two primary classes: Traditional (or Conventional) Machining Processes (TMPs) and Advanced Machining Processes (AMPs) [1]. Whenever a single process (traditional or advanced) cannot satisfactorily perform the desired function, two or more than two processes (TMPs and/or AMPs) are combined to take advantage of the merits of the individual process, and to minimize their unwanted effects. These combined processes are known as *"hybrid processes"* [2, 3].The classification of these processes is given in Figure 1.1. Usually, an individual AMP can perform certain operation(s) but not necessarily all types of operations. In this chapter, an attempt has been made to examine the capabilities of the AMPs with reference to the kinds of operations they can perform, and how they can perform different kinds of operations. To understand the applications of these processes in different kinds of operations, the working principles of these processes are discussed first. It is important to note that there are different types of operations that can be performed by one type of process, say, electrochemical machining, electrochemical drilling, or electrochemical milling, but the principle of electrochemical machining will remain the same (i.e., Faraday's laws of electrolysis) in different types of operations. The scientific analyses of different types of AMPs are discussed in different chapters by the experts in the individual

DOI: 10.1201/9780429160011-1

1

```
┌─────────────────────────────────────────┐
│    Classification of Machining Processes  │
└─────────────────────────────────────────┘
```

┌──────────────────────┐   → Turning
│ Traditional Machining │   → Milling
│   Operations   │   → Drilling
└──────────────────────┘   → Shaping

┌──────────────────────┐   → Mechanical AMPs: USM, AJM, WJM, AWJM
│ Advanced Machining   │   → Thermal AMPs: EDM, LBM, EBM, PAM
│   Processes   │   → Electrochemical & Chemical AMPs: ECM, CHM, BioChM
└──────────────────────┘

┌──────────────────────┐   → ECG, EDG, ECSM, etc.
│   Hybrid Processes   │
└──────────────────────┘

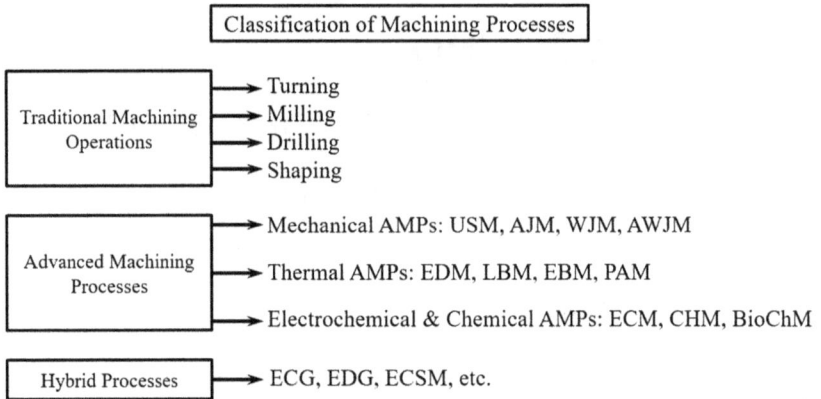

FIGURE 1.1 Classification of machining processes. Here, the abbreviations used are as follows –EDM: Electric discharge machining, LBM: Laser beam machining, EBM: Electron beam machining, PAM: Plasma arc machining, ECM: Electrochemical machining, ChM: Chemical machining, BioChM: Biochemical machining, ECG: Electrochemical grinding, EDG: Electric discharge grinding, and ECSM: Electrochemical spark machining.

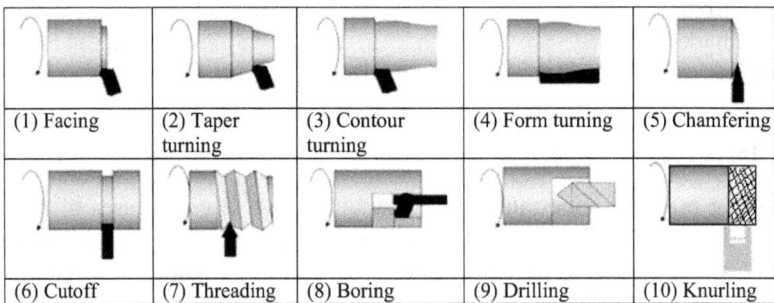

(1) Facing	(2) Taper turning	(3) Contour turning	(4) Form turning	(5) Chamfering
(6) Cutoff	(7) Threading	(8) Boring	(9) Drilling	(10) Knurling

FIGURE 1.2 Different kinds of operations performed using traditional machining process on a lathe machine.

processes, hence they are not elaborated in this chapter. Here, the emphasis is on the fact that different types of operations can be performed by one type of process, but the working principle remains the same.

All three types of AMPs (mechanical, thermal, and electrochemical and chemical) can be suitably employed for performing different kinds of operations. However, the accuracy, tolerances, surface finish, and surface integrity will depend on the type of process, selected process parameters, and the type of workpiece material being used. Different types of the AMPs with certain constraints can be employed to perform different kinds of machining operations. It is possible by providing different combinations of relative motion between the tool and workpiece. As can be seen in Figure 1.2, different kinds of operations like turning, drilling, cutting, threading, milling, countersinking, boring, etc. can be performed by using one type of material removal process, that is a conventional machining process in this case. It is important to note that the mechanism of material removal remains the same in different kinds of operations while using one particular material removal process.

Figure 1.3 shows a classification of machining operations which can be macro- or micro-machining (at least one dimension of the feature is greater than 1 μm but less than 1 mm). It can have a single operation like threading, texturing, deburring, drilling, etc. being performed by a single machining process, namely electrochemical machining, laser beam machining, or a traditional metal cutting process. However, in some cases it is necessary to perform two different operations simultaneously (multiple operations or a hybrid operation), namely machining and finishing: Electrochemical Machining and Finishing (ECMAF).

In AMPs, the mechanism of material removal is quite different from the traditional machining process. In the latter, the material removal takes place by physical interaction between the tool and workpiece, and a prerequisite is that the tool material should be much harder than the workpiece material. This imposes a basic constraint on the applications of TMPs. In the case of AMPs, different types of processes have different energy sources which are responsible for the removal of material from the workpiece in a particular manner. Usually, performance of AMPs does not depend on the mechanical properties of the workpiece material (except in mechanical AMPs); rather than being based on the process being used, it may depend on electrical properties, magnetic properties, chemical properties, optical properties, or some other material properties. In the following sections, after discussing the working principle of a class of processes, a brief review of different AMPs with reference to the applications of an individual process for performing different kinds of operations is presented. This will help engineers working on the shop floor or scientists in the R&D labs to utilize a process for different kinds of operations with

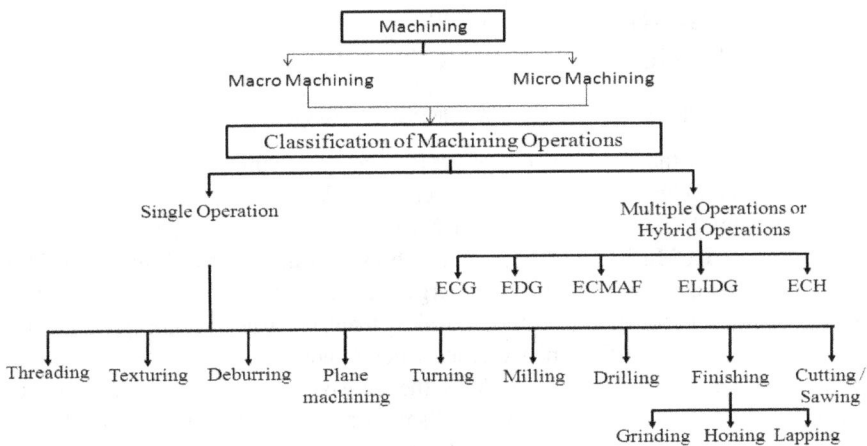

FIGURE 1.3 Fundamental operations and hybrid processes. Here, ECG: Electrochemical grinding, EDG: Electric discharge grinding, and ECMAF: Electrochemical magnetic abrasive finishing, ELIDG: Electrolytic in process dressing and grinding, and ECH: Electrochemical honing.

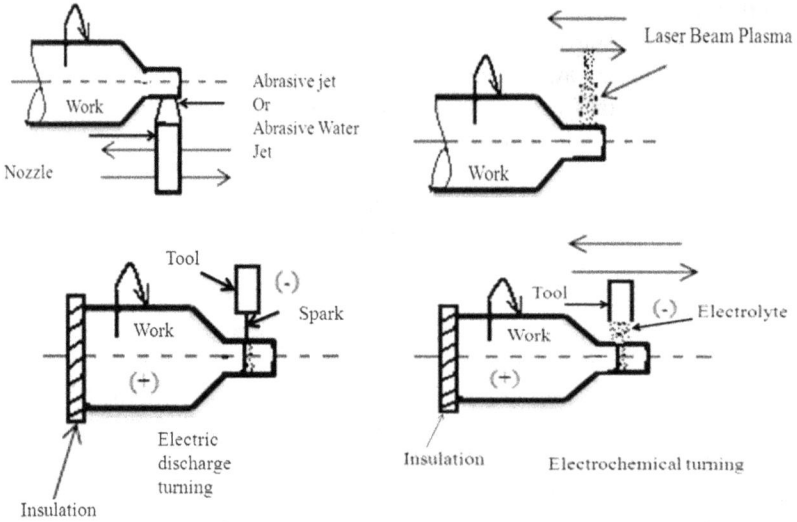

FIGURE 1.4 Relative motion between energy source and workpiece to perform turning operation.

minor changes in the tooling/relative motion. It will also help researchers to develop a process from the point of view of different kinds of operations. Figure 1.4 shows a turning operation being performed using different kinds of AMPs, namely, abrasive jet machining or abrasive water jet machining, laser beam machining, electric discharge machining, and electrochemical machining.

1.2 CLASSIFICATION AND WORKING PRINCIPLES OF ADVANCED MACHINING PROCESSES

1.2.1 MECHANICAL ADVANCED MACHINING PROCESSES

Figure 1.1 shows the classification of AMPs, viz. mechanical, thermal, and electrochemical and chemical AMPs [1]. Mechanical Advanced Machining Processes (MAMPs) mainly include Abrasive Jet Machining (AJM), Ultrasonic Machining (USM), Water Jet Machining (WJM), and Abrasive Water Jet Machining (AWJM). In all these processes, it is the kinetic energy (KE) of water or (water + abrasive particles) which is responsible for the removal of material from the workpiece. In the case of WJM, it is the KE of the water jet which removes/cuts the material from the workpiece [1, 4, 5]. While using MAMPs, the abrasive particle at very high velocity, say, 800–900 m/s in the case of AWJM, hits the workpiece surface [6]. While hitting the workpiece surface, the direction of this high KE abrasive particle may be at 90° to the workpiece surface or at another angle θ ($\theta < 90°$). The angle of the abrasive particle at which it hits the workpiece surface and the properties of the workpiece material (brittle or ductile) decide the mechanism or the way the material is removed from the workpiece. Let us consider an example of brittle workpiece material and

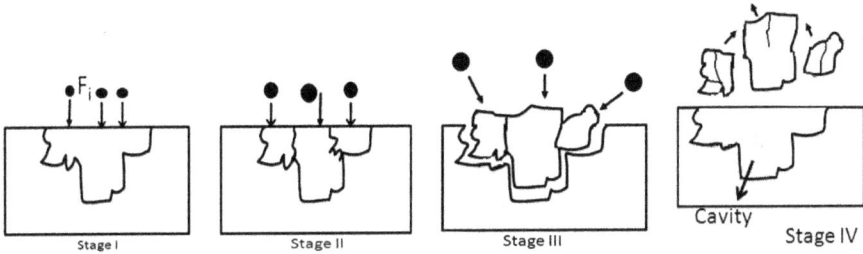

FIGURE 1.5 Stage (I) abrasive particles hitting the workpiece surface, (II) fractured surface, (III) further fracture and separation, (IV) separation of the fractured pieces from the parent material.

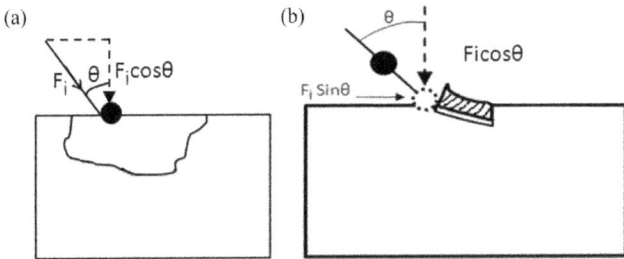

FIGURE 1.6 (a) Abrasive particle hitting the workpiece surface at an angle other than 90°. (b) Formation of micro/nanochip.

abrasive particles hitting the workpiece surface at 90° (Figure 1.5 (Stage I)). In this case, the mechanism of material removal is brittle fracture (Figure 1.5). The fractured layer, when hit subsequently by the abrasive particles, they further break/fracture the material into smaller pieces (Figure 1.5 (Stage II) and (Stage III)) and de-bond (Figure 1.5 (Stage IV)) it from the parent material to separate out in the form of micro-/nano-particles (that are chips).

When the workpiece (brittle material) is hit by an abrasive particle (Figure 1.6) at an angle other than 90°, then the force component ($Fi.Cos\theta$) will be responsible for causing the penetration/fracture and ($Fi.Sin\theta$) will be responsible for the separation of the fractured part or micro-/nanochips from the workpiece. Under these conditions, if the workpiece material is ductile, even then the same two mechanisms of material removal work, that is brittle fracture when the jet angle is 90°, and shear deformation when the angle θ is less than 90°.

When the indentation is in the range of a few nanometer, even brittle material behaves as a ductile material, and it may lead to the removal of very small amount of material by plastic deformation, giving rise to nanosize chips. This would happen when the penetration is in the range of ductile regime deformation of the brittle material [7]. When the workpiece material is ductile then the possible mechanism of material removal is usually plastic deformation. This mechanism of material removal is more prominent when the impact angle $\theta < 90°$ or $\theta > 90°$. However, in the case of impact angle, $\theta = 90°$, the material is being hit repeatedly, and strain hardening of the ductile material takes place. As a result, material removal takes place by brittle fracture and some part may get separated out by shearing of the ploughed material also. The above mechanisms are applicable for the following processes, AJM, USM, and AWJM [8, 9]. However, WJM [10, 11] is usually applicable for comparatively softer materials such as wood, asbestos, polymer composite, and softer metals like aluminum. It is the KE of the water jet which is responsible for shearing/cutting the workpiece material [12].

1.2.2 THERMAL ADVANCED MACHINING PROCESSES (TAMPs)

Thermal Advanced Machining Processes (TAMPs) include Electric Discharge Machining (EDM), Laser Beam Machining (LBM), Electron Beam Machining (EBM), and Plasma Arc Machining (PAM). The mechanism of material removal in all these processes remains almost the same [1, 4, 5, 11]. In these processes, high-intensity heat is generated in the localized area. Usually, the intensity of heat is so high that most of the known workpiece materials are melted, and part of the molten material also gets vaporized.

As shown in Figure 1.7, because of high heat intensity in the localized area, part of the parent material gets melted out and a small part gets vaporized depending upon the temperature of the molten material in the melt pool, the environment surrounding the molten pool (liquid, gas, or vacuum), and the properties of the workpiece material. Part of the molten material is flushed out by the liquid or gas jet and carried away from the melt pool. The remaining molten part in the cavity, on the sides of the cavity, and on the un-machined part in the lateral direction at the top gets resolidified, and it is known as the *recast layer*. The size (thickness) of the recast layer depends on many factors. The resolidified material sometimes obtained is in the form of globules which also indicate that the mechanism of material removal is melting and/or vaporization [13, 14].

The parts machined by any one of the TAMPs, one or more defects are quite commonly observed. These defects include microcrack, recast layer, Heat Affected Zone (HAZ), thermal residual stresses, etc., depending upon the magnitude of the energy input, and the thermal properties of the workpiece material [3–5]. To some extent, the occurrence of these defects also depends on the type of the heat source generated during the process (Gaussian, Uniform, Triangular, or Point heat source) [15, 16]. In the EDM process, only electrically conducting materials can be machined, while

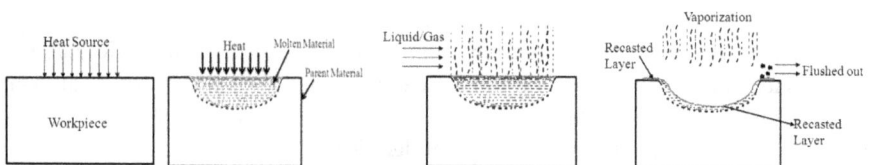

FIGURE 1.7 Different stages involved in the mechanism of material removal in Thermal Advanced Machining Processes.

other TAMPs (LBM, EBM, PAM) can be applied to both electrically conducting as well as electrically non-conductive materials [17–20].

In different TAMPs, the sources of heat generation are of different types. For example, in EDM, the heat is generated by a sparking phenomenon in the dielectric flowing in the Interelectrode Gap (IEG) (between the tool (cathode) and workpiece (anode)). In LBM, when a laser beam hits the workpiece surface it produces such a high intensity of heat that it can melt and even vaporize any material (excluding short pulse lasers such as femtosecond and picosecond lasers). With the help of the optics, the laser beam diameter can be increased or decreased, which also changes heat intensity on the laser beam spot. It is possible to have a laser beam in continuous mode or pulsed mode (nanosecond, picosecond, and femtosecond laser pulses). However, it should be noted that the machining efficiency of the laser beam is extremely low (less than 5% or so). Laser beams can be produced by solid, liquid, or gas as lasing materials [17–20].

With EBM, a beam of electrons at a velocity close to the velocity of light, hits the workpiece surface. The electrons attain very high KE and when they hit the workpiece surface, they are able to produce very high temperature that can melt and even vaporize the workpiece material. For better accuracy and efficiency, a vacuum should be maintained in the machining zone. However, the equipment limits the size of the workpiece that can be machined [21–23].

In PAM, an arc is created between the two electrodes. The anode is the nozzle when electrically non-conducting materials are to be cut, otherwise the workpiece is the anode so that electrically conducting materials can be processed/machined. The latter case is more efficient than the former [1, 4]. The process is good for cutting very thick workpieces but usually the accuracy obtained is comparatively poor.

In beam machining technology, the processes discussed above come under the category of TAMPs. They are good for macro- and micromachining but in general they are not used for nanomachining (or, more correctly for the creation of nanosize features). There is another beam technology process called Ion Beam Machining (IBM) in which material is removed practically atom by atom by a beam of ions. However, in my opinion, it does not fall in to the category of TAMPs because here the mechanism of material removal is not melting and/or vaporizing, but rather it is a kind of mechanical energy (force) of the ions which separates out an atom from the surrounding atoms when the energy of the hitting ion is more than the bonding energy of the atom [24–27].

Focused Ion Beam (FIB) machining is the process in which material is removed atom by atom. This is the only process which can machine at the nano and sub-nano levels. Here, theoretically, a beam of ions removes the material from the top surface of the workpiece atom by atom. Hence, this is the slowest (max. 2–3 µm/hr thickness of the substrate) material removal process. This process is normally recommended for *nanomachining*. Depending upon the type of workpiece material, material removal rate (MRR) depends on the energy of the ion beam (or individual ion). Here, if the KE of an ion is greater than the bonding energy of the atom, it separates out the atom from the workpiece material, and different shapes and sizes of the features can be created but at the nanolevel [26, 27].

1.2.3 Chemical, Electrochemical, and Biochemical Machining Processes

In the third category of AMPs, processes such as Chemical Machining (ChM), Electrochemical Machining (ECM), and biochemical machining are included. In ChM, an appropriate etchant(s) is (are) chosen which can chemically react with the workpiece material and remove the material in the form of reaction product(s). In the case of an alloy, it can form more than one type of reaction product depending upon the elements of the alloy and the etchant(s) used [1]. In this process, the reaction rate (or MRR) keeps on decreasing with time. To overcome this weakness of the process, stirring and heating of the etchant are usually recommended. During ChM, the etchant strength should be maintained by continuously replacing part of the used etchant with fresh etchant after its use for a definite interval of time. This is a unique process in the sense that there are no mechanical forces or any thermal load acting on the workpiece. Hence, the related defects are also not there in the machined workpieces, but it may lead to a corrosive effect if proper post processing (or cleaning) of the machined workpiece is not done [28, 29].

In the case of ECM, only electrically conductive workpiece material can be machined. Secondly, selection of the right kind of electrolyte is a prerequisite for the success of the ECM process. Chemical properties of the workpiece material play a vital role in governing the performance of the process. Another major factor in the process is the proper design of the tool for ECM to achieve the required accuracy of the machined component [30]. Machining of an alloy may require a mixture of electrolytes to obtain better results and higher performance of the process. The ECM process has the second highest MRR (after PAM) in all AMPs, and very good process performance (usually, efficiency >75%). The performance does not depend on the mechanical and physical properties of the workpiece material [31, 32]. For many specific applications, this is the only process which can be employed, and no other traditional or advanced machining process works. For example, drilling of contoured holes of the desired shape and size, and drilling a hole perpendicular to the wall between the two holes (Figure 1.8) [33].

The machining of plastics is also important in some applications. Enzymes are used to machine/eat away the plastics in the desired fashion. Some work is also going on to machine specific metals by this process, known as the *"Biochemical Machining"* process [34].

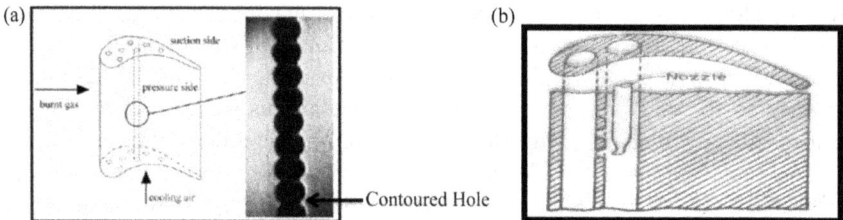

FIGURE 1.8   Contoured hole and hole perpendicular to a wall.

Traditional machining processes fail to perform an operation when the workpiece material becomes harder than the cutting tool material. These processes also fail when complex shaped products, 3-D products, and products with a very high level of surface finish, especially in inaccessible areas, are required. Then it is necessary to search for appropriate alternative machining and finishing processes, say, advanced machining and finishing processes (also called unconventional finishing, non-traditional machining and finishing processes, or hybrid machining and finishing processes [35–37]).

Figure 1.3 shows different types of fundamental operations (usually performed using traditional machining process) and hybrid or multiple operations and processes. Here, multiple operations mean two or more than two operations are being performed at the same time. For example, electrochemical machining and magnetic abrasive finishing, Electrochemical Machining and Grinding (ECG), Chemical Machining and Mechanical Polishing (CMP), and Electrochemical Machining, Grinding, and Dressing (ELID). Performing multiple operations on the same machine and at the same time enhances productivity. However, the basic types of operations are defined by the relative motion between the tool or the source of energy (mechanical, thermal, electrochemical, or chemical) and the workpiece from which the material is to be removed. Figure 1.4 shows an example of the relative motion between the source of energy/tool and workpiece to achieve material removal in a certain manner or in a certain type of operation, say, turning. Depending upon the kind of energy source used to remove the material, the operation is named accordingly, for example, Electrochemical (EC) turning [38–41], Electric Discharge (ED) turning, Laser Beam (LB) turning [42], Abrasive Jet (AJ) turning, Abrasive Water Jet (AWJ) turning, etc. Here, the important point to note is that the workpiece has a rotary motion while the source of energy has a linear motion to perform the turning operation. Further, the working principle of the process remains the same as in simple machining operation for macro- or micromachining. Some researchers have given the rotary motion to the workpiece, but wire as a tool in case of wire electric discharge turning and wire electro chemical turning [38, 39]. In the case of AJ turning, AWJ turning, electric discharge turning, and EC turning, the velocity of the air jet, abrasive water jet, dielectric, and electrolyte, respectively, should be optimized along with other parameters to give the best results. A linear feed rate should be selected so that the working gap in AJT and AWJT, IEG in EDT, and ECT or focal distance in the case of laser beam turning do not fluctuate too much, else productivity will be hampered. On the same lines, chemical turning can be performed by dipping the workpiece in the etchant and then by rotating it. The sole objective being considered here is to reduce the diameter of the workpiece throughout or in the steps by employing a certain type of process.

These different types of operations can be performed by different independent types of traditional machine tools (m/ts), namely, lathe, milling, drilling, grinding, honing m/ts. These operations are normally performed on the individual machines but sometimes more than one type of operation can also be performed on the same machine, say, turning, drilling, boring, and threading on a lathe machine, or milling, drilling, and turning on a computer numerical control (CNC) machining center. To enhance the productivity of traditional m/t, every time to perform two or more than two types of operations on the same traditional m/t is not possible. However, it is possible to perform two or more than two types of operations using the same type of advanced machining process on a single machine, but it may require some modifications in the form of a fixture or attachment in the machine. By introducing such machines, transport time, setup time, lead time, etc. can be reduced.

Here, the important thing to note is that it does not matter which operation (machining, drilling, turning, milling, or even finishing) is being performed using a particular process (ECM, EDM, AWJM, or any other type (but not hybrid)), the mechanism of material removal remains the same, but relative motion between the tool/energy source and workpiece changes according to the type of operation being performed. *Hence, the words "process" and "operation" should not be used synonymously.*

1.3 MICROMACHINING

Macromachining and micromachining are differentiated basically by the size of the feature created on the workpiece, which may be macro or micro in size. If one of the dimensions of the feature created is in micrometer (greater than 1 μm, but equal to or smaller than 999 μm) then the process should be called micromachining, otherwise meso- or macromachining (if all the dimensions of the feature created are equal to or greater than 1 mm). An important point to understand here is that if the parameters of a process are well articulated, all three types of AMPs can be used for micromachining to create microfeatures on the macro or micro size workpiece. However, if a setup designed and fabricated for macromachining is used for micromachining, it may or may not produce the desired results. For this purpose, a separate dedicated miniature size m/t should be designed and fabricated so that the desirable process parameters can be achieved within the desired accuracy. Let us take some examples of the applications of the AMPs in the micro domain.

The major differences required in Abrasive Jet Micromachining (AJMM) are as follows [43–45]. The size of the inner diameter of the nozzle should be reasonably smaller than the size of the hole to be machined or the width of the feature to be created. It requires a much smaller size of the abrasive particles compared to the one used in abrasive jet macromachining. In the case of AJMM, the size of the chips produced is much smaller than those obtained in a normal AJM process. The same arguments hold true for the case of Abrasive Water Jet Micromachining (AWJMM) in which the inner diameter of the nozzle and the size of the abrasive particles should be substantially smaller compared to the case of macromachining to allow it to enter into the micromachining domain. However, it is important to note that the mechanism of material removal remains the same in both cases, that is macromachining and micromachining. More details can be found in various references [46, 47]. In Ultrasonic Micromachining (USMM), in place of a nozzle the dimensions of the tool have to be brought down in the micrometer domain to have microdrilling, microslotting, etc. However, in all the three cases, if the depth of the feature (hole, slot, channel, etc.) is to be made in the micro domain then the machining time has to be controlled separately from the tool design and machining conditions [48–50].

In the case of TAMPs, the processes, namely, EDM, LBM, and EBM are used for micromachining also, but in these cases, the machining parameters have to be brought down to the level so that the crater size created is substantially reduced to produce microfeatures. In the case of Electric Discharge Micromachining (EDMM), the size of the tool, voltage, and pulse duration has to be brought down substantially to create nanometer or micrometer size craters to create the microfeatures on the workpiece [51–53]. The type of dielectric can also be changed to suit the creation of nanosize craters [54]. The selected dielectric should have low viscosity so that it can reach in to the intricate microfeatures. In the case of Laser Beam Micromachining (LBMM), the focused laser beam spot size should lie in the range of nanometers to a few micrometers [54–57]. Short pulse lasers are capable of producing defect-free microfeatures on the workpieces [55]. In the case of Electron Beam Micromachining (EBMM) [58], the electron beam diameter should be only a few micrometers. It can be controlled by changing the machining parameters.

For Electrochemical Micromachining (ECMM), the tool size and shape have to be designed as per the requirements [59]. To create microfeatures, the voltage, IEG, and electrolyte conductivity have to be reduced substantially (usually, voltage 1–5 V, IEG $\cong$ 5–10 μm, electrolyte concentration $\cong$ 5–10%) [12]. These EC micromachined parts do not have any residual stresses, recast layers, and micro/nanocracks as the defects observed in TAMMPs but they may corrode the parts if a proper electrolyte is not used.

1.4   REMARKS

This book deals with some of the AMPs with an emphasis on the science behind the material removal mechanism of these individual processes. It will be clear after reading working principles and their scientific analysis in different chapters that there is a lot more research work that can be done to enhance the capabilities of an individual process in terms of its machining efficiency, accuracy, tolerances, and applications. Each process in this book has been discussed by taking an example of an individual operation(s), as discussed in this chapter. The applications of these processes can be significantly expanded by using them for different kinds of operations by modifying the setup, if required; not only for different kinds of operations, but a lot more can be done by hybridizing the processes as per the products' requirements of individual industries. Some chapters in this book focus on hybridized processes also.

REFERENCES

1. Jain V. K. (2021). *Advanced Machining Processes*, 2nd Edition. Allied Publishers, New Delhi, India.
2. Chavoshi S. Z. and Luo X. (2015). Hybrid Micro-Machining Processes: A Review. *Precision Engineering*, Vol. 41, pp. 1–23.
3. Jain V. K. (Ed.). (2014). *Introduction to Micromachining*, 2nd Edition. Narosa Publishing House, New Delhi, India.

4. Benedict G. F. (1987). *Non-Traditional Machining Processes.* Marcel Dekker Inc., New York.

5. McGeough J. A. (1988). *Advanced Machining Methods.* Chapman and Hall, London.

6. Verma A. P. and Lal G. K. (1985). Basic Mechanics of Abrasive Jet Machining. *Journal of the Institution of Electrical Engineers,* Vol. 66, pp. 74–81.

7. Balasubramaniam R. and Suri V. K. (2014). Diamond Turn Machining. In *Introduction to Micromachining* (Ed. V. K. Jain). Narosa Publishing House, Pvt. Ltd, New Delhi.

8. Miller D. S. (2004). Micromachining With Abrasive Water Jets. *Materials Processing Technology,* Vol. 149(1–3), pp. 37–42.

9. Hashesh M. (1989). A Model for Abrasive Water Jet (AWJ) Machining. *Transactions of the ASME: Journal of Engineering Materials and Technology,* Vol. 111, pp. 154–162.

10. DeMeis R. (1986). Cutting Metal With Water. *Aerospace America,* Vol. 24(3), pp. 22–23.

11. Pandey P. C. and Shan H. S. (1980). *Modern Machining Processes.* Tata McGraw Hill, New Delhi.

12. Bhattacharya B. (2015). *Electrochemical Micromachining for Nanofabrication, MEMS.* Elsevier Publishers, Waltham USA

13. Koshy P., Jain V. K. and Lal G. K. (1993). Experimental Investigations into EDM With Rotating Disc Electrode. *Precision Engineering,* Vol. 15(1), pp. 6–15.

14. Newman S. T. and Ho K. H. (2003). The State of the Art-Electric Discharge Machining. *International Journal of Machine Tools and Manufacture,* Vol. 43(13), pp. 1287–1300.

15. Shankar P., Jain V. K. and Sundararajan T. (1997). Analysis of Spark Profile During EDM. *Machining Science and Technology,* Vol. 1(2), pp. 195–217.

16. Madhu P., Jain V. K., Sundarajan T. and Rajurkar K. P. (1991). Finite Element Analysis of EDM Process. *Processing of Advanced Materials,* Vol. 1(3–4), pp. 161–174.

17. Vasa N. J. (2014). Laser Micromachining Techniques and Applications. In *Introduction to Micromachining* (Ed. V. K. Jain). Narosa Publishing House Pvt. Ltd., New Delhi.

18. Sidpara A. and Jain V. K. (2014). Electron Beam Micromachining. In *Introduction to Micromachining* (Ed. V. K. Jain). Narosa Publishing House Pvt. Ltd., New Delhi.

19. Chryssolouris G. (1991). *Laser Machining: Theory and Practice.* Springer Verlag, New York.

20. Erden A. (1982). Role of Dielectric Flushing on EDM Performance. In *DROE. 23rd IMTDR Conference,* Manchester, UK, pp. 283–289.

21. Utke H. and Melngailis. (2008). Gas-Assisted Focused Electron Beam and Ion Beam Processing. *Journal of Vacuum Science and Technology B,* Vol. 26(4), pp. 1197–1267.

22. MeGeough J. A. (2002). Electron Beam Machining. In *Micromachining of Engineering Materials.* Marcel Dekken Inc., New York.

23. Brown G. and Nichols K. G. (1966). A Review of the Use of Electron Beam Machines for Thermal Milling. *Journal of Materials Science,* Vol. 1(1), pp. 96–111.

24. Kulkarni V. N., Shukla N. and Rajupt N. S. (2013). Micro-and Nanomanufacturing by Focused Ion Beam. In *Micromanufacturing Processes* (Ed. V. K. Jain). Taylor and Francis (CRC Press), New York.

25. Arvindan S., Rao P. V. and Yoshinom. (2014). Focussed Ion Beam Machining. In *Introduction to Micromachining* (Ed. V. K. Jain). Narosa Publishing House, New Delhi.

26. Tseng A. A. (2004). Recent Developments in Micromilling Using Focused Ion Beam Technology. *Journal of Micromechanics Microengineering,* Vol. 14, pp. R15–R34.

27. Tripathi S. K., Shukla N. and Kulkarni V. N. (2008). Correlation Between Ion Beam Parameters and Physical Characteristics of Nano Structures Fabricated by Focused Ion Beam. *Nuclear Instruments and Methods in Physics Research B,* Vol. 266, p. 1468.

28. McGeough J. A. (1974). *Principles of Electrochemical Machining.* Chapman and Hall, London.

29. De Bar A. E. and Oliver D. A. (Editor). (1975). *Electrochemical Machining.* Mc Donald and Co. Ltd., London.

30. Jain V. K. and Pandey P. C. (1980). Tooling Design for ECM. *Precision Engineering,* Vol. 2, pp. 195–206.

31. Gurklis J. A. (1965). Metal Removal by Electrochemical Methods and Its Effects on Mechanical Properties of Metals, Defense Metals Information Center Report 213, Bettle Memorial Institute.

32. Evans J. M., Boden P. J. and Baker A. A. (1971). Effects of Surface Finish Produced During ECM Upon the Fatique Life of Nimonic 80 A. In *Proc. 12th Int. Mach. Tool Des. Res. Cont.,* p. 271.

33. Jain V. K., Chavan A. and Kulkarni A. (2009). Analysis of Contoured Holes Produced Using STED Process. *The International Journal of Advanced Manufacturing Technology,* Vol. 44, pp. 138–148.

34. Hong-Hoching J.-H. C. and Jadhav U. U. (2013). Biomachining Acidithiobacillus-Genus -Based Metal Removal. In *Micromanufacturing Processes* (Ed. V. K. Jain). Taylor and Francis (CRC Press), New York.

35. Jain V. K., Sidpara A., Ravi Sankar M. and Das M. (2012). Introduction. In *Micromanufacturing Processes* (Ed. V. K. Jain). Taylor and Francis Publishers, USA.

36. Das M., Jain V. K. and Ghoshdastidar P. S. *Nanofinishing Process Using Magnetorheological Polishing Medium.* Lambert Academic Publishing, Germany, 2012.

37. Kumara S., Jain V. K. and Sidpara A. (2015). Nanofinishing of Freeform Surfaces (Knee Joint Implant) by Rotational-Magneto Rheological Abrasive Flow Finishing (R-MRAFF). *Precision Engineering,* Vol. 42, pp. 165–178.

38. Ali El-Taweel T. and Gouda S. A. (2011). Performance Analysis of Wire Electrochemical Turning Process—RSM Approach. *International Journal of Advanced Manufacturing Technology,* Vol. 53, pp. 181–190. DOI: 10.1007/s00170-010-2809-x.

39. Yahyavi Zanjani M., Ghattan Kashani H. and Mirahmadi A. (2013). Improvement of Electrochemical Turning for Machining Complex Shapes Using a Simple Gap Size Sensor and a Tubular Shape Tool. *International Journal of Advanced Manufacturing Technology,* Vol. 69(1–4), p. 375.

40. Yan M.-T. and Hsieh P.-H. (2014). Monitoring and Adaptive Process Control of Wire Electrical Discharge Turning. *International Journal of Automation Technology,* Vol. 8(3), pp. 468–469.

41. Hofsted A. and Van Tin B. (1970). Some Remarkgon Electrochemical Turning. *Annals CIRP,* Vol. 18, pp. 93–106.

42. Kibria G., Doloi B. and Bhattacharya B. (2015). Pulsed:Nd YAG Laser Microturning of Alumina Ceramics. In S. N.Joshi and U. S.Dixit (Eds.), *Laser Based Manufacturing: Topics in Mining, Metallurgy and Materials Engineering,* Springer, New Delhi, pp. 343–380.

43. Ko T. J. (2014). Abrasive Jet Micro Machining. In *Introduction to Micromachining* (Ed. V. K. Jain). Narosa Publishing House Pvt. Ltd., New Delhi.

44. Saragih A. S. and Ko T. J. (2009). A Thick SU-8 Mask for Micro Abrasive Jet Machining on Glass. *Journal of Advanced Manufacturing Technology,* Vol. 41, pp. 734–740.

45. Ghobeity A., Getu H., Krajac T., Spelt J. K. and Papini M. (2007). Process Repeatability in Abrasive Jet Micromachining. *Journal of Materials Processing Technology,* Vol. 190, pp. 51–60.

46. Miller D. S. (2014). Micromachining With Abrasive Water Jets. In *Introduction to Micromachining* (Ed. V. K. Jain). Narosa Publishing House, New Delhi.

47. Liu H. T. (2010). Water Jet Technology for Machining Fine Features Pertaining to Micromachining. *Journal of Manufacturing Processes,* Vol. 12, pp. 8–18.

48. Adithan M. and Venkatesh V. C. (1976). Production Accuracy of Holes in Ultrasonic Drilling. *Wear*, Vol. 40(3), pp. 309–318.

49. Dornfeld D., Min S. and Takeuchi Y. (2006). Recent Advances in Mechanical Micromachining. *Annals of the CIRP*, Vol. 55(1), pp. 1–24.

50. Jadoun R. S. (2014). Ultrasonic Micromachining. In *Introduction to Micromachining* (Ed. V. K. Jain). Narosa Publishing House, New Delhi.

51. Joshi S. S. (2014). Micro-Electric Discharge Micromachining. In *Introduction to Micromachining* (Ed. V. K. Jain). Narosa Publishing House, New Delhi.

52. Dhanik S. and Joshi S. S. (2005). Modeling of a Single RC-Pulse Discharge in Micro-EDM. *Transactions of the ASME: Journal of Manufacturing Science & Engineering*, Vol. 127, pp. 759–767.

53. Masu Zawa T. (1997). State of the Art of Micro-Machining. *Annals of the CIRP*, Vol. 29(2), pp. 621–628.

54. Kagaya K., Oishi Y. and Yada K. L. (1990). Micro Electro Discharge Machining Using Water as a Working Fluid2: Narrow Slit Fabrication. *Precision Engineering*, Vol. 12, pp. 213–217.

55. Alemohammad H., Toyserkani E. and Pinkerton A. J. (2008). Femtosecond Laser Micromachining of Fiber Bragg Gratings for Simultaneous Measurement of Temperature and Concentration of Liquids. *Journal of Physic D: Applied Physics*, Vol. 41, pp. 1–9.

56. Powell P. M. (2003). Laser Based Micromachining Gets Practical. *Photonics Spectra*, Vol. 37, p. 70.

57. Rizvi N. H. and Apte P. (2002). Developments in Laser Micromachining Techniques. *Journal of Material Processing Technology*, Vol. 127, pp. 206–210.

58. Utke O. and Mengailis. (2008). Gas-Assisted Focused Electron Beam and Ion Beam Processing. *Journal of Vacuum Science and Technology B*, Vol. 26(4), p. 1197.

59. Reddy M. S., Jain V. K. and Lal G. K. (1988). Tool Design for ECM: Correction Factor Method. *Transactions of ASME: Journal of Engineering for Industry*, Vol. 110, pp. 111–118.

2 Abrasive Water Jet Machining (AWJM)

Gokhan Aydin and Izzet Karakurt

CONTENTS

2.1 INTRODUCTION

Water jet technology is one of the most recently developed non-traditional manufacturing processes [1, 2]. It uses a fine jet of ultra-high-pressure water to cut the workpiece by means of erosion [3, 4]. Water jet machining can be used to cut soft workpieces like plastic, rubber, or wood. In order to cut hard workpieces like metals or rock (e.g., granite), abrasive particles are mixed in the water jet [5]. When the abrasive particles are used in the water for machining purposes then it is called

DOI: 10.1201/9780429160011-2

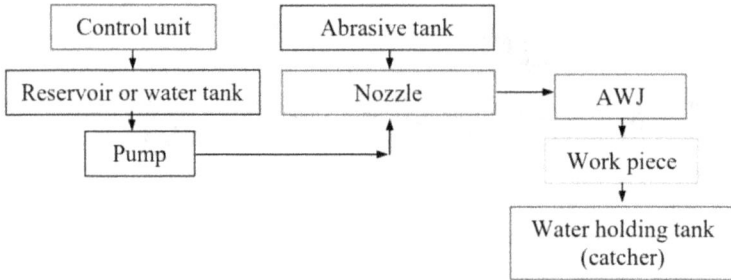

FIGURE 2.1 Major components of AWJM technology.

TABLE 2.1
Superiorities and Limitations of AWJM Technology

Superiorities

No thermal damage or distortion
High machining versatility
Ability to produce contours
Easy integration with mechanical manipulator (e.g., to computer-controlled systems)
Small cutting forces and fast setup
Minimal burrs
Easy availability of water as the working fluid

Limitations

Noise associated with AWJ cutting
High initial capital cost and operating costs
Low energy efficiency
Wear of nozzle
Geometry deficiencies in contouring process

Abrasive Water Jet Machining (AWJM) [6–9]. The major components of AWJM technology are presented in Figure 2.1. As seen, a traditional AWJ system includes components such as a high-pressure water pump, pressure-resistant pipes which are responsible for supplying high-pressure water, an abrasive tank for adding abrasive particles to the water, a cutting head, a control unit which directs the cutting head to the desired coordinates and a water holding tank/table parts [10, 11].

AWJM technology has various superiorities over the other traditional and advanced machining processes [12, 13], for example, its cutting capabilities. However, cutting depth and kerf quality restrict wider applications of the technology. Table 2.1 presents some superiorities and limitations of AWJM technology.

The unique technological capabilities and broad functionalities of AWJM technology have made it a competitive tool in several industrial applications in recent years [14]. Some of the sectors using AWJM technology are presented in Table 2.2.

TABLE 2.2

Sectors Benefiting from AWJM Technology [15–17]

Sector	Workpieces
Aviation–Aerospace	Wing section, turbine blades, landing gear
Automotive	Gear rings, deburring, engine components
Defense	Bullet proof glass, armor plating, machining of composites
Glass	Window glass, shaping of special purposes
Stone–Decoration	Rock and rocklike material
Composite	Polymer composites, metal matrix composites
Others	Metallurgy, electronics, instrumentation, optics, medical, textile

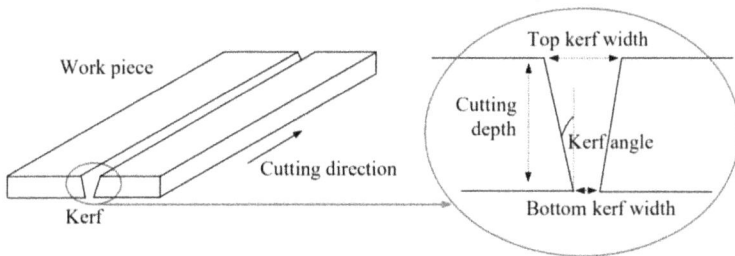

FIGURE 2.2 A typical kerf profile produced by AWJM.

2.2 PERFORMANCE OUTPUTS OF AWJM

The cutting geometry (cut depth and kerf profile), specific cutting energy, and surfaces of the workpiece machined (surface roughness and striation profile) could be considered for assessment of the cutting performance of the AWJM.

2.2.1 CUTTING GEOMETRY

In AWJ cutting, the jet creates a kerf in the workpiece that decreases along the cut depth. The form of the kerf is considered as one of the main indicators indicating the performance of the process [18]. Typical kerf geometry is shown in Figure 2.2. The kerf angle is taken as a characteristic parameter in the kerf profile [19, 20]. It is calculated using Equation 2.1. Another performance evaluation parameter is the cut depth, which can be defined as the vertical distance between the top surface and bottom of the kerf (Figure 2.2). A higher cut depth and lower kerf angle indicate a desirable cutting performance/response [21, 22].

$$\theta = \tan^{-1}\left[\left(W_{top} - W_{bottom}\right)/2h_k\right] \tag{2.1}$$

where θ is kerf angle, W_{top} is top kerf width, W_{bottom} is bottom kerf width, and h_k is kerf depth.

2.2.2 SPECIFIC CUTTING ENERGY

Specific cutting energy refers to the amount of energy required to cut a unit volume of workpiece material. It is one of the most useful methods used to estimate the efficiency of cutting machines [23–25]. In order to calculate the specific cutting energy, it is necessary to know the AWJ energy and the cut volume. Before and after cutting (at this stage, the abrasive particles in the cutting slot are cleaned with compressed air), the weight loss is determined by weighing the workpiece being machined (following the relevant procedure). The cut volume is calculated using the workpiece density and weight loss. Additionally, Bernoulli's equation could be used to determine the AWJ energy (Equation 2.2).

$$p + \rho_w v_w/2 + \rho_w g h = \text{Constant} \tag{2.2}$$

where p is water pressure, v_w is water jet speed, ρ_w is water density, g is acceleration due to gravity, and h is height of the observed points above the reference plane.

$$v_{wj} = (2p/\rho_w)^{0,5} \tag{2.3}$$

$$Q_w v_{wj} + Q_{va} = (Q_w + Q) v_{awj} \tag{2.4}$$

$$v_{awj} = Q_v v_{wj}/(Q_w + Q) \tag{2.5}$$

$$v_{awj} = v_{wj}/(1 + Q/Q_w) \tag{2.6}$$

where Q_w: water flow rate (g/min), Q: abrasive flow rate (g/min), v_{wj}: the speed of pure water jet (mm/min), v_a: velocity of abrasive particles (mm/min) and v_{awj}: speed of abrasive water jet (mm/min).

During the mixing process, some momentum is lost as the abrasive particles strike the water jet and inner surface of the focusing tube.

$$v_{awj} = \eta v_{wj}/(1 + Q/Q_w) \tag{2.7}$$

where, η is the momentum loss factor, whose value lies between 0.65 and 0.85 [26].

The abrasive flow rate determines the number of impacting abrasive particles and the kinetic energy (KE) of AWJ. The AWJ KE (E_{awj}) can be expressed by the following equations:

$$E_{awj} = 0,5 Q v_{awj}^2 \tag{2.8}$$

$$E_{awj} = 0,5 Q \left[\eta v_{wj}/(1 + Q/Q_w) \right]^2 \tag{2.9}$$

The combination of Equations 2.3 and 2.9 gives the AWJ KE required to overcome the fracture energy of the workpiece being cut.

$$E_{awj} = \left[0,5Q\eta^2 2pQ_{wj}^2\right]\Big/\left[\rho_w\left(Q_{wj}+Q\right)^2\right] \qquad (2.10)$$

$$e_c = E_{awj}/V \qquad (2.11)$$

where E_{awj}: the AWJ energy, e_c: specific cutting energy (J/mm³), V: cutting volume (mm³/sec).

As a result, the AWJ energy and specific cutting energy are calculated using Equations 2.10 and 2.11, respectively.

2.2.3 CHARACTERIZATION OF MACHINED SURFACE

Surface roughness is an important parameter for determining the jet–workpiece material interaction and evaluating the damage caused by this interaction on the surface [27, 28]. Low surface roughness is desired in order to avoid secondary processing in AWJM technology, as it is in the other machining technologies [29]. Machined surfaces produced by the AWJ can be divided into two zones: The first along the cutting depth as the *cutting-wear zone* and the second as the *deformation-wear zone* (see Figure 2.3). Uniform surface texture occurs at the cutting-wear zone since abrasive particles impact the kerf wall at shallow angles. This uniform surface is generally assessed by a surface roughness measure [30]. At the lower part of the kerf, which is called the deformation-wear zone, the AWJ forms large striations defined by larger irregularities rather than surface roughness [31]. Striations are formed when the ratio between the available jet energy and the required energy for the removal of material becomes comparatively small [27].

2.2.4 FACTORS AFFECTING AWJM PERFORMANCE

The cutting performance of AWJM is mainly affected by operating variables such as traverse speed, abrasive flow rate, stand-off distance, water pressure, and the workpiece properties (e.g., strength, hardness) [32–34]. AWJ cutting performance decreases with an increase in traverse speed and a decrease in the abrasive flow

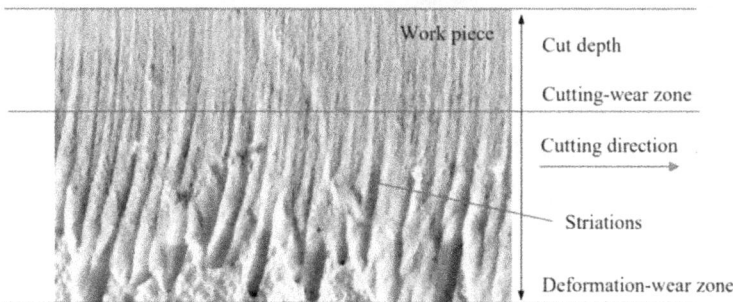

FIGURE 2.3 A representative figure for zones produced by AWJM.

rate due to a decreasing number of abrasive particles impinging on the workpiece [35–37]. As another cutting parameter, the stand-off distance is related to the effective jet diameter. When the jet spreads out of the nozzle, it diverges and loses a part of its KE. Therefore, an increase in the stand-off distance negatively affects AWJ machining performance. Additionally, higher water pressure increases the KE of the individual abrasive particles inside the jet and enhances its capability for material removal [38, 39].

The resistance of the materials to any mechanical breakdown depends on several factors, such as strength and hardness (especially microhardness) of the materials. Among the material properties, the uniaxial compressive strength is one of the important parameters representing bearing capacity of the material. Its variation is explained by a number of factors including material composition. It can also be defined as the threshold value for any material to fail under external forces. In AWJ cutting, the jet cuts the workpiece by a rapid erosion process when its force exceeds the compressive strength of the workpiece. In addition to the strength, the hardness of the workpiece is also an important factor which affects AWJ performance. In an erosion process, a particle is subjected to the surface with a force which results in the deceleration of the particle. The surface reacts to this pressure with an elastic or plastic deformation. The influence of hardness is generally low if the abrasive particles are significantly harder than the hardness of the workpiece surface. However, if the hardness of the workpiece surface is close to or harder than the hardness of the abrasive particles, then lower cutting performance is obtained due to the deformation/fracture of the abrasive particles [40].

For workpieces composed of different minerals (such as rocks), textural properties such as mineral grain size, spatial arrangements of the grains, microfractures, and mineral cleavages inside the minerals could also affect AWJ performance [41]. The initial breakage and/or fracture occurs along the grain boundaries before the mineral/individual grains begin to fracture. These breakages and/or fractures, which initially occur along the grain boundaries, make the grains disintegrate easily [42]. Therefore, fine-grained rocks give better results on the basis of cut depth and kerf angle. On the other hand, there is a relatively lower amount of grain boundaries (it increases the surface roughness) for a specific area in coarse-grained rocks. Therefore, coarse-grained rocks present better results in terms of workpiece surface roughness.

2.3 ABRASIVES

2.3.1 ABRASIVE PROPERTIES

The cutting capability of AWJM can be significantly affected by the type of abrasives that are utilized [43–45]. The abrasive properties presented below should be determined before the selection of an appropriate abrasive.

- **Hardness**: Using a low hardness abrasive will reduce the traverse speed, while extending the nozzle life [40].

- **Specific gravity**: High KE required to cut the harder workpieces. This energy is stored in the abrasive particles which have higher specific gravity than water [20]. Therefore, the ideal abrasive particles should have sufficiently high specific gravity for efficient cutting.
- **Shape**: Spherical and sharp-edged abrasive particles are commonly used in AWJM applications. A sphere is generally considered to be the ideal shape for conveying mass in a water stream. However, the acceleration and fragmentation of the abrasive particles must be balanced for efficient machining. The most suitable particle shape depends on the workpiece properties being cut and edge finish requirements. For instance, sharp-angular grains cut more quickly and offer superior edge finish. Spherical grains are used more for general-purpose standard cutting applications [43, 45].
- **Toughness**: The disintegration of abrasive particles is carried out in the mixing chamber/focusing tube (particle–particle, particle–water jet, and particle–wall collisions) and on the workpiece cutting surface (particle–particle and particle–workpiece collisions). Brittle abrasive particles can be easily fragmented in the focusing tube, and these fragmented particles cannot perform effective cutting [46]. On the other hand, extremely tough abrasive particles dull during the mixing process, causing ineffective cutting conditions. Therefore, the ideal abrasive should have a certain breakdown rate to obtain sharp and angular cutting edges.

The following abrasive properties are controlled in abrasive processing [43, 47–49].

- **Purity**: Materials are mined, milled, and processed for the production of the abrasive particles. During the refining process, high-purity abrasives need more processing stages than low-purity abrasives. As a result, high-purity abrasives cost more, but they present higher cutting performance. Low-purity abrasives may include grains that can negatively affect the AWJ cutting performance.
- **Particle size distribution**: Control of the particle size distribution is extremely important to maximize AWJM performance. The nozzle can be clogged in cases of the use of coarse or oversized abrasive particles. This may even result in the machining process to stop. Conversely, irregular feed or sputtering in the jet may be observed since finer abrasive particles can collect in the feed line/cutting head. It is very difficult to adjust the abrasive feed rate and maintain traverse speed in AWJM applications where abrasive particles with inconsistent size distribution are used [43, 50].

2.3.2 Commonly Used Abrasives

The materials used as abrasives can be broadly classified as either natural or synthetic. Natural abrasives such as diamond, corundum, and emery are found as natural deposits, and can be mined and processed for use with little alteration. Most natural abrasives, other than a natural diamond, vary widely in their properties. Therefore, synthetic abrasives have replaced these natural abrasives as industrial applications require consistent properties. Synthetically manufactured for the purpose of abrasion, many synthetic abrasives are effectively identical to a natural mineral, differing only in that the synthetic abrasives are produced rather than mined [51–53]. In AWJ cutting, the following abrasives are commonly used [49]. Additionally, Tables 2.3 and 2.4 give some physical and chemical properties, respectively, of commonly used abrasives.

- **Brown fused alumina**: It is produced by the smelting of calcined bauxite, which is a source of aluminum oxide, in an electric arc furnace at high temperature. A slow solidification process follows the fusion, to yield blocky crystals. The cooled crude material is further crushed, cleaned for magnetic impurities, and classified.
- **White fused alumina**: It is produced from the fusion of high purity calcined alumina in electric arc furnaces.
- **Mixed alumina**: Mixed alumina is a secondary product, produced from ceramic grinding wheels and reclaimed alumina dust from the blasting and grinding industries.
- **Garnet**: Garnet is a kind of naturally forming material. It is extracted and processed to get abrasive particles.
- **Emery powder**: Natural emery is used as raw material for producing emery powder.
- **Silicon carbide**: Silicon carbide is the third hardest mineral in the world, behind diamond and boron carbide. It is a synthetic abrasive produced through the fusion of high-grade silica sand and finely ground carbon such as petroleum coke, in an electric furnace at high temperature.
- **Glass beads**: They are produced from glass sand. Glass sand is firstly fused into a spherical shape by high-temperature suspension. Then, it is annealed to eliminate internal stress and prevent breakage.

TABLE 2.3
Physical Properties of the Commonly Used Abrasives [49]

Abrasive Type	Hardness (Mohs)	Grain Shape	Specific Gravity	Bulk Density* (g/cm³)
Brown fused alumina	9	Angular	3.9–4.1	1.5–2.1
White fused alumina	9	Angular	3.9–4.1	0.8–2.1
Mixed alumina	9	Angular	3.9	1.5–1.8
Garnet	7.5–8	Angular	3.5–4.3	1.9–2.2
Emery powder	7.5–8.5	Angular	3.5–4	1.6–1.9
Silicon carbide	9–10	An./splintery	3.2	0.75–1.82
Glass beads	6	Round	2.5	1.5–1.6
Glass granules	6	Angular	2–2.6	1.2–1.8
Ceramic beads	7–7.5	Round	3.8	2.1–2.4

*depending on grain size

TABLE 2.4

Chemical Compositions (% By Weight) of the Commonly Used Abrasives [49]

Abrasive Type	Al$_2$O$_3$ (%)	Fe$_2$O$_3$ (%)	SiO$_2$ (%)	CaO (%)	MgO (%)	Other (%)
Brown fused alumina	94.20	0.33	1.34	0.33	-	TiO$_2$: 2.79
White fused alumina	99.69	0.02	0.03	0.05	0.01	TiO$_2$: 0.01; Na$_2$O: 0.20
Mixed alumina	88.10	0.20	6.80	0.30	0.30	SIC: 5.80; TiO$_2$: 0.70
Garnet	23.00	33.00	35.00	1.00	7.00	MnO: 1.00
Emery powder	62.30	25.60	8.40	1.20	0.20	-
Silicon carbide	-	0.20	-	-	-	SIC: 98.00; C-frei: 0.20
Glass beads	<2.50	<0.50	70.00–75.00	7.00–12.00	<5.00	K$_2$O: <1.50; Na$_2$O: 12.00–15.00
Glass granules	<2.50	<0.50	65.00–75.00	7.00–12.00	<5.00	Na$_2$O: 12.00–18.00; K$_2$O: <1.50
Ceramic beads	4.57	0.14	27.77	3.47	-	ZrO$_2$: 61.98; TiO$_2$: 0.34

- **Glass granules**: They are the products of various glass sources, including glass bottles, industrial/construction glass, car windshields, and waste glass from glass manufacturers. These wastes are easily crushed or melted/recycled to produce the final products. It can be stated that glass granules are more abrasive than round glass beads due to their angular grain shape.
- **Ceramic beads**: They are produced by high-temperature electrofusion of oxides. The bead internal structure consists of a compact assembly of a crystalline zirconium network, which is tightly interlocked with the amorphous silica phase.

2.3.3 Abrasive Recycling

Abrasive cost, including disposal cost, is often the largest component in the cost of AWJ cutting. This can be up to 75% of the total cost depending on the parameters, such as abrasive mass flow rate, abrasive price, and the cost of the AWJM system [54, 55]. However, the abrasive particles can be reused, which can reduce the abrasive and the disposal costs. A general idea about the cost of abrasives can be obtained by making an evaluation on a garnet abrasive basis [56]. The price of the garnet ranges from US$0.15–0.30/kg, depending on the purchase volume. The cost of abrasives can be determined if an average abrasive cost of US$0.225 is taken as a reference. Assuming that the abrasive feed rate is 300 g/min in AWJ cutting, the hourly abrasive cost will be calculated as US$4.80 (0.3 kg/min × 60 min × 0.225/kg). If a cutting time of six hours per day is used, the daily abrasive cost can be calculated as US$24.30. Monthly and annual abrasive costs are then calculated as US$729 and US$8748 respectively.

The abrasive particles can be recycled many times in order to reduce the costs of abrasive and disposal in AWJ cutting. Through the recycling system, abrasive particles can be automatically recycled and fed into the cutting system. A schematic illustration of the abrasive recycling process in AWJM is depicted in Figure 2.4. It can be stated that a stirring overflow separator is the core component of the recycling system [52]. Solid particles in liquid have two effects of gravity and buoyancy. Based

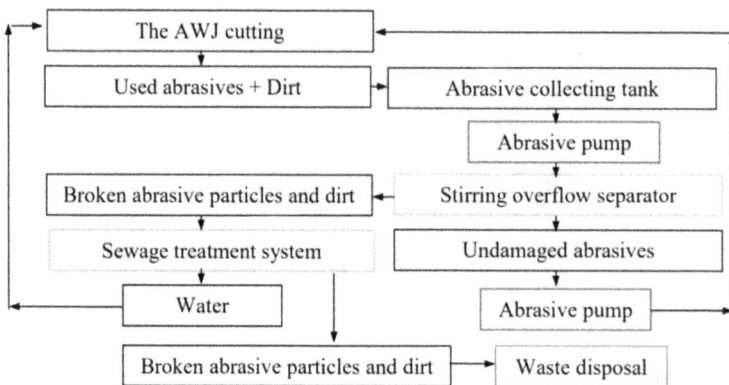

FIGURE 2.4 Schematic illustration of the abrasive recycling process in AWJM.

on mathematical analysis, the following equation could be obtained for spherical solid particle settling velocity μ_s in a fluid as given in Equation 2.12.

$$\mu_s = A\left[4\left(C_s - C\right)gD_s\big/\left(3C_dC\right)\right]^{0.5}\left[1-\left(D_s/D\right)^2\right] \tag{2.12}$$

where μ_s is fall velocity of particles, A is shape factor, C_d is drag coefficient, C_s is the specific weight of particles, C is the specific weight of water, g is gravity due to acceleration, D_s is the equivalent diameter, and D is the diameter of the separator.

When the target (workpiece) is spherical, $A = 1$; and when the target is an irregular shape, $A < 1$,

The drag coefficient C_d is related to the Reynolds number.

$$\text{When } 1 < \text{Re}, C_d = 24/\text{Re} \tag{2.13}$$

$$\text{When } 1 < \text{Re} < 1000, C_d = 3\left(Ds\mu_s\rho_w/L\right)^{-0.625} \tag{2.14}$$

$$\text{When } \text{Re} > 1000, C_d \approx 0.45$$

where, L is fluid viscosity, and ρ_w is the density of water. μ_s can be obtained through the combination of Equations 2.12 and 2.14.

Due to the extensive influence of gravity and shape, and other reasons, objects fall into the water at different rates [57]. Thus, the suspension is divided into different layers, such as upper (the dirt), middle (broken abrasive particles), and bottom layer (integrated abrasive particles). The integrated abrasive particles at the bottom layer are pumped to the AWJ cutting system while dirt and broken abrasive particles are separated away by the separator [52].

2.4 TECHNIQUES FOR IMPROVING AWJ CUTTING PERFORMANCE

Machining applications with AWJM technology are normally required to be carried out quickly, efficiently, and economically. In addition, the surface quality of the material obtained after the process is expected to be in accordance with the desired level. However, some obstructions in terms of the AWJ cutting capability such as cut depth, kerf, and surface qualities limit its widespread use despite its many advantages over the other conventional and advanced manufacturing technologies. Low traverse speed at high pressure can be selected to improve cut depth, kerf, and surface properties. However, this combination is not preferred in practice due to economic reasons. Therefore, various attempts have been made to incorporate new features in the AWJ machining process. Among these multi-pass cutting, increasing the machine capacity (higher water pressure), angling the jet forward in the cutting plane, and controlled nozzle oscillation are found to have some impact in improving the cutting performance without any additional cost to the cutting process [54, 58, 59].

2.4.1 Multi-Pass Cutting

Multiple passes cutting along the same cutting line with an appropriate combination of cutting parameters provide a distinct advantage over single-pass cutting [18, 60]. A multi-pass cutting operation is one where the nozzle travels over the same kerf a number of times. This technique has a positive effect especially on the cut and cutting-wear zone depths. It can be noted that the cut depth and cutting-wear zone depth increase linearly at first, and then a decrease begins when a critical level is reached in a multi-pass cutting application [61–63]. More research work is needed on multi-pass cutting to develop the optimization strategies for choosing the optimum number of passes and cutting parameters in different workpieces.

2.4.2 Angling the Jet Forward in the Cutting Plane

At shallow cut depths, the direction of the jet impacting on the workpiece does not change much. With an increase in cut depth (through the workpiece thickness), the impact angle increases. With the widening of the impact angle, the cutting energy of the abrasive particles in the jet decreases and the cutting direction changes due to the jet tail-back effects, and more waviness/striation structure occurs on the cut surface. As a result, it is not able to cut effectively. For this reason, it is possible to improve the cut surface quality by a method of the jet contacting the workpiece from different angles (angling the jet forward in the cutting plane). Changes in the jet impact angle can lead to a change in the erosion mode effective in material removal, resulting in a reduction of the waviness/striation structure on the cut surface [64]. The optimum jet impact angle to be selected in the method is directly related to the properties of the workpiece [14, 64, 65].

2.4.3 Increasing Machine Capability

Increasing the water pressure can improve some major cutting performance indicators [66]. Similarly, the use of an ultra-high-pressure at 690 MPa shows that high water pressure can increase the cutting efficiency, especially the material removal rate and narrow kerf taper. Since there is a critical value for water pressure depending on other process parameters, increasing water pressure may not always produce efficient results. In addition, the use of ultra-high water pressure can present some technical problems related to the water jet system's availability and high maintenance costs that also limit the successful applications [67].

2.4.4 Controlled Nozzle Oscillation

The controlled nozzle oscillation technique is one of the most effective methods, developed to improve AWJ cutting performance on the basis of the cut and cutting-wear zone depths [58, 61]. The method works on the principle of a pendulum-like forward and backward nozzle motion in the cutting plane at a given angle (amplitude) and frequency. In this method, oscillation angle and frequency are considered as two main factors affecting the cutting performance. In other words, the contact angle of the jet to the workpiece is 90°, and it does not change during cutting in the normal cutting method with the AWJ. In the oscillation technique, the contact angle of the jet can be controlled, and it can be changed depending on the properties of the workpiece [3].

2.5 TECHNOLOGICAL DEVELOPMENTS IN AWJ CUTTING APPLICATIONS

Magnetorheological abrasive water jet, cryogenic abrasive water jet, thermally assisted abrasive water jet, and ice-assisted abrasive water jet are some types of AWJ technology developed in recent years.

2.5.1 MAGNETORHEOLOGICAL AWJ CUTTING/POLISHING

In classic AWJ technology, the growth of the diameter of the jet exiting the nozzle prevents efficient cutting. This is undesirable as it requires secondary treatment, especially where precise machining is required. Magnetorheological fluids can be used in AWJ technology to eliminate this undesirable situation [68, 69]. This method is described as a method obtained by adding some chemical additives and dispersed ferromagnetic particles to the classical abrasive–water mixture. Ferromagnetic particles in the mixture cause the abrasive particles to be arranged in a particular structure with high pressure in the presence of a magnetic field. Under magnetization, particles are organized into a spatial structure, which has higher yield stress, among other things. Magnetorheological liquid inside the nozzle is suppressed and internal jet vacancies are filled. As a result, a highly chained and collimated fluid jet can be obtained inside and beyond the nozzle [70]. In this way, the jet with increased viscosity is maintained at the nozzle exit, and the exit diameter is significantly prevented from growing, depending on the workpiece cutting depth. Maintaining the jet's exit diameter for a long time causes the abrasive particles in the jet to perform more effective or efficient cutting by preserving their KE for a long time [71]. The jet outlet diameter can only be effective up to a few nozzle diameters in classic AWJ technology, while the jet outlet diameter is effective up to ten times nozzle diameter in this method [72]. Despite its significant advantages, its high cost and technical complexity limit the use of the magnetorheological AWJ method [71].

2.5.2 CRYOGENIC AWJ CUTTING

Some soft workpieces cannot be precisely cut with conventional AWJ technology due to losses in their surface integrity. Cutting such workpieces with AWJ technology using coolant mixture can drastically change the material removal mechanism [71]. The cooling process causes some changes in the structure of the workpiece in machining applications. This reduces the resistance of the workpiece to a mechanical external influence. Therefore, the use of the cooling process in cutting/processing technologies can be considered as an alternative method in terms of both

performance enhancement and environmental aspects [73]. There are different coolant alternatives such as oxygen, hydrogen, nitrogen, helium, and carbon dioxide that
can be used as coolants in cutting technologies. Among them, liquid nitrogen is the
most preferred type due to its significant advantages such as not having any harmful effects, easy disposal after cutting, and not being corrosive and flammable [74].
The number of abrasive particles embedded in the surface during cutting has been
significantly reduced as a result of the coolant delivered to the workpiece because it
hardens the surface of the workpiece. This results in a higher quality surface after
cutting [74]. Despite the increase in cutting performance provided by this method, it
increases the total energy consumption and number of microcracks that may occur
due to high level of cooling. Such effects limit its use [71, 75].

2.5.3 THERMALLY ASSISTED AWJ MACHINING

The purpose of the thermally assisted AWJM method is to heat the surface of workpieces that are difficult to cut to reduce their hardness. Thus, hard surfaces are transformed into workpieces that can be cut more easily by deformation. The heat required
during machining is supplied by oxy-acetylene welding equipment connected to the
abrasive water jet, using acetylene as the main fuel. Acetylene is a hydro carbona-
ceous gas that is used as a fuel for welding applications in many industrial areas. A
simple schematic representation of the method is given in Figure 2.5.

2.5.4 ICE-ASSISTED AWJ MACHINING

As is well known, abrasive particles are added to the system in order to increase the
cutting power of the water jet and to cut hard and strong workpieces such as rock by
traditional AWJ technology. Although, there are various alternatives for this abrasive
material, garnet is commonly used [5, 77]. The addition of these abrasive particles,
including garnet, to the system brings along some undesirable situations such as
increased production and operating costs, environmental pollution, and nozzle wear
[78]. Ice particles can be used as abrasive particles for the elimination of these dis-
advantages [79]. A jet including ice particles has lower energy when compared to
AWJ cutting where garnet abrasive is used. However, it can be stated that ice-assisted
AWJ is low cost and it is environment-friendly. Ice-assisted AWJ machining can
offer many advantages in the applications, say, contaminated surfaces are cleaned.

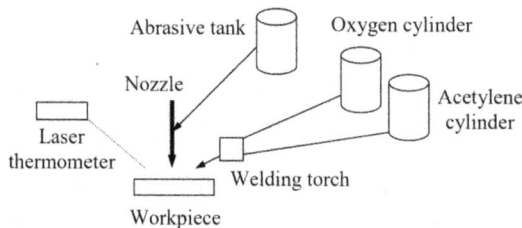

FIGURE 2.5 Schematic diagram of a setup of thermally assisted AWJM [76].

It can also be considered for applications where pollution is not desirable, such as the pharmaceutical and space industries, and food sector, as it is environmentally friendly [78, 80].

2.6 CONCLUDING REMARKS

The cutting geometry, specific cutting energy, and surfaces of the workpiece machined could be used as responses for evaluating the cutting performance of AWJM technology. These outputs are mainly affected by the operating variables and properties of the workpiece. Prior to the use of the AWJM process, the levels of the operating variables and properties of the workpiece should be effectively investigated, and an optimum combination of the operating variables for a specific workpiece should be determined in order to achieve the desirable cutting performance in terms of cut depth and surface roughness.

In AWJ cutting, the control of the cutting is mainly provided by changes in the traverse speed while other operating variables are generally kept constant during the cutting operations. For example, considering the pressure limits of the AWJ machine, the water pressure and abrasive feed rate can be selected as high as possible so as not to cause any problems such as clogging of the nozzle. Additionally, the stand-off distance is commonly adjusted at 3–4 mm. The traverse speed, on the other hand, needs to be determined depending on the properties of the AWJM technology and workpiece that will be cut, unlike other operating variables. It may be advised to keep the traverse speed low to achieve better cutting results. However, in this case, time will be a limiting factor. Therefore, it would be a useful approach to consider time as a limiting factor at the stage of determining the traverse speed.

Affecting the cutting performance and profitability of the AWJM technology, the abrasive properties such as type, hardness, gravity, shape, and toughness should be analyzed in detail prior to the selection of an appropriate abrasive. As an effective method for reducing the abrasive and disposal costs of the AWJM technology, the abrasive particles should be recycled by an appropriate recycling system.

Performance improvement in AWJ cutting technology could be provided with techniques such as increasing the machine capacity, angling the jet forward in the cutting plane, controlled nozzle oscillation and multi-pass cutting applications. With these techniques, it is possible to obtain better results without additional cutting costs. The technique of angling the jet forward toward the cutting plane can lead to the reduction of the waviness/striations on the machined surface by causing a change in the erosion mode. It can also have a positive effect on the kerf angle while the technique of controlled nozzle oscillation allows significant increase in cutting and cutting-wear zone depths. In addition, low surface roughness values can be obtained with the relevant technique. On the other hand, it is possible to get deeper cutting and cutting-wear zone depth and kerf angle with low surface roughness with the multi-pass cutting application by adjusting the relationship between the appropriate traverse speed and the number of passes in low-capacity operations.

Industrial demands for various parts of production have brought different technological searches for diversification of AWJ as well. Magnetorheological AWJ, cryogenic AWJ, thermally and ice-assisted AWJ are some types of AWJ technology developed in recent years. Soft workpieces can be effectively cut by the cryogenic AWJ method. Moreover, while the magnetorheological AWJ method is an effective method where precise surface cuts and/or machining is required to control the distribution of abrasive particles in the jet, the thermally assisted AWJ method is more effective in cutting/processing workpieces with hard surfaces. The ice-assisted AWJ method, which was developed to fill the gap between water and AWJ technologies, is more environmentally friendly than classical AWJ technology, and has higher cutting/processing efficiency than water jet technology.

As a result, AWJ cutting/machining technology to meet different industrial demands has become increasingly common in recent years. Considering the research and development studies on AWJM technology, it is possible to say that the technology will have a much more common use in the near future.

NOMENCLATURES/ABBREVIATIONS

A	Shape factor
AWJ	Abrasive water jet
AWJM	Abrasive water jet machining
C	Specific weight of water
C_s	Specific weight of particles
C_d	Drag coefficient
D	Diameter of the separator
D_s	Equivalent diameter
E_{awj}	Abrasive water jet energy
e_c	Specific cutting energy
g	Acceleration due to gravity
h	The height of the observed points above the reference plane
h_k	Kerf depth
KE	Kinetic energy
L	Fluid viscosity
p	Water pressure
Q	Abrasive flow rate
Q_w	Water flow rate
V	Cutting volume
v_a	Velocity of abrasive particles
v_{awj}	Speed of abrasive water jet
v_w	Water jet speed
v_{wj}	Speed of pure water jet
W_{bottom}	Bottom kerf width
W_{top}	Top kerf width
η	Momentum loss factor
μ_s	Fall velocity of particles
ρ_w	Water density
θ	Kerf angle

REFERENCES

1. Sitek, L., J. Foldyna, P. Martinec, J. Klich, M. Mašláň. 2012. On the preparation of precursors and carriers of nanoparticles by water jet techonology. *Tehničkivjesnik* 19(3):465–474.
2. Srinivasu, D. S., D. A. Axinte, P. H. Shipway, J. Folkes. 2009. Influence of kinematic operating parameters on kerf geometry in abrasive waterjet machining of silicon carbide ceramics. *International Journal of Machine Tools and Manufacture* 49:1077–1088.
3. Chen, F. L., E. Siores, K. Patel. 2002. Improving the cut surface qualities using different controlled nozzle oscillation techniques. *International Journal of Machine Tools & Manufacture* 42:717–722.
4. Engin, C. I. 2012. A correlation for predicting the abrasive water jet cutting depth for natural Stones. *South African Journal of Science* 108(9/10):1–11.
5. Saraçyakupoğlu, T. 2012. Analysis of material, pressure, cutting velocity and waterjet diameter's effect on the surface quality for the waterjet cutting. PhD dissertation, Eskisehir Osmangazi University.
6. Aydin, G., I. Karakurt, C. Hamzacebi. 2014. Artificial neural network and regression models for performance prediction of abrasive waterjet in rock cutting. *International Journal of Advanced Manufacturing Technology* 75:1321–1330.
7. Karakurt, I., G. Aydin, K. Aydiner. 2012a. A study on the prediction of kerf angle in abrasive waterjet machining of rocks. *Proceedings of the Institution of Mechanical Engineers, Part B: Journal of Engineering Manufacture* 226:1489–1499.
8. Karakurt, I., G. Aydin, K. Aydiner. 2012b. An experimental study on the cut depth of the granite in abrasive waterjet cutting. *Materials and Manufacturing Processes* 27(5):538–544.
9. Karakurt, I., G. Aydin, K. Aydiner. 2014. An investigation on the kerf width in abrasive waterjet cutting of granitic rocks. *Arabian Journal of Geosciences* 7(7):2923–2932.
10. Pawar, P. J., U. S. Vidhate, M. Y. Khalkar. 2018. Improving the quality characteristics of abrasive water jet machining of marble material using multi-objective artificial bee colony algorithm. *Journal of Computational Design and Engineering* 5:319–328.
11. Mrudhula, A. V. L., C. V. S. P. Rao, S. M. N. R. Singu. 2017. Experimental evaluation of optimal parameters for abrasive water jet machining process of granite. *International Journal of Engineering Trends and Applications* 4(2):15–19.
12. Zhong, Y. 2008. A study of the cutting performance in multipass abrasive waterjet machining of alumina ceramics with controlled nozzle oscillation. MSc dissertation, The University of New South Wales.
13. Dani, D. N., H. N. Shah, H. B. Prajapati. 2016. An experimental investigation of abrasive water jet machining on granite. *International Journal for Innovative Research in Science & Technology* 3(5):26–31.
14. Shanmugam, D. K., J. Wang, H. Liu. 2008. Minimisation of kerf tapers in abrasive waterjet machining of alumina ceramics using a compensation technique. *International Journal of Machine Tools & Manufacture* 48:1527–1534.
15. Belloy, E., E. Walckiers, A. Sayah, M. A. M. Gijs. 2000. Introduction of powder blasting for sensor and microsystem applications. *Sensors and Actuators A: Physical* 84:330–337.
16. Ergene, B., C. Bolat. 2019. A review on the recent investigation trends in abrasive waterjet cutting and turning of hybrid composites. *Sigma Journal of Engineering and Natural Sciences* 37(3):989–1016.
17. Hallmann, L., P. Ulmer, E. Reusser, C. H. F. Hämmerle. 2012. Effect of blasting pressure, abrasive particle size and grade on phase transformation and morphological change of dental zirconia surface. *Surface and Coatings Technology* 206:4293–4302.

18. Wang, J., D. M. Guo. 2003. The cutting performance in multipass abrasive water-jet machining of industrial ceramics. *Journal of Materials Processing Technology* 133:371–377.

19. Maros, Z. 2012. Taper of cut at abrasive waterjet cutting of an aluminium alloy. *Production Processes and Systems* 5(1):55–60.

20. Aswathy, K., P. Govindan. 2015. Modeling of abrasive water jet machining process. *International Journal of Recent Advances in Mechanical Engineering* 4(3):59–71.

21. Aydin, G., I. Karakurt, K. Aydiner. 2013b. Prediction of cut depth of the granitic rocks machined by abrasive waterjet (AWJ). *Rock Mechanics and Rock Engineering* 46(5):1223–1235.

22. Aydin, G., I. Karakurt, K. Aydiner. 2013c. Wear performance of saw blades in processing of granitic rocks and development of models for wear estimation. *Rock Mechanics and Rock Engineering* 46(6):1559–1575.

23. Kuram, E., B. Ozcelik, M. Bayramoğlu, E. Demirbas, B. T. Simsek. 2013. Optimization of cutting fluids and cutting parameters during end milling by using D-optimal design of experiments. *Journal of Cleaner Production* 42:159–166.

24. Jankovic, P., M. Madic, D. Petkovic, M. Radovanovic. 2018. Analysis and modeling of the effects of process parameters on specific cutting energy in abrasive water jet cutting. *Thermal Science* 22(5):1459–1470.

25. Hlavac, M. L., M. I. Hlavacova, V. Geryk. 2017. Taper of kerfs made in rocks by abrasive water jet (AWJ). *International Journal of Advanced Manufacturing Technology* 88:443–449.

26. Zhang, S., P. Nambiath. 2005. Accurate hole drilling using an abrasive water jet in titanium. *American Watejet Conference*, 2005, Houston, TX, USA.

27. Xu, S. 2005. Modelling the cutting process and cutting performance in abrasive waterjet machining with controlled nozzle oscillation. PhD dissertation, Queensland University of Technology.

28. Abdullah, R., A. Mahrous, A. Barakat. 2016. Surface quality of marble machined by abrasive water jet. *Cogent Engineering* 3(1):1178626.

29. Aydin, G., I. Karakurt, K. Aydiner. 2011. An investigation on the surface roughness of the granite machined by abrasive waterjet. *Bulletin of Materials Science* 34(4):985–992.

30. Aydin, G., I. Karakurt, K. Aydiner. 2013a. Investigation of the surface roughness of rocks sawn by diamond sawblades. *International Journal of Rock Mechanics and Mining Sciences* 61:171–182.

31. Alsoufi, S. M., K. D. Suker, W. M. Alhazmi, S. Azam. 2017. Influence of abrasive water-jet machining parameters on the surface texture quality of Carrara marble. *Journal of Surface Engineered Materials and Advanced Technology* 7:25–37.

32. Oh, T., G. Cho. 2014. Characterization of effective parameters in abrasive waterjet rock cutting. *Rock Mechanics and Rock Engineering* 47:745–756.

33. Oh, T., G. Joo, G. Cho. 2019a. Effect of abrasive feed rate on rock cutting performance of abrasive waterjet. *Rock Mechanics and Rock Engineering* 52:3431–3442.

34. Mıynarczuk, M., M. Skiba, L. Sitek, P. Hlavac, P. Hlavacek, A. Kozusnikova. 2014. The research into the quality of rock surfaces obtained by abrasive water jet cutting. *Archives of Mining Sciences* 59(4):925–940.

35. Gryc, R., L. M. Hlavac, M. Mikolas, J. Sancer, T. Danek. 2014. Correlation of pure and abrasive water jet cutting of rocks. *International Journal of Rock Mechanics and Mining Sciences* 65:149–152.

36. Gupta, V., P. M Garg, K. M. Batra, R. Khanna. 2013. Analysis of kerf taper angle in abrasive water jet cutting of Makrana white marble. *Asian Journal of Engineering and Applied Technology* 2(2):35–39.

37. Gupta, V., P. M. Pandey, M. P. Garg, R. Khanna, N. K. Batra. 2014. Minimization of kerf taper angle and kerf width using Taguchi's method in abrasive water jet machining of marble. *Procedia Materials Science* 6:140–149.

38. Arab, P. B., T. B. Celestino. 2017. Influence of traverse velocity and pump pressure on the efficiency of abrasive waterjet for rock cutting. *Soils and Rocks* 40(3):255–262.

39. Arab, P. B., T. B. Celestino. 2020. A microscopic study on kerfs in rocks subjected to abrasive waterjet cutting. *Wear* 448–449:203–210.

40. Fowler, G., I. R. Pashby, P. H. Shipway. 2009. The effect of particle hardness and shape when abrasive water jet milling titanium alloy Ti6Al4V. *Wear* 266:613–620.

41. Aydin, G., I. Karakurt, K. Aydiner. 2012. Performance of abrasive waterjet in granite cutting: Influence of the textural properties. *Journal of Materials in Civil Engineering* 24(7):944–949.

42. Ehlen, J. 2002. Some effects of weathering on joints in granitic rocks. *Catena* 49(1–2):91–109.

43. Rapple, R. R. 2014. Selecting the right waterjet abrasive. *The Fabricator.* http://www.thefabricator.com/article/waterjetcutting/selecting-the-right-waterjet-abrasive (Accessed April 15, 2016).

44. Oh, T., D. Park, D. Kong. 2019b. Performance analysis of sand abrasives for economical rock cutting using waterjet. *Korean Tunnelling and Underground Space Association* 21(6):763–778.

45. Aydin, G., S. Kaya, I. Karakurt. 2019. Effect of abrasive type on marble cutting performance of abrasive waterjet. *Arabian Journal of Geosciences* 12:357.

46. Long, X., X. Ruan, Q. Liua, Z. Chen, S. Xue, Z. Wu. 2017. Numerical investigation on the internal flow and the particle movement in the abrasive waterjet nozzle. *Powder Technology* 314:635–640.

47. Cha, Y., T. Oh, G. Cho. 2020. Effects of focus geometry on the hard rock-cutting performance of an abrasive waterjet. *Advances in Civil Engineering.* Article ID: 1650914.

48. Cha, Y., T. Oh, H. Hwang, G. Cho. 2021a. A simple approach for evaluation of abrasive mixing efficiency for abrasive waterjet rock cutting. *Applied Sciences* 11(4):1543.

49. Kuhmichel. 2021. Abrasives for experts. https://www.kuhmichel.com/en/ (Accessed April 12, 2021).

50. Cha, Y., T. Oh, G. Joo, G. Cho. 2021b. Performance and reuse of steel shot in abrasive waterjet cutting of granite. *Rock Mechanics and Rock Engineering* 54:1551–156.

51. Mason, T. O. 2021. Abrasives. https://www.britannica.com/technology/abrasive (Accessed April 12, 2021).

52. Dong, Y., W. Liu, H. Zhang, H. Zhang. 2014. On-line recycling of abrasives in abrasive water jet cleaning. *Procedia CIRP* 15:278–282.

53. Oh, T., G. Joo, Y. Cha, G. Cho. 2019c. Effect of garnet characteristics on abrasive waterjet cutting of hard granite rock. *Advances in Civil Engineering Volume.* ID: 5732649.

54. Vu, N. P. 2008. Performance enhancement of abrasive waterjet cutting. MSc. dissertation, Technische Universiteit Delft.

55. Aydin, G. 2014. Recycling of abrasives in abrasive water jet cutting with different types of granite. *Arabian Journal of Geosciences* 7(10):4425–4435.

56. Sicheng. 2021. Garnet sand prize in the world markets. https://blastinggarnet.com/garnet-sand-price-in-the-world-markets/ (Accessed April 28, 2021).

57. Aydin, G. 2015. Performance of recycling abrasives in rock cutting by abrasive waterjet. *Journal of Central South University* 22(3):1055–1061.

58. Wang, J. 2007. Predictive depth of jet penetration models for an abrasive waterjet cutting of alumina ceramics. *International Journal of Mechanical Sciences* 49:306–316.

59. Karakurt, I., G. Aydin, F. Yıldırım, S. Kaya. 2019. Methods for improvement of abrasive waterjet cutting performance. *26th International Coal Congress and Exhibition of Turkey*, April 16–19, 2019, Antalya/Turkey, 1:1340–1345.

60. Momber, A. W., R. Kovacevic. 1998. *Principles of Abrasive Waterjet Machining*. New York: Springer.

61. Wang, J., Y. Zhong, 2009. Enhancing the depth of cut in abrasive waterjet cutting of alumina ceramics by using multipass cutting with nozzle oscillation. *Machining Science and Technology* 13:76–91.

62. Wang, J. 2010. Depth of cut models for multipass abrasive waterjet cutting of alumina ceramics with nozzle oscillation. *Frontiers of Mechanical Engineering in China* 5(1):19–32.

63. Azhari, A., C. Schindler, C. Godard, J. Gibmeier. 2016. Effect of multipasses treatment in waterjet peening on fatique performance. *Applied Surface Science* 388:468–474.

64. Wang, J. 2003. The effect of jet impact angle on the cutting performance in AWJ machining of alumina ceramics. *Key Engineering Materials* 238–239:117–122.

65. Chen, F. L., E. Siores. 2001. The effect of cutting jet variation on striation formation in abrasive waterjet cutting. *International Journal of Machine Tools & Manufacture* 41(10):1479–1486.

66. Momber, A. W., R. Kovacevic. 1997. Test parameter analysis in abrasive jet cutting of rocklike materials. *International Journal of Rock Mechanics and Mining Sciences* 34(1):17–25.

67. Li, H. 2004. A study of the cutting performance in abrasive waterjet contouring of alumina ceramics and associated jet dynamics characteristics. PhD dissertation, Queensland University of Technology.

68. Jha, S., V. K. Jain. 2004. Design and development of the magnetorheological abrasive flow finishing process. *International Journal of Machine Tools and Manufacture* 44:1019–1029.

69. Tricard, M., W. I. Kordonski, A. B. Shorey. 2006. Magnetorheological jet finishing of conformal, freeform and steep concave optics. *CIRP Annal-Manufacturing Technology* 55:309–312.

70. Shorey, A. B., W. I. Kordonski, S. R. Gorodkin, S. D. Jacobs, R. F. Gans, K. M. Kwong, C. H. Farny. 1999. Design and testing of a new magnetorheometer. *Review of Scientific Instruments* 70:4200–4206.

71. Melentiev, R., F. Fang. 2018. Recent advances and challenges of abrasive jet machining. *CIRP Journal of Manufacturing Science and Technology* 22:1–20.

72. Kordonski, W., S. Gorodkin. 2011. Material removal in magnetorheological finishing of optics. *Applied Optics* 50(14):1984–1994.

73. Yuvaraj N., M. P. Kumar 2016a. Experimental study of cryogenically enhanced abrasive waterjet cutting of AISI D2 steel. *Proceedings of 6th International & 27th All India Manufacturing Technology, Design and Research Conference*, December 16–18, Maharashtra/India.

74. Yuvaraj N., M. P. Kumar. 2016b. Cutting of aluminium alloy with abrasive waterjet and cryogenic assisted abrasive waterjet: A comparative study of the surface integrity approach. *Wear* 362–363:18–32.

75. Gradeen, A. G., J. K. Spelt, M. Papini. 2012. Cryogenic abrasive jet machining of polydimethylsiloxane at different temperatures. *Wear* 274–275:335–344.

76. Patel, D., P. Tandon. 2015. Experimental investigation of thermally enhanced abrasive waterjet machining of hard-to-machine metals. *CIRP Journal of Manufacturing Science and Technology* 10:92–101.

77. Aydin, G., S. Kaya, I. Karakurt. 2017. Utilization of solid-cutting waste of granite as an alternative abrasive in abrasive waterjet cutting of marble. *Journal of Cleaner Production* 159:241–247.

78. Geren, N., M. Tutal. 2002. Latest developments and new trends in waterjet cutting systems and its situation in Turkey. *4th GAP Engineering Congress*, June 6–8, Şanlıurfa/Turkey (in Turkish).

79. Jerman, M., H. Orbanic, A. Lebar, I. Sabotin, P. Dresar, J. Valentincic. 2016. Measuring the water temperature changes in ice abrasive water jet prototype. *Procedia Engineering* 149:163–168.

80. McGeough, J. A. 2016. Cutting of food products by ice-particles in a water-jet. *Procedia CIRP* 42:863–865.

3 Electrical Discharge Machining

Mahavir Singh, J. Ramkumar, and V. K. Jain

CONTENTS

DOI: 10.1201/9780429160011-3

3.1   INTRODUCTION

Electrical Discharge Machining (EDM) is a pioneer machining process in the domain of electrophysical and chemical processes [1, 2]. This class of machining is also popularly categorized as the Advanced Machining Process (AMP) [3]. Based on the utilization of energy for the removal of material, it is characterized as an electrothermal type of AMP due to the input power supplied to the electrodes in the form of electrical energy, and subsequent conversion of this energy into thermal/heat energy. Thermal energy conducted to the workpiece (anode) and tool (cathode) from the plasma channel either melts or melts and partially vaporizes the material from both the electrodes [4]. This results in the formation of tiny cavities on the electrodes, known as craters [5]. Repetition of the occurrence of such craters results in the removal of material from the work surface and some material removal from the tool surface as well. In this manner, the tool's approximate reverse image (or approximate inverse replica) is copied on the workpiece with a slight deviation caused by tool wear and overcut. Figure 3.1 illustrates the basic configuration of the EDM process, depicting two electrodes separated by a small spark gap, i.e., Inter Electrode Gap (IEG) occupied by the dielectric fluid. Bubbles and debris particles are generated in the machining process [6]. The gap between the electrodes in the direction of motion is called the *primary discharge gap*, whereas the gap between the electrodes in the direction normal to the tool motion is termed the *secondary discharge gap* (contributes to radial overcut). The EDM process has excelled in the realm of complex micro/macro fabrication owing to its inherent advantages. These include, namely, that the process performance is not limited by the mechanical properties/hardness of the work material, absence of direct physical contact between the electrodes that facilitates deformation-free machining of delicate and thin parts, high dimensional accuracy and repeatability of the machined features, complex miniature

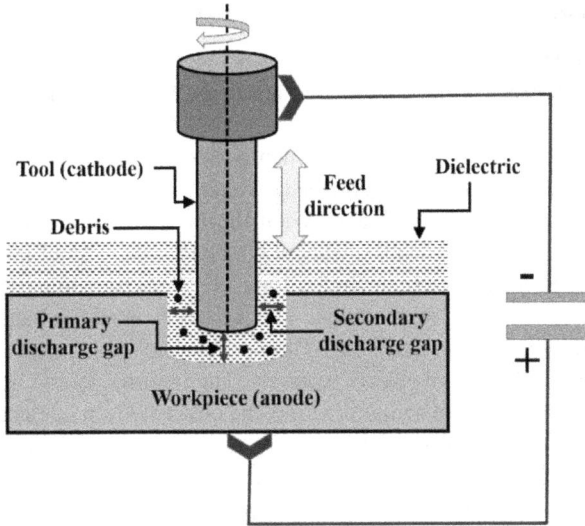

FIGURE 3.1 Schematic illustration of the EDM process.

machining through a simpler cylindrical tool, and high aspect ratio (ratio of a feature's depth to its width/diameter) machining, etc. [3, 7, 8].

Over the decades, several modifications have been accomplished toward improvement in machining accuracy and productivity of the process. These modifications have led to the evolution of various process variants and process performance improvement. Despite the modifications in the process configuration, fundamental understanding of the physical phenomenon occurring in the small IEG and material removal behavior remains of utmost importance for the scientific analysis of the process. This chapter discusses the scientific approach toward the fundamental aspects of the EDM process, beginning from the electronic breakdown of the dielectric fluid to that of the material removal from the electrodes surface, i.e., crater formation.

3.2 FUNDAMENTALS OF THE EDM PROCESS

The working principle of the EDM process is still debatable owing to the differences in material removal methods, which the researchers have postulated due to thermal energy, thermal shocks, and mechanical stresses [9]. At high-energy input to the workpiece (above 10 J), the mechanical impact of particles (fine solid particles or ionized atoms) dominates material removal. However, at a comparatively low energy level, erosion due to thermal effects is considered as a mechanism responsible for material removal through the creation of craters [10]. However, to a great extent, the electrothermal theory provides an insight into the material removal mechanism in the EDM process [11]. A brief working principle of the EDM process based on electrothermal theory is discussed as follows.

3.2.1 Mechanism of Material Removal in EDM

There are three major stages of the overall working of the EDM process, namely, the preparation phase, discharging phase, and power termination phase, which are discussed as follows:

According to electrothermal theory, when a potential is applied across the two electrodes, an electric field is established, which increases with the consequent increase in the instantaneous voltage. When the instantaneous voltage across the electrode (within the IEG) reaches a level at which the intensity of the electric field is sufficiently high (comparable or higher than the dielectric strength of the fluid), electrons are emitted from the cathode. The variation in the voltage and current with time for resistance-capacitance (R-C) and transistor-based power supplies are depicted in Figures 3.2(a) and (b), respectively. The intense electric field causes the cold emission of electrons. In actual practice, the cold emission of electrons requires some time once the voltage touches the breakdown value, and this time duration is termed the "ignition delay period", as indicated in Figure 3.2(b). The ignition delay period depends upon the spark gap, chemical properties of the dielectric fluid, open-circuit voltage, etc. At this moment, electrically non-conducting dielectric fluid breaks down due to ionization of its molecules by the electrons emitted initially and electrons generated subsequently through dielectric ionization. The electron avalanche breaks down the dielectric fluid, which results in the formation of a large number of electrons and other species such as positive ions and neutrals in the spark gap. Soon, the concentration of these ions and electrons becomes substantially high, that this state of matter is approximated as "*plasma*". The electrical

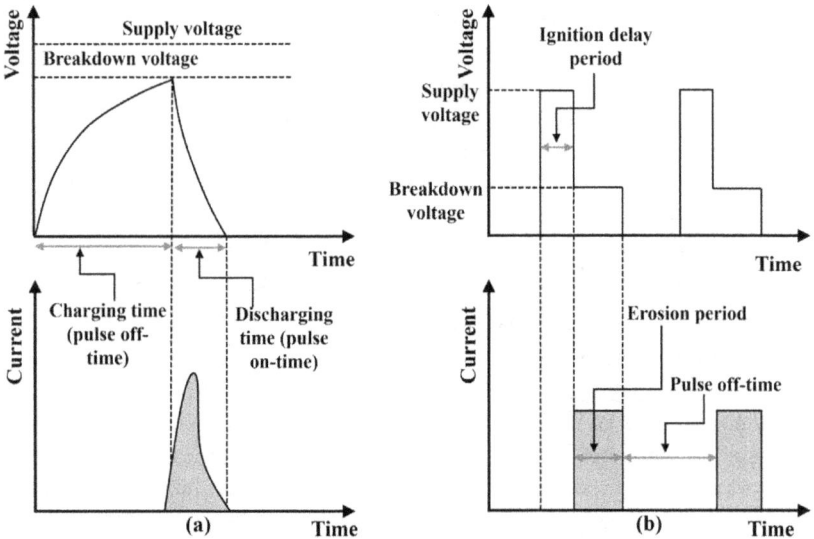

FIGURE 3.2 Typical variation of voltage and current with time (a) for R-C circuit, (b) transistor circuit.

resistance of the plasma during the "pulse on-time" is significantly low that it allows the flow of electrons and positive ions toward the anode and cathode, respectively. This movement could be visualized as a light beam called a *"spark"* [3]. Upon the bombardment of electrons and ions, their Kinetic Energy (KE) is transformed into intense heat energy. The resistive heating caused by the flow of current generates an extremely high temperature (10,000–12,000°C) in the plasma channel [12, 13]. Heat transfer occurs from the plasma channel to both electrodes. Since the electrons move approximately with the speed of light, the KE of electrons is much higher than that of positive ions. Hence, more material removal occurs from the anode compared to the cathode. Evaporation of dielectric and electrodes material produces a bubble encapsulating the plasma column. This generates immense pressure in the plasma channel that prevents the escape of evaporated material. After the pulse duration, the dielectric fluid must recover its dielectric strength for the subsequent sparks. In order to achieve reionization of the dielectric fluid and prevent the workpiece from over-heating, the EDM power is automatically switched off, as shown in Figure 3.2(b). The period during which there is no power supply across the electrodes is called the discharge interval (pulse off-duration). As the power is withdrawn from the plasma column, it explodes and generates a mechanical shockwave which expels molten metal from the crater. This explosion of the plasma channel also creates a partial vacuum that is responsible for the suction of fresh dielectric [14]. Hence, the liquid metal is expelled from the discharge region, and a crater is formed [15]. Inadequate flushing may result in an abnormal discharge or system failure [16].

The crater's size, shape, and overlapping frequency determine the final surface finish of the component [17]. Figure 3.3 demonstrates the steps involved (as discussed) in the plasma–material interaction, and crater evolution phenomenon in a typical EDM process. The generation of sparks at the locations, being equal to or smaller than the breakdown gap, erodes the material from the entire surface in the proximity of the tool. In this manner, approximately the inverse replica of the tool is formed on the workpiece [3]. The crater size (which determines the minimum machinable dimensions) formed in the EDM is a direct function of energy available per pulse, depending upon the machining current. Therefore, electrodes with high electrical resistivity could restrict the maximum flow of current during machining [18].

Generally, straight polarity (workpiece positive and tool negative) is preferred for low Tool Wear Rate (TWR) and high Material Removal Rate (MRR) [19]. It has also been observed that if tool and workpiece materials are alike, the anode erodes at a faster rate. Therefore, the workpiece is generally made an anode [20]. Several researchers have employed reverse polarity (workpiece negative and tool positive), especially for a long pulse on-time, resulting in comparatively high MRR, low TWR, and improved surface characteristics [21]. Still, the clear distinction between low and high pulse duration, as well as the suitability of reverse polarity for similar electrodes (tool and workpiece) material, is debatable. In some instances, reverse polarity has been attempted for the electric discharge deposition process identical to layer-by-layer manufacturing and micro coating for surface modifications [22, 23].

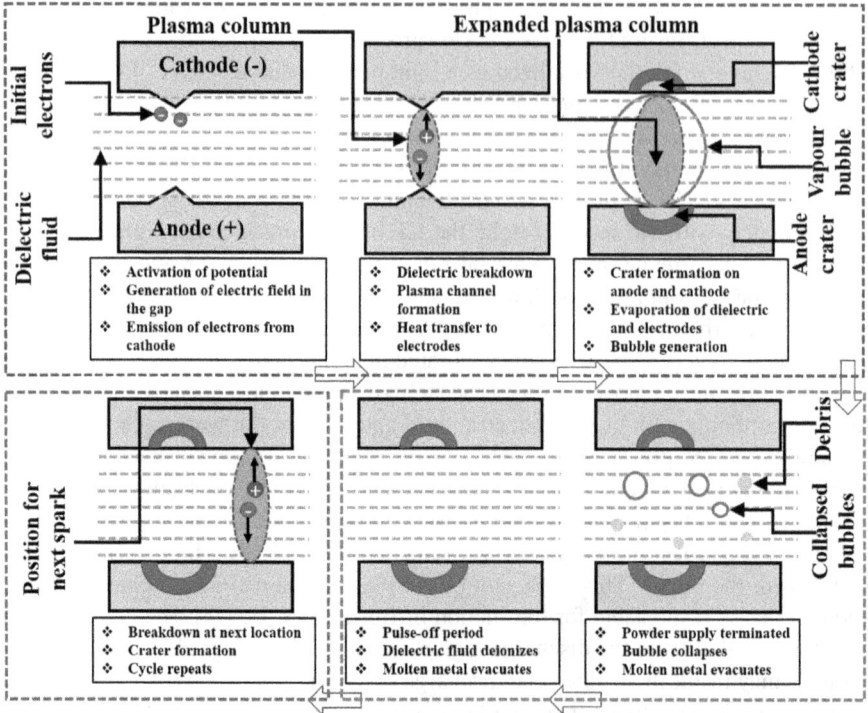

FIGURE 3.3   Plasma material-interaction and crater creation in the EDM process.

3.3   ELECTRON EMISSION FROM THE CATHODE AND PLASMA CHANNEL FORMATION

In the overall functioning of the EDM process, plasma column formation in the IEG, its expansion during the pulse duration, its exploding force as the power is withdrawn from the electrodes, heat/energy transfer from the plasma to the electrodes predominantly determine the nature of the process. Thus, it is essential to understand the mechanism of plasma column formation in the EDM process. The plasma column formation mechanism can be categorized into two major steps, i.e., primary electron emission from the cathode surface and intense ionization of the neutral molecules of the dielectric fluid. For the initiation of dielectric ionization and progression of plasma generation, initial electrons are required, which are supplied by the cathode surface.

3.3.1   Electron Emission from the Cathode

Electron emission from the cathode surface requires a minimum energy to remove an electron from the surface, contrary to only the knocking of electrons from an atom. To remove an electron(s) from a lower energy level (Fermi level) to the medium (dielectric fluid) in the vicinity of the solid surface, the minimum/threshold energy is required, which is termed *"work function"*. Work function is a characteristic of the cathode material. The process of electron emission from the cathode surface is attainable in the following ways:

3.3.1.1 Photoelectric Emission

Energy to the cathode is supplied in the form of photons with the energy of an individual photon given as $E = hv$, where h is Plank's constant and v is the frequency of the photon [24, 25]. There must be a critical frequency above which electrons are emitted from the cathode surface, and the excess energy is imparted as the KE gain of the electrons utilized for ionization of the dielectric fluid.

3.3.1.2 Thermionic Emission

At room temperature or sufficiently low temperature, the electrons in the atoms have inadequate energy to cross the work function barrier and thus remain within the atom [24]. However, as some external energy source heats the work surface, the atoms/electrons gain sufficient energy due to lattice vibration, and if it is sufficient to cross the work function of the material, electron emission occurs. The emission current in thermionic emission is related to the working temperature, given as follows [25]:

$$J = ZT^2 \exp\left(-\frac{\varnothing}{KT} \right) \tag{3.1}$$

where J is the current density, Z is a constant, $\varnothing$ is the work function, K is the Boltzmann's constant and T is the temperature. Equation 3.1 reveals that the electron/current density increases as the work function reduces, however it increases with a rise in the absolute temperature.

3.3.1.3 Field Emission

When an electrostatic field or electric field is induced in the gap, the electrons are removed from the cathode surface provided the electric field is sufficiently high to overcome the potential barrier of the cathode surface [24]. The field emission is also known as the *"tunnel effect"*. The threshold electric field required for inducing a microampere current through the field emission is in the range of 10^9–10^{10} V/m at considerably high applied potential (about 2–5 kV) [25, 26]. However, the magnitude of potential for field emission depends on the level of the IEG, which is significantly small in the EDM process. The effective level of the cathode's work function in the existence of electric field is reduced in the following way [26]:

$$\varnothing_{eff} = \varnothing_a - \sqrt{\frac{eE}{4\pi\varepsilon_0}} \tag{3.2}$$

where $\varnothing_{eff}$ and $\varnothing_a$ are the effective work function and original work function of the cathode material, respectively, e and E are the electronic charge and electric field, respectively and ε_0 is the permittivity of vacuum.

3.3.1.4 Electron Emission by Positive Ions and the Impact of High-Energy Atoms

It is also possible to knock the electrons from the cathode material through the impacts of positive ions and atoms, which are excited during the ionization process from the primary electrons. In fact, this kind of electron emission is termed secondary emission, as the generation of positive ions or excited atoms is achieved by the ionization process due to the primary emitted electrons from the cathode surface or electrons released during the series of ionization [24–26]. It is essential that the ions or atoms must have energy approximately twice that of the work function of the cathode material. It is because a part of the energy is consumed in the absorption of these ions by the cathode, and the remaining amount of energy is utilized to emit secondary electrons.

3.4   DIELECTRIC BREAKDOWN THEORY IN THE EDM

Dielectric fluid is the primary element of the EDM process, which serves various purposes starting from maintaining an electrically non-conductive path and creating a spark upon breakdown, evacuating debris, absorbing the excess heat from the tool, and confining the plasma column [27, 28]. The following sections discuss the breakdown mechanism of gases and liquid dielectric prevalently used in the EDM process.

3.4.1   Gas Breakdown Mechanism

Liquid-based dielectrics are extensively used in the EDM process, due to their ability to contain a plasma column in a smaller area, better flushing ability, and heat-absorbing capacity. But the breakdown mechanisms of the liquid dielectrics are not well established and uniformly accepted. However, the breakdown mechanism of gaseous dielectrics is better known, based on the ionization principle of gas molecules through primary electrons and subsequent avalanche of electrons created. The ionization mechanism in gases is adequately explained by Townsend's first and second ionization coefficient.

3.4.1.1   Townsend's First Ionization Coefficient

To understand the gas dielectric breakdown, two parallel electrodes separated by a gap, "d", is assumed. In the absence of any electric field, the gap filled with gas is in equilibrium condition, and it is disturbed when an electric field of certain intensity is applied across the electrodes. The variation in current flowing in the gap with respect to the applied potential is shown in Figure 3.4. As the potential is applied, the current flow in the gap is proportional to the applied voltage (V) up to the saturation current (i_0) corresponding to the potential V_1. After that, the current becomes almost constant until the potential is increased to V_2. Due to the increased electric field at higher applied potential (beyond V_2), an exponential rise in current is observed. It can be explained by the ionization of gas molecules by the initially liberated electrons from the cathode. The ionization results in a chain reaction and creates a higher number

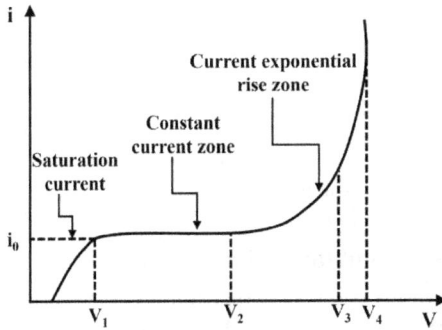

FIGURE 3.4 Variation in current with applied voltage during gases dielectric breakdown.

of ions and electrons in the gap. These electrons attain higher KE at higher potential, i.e., at the higher electric field during the collisions, and create further ionization.

Townsend's first ionization coefficient explains an exponential rise in the current, as depicted in Figure 3.4. Let n_0 be the initial number of electrons emitted from the cathode. These electrons, after moving a distance x from the cathode surface toward the anode, become n. Additional dn electrons are generated in the gap when the n electrons travel a distance dx. Therefore, an increment in the electrons can be represented in the following way [25, 26]:

$$dn = \alpha n dx$$

$$\frac{dn}{n} = \alpha dx$$

$$\ln(n) = \alpha x + A \tag{3.3}$$

Here, A is the constant of integration. At $x = 0, n = n_0$ and $A = \ln(n_0)$. Putting A in Equation 3.3, we get:

$$\ln(n) = \alpha x + \ln(n_0)$$

$$\alpha x = \ln\left(\frac{n}{n_0}\right) \tag{3.4}$$

$$n = n_0 e^{\alpha x}$$

At $x = d$ (IEG), the total number of electrons incident on the anode surface is given as follows:

$$n = n_0 e^{\alpha d} \tag{3.5}$$

Therefore, the current flowing in the IEG is represented as follows:

$$I = I_0 e^{\alpha d} \tag{3.6}$$

Here, α is Townsend's first ionization coefficient, representing the number of electrons produced by a single electron for the unit travel in the electric field direction. $e^{\alpha d}$ depicts the total electrons created in the gap due to the ionization process by a single electron traveling from the cathode to the anode surface (distance d) [25].

3.4.1.2 Townsend's Second Ionization Coefficient

Taking a logarithm of both sides of Equation 3.6 and plotting current (I) versus IEG ($x = d$) for a constant electric field (E) and gas pressure (P), the resultant curves are shown in Figure 3.5. There is a linear variation of logarithm of current ($\ln I$) with IEG ($x = d$) having $\ln I_0$ as intercept and α as the slope. However, it has been established by Townsend that there is a deviation of current (solid line) from the linear variation (dotted line) with an increase in voltage. It is attributed to the emission of secondary electrons from the cathode surface due to the positive ions and photons [25].

Let, n_0 and n_+ represent the number of electrons emitted by the cathode by the primary field and due to positive ion impact, respectively. Therefore, the number of electrons reaching the anode (n), as per Equation 3.5 is given by Equation 3.7 [25, 26]:

$$n = \left(n_0 + n_+\right)e^{\alpha d} \tag{3.7}$$

The total number of electrons released by the gas dielectric are calculated by $\left[n - (n_0 + n_+)\right]$. Corresponding to each electron released by the gas, one positive ion is liberated, and subsequently, this positive ion emits electrons from the cathode. Let Y is the Townsend's second ionization coefficient which depicts the number of

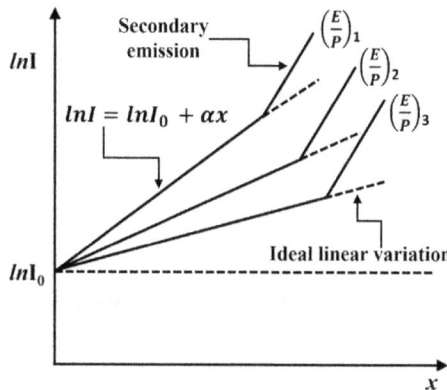

FIGURE 3.5 Plot of logarithmic current vs gap distance (IEG = $x = d$) at different ratios of electric field (E) and dielectric pressure (P).

electrons emitted from the cathode by every incident positive ion. Therefore, the total number of electrons released by the cathode is defined as follows:

$$n_+ = \Upsilon\left[n - \left(n_0 + n_+\right)\right]$$

$$n_+ = \frac{\Upsilon\left(n - n_0\right)}{1 + \Upsilon}$$

(3.8)

Putting Equation 3.8 into Equation 3.7, we get:

$$n = \left(n_0 + \frac{\Upsilon\left(n - n_0\right)}{1 + \Upsilon}\right)e^{\alpha d}$$

(3.9)

Simplifying Equation 3.9, we get:

$$n = \frac{n_0 e^{\alpha d}}{1 - \Upsilon\left(e^{\alpha d} - 1\right)}$$

(3.10)

Expressing Equation 3.10 in terms of current:

$$I = \frac{I_0 e^{\alpha d}}{1 - \Upsilon\left(e^{\alpha d} - 1\right)}$$

(3.11)

Υ is related to a work function of the material. The smaller the work function of the cathode material, the more significant will be the emission. Moreover, Υ increases with the ratio of electric field (E) and gas pressure (P).

3.4.2 BREAKDOWN MECHANISM OF LIQUID DIELECTRIC

The theory of liquid dielectric breakdown is essentially an extension of gas breakdown, which is based on an avalanche of electrons created by the ionization process of dielectric molecules in the existence of an applied electric field. Secondly, the liquid breakdown (particularly impure fluid) is also explained by the presence of foreign particles or impurities such as metallic particles, machined debris, or vapor bubbles. Sometimes it is also known as *"bubble theory"* as the suspended bubbles instigate the breakdown of the gap. Essentially, the inclusion of even a small volume fraction of foreign elements with relative permittivity greater than the base dielectric alters the breakdown phenomenon such that early breakdown of the gap (at lower potential) or at a higher discharge gap can take place.

3.4.2.1 Breakdown Due to Electron Avalanche

It is the most accepted and widely used breakdown mechanism of liquid dielectric, as discussed in the introduction section (thermoelectric theory). This breakdown mechanism is based on the release of initial electrons from the cathode surface, their interaction and subsequent ionization of dielectric molecules. The emitted electrons gain sufficient energy from the applied electric field and partially lose that energy in collision with dielectric molecules. Soon, the ionization process becomes so intense that an avalanche of electrons is created in the IEG. The energy attained by the electrons must be equal to or greater than that lost during the ionization process to achieve the electron avalanche, the condition for which is given as follows [25]:

$$e\lambda E = Ch\nu \tag{3.12}$$

Here, λ depicts the mean free path of electrons (charge e) in electric field E, $h\nu$ is the ionization energy, and C is a constant.

3.4.2.2 Suspended Particles Theory

The suspended particles theory of liquid dielectric applies to a contaminated dielectric. Commercially available dielectric fluids might contain some impurities such as solid particles or may become contaminated during the machining operation as the machined debris is not filtered adequately. These impurities significantly alter the dielectric strength of the fluid and modify the breakdown phenomenon.

Let us assume that the particle is circular having a radius of r and relative permittivity of ϵ_1, is mixed in the liquid dielectric of relative permittivity ϵ_2. Figure 3.6 shows forces acting on the particle in x- and y-directions. Forces acting only in the x-direction, i.e., along the dielectric flow direction, are considered for analysis [29]. As the potential is applied across the electrodes separated by a small gap, an electric field force (F) is exerted on the particle(s), as per the following equation [25]:

$$F = r^3 \frac{(\epsilon_1 - \epsilon_2)}{\epsilon_1 + 2\epsilon_2} E \frac{dE}{dx} \tag{3.13}$$

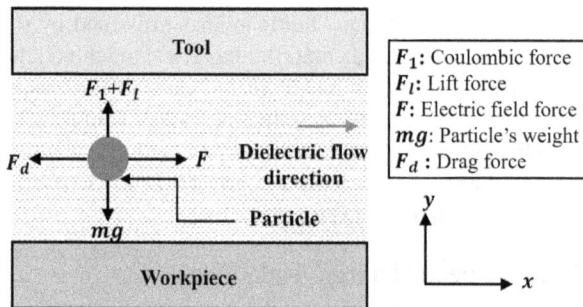

FIGURE 3.6 Schematic demonstration of forces acting on a particle suspended in the liquid dielectric.

Here, E is the applied electric field. This force directs the particles toward the region of the maximum field strength, provided $\epsilon_1 > \epsilon_2$. If ϵ_1 is tending to infinity as it is for metallic particles, then Equation 3.13 is modified as follows:

$$F = r^3 E \frac{dE}{dx} \tag{3.14}$$

This force (F) results in the alignment of particles toward the path of the highest electric field and causes charge concentration. Thus, more particles would migrate to this region such that a chain or particle bridge is formed, which causes intensification of electric field between the particles, and breakdown occurs. The motion of particles moving with a velocity (v) in a fluid of viscosity (η) is opposed by the viscous force, given as follows:

$$F_d = 6\pi\eta r v \tag{3.15}$$

The two forces given in Equations 3.14 and 3.15 are equal in magnitude but act in the opposite direction, thus equating Equations 3.14 and 3.15, we get:

$$v = \frac{r^2}{6\pi\eta} E \frac{dE}{dx} \tag{3.16}$$

Diffusion of particles is also considered, owing to the motion of particles at the higher field region. The drift velocity is expressed as follows:

$$v_d = \frac{-KT}{6\pi\eta r} \frac{dN}{Ndx} \tag{3.17}$$

where K is Boltzmann's constant, T is absolute temperature, and N is concentration of the particles. In the IEG, particles must have a single velocity. Thus, two velocities expressed by Equations 3.16 and 3.17 must be the same, hence we have:

$$\frac{r^2}{6\pi\eta} E \frac{dE}{dx} = \frac{-KT}{6\pi\eta r} \frac{dN}{Ndx}$$

$$\frac{-KT}{r} \frac{dN}{N} = r^2 E dE \tag{3.18}$$

Integrating Equation 3.18, the expression for the required electric field for dielectric breakdown is deduced as follows:

$$E_c^2 = E_0^2 - \frac{2KT}{r^3} \ln N \tag{3.19}$$

Equation 3.19 reveals the required electric field for breakdown, i.e., critical electric field (E_c) when dielectric is contaminated by the particles, is smaller than that required

(E_0) when the dielectric is pure, i.e., there is a reduction in dielectric strength. It is also evident that the electric field required for breakdown is related to the particle's radius (r), concentration (N), and absolute temperature (T). Moreover, it is a function of the relative permittivity of the powder material (ϵ_1) and liquid dielectric (ϵ_2), as per Equation 3.13, which is simplified for $\epsilon_1 \to \infty$.

3.5 PLASMA PROFILE AND ITS EXPANSION

In the previous sections, electron emission, ionization of dielectric fluid, and plasma channel formation are explained with the help of the gas and liquid breakdown mechanisms. Plasma channel formation and its expansion with the progression of a pulse period (pulse on-time) are essential for determining the maximum heat flux, the spot over which the heat flux is influential, and the extent of the machined crater's dimensions (diameter and depth) during a single-spark event. Due to the small IEG and extremely short pulse duration in which the plasma is ignited, the experimental visualization of the progression of the plasma column is a strenuous task. However, with the advent of high-speed imaging techniques, the researchers studied plasma column growth to ascertain the plasma size during various phases of the pulse on-time. Figure 3.7 shows the plasma channel growth during various stages of pulse on-time. Plasma grows during the initial stage of the pulse period, and its size is almost comparable to that of the gap width (IEG), as observed from the captured images after 0.7–2.7 µs. However, it remains invariably constant after that [30]. The size of the molten metal pool, which also determines the size of the crater, increases during the early pulse period and thereafter no significant change is observed due to a reduction in heat flux, which results in insignificant energy for melting the material (refer to Figure 3.8). However, the heat-affected region around the crater increases slightly under the influence of heat flux, as shown in Figure 3.8. This indicates that the diameter of the machined crater is slightly smaller than the plasma column diameter, as the expansion of the plasma column beyond a specific limit results in the Heat-Affected Zone (HAZ) without much change in the molten metal [31]. Figure 3.9 represents the variation in plasma column diameter and machined crater diameter on the copper (anode) and silicon carbide (cathode). The cathode crater is larger than the anode crater, though energy at the anode is high. The greater resistivity of silicon carbide compared with copper contributes to the Joule heating effect, which is negligible for metals (copper) [31]. Based on the assumption

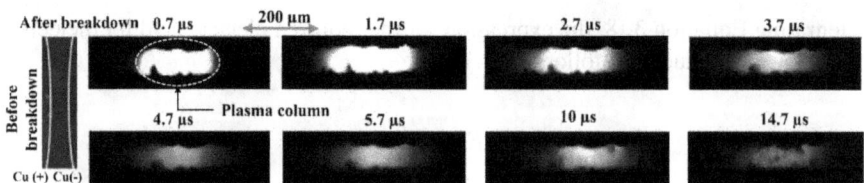

FIGURE 3.7 Captured plasma channel between copper electrodes at different stages of pulse on-time. Reproduced with permission from [30].

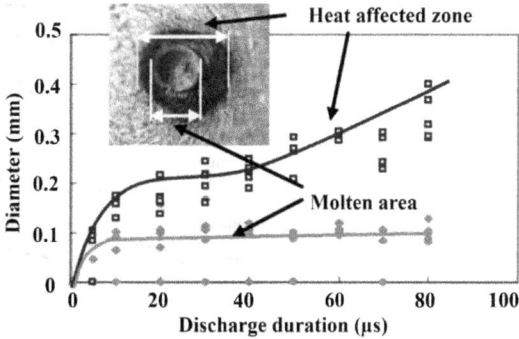

FIGURE 3.8 Variation of molten metal and HAZ with pulse duration for copper as tool and workpiece material. Reproduced with permission from [30].

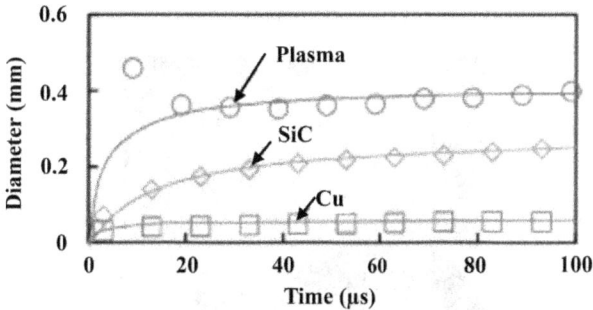

FIGURE 3.9 Comparison of plasma diameter and crater diameter on copper (Cu) anode and silicon carbide (SiC) cathode. Reproduced with permission from [31].

that the heat source radius is equal to the radius of the plasma column (evaluated experimentally by varying pulse on-time) and solving the heat conduction equation in the radial and vertical directions, the surface temperature obtained is less than the melting temperature of the copper workpiece. It reveals that the radius of the *intense heat source* (here, intense means the heat source x-section which has heat intensity capable of melting the workpiece material under its x-section) is less than the radius of the plasma column. Thus, it is indispensable to modify the heat source radius as an *intense heat source* radius so that the size of the molten metal and crater diameter can match [31].

3.5.1 PLASMA COLUMN VARIATION WITH DISCHARGE PARAMETERS

Spark profile, which also depicts the plasma channel profile variation in the IEG and its radius on the anode and cathode during different stages of pulse on-time, is shown in Figure 3.10. The spark radius increases with pulse on-time and current. Its radius is slightly higher at the cathode than at the anode, and at the center of the IEG it is the minimum. The relative variation in the spark radius can be explained

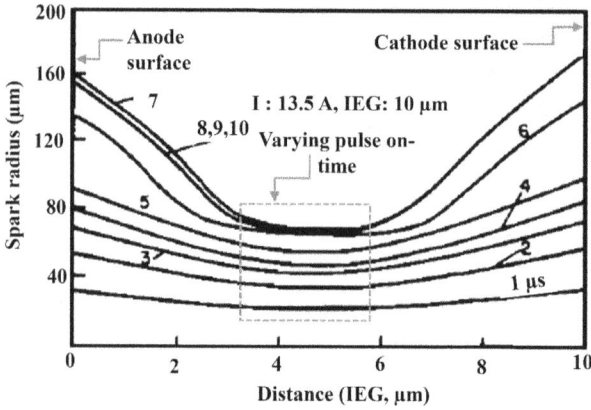

FIGURE 3.10 Plasma channel evolution on anode and cathode and its variation with pulse on-time in µs. Reproduced with permission from [3].

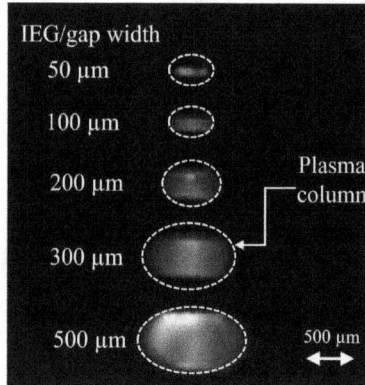

FIGURE 3.11 Plasma channel growth and its relative sizes for varying IEG. Reproduced with permission from [30].

by the electrical conductivity variation with temperature due to which the size of the plasma channel varies for allowing the constant input current [12]. The size of the plasma column is affected by parameters such as current, pulse on-time, dielectric fluid, and IEG. A higher discharge current refers to a higher degree of ionization, i.e., a high concentration of electrons and ions in the gap, necessitating a greater plasma column radius to allow the flow of current through the gap. Pulse on-time dictates the duration of energy supplied to the electrodes, thus assisting in the greater degree of ionization in the gap. Therefore, the plasma column expands with pulse on-time. It is essential to optimize the pulse on-time for higher machining efficiency as, beyond a certain pulse on-time, the increased plasma column radius reduces the heat flux. Liquid dielectrics (or dielectrics with higher viscosity) offer greater resistance to the plasma expansion due to viscous and inertial effects than the gaseous dielectrics, allowing a smaller plasma channel radius for liquid dielectrics.

IEG also determines the size of the plasma column, as is evident from the experimentally captured image of a plasma column at varying IEGs shown in Figure 3.11 [30]. There is a considerable increase in the plasma size when IEG is 500 μm as compared to that at 50 μm IEG or at lower IEGs. It can be attributed to the higher number of electrons, ions, or charge carriers present in the gap when the IEG is high, as discussed by Townsend's ionization theory. Therefore, to allow the flow of a higher number of charge carriers, a larger size plasma column is required. However, the heat flux decreases at a higher level of IEG, which hampers the productivity of the process.

3.5.2 ESTIMATION OF PLASMA RADIUS

Empirical/semi-empirical expressions are available for predicting plasma channel radius as a function of typical EDM process variables such as pulse on-time and current. These expressions are essential in the modeling of the crater formation in the EDM. Based on a variable mass plasma channel, the radius of the plasma column is determined by the underwater experiments given as follows [32]:

$$R = 0.788 \left(t_{on} \right)^{\frac{3}{4}} \qquad (3.20)$$

where R is in micrometers and t_{on} is in microseconds. Similarly, empirical equations are formulated for the plasma radius variation with pulse on-time and current. To determine plasma radius at the cathode surface, it is presumed that the maximum permissible temperature attainable is the evaporation temperature of the cathode material. Thus, the relationship between the cathode plasma radius and thermal properties of the cathode material is expressed as per the following relationship [33]:

$$T_v = \frac{E_d \eta_c 10^6}{K_t R \pi^{\frac{3}{2}}} \tan^{-1} \left(\sqrt{\frac{4 a t_{on} 10^6}{R^2}} \right) \qquad (3.21)$$

where T_v, K_t, and a are the vaporization temperature, thermal conductivity, and thermal diffusivity of cathode material, respectively. E_d and η_c are discharge energy of a pulse and its fraction conducted to the cathode, respectively. R and t_{on} are the plasma radius and pulse on-time, respectively.

3.6 PLASMA CHARACTERIZATION

Plasma composition, its temperature, electron density, and identification of the type of plasma are some of the critical elements by which EDM plasma is characterized. These parameters are related to the extent of material melting/vaporization and material removal in the EDM. There are certain methods for the characterization of the EDM plasma, such as the Langmuir method, microwave, interferometers, and

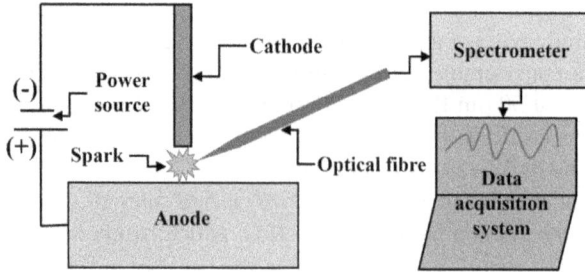

FIGURE 3.12 Schematic representation of optical emission spectroscopy setup for EDM plasma characterization.

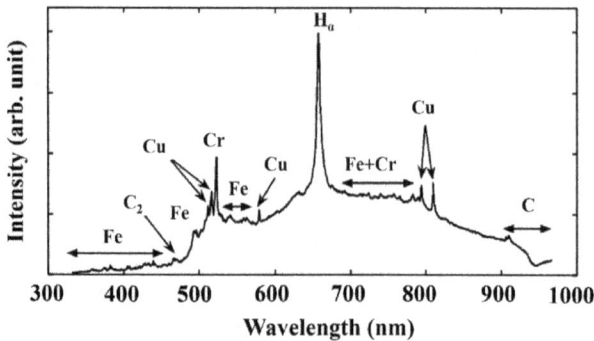

FIGURE 3.13 Typical spectra showing plasma composition (copper tool and steel workpiece, oil dielectric, current 12 A, pulse on-time 2 μs). Reproduced with permission from [36].

optical emission spectroscopy [34]. The Langmuir method necessitates the positioning of a probe in the IEG, which is challenging due to the minute gap between the electrodes. Microwave and interferometer techniques are limited by their lower accuracy due to various problems related to the process. Owing to the limitations associated with some of the above-mentioned techniques, optical emission spectroscopy of plasma characterization is one of the most commonly used and reliable methods [34]. The experimental setup of the optical emission spectroscopy consists of an optical fiber placed precisely in the vicinity of the IEG, a spectrometer to record the emission from the plasma column and a data acquisition system, as shown in Figure 3.12. Sometimes, an integrated system consisting of focusing lenses is utilized to capture the emission with better focus and higher resolution [35].

The composition of plasma is determined by the electrodes material and dielectric. The optical emission spectrometer provides information regarding the typical elemental composition of the plasma, as shown in Figure 3.13. The spectrum shows a significant peak of hydrogen lines (Balmer, H_α line) and some of the atomic carbons due to dissociation of hydrocarbon dielectric during the discharge. Iron and copper lines, due to the evaporation of the electrodes material, are also present in the spectra.

3.6.1 PLASMA TEMPERATURE

Measurement of plasma temperature is on the basis of the intensity ratio of the two spectral lines emitted by the atom of an element in local thermal equilibrium. The spectrum emitted by the EDM plasma can be atomic, ionic, molecular, and continuum emission. At a higher discharge energy level, the relative intensities of the spectral lines are directly related to the plasma temperature. Thus, the plasma temperature is measured by the following equation [34]:

$$T_p = \frac{E_m - E_i}{K} \left(\ln \frac{A_{mn} g_n \lambda_i I_i}{A_{ij} g_i \lambda_m I_m} \right)^{-1}$$

(3.22)

where λ_i and I_i represent the wavelength of the emitted spectrum by the element in plasma during its transition from energy level i to j, and radiant intensity of the emitted spectrum. E_i is the excitation energy at level i, A_{ij} is the probability of transition from energy level i to j. g_i is the statistical weight of energy level i. K is the Boltzmann's constant. Except I_i and I_m, other parameters in Equation 3.22 are obtained from the standard database. Therefore, plasma temperature measurement can be performed by obtaining I_i and I_m, corresponding to two distinct wavelengths.

3.6.2 ELECTRON DENSITY

The flow of current during the discharge in the IEG strongly influences the temperature generation in the plasma channel, as well as the KE transfer of the electrons to the anode. Therefore, electron density, which is a measure of the number of electrons present in the unit volume of IEG, plays a significant part in the EDM process. Plasma characterization allows the determination of electron density while using a pure dielectric fluid (liquid or gas(es)) and mixed dielectric fluid (dissolved electrically conductive substances in the liquid dielectric) [37]. There are methods available for the measurement of electron density based on the emission spectrum from the plasma. The original Saha equation [38], which is amended to derive a "modified Saha equation" is usually adopted for electron density measurement in cases of a dry-EDM process (using any compressed gas as a dielectric fluid), as per the following equation [39]:

$$n_e = 6.6 \times 10^{21} \frac{I_a}{I_i} \frac{A_i g_i}{A_a g_a} \exp\left(-\frac{E^{ion} + E_i - E_a}{T_p} \right)$$

(3.23)

Here, n_e represents electron density (/cm³). I_a and I_i are the radiant intensities, A_a and A_i are transition probabilities from energy levels i to j, g_a, and g_i are the statistical weights, E_a and E_i are the excitation potentials in eV, of the atomic and ionic lines of the same element, respectively. E^{ion} is the ionization energy of the element. Alternatively, the plasma electron density produced in the hydrocarbon-based dielectric, wherein hydrogen is also liberated during the dissociation process, can be evaluated. The Stark broadening and shift of hydrogen Balmer (H_α) line can be utilized for the determination of electron density. It is evaluated by the Full Width at Half Maximum (FWHM) for the H_α spectral line, as follows [37]:

$$n_e = 8.83 \times 10^{22} \left(\Delta\lambda_w\right)^{1.6} \tag{3.24}$$

$\Delta\lambda_w$ is the FWHM of the H_α line in nanometers.

3.6.3 Debye Length

It is defined as the length over which interaction between charges/separation of charges occurs. Therefore, higher the Debye length, the lower the probability of charge separation due to reduced interaction between the molecules. Two essential parameters, namely T_p and n_e are employed for the evaluation of Debye length (λ_d), given as per the following expression [34]:

$$\lambda_d = \sqrt{\frac{\varepsilon_0 K T_p}{n_e e^2}} \tag{3.25}$$

where ε_0 is the permittivity of free space, K is the Boltzmann's constant, and e is the charge of an electron. The Debye length for arc discharges is measured to be around 0.7 μm [34]. For classification of the plasma, a parameter generally termed the plasma parameter ($^\wedge$) is determined. This parameter defines the number of particles in a sphere called the Debye sphere having a radius equal to the Debye length. It is given as follows [34]:

$$^\wedge = 4\pi n_e \lambda_d^3 \tag{3.26}$$

For the value of $^\wedge \ll 1$, the plasma is termed as an ideal plasma. $^\wedge > 1$ refers to a strongly coupled plasma in which the Debye sphere is compactly populous, whereas $^\wedge \leq 1$ depicts a weakly coupled non-ideal plasma in which the Debye sphere is occupied dispersedly [36]. Based on the numerical value of plasma parameter for hydrocarbon dielectric, most of the EDM plasma is identified as weak and non-ideal [36], whereas for gases dielectric (dry-EDM), it is ideal [34].

3.7 ENERGY FRACTION TRANSFERRED TO THE ELECTRODES

Figure 3.14(a) schematically represents the distribution of energy of a single discharge to the tool, workpiece and in the IEG. A certain fraction of the discharge energy is transferred to the anode and cathode from the plasma column. The amount of energy transferred to the respective electrodes controls the extent of material removal from that electrode. A significant part of the heat energy delivered to the workpiece is lost by conduction, convection, and radiation, resulting in very small percentage of the total energy being used for the workpiece erosion. Generally, the

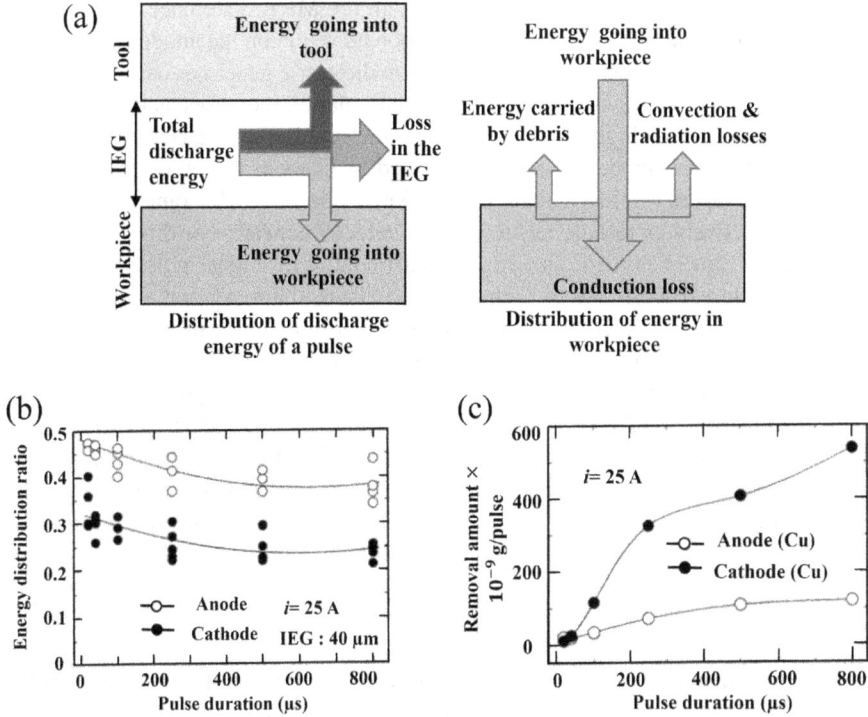

FIGURE 3.14 (a) Distribution of energy of a single discharge in the EDM. (b) energy distribution ratio to the anode and cathode with pulse duration. (c) comparisons of the removal amount from the anode and cathode with pulse duration (electrodes material: copper) [41].

energy transmitted to the anode is higher than to the cathode at all pulse on-time, as shown in Figure 3.14(b), when copper has been used both as the anode and cathode material. The electrons have significantly higher energy upon their impact on the anode surface than that of the positive ions impacting on the cathode. The combined Temperature-Field (T-F) theory explains the relative variation in energy sharing to the electrodes [40]. This is the primary reason for connecting the workpiece at the positive terminal of the power supply (i.e., anode) for higher material removal and tool to the negative terminal to minimize unwanted tool erosion. However, it has been observed that the anode erosion rate is not always higher at all pulse durations than the cathode erosion rate under certain conditions, even though the energy fraction available at the anode is constantly larger than that for the cathode. In fact, the anode erosion rate is higher during the initial pulse duration (about 20 µs). After that, it is smaller than the cathode erosion, as given in Figure 3.14(c) [41]. Therefore, under the prescribed conditions of the same copper as material for both the electrodes, it is not the energy transferred to the electrodes alone, which explains the erosion behavior of the electrode material. Instead, there is a secondary phenomenon which governs the difference in the volumetric removal rate from the cathode and anode, at different pulse durations. This discrepancy in the MRR difference on the anode and cathode is explained with the help of carbon adhesion on the anode at the higher pulse on-time [41]. Dissociation of hydrocarbon dielectric generates carbon particles that adhere to the anode surface, and their accumulation increases as the pulse progresses. Owing to the higher boiling temperature and lower electrical conductivity of the carbon layer, anode erosion is prevented or happens at a slower rate, which eventually hampers the MRR from the copper anode. This phenomenon has been used constructively by researchers to minimize tool wear rate. For this purpose, the tool electrode has been made as an anode, particularly for higher pulse on-time and when the hydrocarbon-based dielectric is used. Thus, the formation of carbon (turbostratic carbon) occurs on the tool (anode), which minimizes or sometimes prevents tool erosion to a larger extent, especially when a higher pulse duration is adopted [42]. Spectroscopic analysis of the plasma column confirms the presence of a lower density copper vapor (copper as a workpiece material), as the adhered carbon layer on the copper (anode) lowers its vaporization [43]. In addition to the phenomenon of carbon adhesion, at higher pulse on-time, the positive ions have adequate time to reach the cathode surface. Thus, owing to the higher mass of the positive ions and their numbers, their KE would be higher. Thus, the positive ions' effect predominates the electrons' effect on the cathode [9]. Therefore, the negative electrode erodes at a slightly higher rate for a longer pulse on-time and vice versa and it is also applicable to dissimilar electrodes materials [21].

Energy distribution to the electrodes is generally measured through the temperature measurement of a particular electrode (anode or cathode). The temperature obtained is compared with the erosion temperature generated under the analytical/numerical analysis while assuming a certain energy fraction going into an electrode. Therefore, following the "trial and error method" under the varying energy fraction, the assumed energy fraction at which a close match between the predicted and the measured temperature is attained, is considered as the fraction of the energy available at that electrode.

Some researchers have predicted the fraction of discharge energy of a pulse delivered to the electrodes by experimental, spectroscopic, analytical and numerical analysis of the EDM process. One of the fundamental theoretical models for the material removal phenomenon at the anode and cathode had assumed Gaussian distribution and point heat flux distribution, respectively. Based on the models, the energy partition factor to the anode and cathode are determined as 8% and 18%, respectively [15, 44]. The energy partition factor for the anode and cathode is evaluated to be 39% and 14%, respectively, by comparison of theoretical and experimental results [45]. The estimation of the energy fraction to the electrodes is based on the assumption that it remains constant for a set of electrode material and dielectric. However, the energy fraction conducted to the electrodes (say, workpiece) also varies with input variables, namely machining current, pulse on-time, etc. The thermal diffusivity of the electrode material also decides the energy fraction delivered to a particular electrode. Therefore, the copper cathode receives a higher percentage of energy than the steel anode. Moreover, the maximum temperature attainable at the copper cathode is lower than the steel anode, as the higher thermal diffusivity of copper conducts the incident heat more easily [12]. Considering a fixed pulse on-time, the energy fraction to the workpiece varies from 0.17–0.23, when current rises from 2–19 A. Conversely, for a constant value of current it varies from 0.07–0.53 when pulse on-time increases from 2–64 µs [46]. Therefore, it can be concluded that the energy fraction conducted to an electrode (anode or cathode) is governed by the combination of tool, workpiece, and dielectric fluid along with input process parameters.

3.8 DISTRIBUTION OF HEAT FLUX TRANSMITTED TO THE ELECTRODES

The primary source of thermal energy necessary for melting and partial vaporization of the electrode materials is the proportion of the discharge energy being transferred from the plasma column to the respective electrodes. The intensity of heat flux over the electrode surface is not uniform across the plasma material-interaction zone; instead, it is maximum at the center and then decreases as the distance from the center increases. Therefore, a portion of the discharge energy available for the erosion of workpiece material is assumed to be distributed as per the following ways:

3.8.1 POINT HEAT FLUX (PHF) DISTRIBUTION

It is assumed that the plasma column is active over a finite small area such that it can be approximated as a point. Point Heat Flux (PHF) distribution has been assumed by the early researchers, especially at the cathode [15], wherein, due to the smaller radius of the plasma column, the PHF distribution can be implemented. The heat flux distribution on the anode is represented by the following equation [47]:

$$q = \frac{\eta E_d}{2\pi \times r^2} \tag{3.27}$$

where η is the proportion of the total discharge energy (E_d) of a pulse transmitted to the workpiece and r is the distance from the center. q denotes the intensity of heat flux on the electrodes.

3.8.2 UNIFORM HEAT FLUX (UHF) DISTRIBUTION

Uniform heat flux (UHF) distribution considers the constant magnitude of heat over the assumed radius of the heat source. The heat flux distribution is represented by the following equation [47]:

$$q = \frac{\eta E_d}{\pi \times R^2} \tag{3.28}$$

where η is a proportion of the total discharge energy (E_d) of a pulse transmitted to the workpiece and R is the plasma radius. Analysis of a machined crater profile depicts the maximum depth at the center, and it reduces away from the center. It shows that the intensity of heat is maximum at the center of the plasma channel, and it reduces as the distance from the center increases.

3.8.3 TRANSIENT HEAT FLUX (THF) DISTRIBUTION

Heat flux is assumed to vary with the advancement in the pulse period. As the plasma channel grows with pulse on-time, there is a subsequent reduction in the heat flux to the electrodes. Thus, the variation of heat flux is transient in nature rather than the uniform intensity over the entire pulse on-time.

3.8.4 GAUSSIAN HEAT FLUX (GHF) DISTRIBUTION

Gaussian distribution of input heat flux to the workpiece accounts for the heat flux intensity variation with the radius of the plasma column. The heat flux distribution as a function of plasma radius and the distance from the center of the plasma column is given as follows [48]:

$$q(r) = Q_{max} \exp\left\{-4.5\left(\frac{x}{R(t)}\right)^2\right\} \tag{3.29}$$

where $R(t)$ is the time-dependent plasma channel radius, Q_{max} is intensity of the maximum heat flux at the center of the plasma channel, which is evaluated as follows [49]:

$$Q_{max} = \frac{(\eta \times 4.457 \times E_d)}{t_{on} \times \pi \times R(t)^2} \tag{3.30}$$

where η is the proportion of the total discharge energy (E_d) of a pulse transmitted to the workpiece and t_{on} is the pulse on-time. Figure 3.15 illustrates a typical crater profile (crater radius and depth) adopting Gaussian distribution of the heat flux and

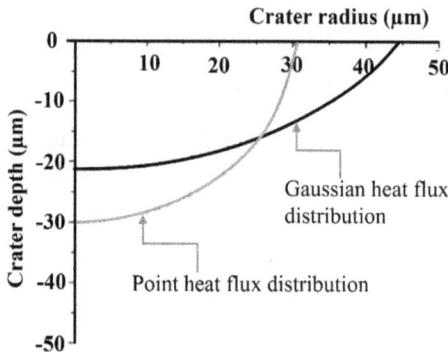

FIGURE 3.15 Simulated crater profile with Gaussian distribution of the heat flux and point heat flux distribution.

PHF distributions [50]. A shallower crater is obtained with Gaussian distribution than a point heat source, and a deeper crater is obtained with the PHF distribution.

3.8.5 COMPARISONS OF THE CRATER FORMED WITH DIFFERENT TYPES OF HEAT FLUX DISTRIBUTIONS

Plasma–material interaction results in the melting and evaporation of workpiece and tool material under the heat flux. This generates a small cavity known as a crater on both the electrodes. A crater is the unit removal in the EDM process, determining the nature of material removal, tool erosion, surface roughness, and thermal defects in the machined components. Therefore, to understand the fundamentals of the EDM process, the study of craters formed on the workpiece surface becomes essential. Figures 3.16(a–f) show the simulated crater profile for various heat flux distributions. The resultant crater is nearly circular in shape for PHF, as shown in Figure 3.16(a). Crater under UHF has nearly constant depth along its radius (refer to Figure 3.16(b)) due to the uniformity of heat flux. Time-varying heat source, as

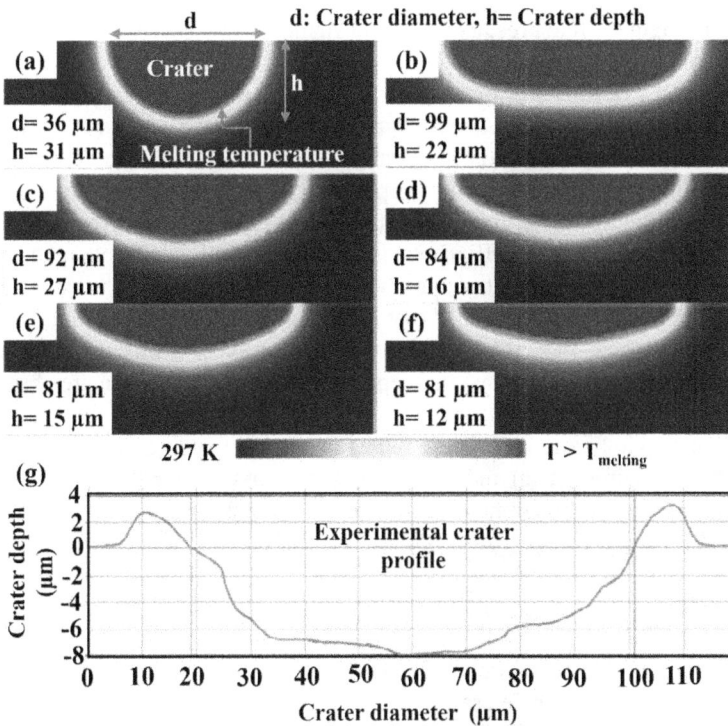

FIGURE 3.16 Simulated single crater profiles on the workpiece surface under various heat flux distribution: (a) point heat source, (b) uniform/disc heat source, (c) time-varying heat source, (d)+temperature-dependent material properties, (e)+latent heat of melting, (f)+latent heat of vaporization, (g) profile of an experimental crater. Reproduced with permission from [52].

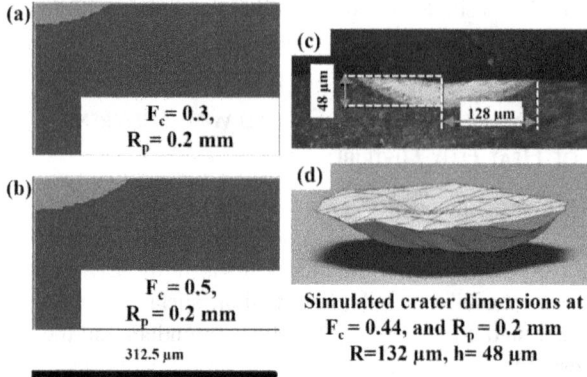

(a)

F_c= 0.3,
R_p= 0.2 mm

(b)

F_c = 0.5,
R_p = 0.2 mm

312.5 μm

(c)

48 μm   128 μm

(d)

Simulated crater dimensions at
F_c = 0.44, and R_p = 0.2 mm
R=132 μm, h= 48 μm

FIGURE 3.17   (a) and (b) Simulated crater profile using Gaussian distribution of the heat flux. (c) cross-sectional view of a single-machined crater. (d) 3-D image of the machined crater. (F_c and R_p are the fraction of energy and plasma radius, respectively. R and h are the crater radius and crater depth, respectively). Reproduced with permission from [51].

depicted in Figure 3.16(c), results in plasma channel expansion with time, thus generating a crater slightly bowl shaped. Inclusion of the temperature-dependent material properties, latent heat of melting and evaporation in the time-varying heat flux distribution results in a subsequent change in crater shape and size, as shown in Figures 3.16(d–f). Figure 3.16(f) shows a shallower crater, having a depth relatively shallower than the diameter owing to plasma channel expansion. The shape and dimensions of the simulated crater in Figure 3.16(f) are closer to the machined crater, as shown in Figure 3.16(g). PHF results in a considerably higher molten volume and unrealistic workpiece temperature, probably due to the higher energy density applied [51]. Further, it has been established that among all the aforementioned distributions of the available heat flux, the Gaussian distribution of heat flux to the electrodes generates a crater whose dimensions (depth and diameter) and shape are observed to be the closest to those of a single-machined crater at the same machining parameters [51], as depicted in Figure 3.17. The simulated craters at two distinct energy fractions are shown in Figures 3.17(a) and (b), whereas the cross-sectional and 3-D profile of the machined crater at the same input parameters is shown in Figures 3.17(c) and (d), respectively.

3.9   MODELING OF CRATER FORMATION IN THE EDM PROCESS

As discussed earlier, the nature of the EDM process is predominantly appraised via the understanding of a single crater machined on the workpiece surface. Moreover, the crater generated on the tool surface is equally important for the complete analysis of the EDM process, as the crater on the tool surface determines tool erosion behavior. Thus, for the scientific analysis of the EDM process, a single crater is formed on the workpiece surface based on the distribution of discharge energy of a single discharge from the plasma column, plasma/heat source radius, and pulse on-time. On analyzing a single crater, it is easier to understand the material removal mechanism of the actual EDM process as the machined surface is the amalgamation of a series of such single craters with certain overlaps. Over the years, analytical modeling, numerical simulation, and Molecular Dynamic Simulation (MDS) have been proposed for the modeling of a single crater event in the EDM process [53–57]. The theoretical modeling approaches rely on the solution of the heat conduction equation based on the initial and boundary conditions [58]. This provides the temperature distribution in the workpiece, which can be used for determining the crater dimensions.

3.9.1 ELECTROTHERMAL MODELING OF THE EDM

Electrothermal modeling of the EDM process consists of the incidence of the heat of discharge from the plasma to the workpiece, conduction of the input heat in the workpiece, and melting or partial evaporation of workpiece material based on the temperature distribution. In this section, a fundamental electrothermal model of the EDM process is discussed. The following assumptions are considered for electrothermal modeling [54]:

1. A circular heat source is assumed to be effective on the workpiece, having uniform intensity over the entire radius. The radius of the heat source is constant during discharge.
2. The heat transfer from the plasma channel to the workpiece is through conduction mode only.
3. Except for the portion of the workpiece wherein the heat source is incident, all other surfaces are assumed to be insulated.
4. Thermal properties of the workpiece material are temperature independent.
5. Mode of material removal is due to melting only. Vaporization and other modes of material removal are neglected.

Figure 3.18 shows a schematic diagram for thermophysical analysis of the EDM process. Assuming the workpiece to be a semi-infinite body, the conduction equation without heat generation for heat transfer in the workpiece is given as follows [54]:

$$\frac{\partial^2 T}{\partial r^2} + \frac{1}{r}\frac{\partial T}{\partial r} + \frac{\partial^2 T}{\partial z^2} = \frac{1}{a}\frac{\partial T}{\partial t} \tag{3.31}$$

where T is temperature, t represents time, r and z are the coordinates of the cylindrical coordinate system, and a is thermal diffusivity of the workpiece material.

The Initial Conditions (ICs) and Boundary Conditions (BCs) are defined as follows [45, 54]:

$$T(r,z,0) = 0 \tag{3.32}$$

$$-K_t \frac{\partial T}{\partial z}(r,0,t) = \begin{cases} Q, & (0 < r < R) \\ 0, & (r > R) \end{cases} \tag{3.33}$$

$$T(\infty,\infty,t) = 0 \tag{3.34}$$

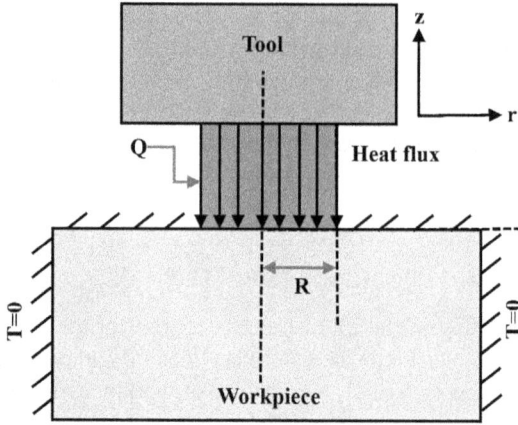

FIGURE 3.18 Schematic diagram of an idealized domain adopted for thermophysical modeling.

Applying ICs and BCs in Equation 3.31, the solution for temperature distribution can be represented by assuming a number of instantaneous point heat sources distributed within a circle of radius $(r = r')$ [54]:

$$T(r,z,t) = \frac{E_d}{4\rho c (\pi a t)^{3/2}} \exp\left[-\frac{r^2 + r'^2 + z^2}{4at}\right] \times I_0\left(\frac{rr'}{2at}\right) \qquad (3.35)$$

where E_d is the discharge energy per pulse (VIt'), ρ and c are the density and specific heat of the workpiece material, respectively. I_0 is the Bessel function (modified) of zero order. The solution for the workpiece under the influence of the heat flux, i.e., $(0 < r < R)$ can be obtained by integrating Equation 3.35.

$$T(r,z,t) = \frac{Q}{2\rho c \pi^{1/2} (at)^{3/2}} \exp\left[-\frac{r^2 + z^2}{4at}\right] \times \int_0^\infty \exp\left[-r'^2\left(\frac{1}{4at} + \frac{1}{r^2}\right)\right]$$

$$\times I_0\left(\frac{rr'}{2at}\right) r' dr' \qquad (3.36)$$

where Q is the discharge energy density expressed as follows [45]:

$$Q = \frac{\eta VI}{\pi R^2} \qquad (3.37)$$

where η is the fraction of the discharge energy going to a particular electrode (anode or cathode), V and I are the discharge voltage and current, respectively. Equation 3.36 is reduced to the following form by evaluating the integral by Laplace transformation:

$$T(r,z,t) = \frac{QR^2}{\rho c(\pi at)^{1/2}} \frac{1}{(4at+R^2)} \exp\left[-\frac{r^2}{(4at+R^2)} - \frac{z^2}{4at}\right] \qquad (3.38)$$

where R is the heat source radius, i.e., plasma channel radius. Temperature distribution in the workpiece for a finite pulse on-time (t') is given by the following equation:

$$T(r,z,t) = \frac{QR^2 a^{1/2}}{K_t \pi^{1/2}} \int_0^t \frac{dt'}{t'^{\frac{1}{2}}(4at'+R^2)} \times \exp\left[-\frac{r^2}{(4at'+R^2)} - \frac{z^2}{4at'}\right] \qquad (3.39)$$

where K_t is thermal conductivity of the workpiece material. Equation 3.39 is valid for a temperature lower than the melting temperature of the workpiece (T_m). To extend this equation for temperature exceeding the melting temperature, the specific heat term is modified to account for the latent heat of melting (H). Therefore, the modified thermal diffusivity (a') is defined as follows:

$$a' = \frac{K_t}{\rho c'}$$

$$\text{Where, } c' = c + \frac{H}{T_m}$$

The numerical solution of Equation 3.39 provides crater diameter, crater depth on the workpiece and tool surfaces at varying energy density (ratio of power per pulse to that of the heat source area), plasma radius, and pulse on-time. Other theoretical models have predicted the energy fraction to the cathode and anode, plasma flushing efficiency, cathode and anode erosion rate, etc. [15, 44]. However, in these models, point and Gaussian heat flux distributions are assumed for cathode and anode, respectively.

3.9.2 NUMERICAL SIMULATION OF THE EDM

Theoretical modeling of the EDM process approximates the crater dimensions based on the temperature distribution in the electrode material. However, the predicted results are less accurate than the actual EDM conditions due to the simplifying assumptions made in the analysis. Moreover, the melt flow, crater evolution, effect of plasma column on the crater, etc., are challenging to consider in the theoretical modeling. Therefore, numerical simulation becomes one of the most reliable and versatile methods of the advanced understanding of the EDM process through the analysis of a single crater. A basic numerical model consists of the input heat flux to the workpiece material being processed and its distribution (generally Gaussian distribution), radius of the plasma channel, pulse on-time, material properties (temperature-dependent or independent), etc. Figure 3.19 illustrates a typical 2-D domain adopted for the numerical analysis of EDM process (boundary 1 receives heat of the plasma and subjected to convection

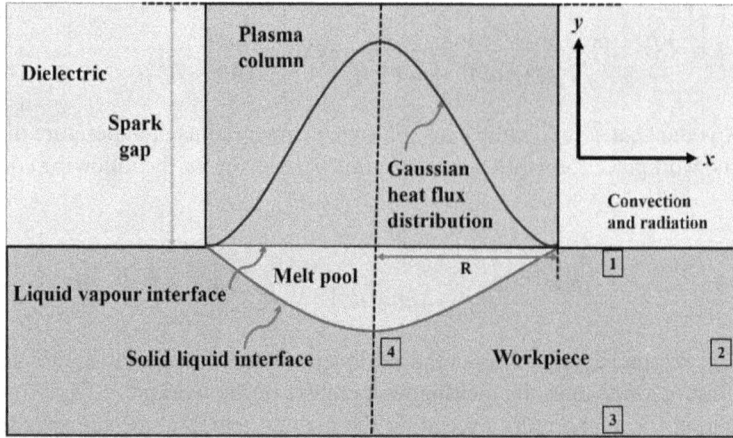

FIGURE 3.19 A typical 2-D physical domain used for numerical simulation. Reproduced with permission from [5].

and radiation losses, boundaries 2, 3, and 4 are insulated). Finite Element Modeling (FEM) solvers such as COMSOL Multiphysics, ANSYS, etc., are utilized to solve the numerical model, which results in creating a crater on the surface of interest. The simulated craters are compared with the machined crater for appraising the prowess of the model. The numerically simulated crater dimensions (diameter and depth) show a much closer match with the single-machined crater when the latent heat of melting and temperature dependency of the workpiece material thermal properties are accounted for [50]. Crater evolution during the pulse on-period (workpiece heating phase) and that during the subsequent bubble-exploding phase shows a significant role of initial bubble pressure (during the pulse off-period) in the crater formation. For different discharge energy conditions, different levels of initial bubble pressures result in a close agreement between the numerically simulated crater and experimental crater [59]. Similarly, there is a dominant role of plasma column pressure on the melt pool during the workpiece heating, i.e., during the pulse on-time [3]. The plasma pressure results in the delay in evaporation of the workpiece material, as is evident from the increased liquid-vapor interface temperature. Higher plasma pressure during the pulse on-period results in the evacuation of the molten material, violent evaporation, and deteriorated crater profile, thus necessitating the importance of initial plasma pressure in the EDM process [5].

It is also necessary to consider various forces acting on the melt pool during melting. These forces include normal plasma pressure acting on the melt pool, recoil pressure exerted by the escaping vapors, normal stress, and tangential shear stress due to melt pool curvature, and Marangoni convection caused by the surface tension gradient [5, 60]. Therefore, energy, momentum and mass conservation equations are solved simultaneously to show temperature distribution, melt flow, and velocity profile in the melt and vapor phases [60]. Figures 3.20(a–d) show the crater evolution on the Ti-6Al-4V workpiece during different pulse on-times. The simulated images represent material evaporation (mushroom plume),

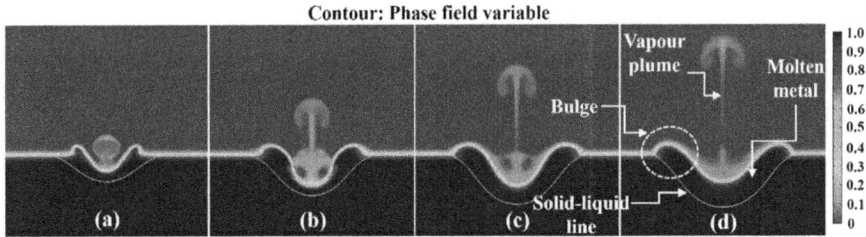

FIGURE 3.20 Crater evolution on Ti-6Al-4V alloy during different stages of pulse on-time: (a) at 6 μs, (b) at 8 μs, (c) at 10 μs, (d) at 12 μs. Reproduced with permission from [5].

outward melt flow due to recoiling effect, bulge formation around the crater, and backflow at later stages (as is evident by a slight reduction in the crater depth at 12 μs) [5]. Transient temperature and pressure variation in the molten metal during the workpiece heating phase is assessed along with the effect of molten metal flow in crater growth, bulge formation around the crater, etc. [61]. Numerical simulation enables the determination of material flushing efficiency, which is the ratio of the volume of the crater formed to that of the actual crater volume when the entire molten metal is evacuated. It is observed to be significantly low in the EDM process, thus leaving the majority of the molten metal on the crater surface in the form of a recast layer [5, 62]. However, the absence of dielectric flow and plasma-exploding force during the pulse off-period could explain the lower flushing efficiency obtained in the numerical simulation. Nevertheless, the lower machining efficiency of the EDM process is ascribed to the inadequate evacuation of molten metal from the crater, which is validated by the formation of the recast layer.

3.9.3 MOLECULAR DYNAMICS SIMULATION OF THE EDM

Molecular Dynamics Simulation (MDS) has also been performed to realize the mechanism of the EDM process at molecular level. MDS focuses on material removal at femtosecond level pulse duration for monocrystal material of nanometer dimensions which are extremely smaller than the actual EDM conditions. Thus, the spatiotemporal scale used in MDS is such that these results cannot be validated with the experimental values; however, they can provide a qualitative estimation of the physical events occurring in the narrow gap. The inter-atomic potential is adopted to estimate the energy of the system in MDS. The effect of the incident energy on material melting/evaporation is visualized as the atoms are leaving the surface. Based on the number of atoms above the melting and vaporization temperature, the contribution of vaporization and melting on a crater formation is evaluated [63]. The influence of the overlapping between the adjacent craters on material removal behavior is studied using MDS [63]. During overlapping of the craters, the depth and diameter of the previously machined craters decrease, but the evaporation rate increases for the subsequent discharges as the amorphous layer, which is generated due to the evolution of the previous crater, assists in

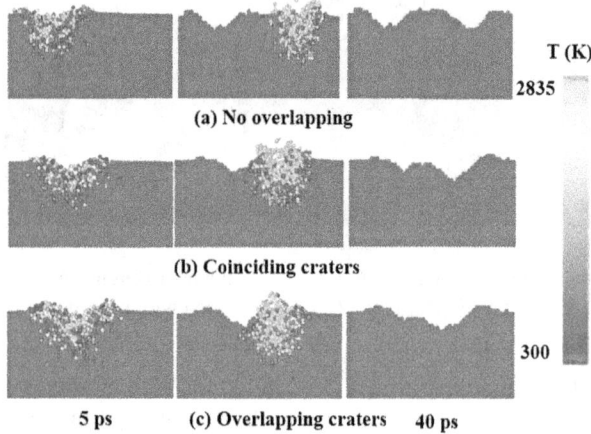

FIGURE 3.21 MDS images of crater formation in EDM at varying craters overlapping. Reproduced with permission from [63].

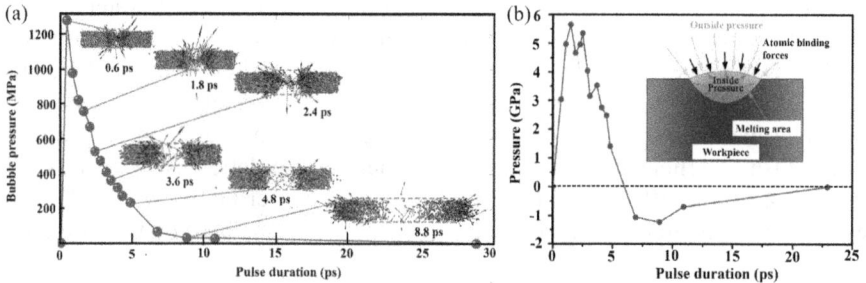

FIGURE 3.22 (a) Variation of bubble pressure with pulse duration; the arrows in the insets show velocity vectors and dotted lines show the bubble region. Reproduced with permission from [64]. (b) variation of pressure inside the melt pool with pulse duration. Inset shows various forces acting on a typical melt pool [65].

the ease of evaporation in the subsequent craters. Figure 3.21 shows the MD simulated craters for different overlapping conditions. Extremely low removal efficiency (below 5%) has resulted, due to resolidification of the molten metal [53]. The presence of a bulge around the crater edges due to resolidified molten metal and shear flow due to high pressure in the superheating material is observed.

MDS enables pressure estimation in the bubble generated around the plasma column and that in the workpiece region wherein the material removal occurs due to melting or partial evaporation and ejection of the molten metal. Figure 3.22(a) depicts the pressure variation in the bubble during different pulse duration [64]. It is necessary to analyze the variation of pressure in the workpiece molten region during pulse duration, as this pressure tries to overcome the pressure of the bubble for the evacuation of the liquid metal or evaporation of the workpiece material [65]. Figure 3.22(b) shows a typical pressure variation in an assumed cylindrical region (wherein melting and evaporation of material occur). As discharge takes place, conduction of heat energy to the workpiece results in high average pressure in the computational domain of the workpiece. With the progression of pulse duration, the average pressure decreases. It becomes negative and approaches zero toward the end of the pulse. Relative magnitude of the maximum bubble pressure and that within the molten metal reveal that higher pressure within the melt pool is responsible for material removal and melt ejection in the EDM.

3.10   EDM: POTENTIAL APPLICATIONS

The EDM process has been extensively utilized for micro/macro fabrication on difficult-to-process materials. The prominent and one of the early variants of the EDM process, i.e., die-sinking EDM, is suitable for die, mold, and complex fabrication by replicating the tool's mirror image (approximately) on the work material [66]. Microhole drilling for the plethora of applications, namely an array of microholes for the textile and filter industries, precise microholes for fuel injector nozzles in diesel engines and that for printers to dispense the ink, higher aspect-ratio cooling holes in turbine blades, etc., are machined accurately using Electric Discharge-Drilling (ED-Drilling) process. One of the recently developed and promising variants of the EDM process for complex 3-D fabrication using a simpler cylindrical tool electrode, i.e., Electric Discharge-Milling (ED-Milling) operation [67, 68] is apposite for the fabrication of µ-channels, µ-pillars, blind µ-pockets, turbine impeller machining, to name just a few. Texturing using the EDM process and its variants are also receiving substantial attention [69]. Wire Electric Discharge Machining (Wire-EDM) [70, 71] and microtool fabrication methods such as Wire Electric Discharge Grinding (Wire-EDG) [72] are versatile processes for the slicing of thick materials, and precise in situ fabrication of microtools for ED-Drilling and ED-Milling operations.

3.11   RESEARCH OPPORTUNITIES

The EDM process has shown its usefulness in the advanced machining regime, and it is regarded as one of the most versatile AMPs. Owing to its material removal behavior and control of the tool feed mechanism, the accuracy and precision achievable from the process are superior compared to other AMPs. Moreover, automation of the machine tools and their operation has brought innovation in the overall versatility of the process. Despite the number of benefits offered by the EDM process, there are some inherent limitations associated with the process, due to which incessant research and advancements are carried out to further improve the competency of the process. Keeping in view the shortcomings of the process, research opportunities in the following domains of the EDM process are encouraged:

3.11.1   Lower Machining Yield

The machining yield, i.e., the machining rate of the EDM process, is relatively low, which is attributed to the several avenues of energy losses that occur in the IEG. Besides energy losses, the non-productive discharges such as short circuits and arcing due to inadequate removal of debris also contribute to the poor machining rate. The inefficient exclusion of debris particles and liquid metal from the narrow IEG is one of the critical elements that bridges the gap and results in arcing and frequent short circuits. Over the years, several modifications in the process have been accomplished to assist efficient flushing in the IEG, including ultrasonic vibration assistance, magnetic field assistance, tool rotation, microslots and machining holes on the tools, gravity assistance for debris evacuation, powder mixed dielectric, etc. The aforementioned process modifications have improved the machining efficiency of the process; still, there are substantial opportunities to develop more suitable and well accepted techniques for enhancing the machining rate of the EDM process.

3.11.2 TOOL WEAR

Wear of the tool electrode is inevitable in the EDM process, as the discharge energy of a pulse is conducted to the workpiece and a tool in different proportions. Unwanted wear of the tool electrode deteriorates the dimensional accuracy of the machined features. Tool wear in the EDM process cannot be eradicated entirely. However, it can be minimized to an extent by suitable choice of tool material, process parameters, dielectric fluid, etc. Mitigation of tool wear and its adverse impact on dimensional accuracy is one of the challenges of the EDM process.

3.11.3 THERMAL DAMAGE

Thermal defects in the machined components such as residual thermal stresses, microcracks, HAZ, and yield strength reduction are prevalent due to high-temperature gradient developed during machining in the workpiece materials. These defects are detrimental to the functioning of any component. Thus, there is a need to analyze the nature of thermal damage and its minimization in the EDM process for its better performance and widespread acceptability.

3.11.4 HIGH SURFACE ROUGHNESS

Surface roughness of EDMed machined features is related to the depth of a single crater, which is a function of the discharge energy of a pulse. Thus, for a better surface finish of the machined components, the energy of a pulse needs to be low. However, it essentially reduces the machining rate. In addition to the depth of a crater, resolidified material on the crater surface known as the recast layer influences the roughness of the features. Research in the direction of minimization of recast layer formation and lowering the surface roughness using some appropriate techniques are being carried out for EDM performance enhancement.

3.11.5 MACHINING OF ELECTRICALLY NON-CONDUCTIVE MATERIALS

One of the limitations of the EDM process is its suitability to machine only to electrically conductive materials [73]. Electrically non-conductive (insulating) materials can be processed through the EDM process by the additive or assistive layer method, wherein a conductive layer is coated on the insulating material [74]. Primary sparks occur between the conductive layer and tool; simultaneously, due to the dissociation of hydrocarbon dielectric, a carbon layer is formed on the workpiece surface, which is electrically conductive. Thus, subsequent sparking between the tool and the formed carbon layer results in the machining of insulating materials such as ceramics. However, the assistive layer method is still an evolving technique and has certain limitations, which demand the development of advanced techniques for the efficient machining of insulating materials by the EDM process.

3.11.6 Green Manufacturing

Emission of aerosols and harmful gases from the machining zone can be potential health hazards to the operators and surroundings. Toward the environmental sustainability of the EDM process, the development of new dielectric fluids can be carried out besides dry-EDM, mist-EDM, and spray-EDM.

3.12 SUMMARY

This chapter focuses on the fundamentals of material removal phenomenon in the EDM process. The plasma channel, a primary heat source for material removal from the workpiece, is the most critical element in the EDM process. Therefore, the formation of the plasma channel, its behavior during the pulse on-time, plasma characterization, energy transfer to the electrodes, material melting and evaporation, melt flow during the crater evolution, and crater formation are discussed to understand the EDM process. Potential applications and prospective opportunities for research in some selected domains of the EDM process are highlighted.

ACRONYMS/ABBREVIATIONS

AMPs	Advanced Machining Processes
EDM	Electrical Discharge Machining
ED-Drilling	Electrical Discharge-Drilling
ED-Milling	Electrical Discharge-Milling
FEM	Finite Element Modeling
FWHM	Full Width at Half Maximum
GHF	Gaussian Heat Flux
HAZ	Heat-Affected Zone
IEG	Inter-Electrode Gap
KE	Kinetic Energy
MRR	Material Removal Rate
MDS	Molecular Dynamic Simulation
PHF	Point Heat Flux
R-C	Resistance-Capacitance
T-F	Temperature-Field

TWR	Tool Wear Rate
THF	Transient Heat Flux
UHF	Uniform Heat Flux
Wire-EDG	Wire Electric Discharge Grinding
Wire-EDM	Wire Electric Discharge Machining

SYMBOLS

K	Boltzmann's constant
N	Concentration of particles
E_c	Critical electric field
I	Current
j	Current density
λ_d	Debye length
ρ	Density
E_d	Discharge energy
Q	Discharge energy density
η	Discharge energy fraction conducted to the anode
η_c	Discharge energy fraction conducted to the cathode
V	Discharge voltage
v_d	Drift velocity
$\varnothing_{eff}$	Effective work function
E	Electric field
F	Electric field force
e	Electronic charge
ne	Electrons density
E_i	Excitation energy at level i
E_a	Excitation potential
υ	Frequency of light
$\Delta\lambda_w$	FWHM of the H_α line
P	Gas pressure
E^{ion}	Ionization energy
q	Intensity of heat flux
Q_{max}	Intensity of maximum heat flux
d	Interelectrode gap
H_α	Hydrogen Balmer line
H	Latent heat of melting
λ	Mean free path
T_m	Melting temperature
μ	Micro
c'	Modified specific heat
a'	Modified thermal diffusivity
n	Number of electrons
r	Particle's radius
v	Particle's velocity

ε_0	Permittivity of vacuum
h	Planck's constant
R	Plasma channel radius
$\wedge$	Plasma parameter
T_p	Plasma temperature
A_{ij}	Probability of transition from energy level i to j
t_{on}	Pulse on-time
I_i	Radiant intensity of the emitted spectrum
ε_2	Relative permittivity of dielectric
ε_1	Relative permittivity of powder particles
c	Specific heat
g_i	Statistical weight of energy level i
T	Temperature
K_t	Thermal conductivity
a	Thermal diffusivity
α	Townsend's first ionization coefficient
Υ	Townsend's second ionization coefficient
T_v	Vaporization temperature
F_d	Viscous force
η	Viscosity of fluid
λ_i	Wavelength of the emitted spectrum
$\varnothing$	Work function

REFERENCES

1. Ho KH, Newman ST. State of the art electrical discharge machining (EDM). *Int J Mach Tools Manuf* 2003;43:1287–1300. DOI: 10.1016/S0890-6955(03)00162-7.

2. Singh M, Saxena P, Ramkumar J, Rao RV. Multi-spark numerical simulation of the micro-EDM process: An extension of a single-spark numerical study. *Int J Adv Manuf Technol* 2020;108:2701–2715. DOI: 10.1007/s00170-020-05566-6.

3. Jain VK. *Advanced Machining Processes*. Allied Publishers Private Limited; New Delhi; 2007.

4. Mujumdar SS, Curreli D, Kapoor SG, Ruzic D. A model of micro electro-discharge machining plasma discharge in deionized water. *J Manuf Sci Eng* 2014;136:031011-1-031011–2. DOI: 10.1115/1.4026298.

5. Singh M, Sharma S, Ramkumar J. Numerical simulation of melt-pool hydrodynamics in μ-EDM process. *Procedia CIRP* 2020;95:226–231.

6. Kunieda M, Lauwers B, Rajurkar KP, Schumacher BM. Advancing EDM through fundamental insight into the process. *CIRP Ann – Manuf Technol* 2005;54:64–87. DOI: 10.1016/S0007-8506(07)60020-1.

7. Papazoglou EL, Obratański PK, LeszczyńskaMadej B, Markopoulos AP. A study on electrical discharge machining of titanium grade2 with experimental and theoretical analysis. *Sci Rep* 2021:1–22. DOI: 10.1038/s41598-021-88534-8.

8. Lin YC, Hwang LR, Cheng CH, Su PL. Effects of electrical discharge energy on machining performance and bending strength of cemented tungsten carbides. *J Mater Process Technol* 2008;206:491–499. DOI: 10.1016/j.jmatprotec.2007.12.056.

9. Schumacher BM. After 60 years of EDM the discharge process remains still disputed. *J Mater Process Technol* 2004;149:376–381. DOI: 10.1016/j.jmatprotec.2003.11.060.

10. Heuvelman C. Some aspects of the research on electro-erosion machining. *Ann CIRP* 1969;17:195–199.

11. Mamalis AG, Vosniakos GC, Vaxevanidis NM. Macroscopic and microscopic phenomena of electro-discharge machined steel surfaces: An experimental investigation. *J Mech Work Technol* 1987;15:335–356.

12. Shankar P, Jain VK, Sundararajan T. Analysis of spark profiles during EDM process. *Mach Sci Technol* 1997;1:195–217. DOI: 10.1080/10940349708945647.

13. Abulais S. Current research trends in electric discharge machining (EDM): Review. *Int J Sci Eng Res* 2014;5:100–118.

14. Davim JP, editor. *Nontraditional Machining Processes.* Springer; London; 2013.

15. DiBitonto DD, Eubank PT, Patel MR, Barrufet MA. Theoretical models of the electrical discharge machining process. I. A simple cathode erosion model. *J Appl Phys* 1989;66:4095–4103. DOI: 10.1063/1.343994.

16. Bojorquez B, Marloth RT, Said. Formation of a crater in the workpiece on an electrical discharge machine. *Eng Fail Anal* 2002;9:93–97.

17. Chen Y, Mahdivian SM. Analysis of electro-discharge machining process and its comparison with experiments. *J Mater Process Technol* 2000;104:150–157.

18. Koyano T, Sugata Y, Hosokawa TFA. Micro electrical discharge machining using high electric resistance electrodes. *Precis Eng* 2017;47:480–486. DOI: 10.1016/j.precisioneng.2016.10.003.

19. Lee SH, Li XP. Study of the effect of machining parameters on the machining characteristics in electrical discharge machining of tungsten carbide. *J Mater Process Technol* 2001;115:344–358. DOI: 10.1081/AMP-200060412.

20. Ghosh A and Mallik AK. *Manufacturing Science* (2nd ed.). Affiliated East-West Press Private Limited; New Delhi; 2010.

21. Liu YH, Ji R, Li Q, Yu L, Li X. Electric discharge milling of silicon carbide ceramic with high electrical resistivity. *Int J Mach Tools Manuf* 2008;48:1504–1508. DOI: 10.1016/j.ijmachtools.2008.03.012.

22. Chakraborty S, Kar S, Ghosh SK, Dey V. Parametric optimization of electric discharge coating on Aluminium-6351 alloy with green compact silicon carbide and copper tool: A Taguchi coupled utility concept approach. *Surf Interfaces* J 2017;7:47–57. DOI: 10.1016/j.surfin.2017.02.001.

23. Jain, VK, Seshank S, Sidpara A, Jain H. Some aspects of micro-fabrication using electro discharge deposition process. *Proc. ASME/ISCIE 2012 Int. Symp. Flex. Autom. ISFA2012*, June 18–20, 2012, St. Louis, Missouri, USA, 2012, pp. 1–6.

24. Naidu MS, Kamaraju V. *High Voltage Engineering.* McGraw-Hill; United States of America; 1996.

25. Wadhwa CL. *High Voltage Engineering.* New Age International Publsihers; New Delhi; 2007.

26. Kuffel E, Zaengl WS, Kuffel J. *High Voltage Engineering Fundamentals*; Butterworth-Heinemann; Great Britain; 2000.

27. Chakraborty S, Dey V, Ghosh SK. A review on the use of dielectric fluids and their effects in electrical discharge machining characteristics. *Precis Eng* 2015;40:1–6.

28. Zhang Y, Liu Y, Shen Y, Ji R, Li Z, Zheng C. Investigation on the influence of the dielectrics on the material removal characteristics of EDM. *J Mater Process Technol* 2014;214:1052–1061. DOI: 10.1016/j.jmatprotec.2013.12.012.

29. Jahan MP, Rahman M, Wong YS. Modelling and experimental investigation on the effect of nanopowder-mixed dielectric in micro-electrodischarge machining of tungsten carbide. *Proc Inst Mech Eng Part B J Eng Manuf* 2010;224:1725–1739. DOI: 10.1243/09544054JEM1878.

30. Kojima A, Natsu W, Kunieda M. Spectroscopic measurement of arc plasma diameter in EDM. *CIRP Ann – Manuf Technol* 2008;57:203–207. DOI: 10.1016/j.cirp.2008.03.097.

31. Kitamura T, Kunieda M. Clarification of EDM gap phenomena using transparent electrodes. *CIRP Ann – Manuf Technol* 2014;63:213–216. DOI: 10.1016/j.cirp.2014.03.059.

32. Eubank PT, Patel MR, Barrufet MA, Bozkurt B. Theoretical models of the electrical discharge machining process. III. The variable mass, cylindrical plasma model. *J Appl Phys* 1993;73:7900–7909. DOI: 10.1063/1.353942.

33. Pandey PC, Jilani ST. Plasma channel growth and the resolidified layer in EDM. *Precis Eng* 1986;8:104–110. DOI: 10.1016/0141-6359(86)90093-0.

34. Kanmani S, Karthikeyan G, Ramkumar SDJ. Plasma characterization of dry μ-EDM. *Int J Adv Manuf Technol* 2011;56:187–195.

35. Nagahanumaiah RJ, Glumac N, Kapoor SG, Devor RE. Characterization of plasma in micro-EDM discharge using optical spectroscopy. *J Manuf Process* 2009;11:82–87. DOI: 10.1016/j.jmapro.2009.10.002.

36. Descoeudres A, Hollenstein C, Demellayer R, Wälder G. Optical emission spectroscopy of electrical discharge machining plasma. *J Mater Process Technol* 2004;149:184–190. DOI: 10.1016/j.jmatprotec.2003.10.035.

37. Mujumdar SS, Curreli D, Kapoor SG. Effect of dielectric conductivity on micro-electrical discharge machining plasma characteristics using optical emission spectroscopy. *J Micro Nano-Manuf* 2018;6:1–6. DOI: 10.1115/1.4039508.

38. Saha MN. LIII . Ionization in the solar chromosphere. *Philos Mag Ser* 1920;6:472–488.

39. Djilianova O, Sadowski MJ, Skladnik-Sadowska E, Malinowski K, Scholz M, Blagoev A, et al. The Cu spectra as a tool for late plasma focus diagnostics. *J Phys Conf Ser* 2006;44:175–178. DOI: 10.1088/1742-6596/44/1/024.

40. Lee TH. T-F theory of electron emission in high-current arcs. *J Appl Phys* 1959;30:166–171. DOI: 10.1063/1.1735127.

41. Xia H, Kunieda M, Nishiwaki N. Removal amount difference between anode cathode in EDM process. *Int J Electr Mach* 1993;45–52.

42. Mohri N, Suzuki M, Furuya M, Saito N, Kobayashi A. Electrode wear process in electrical discharge machinings. *Ann CIRP* 1995;44:165–168. DOI: 10.1016/S0007-8506(07)62298-7.

43. Kunieda M, Kobayashi T. Clarifying mechanism of determining tool electrode wear ratio in EDM using spectroscopic measurement of vapor density. *J Mater Process Technol* 2004;149:284–288. DOI: 10.1016/j.jmatprotec.2004.02.022.

44. Patel MR, Barrufet MA, Philip T. Eubank A, DiBitonto DD. Theoretical models of the electrical discharge machining process. II. The anode erosion model. *J Appl Phys* 1989;66:4104–4111.

45. Yeo SH, Kurnia W, Tan PC. Electro-thermal modelling of anode and cathode in micro-EDM. *J Phys D Appl Phys* 2007;40:2513–2521. DOI: 10.1088/0022-3727/40/8/015.

46. Algodi SJ, Adam T, Clare PDB. Modelling of single spark interactions during electrical discharge coating. *J Mater Process Technol* 2017;252: 760–772

47. Yeo SH, Kurnia W, Tan PC. Critical assessment and numerical comparison of electro-thermal models in EDM. *J Mater Process Technol* 2008;203:241–251.

48. Yadav V, Jain VK, Dixit PM. Thermal stresses due to electrical discharge machining. *Int J Mach Tools Manuf* 2002;42:877–888.

49. Kansal HK, Singh S, Kumar P. Numerical simulation of powder mixed electric discharge machining (PMEDM) using finite element method. *Math Comput Model* 2008;47:1217–1237. DOI: 10.1016/j.mcm.2007.05.016.

50. Joshi SN, Pande SS. Thermo-physical modeling of die-sinking EDM process. *J Manuf Process* 2010;12:45–56. DOI: 10.1016/j.jmapro.2010.02.001.

51. Zhang Y, Liu Y, Shen Y, Li Z, Ji R, Wang F. A new method of investigation the characteristic of the heat flux of EDM plasma. *Procedia CIRP* 2013;6:450–455. DOI: 10.1016/j.procir.2013.03.086.

52. Weingärtner E, Kuster F, Wegener K. Modeling and simulation of electrical discharge machining. *Procedia CIRP* 2012;2:74–88. DOI: 10.1016/j.procir.2012.05.043.

53. Yang X, Guo J, Chen X, Kunieda M. Molecular dynamics simulation of the material removal mechanism in micro-EDM. *Precis Eng* 2011;35:51–56.

54. Jilani ST, Pandey PC. Analysis and modelling of EDM parameters. *Precis Eng* 1982;4:215–221.

55. Das S, Klotz M, Klocke F. EDM simulation: Finite element-based calculation of deformation, microstructure and residual stresses. *J Mater Process Technol* 2003;142:434–451. DOI: 10.1016/S0924-0136(03)00624-1.

56. Madhu P, Jain VK, Sundararajan T, Rajurkar KP. Finite element analysis of EDM process. *Process Adv Mater* 1991;1:161–174.

57. Bhattacharya R, Jain VK, Ghoshdastidar PS. Numerical simulation of thermal erosion in EDM process. *J Inst Eng (I) Prod Engg Div* 1996;77:13–19.

58. Van Dijck FS, Dutré WL. Heat conduction model for the calculation of the volume of molten metal in electric discharges. *J Phys D Appl Phys* 1974;7:899–910. DOI: 10.1088/0022-3727/7/6/316.

59. Tao J, Jun Ni, Shih AJ. Modeling of the anode crater formation in electrical discharge machining. *J Manuf Sci Eng* 2018;134:1–11. DOI: 10.1115/1.4005303.

60. Mujumdar SS, Curreli D, Kapoor SG, Ruzic D. Modeling of melt-pool formation and material removal in micro-electrodischarge machining. *J Manuf Sci Eng Trans ASME* 2015;137:1–9. DOI: 10.1115/1.4029446.

61. Tang J, Yang X. A novel thermo-hydraulic coupling model to investigate the crater formation in electrical discharge machining. *J Phys D Appl Phys* 2017;50:1–12.

62. Tang J, Yang X. A Thermo-hydraulic modeling for the formation process of the discharge crater in EDM. *Procedia CIRP* 2016;42:685–690. DOI: 10.1016/j.procir.2016.02.302.

63. Roy T, Sharma A, Datta D, Balasubramaniam R. Molecular dynamics study on the effect of discharge on adjacent craters on micro EDMed surface. *Precis Eng* 2018;52:469–476. DOI: 10.1016/j.precisioneng.2018.02.005.

64. Yue X, Yang X. Molecular dynamics simulation of the material removal process and gap phenomenon of nano EDM in deionized water. *RSC Adv* 2015;5:66502–66510 l. DOI: 10.1039/c5ra11419e.

65. Yue X, Yang X. Molecular dynamics simulation of single pulse discharge process: Clarifying the function of pressure generated inside the melting area in EDM. *Mol Simul* 2017;43:935–944. DOI: 10.1080/08927022.2017.1306649.

66. Uhlmann E, Piltz S, Doll U. Machining of micro/miniature dies and moulds by electrical discharge machining – Recent development. *J Mater Process Technol* 2005;167:488–493. DOI: 10.1016/j.jmatprotec.2005.06.013.

67. Singh M, Jain VK, Ramkumar J. Micro-electrical discharge milling operation. In: Golam Kibria, Muhammad P. Jahan, Bhattacharyya B, editors. *Micro-Electrical Disch. Mach. Process. Technol. Appl.*, Springer Nature, Singapore Pvt Ltd.; 2019, pp. 23–51.

68. Karthikeyan G, Ramkumar J, Dhamodaran S, Aravindan S. Micro electric discharge milling process performance: An experimental investigation. *Int J Mach Tools Manuf* 2010;50:718–727. DOI: 10.1016/j.ijmachtools.2010.04.007.

69. Singh M, Jain VK, Ramkumar J. Micro-texturing on flat and cylindrical surfaces using electric discharge micromachining. *J Micromanufacturing* 2020:1–11. DOI: 10.1177/2516598420980404.

70. Singh M, Singh A, Ramkumar J. Thinwall micromachining of Ti–6Al–4V using microwire electrical discharge machining process. *J Brazilian Soc Mech Sci Eng* 2019;41(338):1–12.

71. Singh M, Ramkumar J, Rao RV, Balic J. Experimental investigation and multi-objective optimization of micro-wire electrical discharge machining of a titanium alloy using Jaya algorithm. *Adv Prod Eng Manag* 2019;14:251–263.

72. Asad ABMA, Masaki T, Rahman M, Lim HS, Wong YS. Tool-based micro-machining. *J Mater Process Technol* 2007;192–193:204–211. DOI: 10.1016/j.jmatprotec.2007. 04.038.

73. Mohri N, Fukuzawa Y, Tani T, Sata T. Some considerations to machining characteristics of insulating ceramics – Towards practical use in industry. *CIRP Ann – Manuf Technol* 2002;51:161–164. DOI: 10.1016/S0007-8506(07)61490-5.

74. König W, Dauw DF, Levy G, Panten U. EDM-future steps towards the machining of ceramics. *CIRP Ann – Manuf Technol* 1988;37:623–631. DOI: 10.1016/S0007-8506 (07)60759-8.

4 Micro-EDM
Modeling and Optimization

Ranajit Mahanti and Manas Das

CONTENTS

4.1 INTRODUCTION

Over the last few decades, the technology boom has been linked to the miniaturization of instruments, parts, and machines due to the advances in micromanufacturing techniques. Microfabrication techniques have a huge demand in the biotechnology, automotive, communication, electronics, and avionics industries. Micromachining techniques – both traditional and non-traditional – are widely used to meet these requirements and to produce miniature products. Microfabrication using conventional machining processes is limited due to tool size, material and various forms of wear, and heat generation owing to the friction force between the tool and workpiece

DOI: 10.1201/9780429160011-4

during machining [1]. Non-traditional (also called advanced, unconventional, or new) micromachining processes are suggested to overcome these problems where no contact between the microtool and workpiece is involved. Among various advanced micromachining technologies, micro electro discharge machining (μ-EDM/micro-EDM) is one of the economic thermoelectric type micromachining processes capable of machining conductive materials regardless of the workpiece's mechanical properties as hardness and strength, adding to facilitate the fabrication of complex microfeatures on hard materials. Microcomponent inaccuracy, on account of vibration, chatter, mechanical stress, deformation of tool electrode, etc., is absent in EDM due to the lack of physical contact between the tool and workpiece [2].

Kurafuji and Masuzawa [3] have realized a microhole in a 50 μm-thick plate of carbide material by micro-ED drilling (micro-EDD). Many researchers have then performed numerous studies to expand and improve the process capabilities of μ-EDM along with various configurations and applications. The material removal mechanism is complex in the μ-EDM process, and it is time-dependent. Understanding the mechanics of material erosion in the narrow interelectrode gap (IEG) and selection of process parameters are necessary for μ-EDM modeling. Low material removal rate (MRR), higher relative tool wear (RTW), inaccuracy in machined features, etc. are the significant problems associated with μ-EDM, and they increase the fabrication cost of the products. Selection of optimum process parameters or a suitable machining environment is an essential aspect of cost-effective fabrication. So, experimental studies and optimization of the μ-EDM process are other research fields to enhance process performance, such as higher MRR or productivity, and better surface qualities and dimensional accuracy.

This chapter presents a comprehensive overview of the numerous developed models (empirical, analytical, and numerical) of the μ-EDM process and experimental studies (mostly optimization approaches). The chapter's contents are divided into two sections. The first section gives an overview of the μ-EDM process, including a brief explanation of the fundamental concept of material removal mechanism, setup information, machining performance with parameters, and μ-EDM variants with applications. The second part discusses the numerous models that have been constructed, including empirical models based on the experimental research, analytical models, and numerical models.

4.2 OVERVIEW OF MICRO-EDM

This section covers the fundamentals and compares macro (traditional) and micro versions, setup details with subsystems, machining performance with parameters, and various configurations of μ-EDM with applications.

4.2.1 BASIC PRINCIPLE OF MICRO-EDM

Micro-EDM works on the same principle as traditional EDM. In μ-EDM, the micro tool and the workpiece are separated by a small distance and submerged in a suitable dielectric fluid. Controlled discharges, resulting in melting and vaporization, are used to remove material from the workpiece. The common dielectrics used are deionized water, and hydrocarbon oil (kerosene, paraffin oil, etc.). A high electric field is generated in regions where the distance between the cathode and anode is smaller when the requisite potential difference from the pulse power supply is applied between the tool (cathode) and workpiece (anode). This strong electric field propels electrons from the cathode to the anode. The electrons collide with neutral dielectric molecules during movement, leading to dielectric breakdown, i.e., splitting into ions and electrons, which lowers the effective resistance in the gap, and enhances the electrical conductivity of the fluid medium. Consequently, the conductive state between the two electrodes is formed by a continuous motion of the electrons. This electron avalanche collides with more dielectric fluid molecules, which leads to a more significant number of electrons and ions being produced. The formation and expansion of bubbles resulting from the dielectric fluid's vaporization cover the area and decrease the effective resistance [4]. All these events, resulting in a plasma channel at minimum IEG and a continuous flow of electrons, is known as discharge.

The collision of high-velocity electrons on the workpiece surface results in a very high temperature due to the conversion of kinetic energy into heat energy. This sudden heat energy melts the workpiece material locally, and some portion of it evaporates immediately. A high-pressure plasma channel removes the molten material and carries it out with the dielectric from the machining zone. After removing molten material, the minimum gap point changes, and the plasma channel shifts to the next minimum spark gap leading to a crater formation on that specified area. The pulse power supply enables the uniform discharge distribution over the workpiece surface, resulting in uniform erosion from the workpiece material. Tool wear is the material erosion from the tool surface caused by ions traveling toward the cathode surface. Tool wear is always associated with μ-EDM, but the tool's erosion is comparatively less than the workpiece, as explained well by Ghosh and Mallik [5]. Figure 4.1 shows the plasma channel in the IEG, the evaporated portion (superheated cavity), the molten cavity, and the crater formed on both electrodes during the μ-EDM process.

4.2.2 Macro- vs Micro-EDM

The general principle of μ-EDM is the same as that of conventional (macro) EDM, and it is simply a smaller version of traditional EDM. However, using lower discharge energy levels, smaller electrode sizes, shorter pulse on-time, and submicron IEG, μ-EDM differs from conventional EDM. The tool electrode may burn if excessive sparking happens, which restricts the energy intensity to a low level. The tool can deflect or break if extreme flushing pressure is applied. The tool electrode wear rate is comparatively higher in μ-EDM, demanding a highly developed tool wear compensation technique [6]. In the micro regime, empirical relationships that predict the pattern of MRR or tool wear rate (TWR) in macro-EDM are invalid. Because of these facts and understanding the mechanism behind material removal and factors affecting the machining efficiency, μ-EDM has to be viewed as a separate machining strategy.

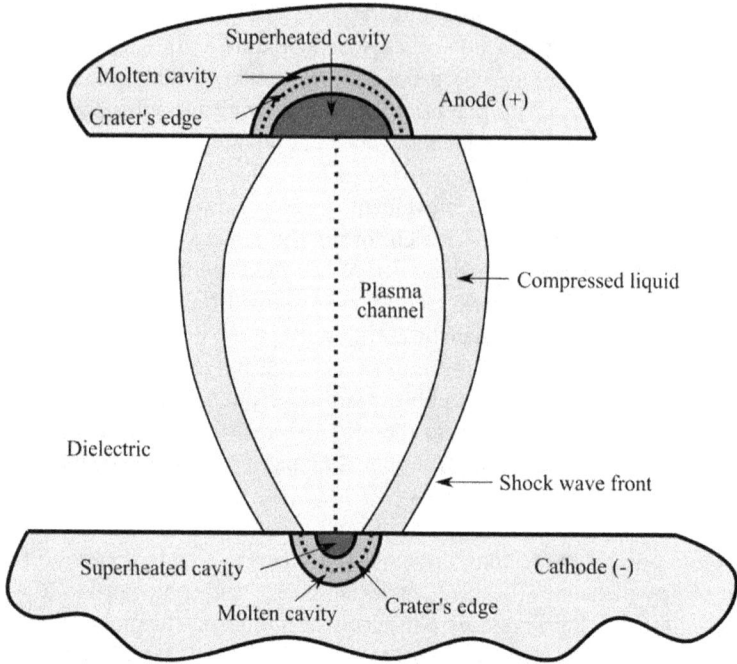

FIGURE 4.1   Schematic representation of μ-EDM process.

The discharge energy varies from 5–11,000 μJ [7–9]; the value of pulse on-time ranges from 0.05–100 μs [10] and the IEG equal to 10 μm. Other significant differences between micro and conventional EDM are the plasma radius and crater size. In μ-EDM, the plasma channel radius maybe equal to the tool diameter [11], while in macro-EDM, it is not so. The crater is tiny due to low discharge energy in μ-EDM, reflecting better surface quality than macro-EDM. Also, MRR is very low compared to conventional EDM. In addition, Table 4.1 demonstrates the differences between the traditional EDM and μ-EDM for better knowledge of the process parameter ranges and capabilities of both versions of EDM.

4.2.3   MICRO-EDM SETUP

The μ-EDM machine tool consists of a machining chamber, a servo control system, a dielectric fluid supply system, a positioning system, an online measurement system, and a pulse generator. Figure 4.2 shows the schematic diagram of a typical μ-EDM setup. In traditional EDM, where a higher MRR is required, a transistor-based pulse generator is often utilized, but in μ-EDM, a resistor-capacitor (RC) pulse generator is generally employed [13]. The machining chamber consists of a workpiece, fixture, and dielectric. The workpiece is rigidly clamped with a fixture, and the machining region is submerged within the dielectric fluid to avoid fire hazards. A dielectric circulating system is used to provide dielectric supply inside the machining chamber.

TABLE 4.1
Comparison of Macro- and Micro-EDM [10, 12]

Characteristics	Macro-EDM	Micro-EDM
Power source	Transistor-based pulse generator	RC-based pulse generator
Power condition	Stable gap voltage and current	Voltage and current varied along the spark gap
Dielectric	Hydrocarbon-based oil and deionized water	Mainly deionized water
Current	>10 A	<5 A
IEG	10–500 μm	<10 μm
Source voltage	40–400 V	10–120 V
Pulse duration	User-defined constant	0.05–100 μs
Pulse frequency	2–200 kHz	100–500 kHz
Discharge energy	0.1–4 J	10^{-6}–10^{-7} J
MRR	0.3–30 mm³/min	0.000125–0.02 mm³/min
Roughness (R_a)	0.8–3.1 μm	0.015–0.8 μm
Feature accuracy	±25–127 μm	±1 μm
Aspect ratio	Up to 100	Up to 50
Recast layer thickness	1–8 μm	<1 μm

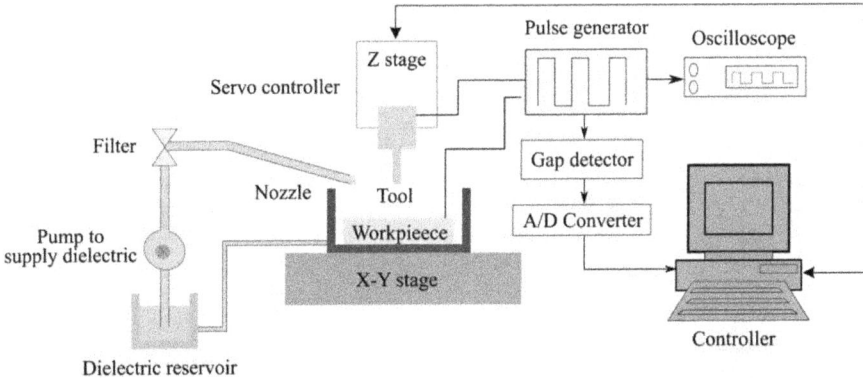

FIGURE 4.2 Schematic diagram of a typical μ-EDM setup.

The dielectric fluid delivery system consists of a reservoir, pump, nozzle, filters, and pipes. Hydrocarbon-based oil (kerosene, paraffin, transformer oil, etc.) and deionized water are mainly used as a dielectric.

The positioning system (CNC actuation system) controls the workpiece's X-Y motion as well as the tool electrode's Z-direction motion. The servocontrol system maintains the optimum interelectrode distance by pulling the electrode from time to time to avoid a short circuit. Analog to a digital (A/D) converter, a pulse-width modulation (PWM) module with a feedback system is used along with a microcontroller to control the servomotor and maintain the optimum spark gap [14]. The gap detector is used to monitor the spark gap during discharges. The oscilloscope is used to observe the pulse characteristics. A charge-coupled device (CCD) is integrated into an online measurement system to track and measure the microfeatures and irregularities present in the fabricated component without removing it from the spindle or fixture.

4.2.4 MACHINING/PERFORMANCE PARAMETERS

The performance or quality characteristics of μ-EDM can be measured using MRR, TWR, thickness of the recast layer, circularity, overcut, dimensional accuracy, etc. The selection of process parameters is critical to control the quality characteristics of μ-EDM.

The process parameters of μ-EDM are mainly divided into two subgroups. These are the electrical parameters and non-electrical types. The electrical categories are source voltage, peak current, capacitance (for RC-based generator), on/off-pulse duration, gap voltage, and current (for transistor-based generator). The non-electrical parameters are tool rotation speed, tool feed rate, type of dielectric and flow rate, flushing pressure, types of workpiece material, tool electrode, etc. The fishbone diagram of process parameters influencing the μ-EDM performance is shown in Figure 4.3 [15].

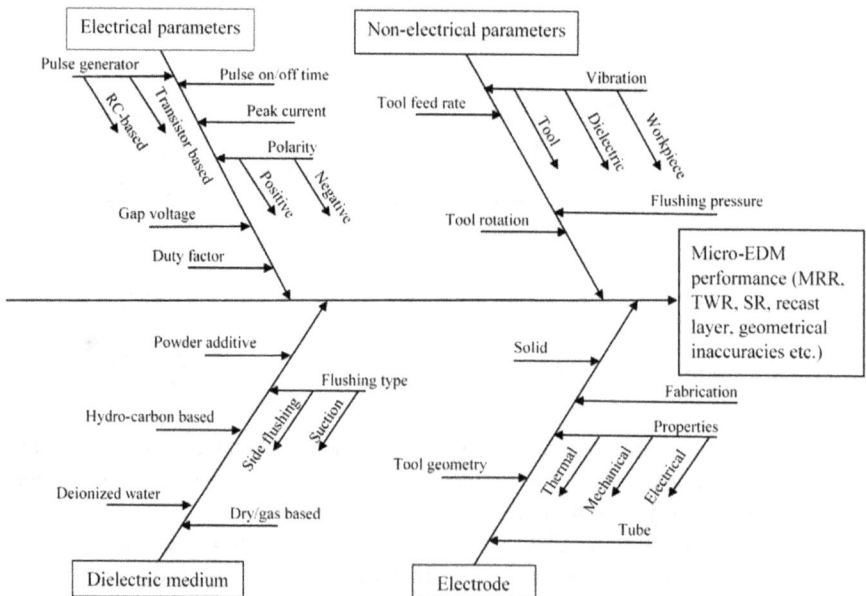

FIGURE 4.3 Fishbone diagram of process parameters that affect the μ-EDM performance [15]. With kind permission from Elsevier, License No. 5158630404924.

4.2.5   Micro-EDM Variants and Applications

Micro-EDM is divided into numerous versions or configurations based on tool electrode geometries, relative motion between microtool and workpiece, and purpose of operation or application. These variants are:

1. *Die-sinking micro-EDM*: In micro die-sinking EDM, the geometry of the tool is an approximate negative replica of the desired product. In this process, plunged motion of the tool is given without rotation of the tool.
2. *Micro-ED milling*: The simple shape (cylindrical rod) tool follows a pre-programmed complex path with rotation to fabricate the complex microfeatures using a suitable tool wear compensation system and layerwise machining [16].
3. *Micro-ED drilling*: The rotating tool moves vertically downward in a continuous dielectric flow and pulsed DC supply conditions to make microholes in the workpiece.
4. *Micro-wire EDM*: In this variant, the continuously moving thin wire (normal value of diameter used is 20 μm) is used as a tool electrode to cut the workpiece throughout [12]. The tool electrode is guided by a computer numerical control (CNC) facility to follow a programmed path to produce complex profiles.
5. *Micro electro discharge grinding* (micro-EDG): To form the tool electrode for subsequent ED milling or drilling, sacrificial electrodes (negative polarity) of various shapes, such as block, wire, and disc are used. According to this, micro-EDG is classified into three configurations, i.e., block EDG (stationary and moving), wire EDG, and disc EDG.
6. *Reverse micro-EDM*: A tool electrode (plate type) with single or multiple holes is used instead of the standard EDM tool electrode. When an EDM tool electrode with holes is plunged into the workpiece, material erosion produces certain micropillars to appear on the workpiece. Such machined parts with a series of micropillars can be used as a tool for ED microdrilling and electrochemical microdrilling of multiple holes simultaneously for enhancing the process performance.

All these μ-EDM variants are restricted in terms of process capabilities such as microfeature size, complexity, maximum aspect ratio, and surface quality. The process capabilities of μ-EDM variants, along with the potential applications, are shown in Table 4.2.

4.2.6   Remarks

The μ-EDM is a scaled down version of macro-EDM. The process parameters of μ-EDM are mainly classified as electrical and non-electrical types. The parameters and quality characteristics of micro and macro-EDM are presented in the

TABLE 4.2
Process Capabilities of Micro-EDM with Applications [12, 17]

Types of Micro-EDM	Machining Capability	Surface Roughness R_a (μm)	Aspect Ratio	Min. Feature Size (μm)	Applications
Micro die-sinking EDM	3D	0.05–0.3	15–20	20	Embossing molds, microinjection molds, 3D microstructures, etc.
Micro-ED milling	3D	0.2–1	10	20	3D microfeatures, micromolds, microfluidic channel, micropillars, etc.
Micro-ED drilling	2D	0.05–0.3	25–50	5	Blind or through microholes, reverse EDM tool electrodes, etc.
Micro-WEDM	21/2 D	0.07–0.2	100–125	30	Forming tools, stamping tools, spinning nozzles, etc.
Micro-EDG	Axial or rotational symmetrical	0.015–0.8	30	3	On-machine fabrication of microelectrodes for μ-ED drilling and milling

comparison table, which helps to select the process parameter ranges in the experiments for different μ-EDM processes. The comparison of various μ-EDM variants with applications is given in a table to understand the process capabilities and potential application fields. Micro-EDM can produce surface quality with submicron roughness parameters, high aspect ratio (depth/width), minimum feature size (up to 3 μm), and 3D complex microfeatures. The basic principle of μ-EDM is also discussed in this section, which helps in developing analytical or theoretical models.

4.3 MODELING AND OPTIMIZATION OF MICRO-EDM

Modeling is a scientific method for investigating machining behavior. In μ-EDM, the working zone is minor (<10 μm) compared to macro-EDM. In this region, the solid (debris), liquid, and plasma states coexist for a very brief period (in the microsecond to nanosecond range) during machining. So, the modeling of μ-EDM is complicated, and machining behavior is dynamic and completely different from conventional EDM. The models allow one to understand the material erosion mechanism, process variables and its performance. The correlation between input and output parameters of the μ-EDM process is described in a mathematical model using equations.

Various modeling approaches of the μ-EDM process are shown in Figure 4.4. Models are mainly classified into three groups depending on their origins: Experimental or empirical models, analytical models, and numerical models. Because of the complexity of machining behavior, researchers have to use discrete and continuous parameter intervals with the multiple models and differential or non-differential objective functions to find optimum solutions. Seeking optimal solution(s) using a suitable optimization strategy based on an objective function derived from the model is time-consuming and challenging. As a result, researchers have developed a wide variety of approaches for solving these optimization problems.

This section presents some empirical models developed based on experimental studies, analytical models, and finite element method (FEM) based numerical models for the μ-EDM process.

4.3.1 EXPERIMENTAL STUDIES AND EMPIRICAL MODELS

Empirical models are based on experimental studies. The quality characteristics (MRR, TWR, surface quality, machining inaccuracy, etc.) are measured in this model by varying the essential process parameters (electrical and non-electrical) in a specific range. Mathematical equations/models are developed for a dataset where various input and output parameters are obtained using experimental runs. These mathematical models are valid only in the selected range of input parameters. Regression analysis or curve fitting techniques are used to form a polynomial relationship of process parameters and responses. Unlike analytical models, these are not assumption-based, and they give a real-time solution. However, the accuracy of these models is based on the machining conditions, response measuring the instrument's accuracy, machine tool, and experimentation.

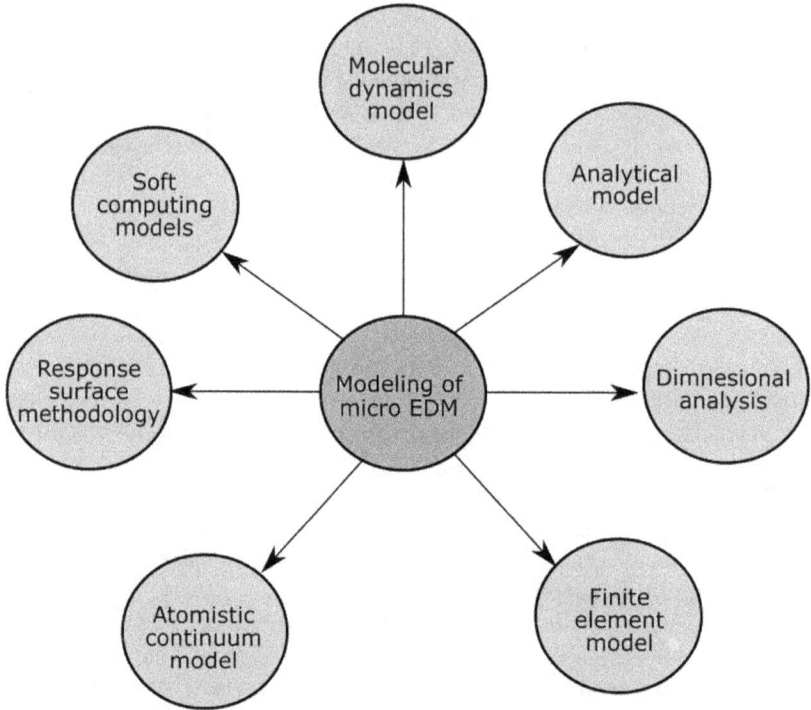

FIGURE 4.4   Various modeling approaches in μ-EDM.

A design of experiment (DOE) plan is used for data collection during experimentation, which is superior to the unplanned methods. It is a comprehensive and scientific method of planning research, data collection, and study with accessible resources. The most commonly used DOE techniques in the literature are factorial design, central composite design of response surface methodology (RSM), and Taguchi DOE. The significant process parameters are determined using statistical means such as the analysis of variance (ANOVA) technique. Various researchers utilize soft computing models to predict the machining characteristics and optimize the process parameters to better understand machining performance. A detailed review of different soft computing techniques and their comparison is available in the literature review [18]. Prior parametric analyses and empirical models of several μ-EDM processes are addressed to understand the relationship between process parameters and machining performance.

Tiwary et al. [19] used μ-EDD on Ti-6Al-4V super alloy to analyze the impact of various input parameters (pulse duration, peak current, gap voltage, and flushing pressure) on the responses such as MRR, taperness, overcut (OC), and TWR. CCD is used to design the experiments taking five levels for each parameter, and RSM is used to construct the correlation between the input parameters and responses. Parameters such as pulse on-time (t_{on}), peak current (I_p), gap voltage (V_g), and flushing pressure (F_p) are selected in the range 1–16 µs, 0.5–2.5 A, 30–50 V, and 0.15–0.35 kg/cm², respectively. The RSM models are used to perform multi-objective optimization in Minitab® software for the maximum MRR and minimum TWR, OC, and taper. Finally, the experimental validation is done using the optimum process parameters. Another experimental analysis of µ-ED drilling of tool steel has been performed using RSM-based CCD by Lin et al. [20]. In this study, the process characteristic such as MRR, tool wear (TW), and OC are evaluated by varying the process variables peak current (0.3–1 mA), pulse duration (6–25 µs), pulse off-time or interval (3–13 µs), and tool rotation speed (100–500 rpm). Regression analyses are performed to develop mathematical models of the responses, and these models are used as objective functions for optimization. According to ANOVA, peak current and its combination with pulse off-time are the most important parameters for model construction. It has been observed that MRR increases with the increasing pulse on-time, peak current, and tool rotation. But the TW and OC increase with the increase of peak current [20]. Ay et al. [21] utilized a soft computing technique (grey relational analysis) for optimization of the parameters such as pulse on-time (3–50 µs) and discharge current (100–1000 mA), in µ-ED drilling of Inconel 718 for evaluating hole performance (taper ratio and hole dilation). The discharge current has more impact on process performance than pulse on-time, stated by investigating the optimum results. The drilled surface's characteristics and tool condition are measured using an optical and scanning electron microscope (SEM). MRR is determined by the discharge energy regulated by the peak current and pulse duration/pulse on-time. The rise in pulse on-time implies that the energy in a single discharge has increased, removing more material from the workpiece and the tool. As a consequence, MRR and TW are increased. The same effect is shown with increasing peak current. The surface quality of the holes has degraded with the rise in peak current (energy density) and pulse width. High discharge energy resulting in intense sparks and deeper craters on the surfaces leads to recast layer formation at the rims of the holes [21].

The combination of soft computing techniques such as artificial neural network (ANN) and genetic algorithm (GA) has been employed to establish a model for optimization of MRR in µ-EDM [22]. The gap voltage in the range of 80–150 V, capacitance in the range of 0.01–0.4 µF of RC-based pulse generator, feed rate in the range of 3–9 µm/s, and rotation speed in the range of 300–600 rpm are considered as process input parameters to develop an ANN-based model for better prediction of MRR. The generated model is employed in GA to optimize the process parameters as an objective function. Dutta et al. (2019) used an assisting electrode (conductive layer coating) and a rotating tool and figured out the material removal process from carbon-fiber-reinforced polymers (CFRP) when fabricating blind holes on CFRP using µ-EDM [23]. The effect of process inputs such as voltage, microtool rotational speed (rpm), and pulse duration on surface characteristics of blind holes and MRR are investigated. The most crucial factor influencing the difference in machining time is discovered to be voltage using ANOVA analysis. The optimum process parameter is the combination of V3/T1/S3 (i.e., 170 V/10 µs/800 rpm). Dutta et al. (2020) also conducted another research to drill microholes on CFRP using an assisting electrode responsible for spark generation by µ-EDM [24]. For hole dilation at the inlet and exit, the grey relational analysis (GRA) methodology is employed to optimize parameters such as voltage (100–190 V), pulse length (10–40 µs), and tool rotation speed (200–500 rpm). The relationship between input variables (voltage, pulse on-time, and tool rotation speed) and grey relational grade (G_i) are presented using regression analysis which is shown in Equation 4.1.

$$G_i = 1.5072 - 0.005157V - 0.001958T - 0.000393S \tag{4.1}$$

The mean of G_i for each input variable to each level is used to determine the optimum level of the input variables. A high value of G_i indicates high process performance. As a result, the best combination of process parameters, such as the voltage of 100 V, pulse on-time of 10 µs, and rotating speed of 200 rpm, has been obtained. The most significant process parameter is the voltage which has the highest contribution (90.95%) compared to pulse on-time and tool speed, contributing 5.91% and 1.48%, respectively. Singh and Singh [25] proposed an experimental study for multi-objective optimization using the desirability function approach in µ-EDD. AISI D2 die steel and copper are used as workpiece and tool materials, respectively. The various input variables such as pulse duration (2–53 µs), pulse interval (6–28 µs), gap voltage (30–62 V), peak current (0.3–2 A), pulse peak voltage (170–205 V), and working time (0.75–1.75 s) are considered in three levels for optimization. MRR and TWR have estimated optimal values of 1663.46 and 166.79 µg/min, respectively, while the experimental values are shown to be similar.

Mehfuz and Ali [26] developed empirical models of MRR and parameters of surface roughness such as average roughness (R_a) and maximum peak to valley height (R_z), and electrode wear ratio (EWR) to explore the effect of input variables which are feed rate (f = 2,4,6 µm/s), capacitance (C = 0.10,1.00,10 nF), and voltage (V = 80,100,120 V) of Be-Cu alloy using tungsten electrode in micro-ED milling [26]. Using ANOVA, second-order quadratic models have been developed for all the response variables and considered only the significant input parameters and their interactions. The developed models for MRR, EWR, and R_a are shown in Equations 4.2, 4.3, and 4.4, respectively. The desirability function approach is used to discover the best process parameters for the maximum MRR and minimum Ra, Rz, and EWR in multi-objective optimization.

$$\frac{1}{\sqrt{MRR}} = 16.85 - 3.68f - 1.24C - 0.070V + 0.28f^2 + 0.09C^2 + 0.04\,fC$$
$$+ 0.01fV \tag{4.2}$$

$$EWR = 0.285 - 0.203f + 0.031C + 0.002V + 0.029f^2 - 0.003C^2$$
$$+ 0.0002\,fC - 0.0003\,fV \tag{4.3}$$

$$R_a = -7.956 - 0.044f + 1.423C + 0.087V - 0.111C^2 - 0.0004V^2$$
$$+ 0.0006\,fV - 0.0007CV \tag{4.4}$$

Karthikeyan et al. [27] conducted an experimental study of µ-ED milling to evaluate performance on EN 24 alloy with tungsten tool (dia. 500 µm) and also to investigate the effect of electrical and non-electrical parameters such as tool rotation speed (S = 100, 500, and 800 rpm), feed rate (F = 10, 25, 45, and 60 µm/sec in Z-axis), aspect ratio (A = 0.5, 1.0, 1.5, and 2.0), and discharge energy (E = 500 and 2000 µJ) on MRR and TWR [27]. It has been reported that rotational speed has the highest impact on the responses. With increased discharge energy, the TWR and MRR rise dramatically. A high rotational speed of the tool electrode results in increased MRR, irrespective of the feed rate and AR. Regression analysis is also provided to understand the degree of interdependence of MRR and TWR on the process parameters. The developed regression equations of MRR and TWR are shown as Equations 4.5 and 4.6, respectively. It has also been reported that multiple passes are required to get a high aspect ratio with desired surface quality.

$$MRR = 5.6e - 3 \times E^{0.318} \times F^{0.34} \times S^{0.439} \times A^{0.246} \qquad (4.5)$$

$$TWR = 1.5e - 4 \times E^{0.435} \times F^{0.428} \times S^{0.435} \times A^{0.383} \qquad (4.6)$$

Increased aspect ratios in microholes, reduced difference in entrance and exit diameters, improved surface quality, and improved geometrical shape have been the focus of recent µ-EDD technological developments. Various researchers have proposed hybrid µ-EDM processes such as vibration-assisted [28], magnetic-field assisted [29–31], sequential machining [32], powder-mixed µ-EDM [33], and high-speed machining to improve the machining performance. The superior responses such as high MRR, low surface roughness, high aspect ratio, and low tool wear ratio with the high spindle rotational speed in the range of 50,000–60,000 rpm have been reported [34, 35].

The debris particles are propelled out of the narrow IEG by high-speed machining, which improves sparking efficiency and reduces anomalous discharge and short circuit conditions. In µ-EDD with high spindle speed, however, the increase in linear velocity at the electrode edge leads the discharge sites to become more dispersed. However, the linear velocity of each point on the electrode's bottom is not uniform. Linear velocity decreases as one gets closer to the electrode's center. The experimental investigations of vibration-assisted µ-EDD have been carried out to drill deeper microholes on tungsten carbide [36], [37]. When vibration is applied to a tungsten carbide workpiece, enhanced performance and quality attributes such as high MRR, low EWR, low surface roughness, and high dimensional accuracy of microholes are achieved. Finally, 50 µm diameter holes with a high aspect ratio (equal to 16.7) and excellent surface quality at the microhole's rim are created using microtools produced online [37]. Also, the vibration-assisted µ-EDD system improves the dielectric flushing condition by improving the debris-expelling behavior, IEG change, and dielectric flow characteristics [38]. The effect of electrical parameters, i.e., gap voltage and capacitance of RC-based pulse generator, and vibration characteristics such as amplitude (0–2.5 µm) and frequency (0–750 Hz) of vibration have also been studied on normal and abnormal (short circuit, effective arc, and complex) discharge

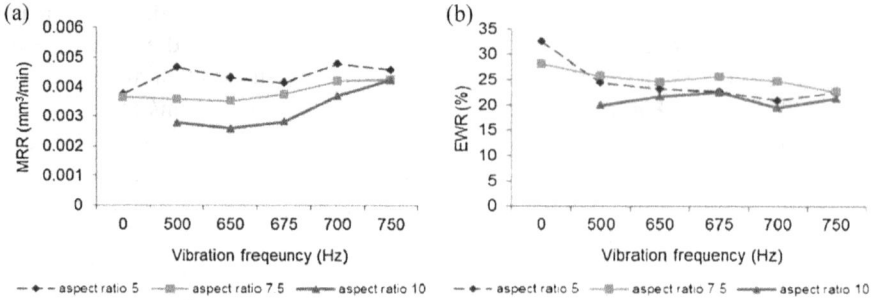

FIGURE 4.5 Effect of vibration frequency on (a) MRR and (b) EWR [36]. With kind permission from Elsevier, License No. 5090081402076.

pulses. It has been found that the percentage of abnormal discharges (mainly percentage of short circuit condition) is reduced with the application of vibration, which indicates stable machining and improved machining characteristics such as better MRR, lower EWR, and better surface quality (smoother and burr-free). Figure 4.5 depicts the relationship between MRR and vibration frequency, and EWR and vibration frequency [36].

An increase in vibration frequency up to an optimum value results in an improvement in MRR, and a reduction in longitudinal electrode wear (LWR) in μ-ED milling of Inconel 718 [39]. Gap voltage (80–125 V), capacitance (0.01–0.3 μF), and vibration frequency (0–280 Hz) are all considered in this parametric analysis. Uniform microchannels with a smooth surface are obtained at a low level of discharge energy and vibration frequency due to the fewer and small size globules formed [39]. Recently, an ultrasonic circular vibration (UCV) electrode with a high frequency of 32.85 kHz has been designed and used to enhance the machining quality of microholes [40]. With the UCV electrode, the morphologies of the microhole array at entry and exit have improved, specifically the diameter accuracy at entry and exit, by 22% and 2.8%, respectively. The benefits of the UCV electrode in improving microhole machining quality can be primarily due to the significantly increased dielectric fluid flow velocity and increased expelling of debris particles in dielectric fluid from the working gap.

4.3.1.1 Remarks

The objective of the experimental modeling-based research in various μ-EDM processes is similar, but the techniques are different. The empirical input/output findings have been utilized to create a first- or second-order regression model using the least square fit to illustrate the relationship between process parameters and performance. In most of the cases, RSM-based regression models are developed. ANOVA analysis is done to find out the most significant parameters. Soft computing models, such as ANN-based models, are also meant to realistically anticipate the correlation between input and response parameters. The optimization of process parameters using soft computing models such as RSM, Taguchi-based method, GA, GRA, desirability function approach, etc. are developed in various μ-EDM variants for cost-effective production. Few literature works have explored the area of a parametric study or empirical modeling of μ-ED milling and hybrid μ-ED milling to fabricate complex-shaped profiles, microslots, grooves etc. Most of the parametric studies are concentrated on μ-EDD. So, extensive parametric studies are required to explore the μ-ED milling process to fabricate complex microstructures (3D).

4.3.2 ANALYTICAL MODELS

Analytical models are created by analyzing the material removal theory using fundamental laws or theoretical study. These models necessitate a thorough understanding of the mechanism. By simplifying the method and choosing suitable assumptions, analytical models may be expressed in a mathematically interpretable manner. The accuracy of analytical models is determined by the hypotheses used. Oversimplification of the mechanism leads to a less reliable model while accounting for the assumptions makes the model more challenging and lengthens the time required to solve it.

In μ-EDM, the material removal procedure is complex and probabilistic. Electrodynamics, hydrodynamics, thermodynamics, and electromagnetics are the disciplines involved with this process. It is almost impossible to provide a straightforward and systematic theory that can fully clarify the essence of the mechanism. However, it is important to provide a valid analytical model that can provide a reasonable estimate of material removal. This estimate can serve as a helpful reference for selecting the necessary machining process parameters. It would also aid in the process planning and monitoring by predicting machining responses such as the MRR, TWR, and surface integrity. The stochastic model is probabilistic, which implies that the result for a given set of inputs may lie within a specified range, and determining the best solution is extremely difficult. There are the analytical models in EDM and also μ-EDM that predict system behavior under various operating conditions. The analytical models of μ-EDM are divided into two parts according to the process characteristics, i.e., plasma channel models and electrode erosion or material removal models. Some of these developed models are discussed in this section.

4.3.2.1 Plasma Channel Modeling

The plasma channel model is utilized to approximate the plasma state following an electrical discharge. Understanding the state of plasma may provide approximated boundary conditions (BCs) for the anode and cathode model. Dhanik et al. [41] described two main approaches to the modeling of the plasma channel in EDM: Fluid dynamics and kinetics.

4.3.2.1.1 Fluid Dynamics Approach

The fluid dynamics approach has used the equation of fluid dynamics (such as the continuity and momentum equation), heat transfer, and state at a particular moment to get the depiction of plasma. Cylindrical and spherical shape plasma have mostly been considered during modeling using this approach [42–44]. However, Shankar et al. [45] reported a non-cylindrical spark shape in the IEG, where the sparks deviate toward electrodes from the IEG's center. While most models are based on a fixed mass of plasma during expansion, Eubank et al. [41] have introduced the variable mass plasma. The assumptions and simplification of the fluid dynamics approach are described in the literature.

Eubank et al. [41] modeled the plasma by including the unsteady state energy balance, fluid dynamics, and state equation of plasma. They have used a basic unsteady state energy balance equation, (Equation 4.7), to account for variations in plasma temperature, mass, and pressure.

$$VIF_p = (H - H_0)\left(\frac{dm}{dt}\right) + m\left(\frac{dH}{dt}\right) \tag{4.7}$$

This equation balances the energy needed to vaporize plasma mass (m_p) using the radiation mode of heat transfer and the energy required to change the enthalpy (H) of plasma by raising the temperature by using the fixed percentage (F_h) of the total energy going into the discharge per certain gap voltage (V) and current (I). The ambient dielectric's enthalpy is H_0. The first part of the equation represents the radiation heat transfer, which can be expressed as Equation 4.8 if the plasma is considered as a blackbody.

$$(H - H_0)\left(\frac{dm_p}{dt}\right) = (2\pi R_p b)(a_R \sigma T^4) \tag{4.8}$$

where a_R is the absorptivity (equal to one), b is the IEG, T is the plasma temperature in K, and σ is the Stefan–Boltzmann constant.

Since plasma spreads over time [45], fluid dynamic properties must be used in the modeling of plasma [44]. Because of the pressure differential, the fluid dynamic equations such as continuity and momentum equation, shown in Equations 4.9 and 4.10 respectively, are employed to represent the liquid dielectric flow.

$$\left(\frac{\partial \rho_0}{\partial t}\right) + \frac{1}{r}\frac{\partial}{\partial r}(\rho_0 r v_r) = 0 \tag{4.9}$$

$$\frac{\partial v_r}{\partial t} + v_r \frac{\partial v_r}{\partial r} + \frac{1}{\rho_0}\frac{\partial P}{\partial r} = 0 \tag{4.10}$$

The above continuity and momentum equations' solution in the cylindrical coordinate system resulting in the relationship between the plasma radius and time, is represented in Equation 4.11.

$$\left(\frac{dR_p}{dt}\right)^2\left(\ln\left(\frac{R_f}{R_p}\right) + \frac{1}{2}\left(\frac{R_p^2}{R_f^2} - 1\right)\right) + R_p\left(\ln\left(\frac{R_f}{R_p}\right)\right)\left(\frac{d^2R_p}{dt^2}\right) = \left(\frac{P - P_f}{\rho_0}\right) \tag{4.11}$$

where R_p = radius of plasma channel, R_f = outer radius of fluid in the spark gap, P_f = ambient pressure, P = plasma pressure and ρ_0 = density of the dielectric liquid.

During a discharge, the temperature and pressure within the plasma reach extremes, causing the dielectric fluid to break and ionize, resulting in a mixture of ionic and neutral species. The ion percentage in the plasma varies depending on the plasma state. The following equation describes the plasma state [44].

$$P = \frac{\lambda T \rho}{216.70} \tag{4.12}$$

where λ is the average number of dissociated species from dielectric molecules, which varies with the plasma state pressure (P) and temperature (T).

4.3.2.1.2 Kinetics Model

The kinetic approach is used to model the association of plasma with the cathode and anode by looking at the distribution of different entities such as electrons, ions, and neutral molecules. This model is based on the consideration of one-dimensional, non-uniform electric field in the IEG, constant heat input, and the absence of expansion of plasma state [46]. The modeling method is split into two phases in this case: Plasma analysis and sheath analysis. A thin layer is formed due to a potential decrease between the plasma and electrodes, called a plasma sheath. The plasma investigation determines the BCs for the sheath analysis. The sheath's relations with the electrodes are modeled in sheath analysis for getting stock removal at the respective electrodes. The plasma region is assumed to be quasi-neutral, and the distribution of ions and electrons ensuring the Boltzmann distribution is as follows,

$$n_e\left(x'\right) = n_0 \exp\left(\frac{e\phi\left(x'\right)}{kT_e}\right) \tag{4.13}$$

where n_0 is the electron density at infinity taken as reference, n_e is the density of electrons, which is the function of x' in the potential gradient direction and T_e is the temperature of the electrons.

To get the reference potential $\left(\phi_0\right)$ and typical ion and electron density (n) in the plasma–sheath region, Singh and Ghosh [46] have used 1D conservation equations of mass and momentum for ions and electrons. The potential distribution function across the IEG is presented in Figure 4.6.

$$\phi_0 = -\left(\frac{kT_e}{e}\right)\ln 2 \tag{4.14}$$

$$n = \frac{1}{2n_0} \tag{4.15}$$

They have also used the momentum expression for separate elements, Poisson's equation (Equation 4.16) for the electric field in the sheath area (Equation 4.17), and

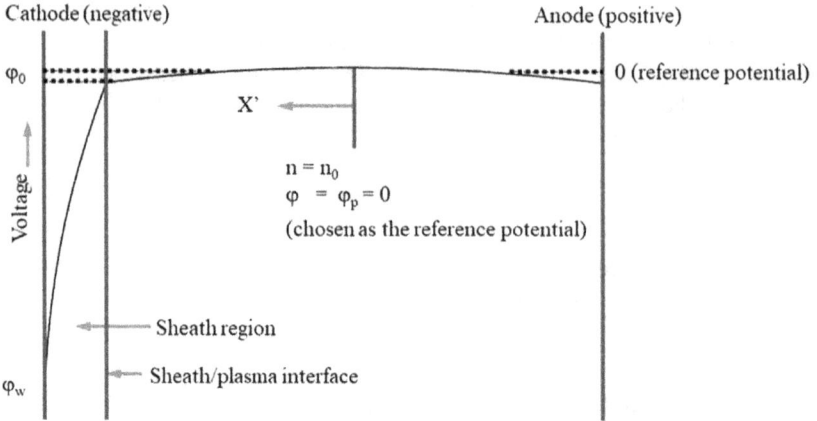

Cathode (negative)   Anode (positive)

φ_0 0 (reference potential)

X'

n = n$_0$
φ = φ_p = 0
(chosen as the reference potential)

Voltage

Sheath region

Sheath/plasma interface

φ_w

FIGURE 4.6   Voltage distribution at IEG in EDM [46]. With kind permission from Elsevier, License No. 5158250671531.

the potential boundary state derived from the plasma zone analysis to analyze the sheath region.

$$\nabla^2\phi = -\frac{e}{\epsilon_0}\left(n_i - n_e\right) \tag{4.16}$$

$$E^2\left(x\right) = \frac{4}{\epsilon_0}n_{se}e\left(\phi_0\phi_w\right)^{\frac{1}{2}} \tag{4.17}$$

where ϕ_w is the voltage at cathode and n_{se} is the typical ion and electron density at the sheath–plasma interface region.

The fluid dynamics method is concerned with changes in bulk plasma characteristics, as can be deduced from the discussion. On the other hand, plasma kinetics is concerned with the distribution of different species within the plasma. Electrothermal erosion is analyzed using the first method (fluid dynamics approach), while electromechanical material removal is studied using the second method (kinetic approach).

Most of the developed models [42–44, 46] have been discussed for macro-EDM. While the physical and chemical behavior during sparking/discharging are the same at the macro and micro scales, and current macro-EDM models must be modified for the use at the microscale. The EDM plasma is expected to disseminate a steady heat flux to the electrodes in the models discussed above to enable material removal. This assumption is invalid in the case of μ-EDM owing to non-uniform discharge using an RC-based pulse generator [47]. Although, the spark gap is small compared to macro-EDM, it might affect the characteristics of the material removal process. The models mentioned above can be compared with the experimental results for a long pulse on-time and high discharge current. In comparison, these models are less accurate with short pulse duration and low current values. Therefore, further analysis is required to model the plasma during discharge in μ-EDM. There has not been much work put into modeling μ-EDM plasma so far. A few researchers have attempted the modeling of μ-EDM plasma [47, 48] which is explained as follows.

Dhanik and Joshi [47] employed a combination of fluid dynamics [44] and kinetics [46] to describe μ-EDM plasma, including breakdown, plasma, and heating phases that are responsible for material erosion. The bubble mechanism and ionization due to the electron impact mechanism are used to explain the breakdown phase of plasma which is unique compared to other models. The bubble mechanism (Figure 4.7) included the small bubble formation in the cathode (tool electrode) due to heating by transferring pre-breakdown current through the micropeak on the tool electrode surface. The bubble rapidly grows and expands toward the workpiece (anode) to form a plasma channel. The Fowler and Nordheim [49] equation is used to quantify the pre-breakdown current due to the cathode's electrons emission. The equation for the bubbles' nucleation rate per unit volume has utilized the expression of nucleated bubble pressure (the Clausius–Clapeyron equation) and the energy density at micro-peaks to get the spherical bubble growth at IEG. A variable mass model is utilized to represent the heating phase, similar to that employed by Eubank et al. [44] to represent μ-EDM plasma during the heating phase. Radiation heat transfer and ion/electron collision with the electrodes responsible for transmitting energy from the μ-EDM plasma to the cathode and anode. A similar analysis like that of Singh and Ghosh [46] of plasma and sheath region has been carried out to quantify the percentage of heat energy transfer to both electrodes by ion and electron bombardment. The time-varying radius, density, pressure, enthalpy, and temperature of the plasma, as well as time-varying electrons density, are all represented in the model. The embryonic plasma properties are also shown, taken with spark gap variation up to 3.5 μm.

Chu et al. [48] have developed a comprehensive μ-EDM plasma model that includes the breakdown and expansion phases considering the magnetic pinch, surface tension, and viscous force of the fluid, which are not considered in the previous

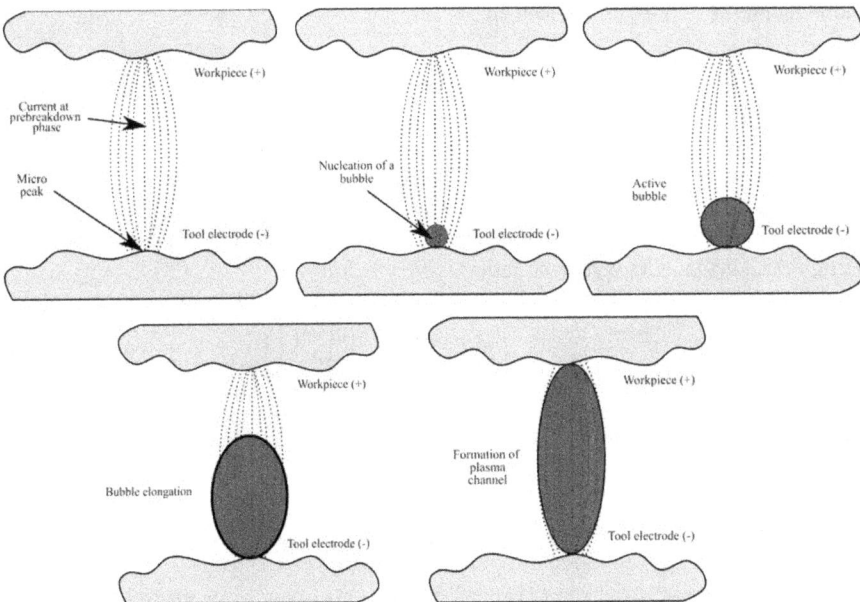

FIGURE 4.7 Schematic diagram of proposed breakdown mechanism.

plasma model. The Dhanik and Joshi [47] models, which are regarded as the initial conditions for the plasma expansion stage, are used to compute plasma parameters such as plasma pressure, temperature, and radius at the breakdown moment. The nucleation temperature of the bubble (T_{nu}) is evaluated using Equation 4.18.

$$Ne^{-\frac{\lambda}{kT_{nu}}}\sqrt{\frac{2\sigma_s}{\pi m}}\exp\left(-\frac{16\pi\sigma_s^3 T_{sat}^2}{3kT_{nu}\left(T_{nu}-T_{sat}\right)^2\left(\rho h_{fg}\right)^2}\right)r_t^3\left(\frac{c_p(l)}{jE_p}\right)$$

$$\times\left(\int_{T_o}^{T_{sat}}\rho_l+\int_{T_{sat}}^{T_{nu}}\rho_m\right)dT - 1 = 0$$

(4.18)

where λ is the per molecule heat of vaporization, m is the mass of a molecule of dielectric, σ_s is the surface tension, k is the Boltzmann constant, T_{sat} is the saturation temperature, ρ_l is the liquid state density, ρ_m is the meta-state density, r_t is the tip radius of micropeak, j is the current density, $c_p(l)$ is the liquid phase heat capacity, and E_p is the micropeak's electric field.

The nucleation bubble radius r_a after the gap breakdown can be calculated as follows:

$$r_a = \frac{2\sigma_s T_{sat}}{\left(T_{nu}-T_{sat}\right)\rho_v h_{fg}}\sqrt[3]{\left(\rho_v/\rho_a\right)}$$

(4.19)

where ρ_v is the vapor state density, ρ_a is the threshold density of vapor, and h_{fg} is the heat of vaporization.

The initial temperature (T_a) at the start of the expansion stage of plasma can be found using the expression as follows:

$$\rho_{avg}\int_{T_{nu}}^{T_a}C_a(T)dT = Ej\tau$$

(4.20)

where growth time $\tau = 3\times\left(b/v_e\right)$, E is the electric field at the spark gap, b is the IEG, and v_e is the electron drift velocity.

The initial pressure (P_a) of the plasma channel after the broken gap is obtained using the Clausius–Clapeyron equation, given as follows:

$$P_a = P_{atm}\exp\left[-\frac{\Delta H_{vap}}{R}\left(\frac{1}{T_a}-\frac{1}{T_{sat}}\right)\right]$$

(4.21)

where, R is the gas constant, ΔH_{vap} is the heat of vaporization, and P_{atm} is the atmospheric pressure.

In the plasma expansion stage, the mathematical expression for the magnetic pinch force is obtained and expressed by the hydrodynamic and energy balance equation considering the magnetic pinch force, viscosity, and surface tension of the dielectric. The model of plasma expansion has been enhanced by using both of these forces, bringing it closer to reality.

4.3.2.2   Modeling of Material Removal or Electrode Erosion

After establishing the plasma channel in the IEG, the material erosion of the cathode and anode will occur due to the heating process. There are various models in the literature [42, 46, 50–55] to estimate the material removal of electrodes in EDM, and they are also applicable to μ-EDM. Based on material removal mechanisms, these models are classified as electrothermal and electromechanical models. This section discusses previous research that has been done on these models.

4.3.2.2.1   Electrothermal Model

The electrothermal model is most common and widely used to recognize material erosion mechanisms in EDM or μ-EDM. According to the electrothermal process, material removal in EDM/μ-EDM occurs due to melting or evaporation due to exceedingly high temperatures achieved at the electrode surfaces. The plasma generated in the IEG is at a high-temperature state and carried a high-intensity current. An electrothermal model is a generic approach for finding the temperature distribution and heat flow in the electrode region by solving a heat transfer equation in the electrode domain. Typically, the plasma heat flux is computed using mathematical relations or a percentage of the electrical power supplied into a discharge. Furthermore, it is considered that the material heated beyond the melting point is removed after the discharge, resulting in material removal from the electrodes.

Yeo et al. [55] attempted to model material removal at cathode and anode in μ-EDM first time. They have used the model developed by Dijck and Snoeys (1972) to develop temperature distribution expression. The uniformly distributed heat flux radius (in μm) is computed at the anode and cathode using experimental data for discharge energies ranging from 5 to 150 μJ, as follows:

$$R_a = 0.0284\, t_{on}^{0.9115} \tag{4.22}$$

$$R_c = 0.0425 t_{on}^{0.0895} \tag{4.23}$$

where R_c and R_a are plasma heat flux radius at the tool (cathode) and workpiece (anode), and t_{on} is the pulse duration (in nanoseconds).

The temperature distribution is derived by employing BCs to evaluate the transient heat conduction equation, i.e., partial differential equation (PDE). The solution of PDE is shown in Equation 4.24.

$$T(r,z,t) = \frac{q \cdot R}{2K_t} \int_0^\infty J_o(\lambda r) J_1(\lambda R)$$

$$\times \left[e^{-\lambda z} erfc\left(\frac{z}{2\sqrt{\alpha t_{on}}} - \lambda\sqrt{\alpha t_{on}} \right) \right. \tag{4.24}$$

$$\left. - e^{\lambda z} erfc\left(\frac{z}{2\sqrt{\alpha t_{on}}} + \lambda\sqrt{\alpha t_{on}} \right) \right] \frac{d\lambda}{\lambda}$$

where J_0 and J_1 are the first kind Bessel functions of zero and first order, respectively, K_t is the average thermal conductivity (W m^{-1} K^{-1}), a is the average thermal diffusivity (m^2 s^{-1}), R is the radius of heat flux at anode or cathode (in µm), which are calculated from the previous expression, already shown in Equation 4.22 or Equation 4.23 and *erfc* is the complementary error function.

Figure 4.8 represents the schematic diagram of the electrothermal model where heat flux q is applied on the bulk material, and R_p is the heat flux radius or plasma radius. The uniformly distributed heat flux (q) over the electrodes is determined using the average gap voltage (V_a), discharge current (I_a), and fraction of energy (C_e) transfer to the electrodes, as in Equation 4.25. The fraction of energy transferred to anode (C_a) and cathode (C_c) are calculated as 39% and 14%, respectively, using experimental results.

$$q = \frac{C_e \cdot V_a \cdot I_a}{\pi R_p^2} \tag{4.25}$$

Dimensionless parameters are used to create temperature distribution models in the vertical and radial directions, which are then displayed to predict crater shape. During a discharge, input variables like voltage and current waveforms, pulse duration, and material characteristics of AISI 4140 alloy steel, such as thermal conductivity, specific heat, density, and melting temperature are used to predict microcrater geometry and dimensions like crater radius and depth. A single discharge experiment is also carried out to validate the models created. The theoretical models are well accepted at low discharge energy levels.

Kurnia et al. [9] have proposed a model for the analytical prediction of µ-EDM responses such as MRR and EWR. They used the concept of Yeo et al. [55] to predict the crater size on both electrodes. It has been assumed that the microcrater formed is of a spherical shape. Equation 4.26 is used to estimate the crater volume (V_c) using crater diameter (d) and depth (h).

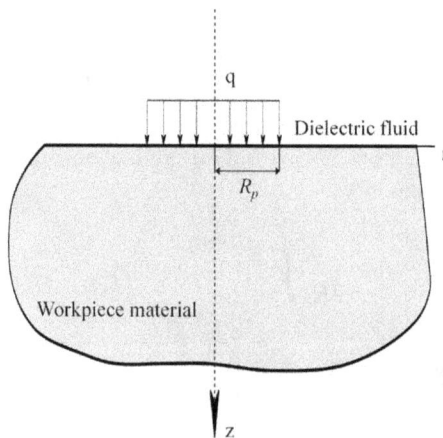

FIGURE 4.8   Schematic representation of the electrothermal model.

$$V_c = \frac{\pi d^2 h}{8} + \frac{\pi h^3}{6} \qquad (4.26)$$

The experimental MRR and EWR are calculated using the measurement of crater volume on both cathode and anode, and total machining time. Then analytical MRR and EWR are modeled using these experimental data considering the removal factor and wear factor.

$$MRR_{analytical} = S_{MRR} \times MRR_{theoretical} = S_{MRR} \times \frac{V_w}{t_m} \qquad (4.27)$$

$$EWR_{analytical} = S_{EWR} \times EWR_{theoretical} = S_{EWR} \times \frac{V_e}{V_w} \qquad (4.28)$$

where V_w is the volumetric material erosion from the workpiece (anode), t_m is the total machining time, S_{MRR} is the material removal factor which takes account of abnormal discharges, and V_e is the volumetric material erosion from the tool electrode (cathode). S_{EWR} is the wear factor that accounts for the carbon deposition on the tool electrode surface and acts as a protective layer on the electrode surface.

Ni et al. [56] have exploited electrothermal models to predict the temperature gradient at the electrode and utilized Hankel transformation and Laplace transformation to solve the heat transfer equation. The temperature gradient at the electrodes is given by Equation 4.29, which depends on the heat flux of plasma (Q_f). Further, the heat flux follows a linear relationship between the discharge current and voltage. So, the temperature gradient is dependent on the discharge current and voltage for constant thermal properties of material. The high amount of discharge current and voltage result in high temperature gradient and causes more material erosion.

$$T(r,z,t) = \frac{2Q_f \sqrt{\alpha t}}{K_t} \left[ierfc\left(\frac{z}{2\sqrt{\alpha t}}\right) - ierfc\left(\frac{\sqrt{z^2 + R^2}}{2\sqrt{\alpha t}}\right) \right] e^{-\left(\frac{r^2}{4\alpha t}\right)} \qquad (4.29)$$

where a is thermal diffusivity of the workpiece, K_t is thermal conductivity of the workpiece, and R is the plasma radius during discharge. Finally, a single discharge experiment has been carried out using RC-based pulse generator to study the influence of open-circuit voltage and capacitance or discharge energy on crater size.

4.3.2.2.2 Electromechanical Model

Very few researchers [46, 57, 58] have proposed the material erosion mechanism using the electromechanical model in conventional EDM for short pulse duration, while no attempt has been made to explain the material removal mechanism using this model in μ-EDM.

Singh and Ghosh [46] have explained the electromechanical theory for material erosion in EDM. The material loss by yielding is caused by the mechanical stress imposed by the electric field on the electrode surface. The plasma and sheath areas are analyzed to determine surface stress (σ), which is shown in Equation 4.30.

$$\sigma = 4n_{se}e\sqrt{\phi_o\phi_w}$$ (4.30)

where n_{se} is the density of ion/electron at plasma and sheath interface, ϕ_o is the reference voltage at infinity, and ϕ_w is the cathode surface's potential.

This term of stress is used to estimate the stress distribution within the electrode, and the crater depth is calculated as the depth over which the highest stress exceeds the yield strength. According to the model, material erosion is caused by electrostatic force for short pulses, whereas melting takes precedence over electrostatic force for longer pulses. The model approximated the crater depth for the short pulse duration, which has a proportionate relationship with discharge current ($\propto I^{1/2}$) and is validated by Williams's experiment [57].

4.3.2.3 Remarks

The plasma channel modeling is divided into a fluid dynamic approach and kinetics approach. In the fluid dynamics approach, the plasma model includes the unsteady energy balance, fluid dynamics (continuity and momentum equations), and the plasma state equation (defined by the plasma parameters, i.e., plasma pressure, temperature, etc.). In the kinetic approach, the plasma region is determined by the distribution of ions and electrons. The fluid dynamics approach is used in most of the literature, and the kinetic method is seldom used due to the analytical complexity of the measurement of electron/ion density. The material removal or crater profile is evaluated mainly by generating the temperature distribution at the electrodes from the electrothermal model, where the plasma channel of the fluid dynamics model is used as BCs. Material removal using an electromechanical model, where the plasma model of kinetics is utilized as the BCs, is not explored in the µ-EDM process. In comparison to empirical models, the literature has rarely found an analytical model of µ-EDM. As a result, analytical modeling for different dielectric fluids (plasma modeling and electrode erosion modeling) is a potential µ-EDM research area.

4.3.3 Numerical Models

Recently, most of the analytical studies have used numerical methods such as finite element (FE), boundary element (BE), finite volume (FV), and finite difference (FD) methods to find the adequate and accurate solution of the governing equations of the process. Most of the µ-EDM modeling has used the FE method, which analyzes the plasma and electrode (cathode/anode) interaction to get temperature distribution on both electrodes using the transient heat conduction model and BCs. The BCs can be obtained using experimentation or empirical models and analytical models. Some of the prior research work based on numerical studies of µ-EDM is depicted in this portion.

4.3.3.1 Single-Spark Modeling

The generalized steps for crater profile generation in single-spark modeling are discussed below. The heat flux impacted region is modeled using a portion of energy distributed to both the electrodes and dielectric, plasma radius, energy intensity, and other BCs such as material properties. The formed plasma channel is believed to be a heat source. Then, melt pool simulation is performed to obtain temperature distribution on both electrodes. Finally, the crater profile is achieved on both electrodes using the temperatures above the electrode material's melting or boiling temperature. The previous research works related to the single-spark model are reported in this section.

Allen and Chen [59] have used a thermal-based numerical model to analyze the material erosion and residual stress of molybdenum in a single-spark process using ANSYS software. An axisymmetric model and constant heat flux such as Gaussian heat flux equation [60] was proposed to evaluate the process performance by process simulation. Thermal analysis means solving of the heat conduction equation based on Fourier's law (Equation 4.31) using the thermal BCs. It is performed to obtain the temperature variation on the workpiece, which is further used to calculate the crater volume or material removal. Figure 4.9 depicts the thermal BCs, and Gaussian heat flux, which are used at the top surface up to the plasma radius. It is assumed that convective heat loss to the rest of the top surface and bottom surface take place, and the other two walls are thermally insulated.

$$K\left[\frac{1}{r}\frac{\partial}{\partial r}\left(r\frac{\partial T}{\partial r}\right)+\frac{\partial^2 T}{\partial z^2}\right]=\rho c\frac{\partial T}{\partial t} \tag{4.31}$$

where T is the temperature as a function of space (r and z) and time (t). K, c, and ρ are the workpiece's thermal conductivity, specific heat, and density, respectively. Parametric studies such as the influence of pulse duration and different work material (molybdenum and steel) based on the numerical model are conducted, and it has

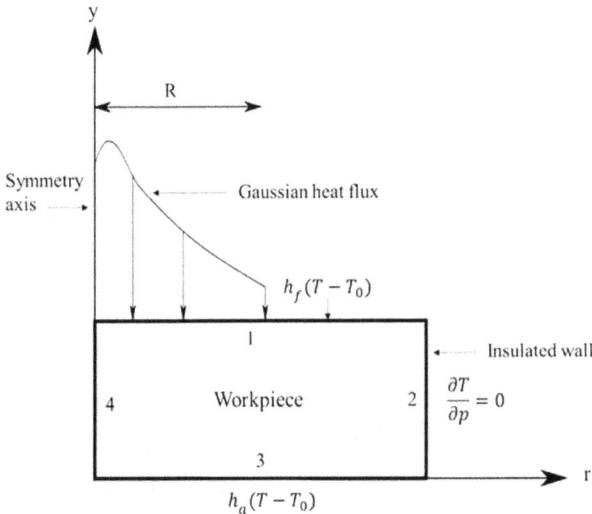

FIGURE 4.9   The axisymmetric model with assumed BCs for process simulation.

been reported that the crater size on molybdenum is smaller compared to steel. The variation of relative tool wear with pulse duration in case of molybdenum work material is higher as compared to steel owing to the superior conductivity and melting point of molybdenum.

Murali and Yeo [61] have utilized the same concept as that by Allen and Chen [59] to estimate the temperature distribution and residual stress in Ti-6Al-4V alloy. Using a heat source (Gaussian heat flow) and temperature-dependent material characteristics, transient thermal studies have been done to estimate crater contour, temperature gradient on titanium alloy, and residual stress on the crater or its neighbors. The simulation results of crater profile dimensions and residual stress are related to the single discharge experimental outcomes using atomic force microscopy (AFM) and nanoindentation methods. The crater's diameter to depth ratios from simulation and experiments are 3.45 and 3.99, respectively. Near the spark core, the simulated stress approaches the material's ultimate tensile strength and eventually decreases as the distance from the center increases.

The finite element approach has been exploited by Kumar and Yadava [62] to evaluate the temperature distribution and crater size in the impacted zone of a single discharge. The parametric studies such as depth and diameter of crater, energy partition, and MRR with varying spark radii have also been performed. It has been observed that the input energy and heat flux have the greatest impact on the energy fraction.

Liu et al. [63] modeled crater formation on metallic glass to predict the crater geometry. They showed various phases in the single crater, including the molten stage, thin crystalline layer, and supercooled liquid layer using ABAQUS software. The simulation is performed under consideration of a single discharge and 100% flushing efficiency, which is defined as the ratio between the actual volumetric removal of material from the crater and the molten pool's volume. The increasing trend of crater dimensions is observed with the discharge voltage, capacitance, and discharge energy both in the simulation and experimentation.

Mujumdar et al. [64] employed the μ-EDM plasma model, including the plasma pressure, radius, and heat flux model for developing a melt pool model to approximate the workpiece's erosion using COMSOL Multiphysics software in a single-spark process [65]. The heat transfer equation (Equation 4.32) and hydrodynamics such as the Navier–Stokes equation (Equation 4.33) and continuity equation (Equation 4.34) are resolved in the area covering the dielectric fluid and workpiece. The melt pool model predicts a crater diameter of 78–96 μm and an overall crater depth of 8–9 μm for a pulse duration of 2 μs based on the crater's profile.

$$\rho_o c \frac{\partial T}{\partial t} + \rho_o c u \cdot \nabla T = \nabla \cdot \left(k \nabla T \right) + Q_{in} \delta \qquad (4.32)$$

$$\rho_o \frac{\partial \mathbf{u}}{\partial t} + \rho_o \left(\mathbf{u} \cdot \nabla \right) \mathbf{u} = \nabla \cdot \left[-p\mathbf{I} + \mu \left(\nabla \mathbf{u} + \left(\nabla \mathbf{u} \right)^T \right) \right] + \mathbf{F}_{sur} + \mathbf{F}_m \qquad (4.33)$$

$$\nabla \cdot \mathbf{u} = 0 \qquad (4.34)$$

where **u** is the velocity vector in m/s, **I** is the identity tensor, ρ_o is the density of dielectric in kg/m³, c is the specific heat capacity in J/kg-K, T is the temperature in K, k is the thermal conductivity in W/m-K, Q_{in} is the plasma heat flux, δ is the function used to transform surface forces into volumetric forces, p is the pressure of fluid in Pa, F_{sur} is the surface forces including the force resulting from the plasma pressure, surface tension and Marangoni force, and F_m is the imaginary damping force.

Shao and Rajurkar [66] simulated crater formation in μ-EDM using the developed electrothermal model. This model has used the heat flux as Gaussian distribution, time-varying plasma radius, and temperature-dependent thermal properties. The analytical model for temperature distribution at electrodes, the plasma radius model, and experimentally measured crater sizes are used to calculate the energy distribution at electrodes which have been reported as 7.37% for the anode and 6.78% for the cathode. The crater profile is experimentally measured and compared with the simulation results. The crater radius and depth (or crater volume) are higher at a higher value of open-circuit voltage and capacitance due to the higher amount of discharge energy. Figure 4.10 shows the temperature distribution along with the crater dimensions for the single discharge μ-EDM process.

Simulation of the crater profile was also carried out by the authors [67] taking into account the combined effects of heat transfer and fluid flow, which is one step forward from previous work. The Marangoni effect (stress-induced due to the presence of surface tension gradient) has been considered as the main contributor to crater development. This model has been used with three governing equations, i.e., Navier–Stokes, continuity, and energy equations, like in work done previously by Mujumdar et al. [65]. This model displays the mechanism of crater formation on stainless steel alloy along with temperature profile and velocity fields. Based on the principle of crater depth reduction after cooling, this model can also estimate the recast layer thickness.

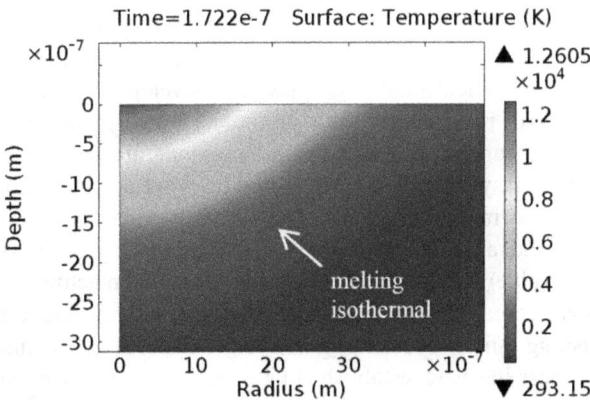

FIGURE 4.10 Simulated temperature profile and crater geometry for single spark [66]. With kind permission from Elsevier, License No. 5090241140349.

The influence of fluid motion in the mushy zone (where the solid and liquid phases coexist) of the workpiece on crater geometry prediction is not considered in any of the models discussed above. Dilip et al. [68] used ANSYS FLUENT simulation to describe this impact on Inconel 718 alloy. They identified that irregular crater geometry is caused by relative motion between the top and successive layers of the molten pool. Temperature-dependent material properties (viscosity, density, specific heat, and thermal conductivity), the heat transfer equation, and fluid flow equation are used for simulation. The thermal BCs such as Gaussian heat flux up to spark radius, R, is taken as 20% of the tool electrode's radius, and convective heat flux for the remaining portion of the top surface is used during pulse on-time. The velocity boundary condition such as Marangoni stress at the top surface and zero velocity gradient in the perpendicular direction to the other three boundary surfaces are considered. In contrast to the experimental performance, the surface profile by simulation differs, approximately from 8.57% to 13.66%. The variation between the experimental and simulated crater geometry increases as the discharge energy is increased, so further melting leads to more resolidification (thickened recast layer) and, therefore, more irregularities on the crater surface. Also, the effect of Marangoni stress on crater profile has been explained using computational fluid dynamics (CFD) simulation of a single discharge experiment by Kliuev et al. [69]. The simulation revealed that while the Marangoni effect does not affect the material removal, it impacts the crater's shape.

4.3.3.2 Multi-Spark Modeling

The spark discharge phenomenon is stochastic and identifying the specific position of sparks and crater overlapping is very difficult in the μ-EDM process. Very few researchers have attempted to develop numerical models using multi-sparks to take this complexity into account.

The concept of an overlapped crater or the effect of multiple discharges on the machined surface has been established by Tan and Yeo [70]. Maximum asperity and lowest asperity conditions are used to assess the topography of the machined surface. Two adjacent craters at the crater rim form a surface asperity with a pointed tip, resulting in the highest surface roughness value in the case of maximum asperity conditions. A two-dimensional model has been developed to solve the shallowest (minimum asperity or roughness) profile on the machined surface using a uniform heat flux with time-varying plasma. The assumptions are made that the surface profile obtained is axisymmetric about the middle point of the separating distance between two neighboring craters, and this distance is equal to the diameter of the heat source. Thus, the range of surface roughness (difference between the highest and lowest roughness value) is determined, revealing the actual machined surface conditions. This range increases with increasing discharge energy due to more melting and plasma flushing efficiency resulting in a larger crater size. An analytical model of surface roughness has been established by Kurnia et al. [71] without considering overlapping craters, reattachment of debris particles, developed microcracks, and pockmarks. Another theoretical model of surface roughness under consideration of debris particles has been presented to establish a realistic model [72]. The debris particles change the gap behavior by affecting the electric field and composition of plasma and melt pool radius in a single discharge. These results are further used in a multi-discharge model to modify the roughness model.

A multi-spark numerical model based on a finite volume approach has been established by Somashekhar et al. [73] to emphasize the effect of the spark discharge ratio (pulse on-time/pulse off-time) on the temperature profile of both electrodes. The simulation findings demonstrate that the temperature in the axial and normal direction increases with the pulse duration and rapidly declines with pulse off-time. The lower oscillation of temperature is shown with the high spark ratio. After each pulse-off time, several sparks are permitted to be fired at the same spot on the workpiece surface in a multi-spark model to increase the depth of the formed crater [74]. This method is shown to be ineffective in estimating the anode profile and depth of layer loss since it entirely ignores the discharges at other sites. Recently, Singh et al. [75] proposed a multi-sparks model based on randomly generated discharges at different locations on the anode (workpiece) surface under consideration of IEG variation. They have extended their work from single spark to multi-sparks. The crater profile (depth and radius) is found using single discharge simulation, and these are input to the multi-discharge models. The steps behind the single spark and multi-sparks simulation in μ-EDM are illustrated in Figure 4.11. In this work, the arbitrary surface roughness values are assigned to the cathode (in the range from −1 to +1 μm) and anode (in the range from −2 to +2 μm) at various points (equal to the number of sparks required) on 0.5 mm length of both electrodes in a 2-D field. A MATLAB® code is generated to store the IEG data at all positions and arrange them in gradually

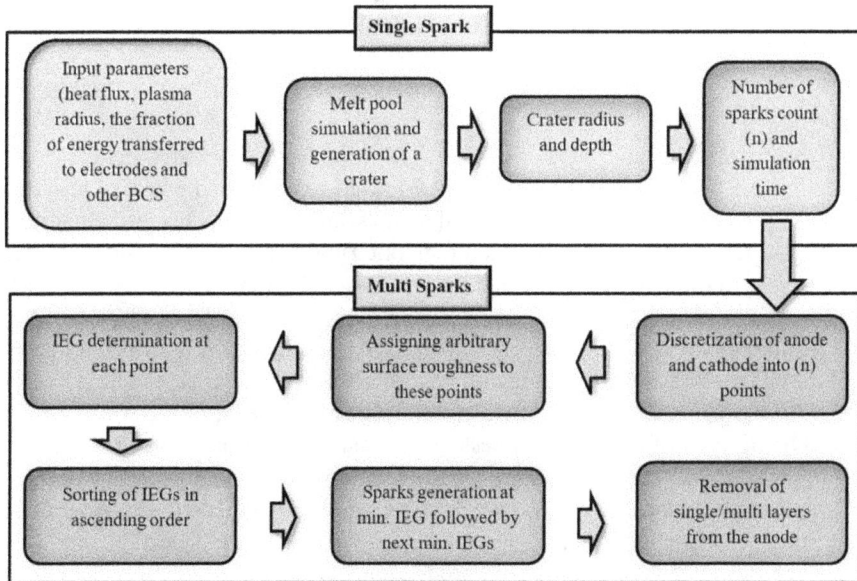

FIGURE 4.11 Steps involved in single spark and multi-spark simulation.

increasing order. Then, this code is combined with COMSOL Multiphysics and acts as an input for controlling the spark location and duration. The first spark is generated at minimum IEG, and the spark is shifted to the next minimum IEG. Thus, successive sparks have occurred at different positions on the workpiece surface, and the machined surface is generated after the removal of material as a layer.

A summary of the various heat flux and plasma radius models used in single-spark and multi-spark simulation of μ-EDM is presented in Table 4.3.

4.3.3.3 Remarks

Numerical simulation based on a single discharge is carried out by solving the heat conduction equation (Fourier equation). Less deviation between the simulation and experimental results is found after using temperature-dependent material properties (mainly thermal properties). A more realistic model is developed after considering heat source as a Gaussian-distributed, expanding plasma model (time-varying), and a fixed fraction of energy transfer to both electrodes. Single-spark models are capable of predicting the crater geometry, size, and, successively, material removal rate.

Still, single-spark modeling of μ-EDM is a potential research field for different combinations of workpiece and tool electrode materials and process parameters. The impact of various heat source models on crater profiles can be a research topic in a single discharge model. The actual surface condition or asperity is evaluated using the multi-spark models, and it also depends on the discharge energy and plasma flushing efficiency of the process. However, measuring overlapping craters in minor IEGs is a complex procedure. Most multi-spark models produced so far are simplistic and fail to capture the accurate topography of the machined surface. This necessitates the development of additional multi-spark models with the real-life assumptions to achieve good matching between the experimental and simulated results.

4.4   CONCLUSIONS

This chapter presents an overview of the μ-EDM process and research in the field of modeling and optimization studies of μ-EDM. The summary and potential research areas of experimental studies or empirical models, analytical models, and numerical models are discussed in the subsection remarks. However, from the discussion, the following key conclusions can be made.

- Micro-EDM is a thermoelectric type advanced machining process. It can produce a variety of microholes, micromolds for injection molding, hot embossing, microslots, microgrooves, micropillars, etc., along with various 3D microfeatures regardless of the material hardness and strength. The study of various developed models such as empirical models, analytical models, and numerical models is required to understand the material removal from the tool electrode and workpiece, and proper selection of process parameters for economical production.
- Empirical modeling has used the experimental results to develop the input/output relationship in the μ-EDM process. Various DOE approaches

TABLE 4.3
Heat Flux and Plasma Radius Models Used in Micro-EDM

Authors	Heat Flux Type	Heat Flux Model, Q (w/m²)	Plasma Radius Model R(t) (μm)
Allen and Chen [59], Singh et al. [75]	Gaussian heat flux	$\dfrac{4.45FUI}{\pi R(t)^2}\exp\left(-4.5\left(\dfrac{r}{R(t)}\right)^2\right)$	Constant radius (5 μm) [59], Empirical equation: $0.0284_{on}^{0.9115}$ [75].
Murali and Yeo [61], Shao and Rajurkar [66, 67], Dilip et al. [68], Somashekhar et al. [73]	Gaussian heat flux	$\dfrac{3.1572FUI}{\pi R(t)^2}\exp\left(-3\left(\dfrac{r}{R(t)}\right)^2\right)$	Constant spark radius (2.1 μm) [61], Empirical equation: $0.788_{on}^{0.75}$ [66, 67], Constant spark radius (20% of the tool radius) [68], Constant spark radius [73].
Kumar and Yadava [62], Liu et al. [63]	Gaussian heat flux	$\dfrac{FUI}{\pi R(t)^2}\exp\left(-2\left(\dfrac{r}{R(t)}\right)^2\right)$	The plasma channel model of Dhanik and Joshi (2005) used [62], $$R(t)=\begin{cases}C_1*t^{\frac{3}{4}} & 0\le t\le t_c \\ C_2 & t\ge t_c\end{cases}$$ Where t is the discharge time, t_c is critical time (taken as 100 μs) and C_1, C_2 are the coefficient selected based on experimental data [63].
Mujumdar et al. [65]	Gaussian heat flux	$\dfrac{q_0(t)R(t)^2}{2\sigma_R^2\left[1-\exp\left(-\dfrac{R(t)^2}{\sigma_R^2}\right)\right]}\exp\left(-\dfrac{r^2}{2\sigma_R^2}\right)$ where $\sigma_R=\dfrac{\text{cathode's radius}}{3}$	Utilized the plasma channel model [64] to predict plasma radius.
Tan and Teo [70]	Uniform heat flux	$\dfrac{FUI}{\pi R(t)^2}$	Empirical equation: $0.059_{on}^{0.79}$

F is a fraction of energy transferred to the electrodes calculated by experimental work, U is the discharge voltage, I is the discharge current, r is the radial distance from the origin, $R(t)$ is the time-varying plasma radius, $q_0(t)$ is the heat flux at $r = 0$ and t_{on} is the pulse on–time or discharge duration.

(CCD-based RSM, Taguchi-based, etc.) are used to collect the data scientifically. ANOVA is used to find the significant process parameters, and these parameters are used to develop more valid and accurate models. Empirical models are further used for process optimization. The range of the parameters is changed for a different combination of tool–workpiece material and variants. Most of the developed models and parametric studies are based on μ-EDD. So, extensive parametric studies or empirical models need to be developed for other variants.

- The plasma channel model evaluates the plasma state during discharge/ spark in analytical modeling. The BCs for the material erosion model are evaluated using the plasma channel model. The material erosion models are mainly based on the electrothermal process. Material erosion occurs at both electrodes attributing to high temperature due to high intensity of current flowing within the plasma channel. The exact solution-based analytical models are not flexible as numerical models, and the solution is time-consuming, demanding the development of numerical models.

- Most of the numerical models developed are based on finite element analysis using ANSYS, COMSOL Multiphysics, and ABAQUS commercial software. Also, several studies are based on a single-spark model for crater formation. The development of the multi-spark model has rarely been used because of the complexity of finding the exact spark locations and crater overlapping. The development of numerical models for different tool–workpiece combinations, various heat flux along with experimental conditions (different combination of process parameters), and hybrid processes (vibration-assisted, magnetic-field assisted, and powder-mixed dielectric) are potential research fields in μ-EDM.

NOMENCLATURE

ANN	Artificial Neural Network	**OC**	Overcut
ANOVA	Analysis of Variance	*P*	Plasma state pressure (Pa)
AR	Aspect Ratio	P_f	Ambient pressure (Pa)
b	IEG (μm)	*PWM*	Pulse-width modulation
BCs	Boundary conditions	*R*	Gas constant (J/mol-K)
C	Capacitance value of RC pulse generator (μF)	R_a	Plasma heat flux radius at anode (μm)
C_a	Fraction of energy at anode	r_a	Nucleation bubble radius after dielectric breakdown (μm)
C_c	Fraction of energy at cathode	R_c	Plasma heat flux radius at cathode (μm)
CCD	Central Composite Design	R_f	outer radius of dielectric fluid in the spark gap (mm)
C_e	Fraction of total energy available at electrodes	R_p	Plasma radius (μm)
CFRP	Carbon-Fiber-Reinforced Polymers	**RSM**	Response Surface Methodology
$c_{p(l)}$	Liquid phase heat capacity (J/kg K)	r_t	Tip radius of micropeak (μm)

DOE	Design of Experiment	**RTW**	Relative Tool Wear
EDD	Electro Discharge Drilling	R_z	Maximum peak to valley height (μm)
EDG	Electro Discharge Grinding	**SEM**	Scanning Electron Microscope
E_p	Micropeak's electric field (V/m)	T_a	Initial temperature at the starting of the plasma expansion stage (K)
EWR	Electrode Wear Ratio	T_e	Temperature of electrons (K)
f	Feed rate of tool electrode (μm/s)	T_{nu}	Nucleation temperature of the bubble (K)
F_p	Flushing pressure (kg/cm²)	t_{on}	Pulse on-time/Pulse duration (μs)
GA	Genetic Algorithm	**TWR**	Tool Wear Rate
Gi	Grey relational grade	**UCV**	Ultrasonic Circular Vibration
GRA	Grey Relational Analysis	V_c	Crater volume (μm³)
h_{fg}	Latent heat of vaporization (J/kg)	V_e	Electron drift velocity (m/s)
IEG	Interelectrode Gap	V_g	Gap voltage (V)
I_p	Peak current (A)	σ	Stefan–Boltzmann constant
j	Current density (A/m²)	w_o	Density of the dielectric liquid (kg/m³)
k	Boltzmann constant	w_l	Liquid state density (kg/m³)
K_t	Average thermal conductivity (W/m K)	w_m	Meta-state density (kg/m³)
LWR	Longitudinal Electrode Wear	w_a	Threshold density of vapor (kg/m³)
m	Mass of a molecule of dielectric (g/mol)	w_v	Vapor state density (kg/m³)
m_p	Mass of the plasma	x_s	Surface tension of dielectric fluid (N/m)
MRR	Material Removal Rate	*a*	Average thermal diffusivity (m²/s)
n_e	Electron density in plasma	φ_o	Reference voltage at infinity (V)
n_i	Ion density in plasma	φ_w	Cathode surface's potential (V)
n_{se}	Typical ion and electron density at the sheath-plasma region		

REFERENCES

1. T. Masuzawa and H. K. Tönshoff. "Three-dimensional micromachining by machine tools." *CIRP Ann. – Manuf. Technol.*, vol. 46, no. 2, pp. 621–628, 1997. DOI: 10.1016/s0007-8506(07)60882-8.

2. K. H. Ho and S. T. Newman. "State of the art electrical discharge machining (EDM)." *Int. J. Mach. Tools Manuf.*, vol. 43, no. 13, pp. 1287–1300, 2003. DOI: 10.1016/S0890-6955(03)00162-7.

3. H. Kurafuji and T. Masuzawa. "Micro-EDM of camented carbide alloys." *J. Jpn. Soc. Elect. Mach. Eng.*, vol. 2, no. 3, pp. 1–16, 1968.

4. E. C. Jameson. *Electrical Discharge Machining*. Dearborn, MI: Society of Manufacturing Engineers, 2001.

5. A. Mallik and A. K. Ghosh. *Manufacturing Science*. Harlow, England: Ellis Horwood Ltd, 1986.

6. M. P. Jahan, M. Rahman, and Y. S. Wong. *Micro-Electrical Discharge Machining (Micro-EDM): Processes, Varieties, and Applications*, vol. 11. Elsevier, 2014.

7. J. Tao, J. Ni, and A. J. Shih. "Modeling of the anode crater formation in electrical discharge machining." *J. Manuf. Sci. Eng. Trans. ASME*, vol. 134, no. 1, 2012. DOI: 10.1115/1.4005303.

8. Y. S. Wong, M. Rahman, H. S. Lim, H. Han, and N. Ravi. "Investigation of micro-EDM material removal characteristics using single RC-pulse discharges." *J. Mater. Process. Technol.*, vol. 140, no. 3, pp. 303–307, 2003. DOI: 10.1016/S0924-0136(03)00771-4.

9. W. Kurnia, P. C. Tan, S. H. Yeo, and M. Wong. "Analytical approximation of the erosion rate and electrode wear in micro electrical discharge machining." *J. Micromech. Microeng.*, vol. 18, no. 8, p. 085011, 2008. DOI: 10.1088/0960-1317/18/8/085011.

10. L. Raju and S. S. Hiremath. "A state-of-the-art review on micro electro-discharge machining." *Procedia Technol.*, vol. 25, pp. 1281–1288, 2016. DOI: 10.1016/j.protcy.2016.08.222.

11. Z. Katz and C. J. Tibbles. "Analysis of micro-scale EDM process." *Int. J. Adv. Manuf. Technol.*, vol. 25, no. 9, pp. 923–928, 2004. DOI: 10.1007/S00170-003-2007-1.

12. B. Bhattacharyya and B. Doloi. *Micromachining Processes*, In *Modern Machining Technology*. Elsevier, 2020, pp. 593–673

13. Jahan, M. Pervej, A. B. M. Ali Rahman, Asad., M., Wong, Y. S., and Masaki, T. "Micro-Electro Discharge Machining (µEDM)." In *Micro-Manufacturing*. Hoboken, NJ, USA: John Wiley & Sons, Inc., 2011, pp. 301–346.

14. G. Yang, L. Feng, and H. Lin. "Research on an embedded servo control system of micro-EDM." *Appl. Mech. Mater.*, vol. 120, pp. 573–577, 2012. DOI: 10.4028/www.scientific.net/AMM.120.573.

15. Q. Liu, Q. Zhang, M. Zhang, and J. Zhang. "Review of size effects in micro electrical discharge machining." *Precis. Eng.*, vol. 44, pp. 29–40, 2016. DOI: 10.1016/j.precisioneng.2016.01.006.

16. D. T. Pham, S. S. Dimov, S. Bigot, A. Ivanov, and K. Popov. "Micro-EDM – Recent developments and research issues." *J. Mater. Process. Technol.*, vol. 149, nos. 1–3, pp. 50–57, 2004. DOI: 10.1016/j.jmatprotec.2004.02.008.

17. A. M. Sidpara and G. Malayath. "Introduction to micro EDM." In *Micro Electro Discharge Machining*. CRC Press, 2019, pp. 1–35.

18. M. Chandrasekaran, M. Muralidhar, C. M. Krishna, and U. S. Dixit. "Application of soft computing techniques in machining performance prediction and optimization: A literature review." *Int. J. Adv. Manuf. Technol.*, vol. 46, nos. 5–8, pp. 445–464, 2010. DOI: 10.1007/s00170-009-2104-x.

19. A. P. Tiwary, B. B. Pradhan, and B. Bhattacharyya. "Study on the influence of micro-EDM process parameters during machining of Ti–6Al–4V superalloy." *Int. J. Adv. Manuf. Technol.*, vol. 76, nos. 1–4, pp. 151–160, 2014. DOI: 10.1007/s00170-013-5557-x.

20. Y. C. Lin, C. C. Tsao, C. Y. Hsu, S. K. Hung, and D. C. Wen. "Evaluation of the characteristics of the microelectrical discharge machining process using response surface methodology based on the central composite design." *Int. J. Adv. Manuf. Technol.*, vol. 62, nos. 9–12, pp. 1013–1021, 2012. DOI: 10.1007/s00170-011-3745-0.

21. M. Ay, U. Çaydaş, and A. Hasçalik. "Optimization of micro-EDM drilling of inconel 718 superalloy." *Int. J. Adv. Manuf. Technol.*, vol. 66, nos. 5–8, pp. 1015–1023, 2013. DOI: 10.1007/s00170-012-4385-8.

22. K. P. Somashekhar, N. Ramachandran, and J. Mathew. "Optimization of material removal rate in micro-EDM using artificial neural network and genetic algorithms." *Mater. Manuf. Process.*, vol. 25, no. 6, pp. 467–475, 2010. DOI: 10.1080/10426910903365760.

23. H. Dutta, K. Debnath, and D. K. Sarma. "A study of material removal and surface characteristics in micro-electrical discharge machining of carbon fiber-reinforced plastics." *Polym. Compos.*, vol. 40, no. 10, pp. 4033–4041, 2019. DOI: 10.1002/pc.25264.

24. H. Dutta, K. Debnath, and Deba, and K. Sarma. "Multi-objective optimization of hole dilation at inlet and outlet during machining of CFRP by µEDM using assisting-electrode and rotating tool." *Int. J. Adv. Manuf. Technol.*, vol. 110, pp. 2305–2322, 2020. DOI: 10.1007/s00170-020-05987-3/Published.

25. A. P. Singh and D. K. Singh. "Multi response optimization for micro-EDM machining of AISI D2 die steel using RSM and neural network." *Mater. Today Proc.*, vol. 43, pp. 1449–1455, 2021. DOI: 10.1016/j.matpr.2020.09.183.

26. R. Mehfuz and M. Y. Ali. "Investigation of machining parameters for the multiple-response optimization of micro electro discharge milling." *Int. J. Adv. Manuf. Technol.*, vol. 43, nos. 3–4, pp. 264–275, 2009. DOI: 10.1007/s00170-008-1705-0.

27. G. Karthikeyan, J. Ramkumar, S. Dhamodaran, and S. Aravindan. "Micro electric discharge milling process performance: An experimental investigation." *Int. J. Mach. Tools Manuf.*, vol. 50, no. 8, pp. 718–727, 2010. DOI: 10.1016/j.ijmachtools.2010.04.007.

28. M. Choubey and M. Rawat. "A review on various methods to improve process capabilities of electrical discharge machining process." *Mater. Today Proc.*, 2021. DOI: 10.1016/j.matpr.2021.03.169.

29. S. H. Yeo, M. Murali, and H. T. Cheah. "Magnetic field assisted micro electro-discharge machining." *J. Micromech. Microeng.*, vol. 14, no. 11, pp. 1526–1529, 2004. DOI: 10.1088/0960-1317/14/11/013.

30. J. M. Jafferson, P. Hariharan, and J. Ram Kumar. "Effects of ultrasonic vibration and magnetic field in micro-EDM milling of nonmagnetic material." *Mater. Manuf. Process.*, vol. 29, no. 3, pp. 357–363, 2014. DOI: 10.1080/10426914.2013.872268.

31. P. Sivaprakasam, P. Hariharan, and G. Elias. "Experimental investigations on magnetic field-assisted micro-electric discharge machining of inconel alloy." *Int. J. Ambient Energy*, pp. 1–8, 2020. DOI: 10.1080/01430750.2020.1758782.

32. Z. Zeng, Y. Wang, Z. Wang, D. Shan, and X. He. "A study of micro-EDM and micro-ECM combined milling for 3D metallic micro-structures." *Precis. Eng.*, vol. 36, no. 3, pp. 500–509, 2012. DOI: 10.1016/j.precisioneng.2012.01.005.

33. S. H. Yeo, P. C. Tan, and W. Kurnia. "Effects of powder additives suspended in dielectric on crater characteristics for micro electrical discharge machining." *J. Micromech. Microeng.*, vol. 17, no. 11, p. N91, 2007. DOI: 10.1088/0960-1317/17/11/N01.

34. Y. Yahagi, T. Koyano, M. Kunieda, and X. Yang. "Micro drilling EDM with high rotation speed of tool electrode using the electrostatic induction feeding method." *Procedia CIRP*, vol. 1, no. 1, pp. 162–165, 2012. DOI: 10.1016/j.procir.2012.04.028.

35. G. L. Feng, X. D. Yang, and G. X. Chi. "Study on machining characteristics of micro EDM with high spindle speed using non-contact electric feeding method." *Int. J. Adv. Manuf. Technol.*, vol. 92, nos. 5–8, pp. 1979–1989, 2017. DOI: 10.1007/s00170-017-0290-5.

36. M. P. Jahan, Y. S. Wong, and M. Rahman. "Evaluation of the effectiveness of low frequency workpiece vibration in deep-hole micro-EDM drilling of tungsten carbide." *J. Manuf. Process.*, vol. 14, no. 3, pp. 343–359, 2012. DOI: 10.1016/j.jmapro.2012.07.001.

37. M. P. Jahan, M. Rahman, Y. S. Wong, and L. Fuhua. "On-machine fabrication of high-aspect-ratio micro-electrodes and application in vibration-assisted micro-electrodischarge drilling of tungsten carbide." *Proc. Inst. Mech. Eng. Part B J. Eng. Manuf.*, vol. 224, no. 5, pp. 795–814, 2010. DOI: 10.1243/09544054JEM1718.

38. M. P. Jahan, T. Saleh, M. Rahman, and Y. S. Wong. "Development, modeling, and experimental investigation of low frequency workpiece vibration-assisted micro-EDM of tungsten carbide." *J. Manuf. Sci. Eng. Trans. ASME*, vol. 132, no. 5, pp. 1–8, 2010. DOI: 10.1115/1.4002457.

39. D. R. Unune and H. S. Mali. "Experimental investigation on low-frequency vibration-assisted μ-ED milling of Inconel 718." *Mater. Manuf. Process.*, vol. 33, no. 9, pp. 964–976, 2018. DOI: 10.1080/10426914.2017.1388516.

40. Z. Li, J. Tang, and J. Bai. "A novel micro-EDM method to improve microhole machining performances using ultrasonic circular vibration (UCV) electrode." *Int. J. Mech. Sci.*, vol. 175, p. 105574, 2020. DOI: 10.1016/j.ijmecsci.2020.105574.

41. S. Dhanik, S. S. Joshi, N. Ramakrishnan, and P. R. Apte. "Evolution of EDM process modelling and development towards modelling of the micro-EDM process." *Int. J. Manuf. Technol. Manag.*, vol. 7, nos. 2–4, pp. 157–180, 2005. DOI: 10.1504/IJMTM.2005.006829.

42. F. V. Dijck and R. Snoyes. "Plasma channel diameter growth affects stock removal in EDM." *CIRP Ann.*, vol. 21, no. 1, pp. 39–40, 1972.

43. P. K. Eckman and E. M. Williams. "Plasma dynamics in an arc formed by low-voltage sparkover of a liquid dielectric." *Appl. Sci. Res. Sect.* B, vol. 8, no. 1, pp. 299–320, 1960.

44. P. T. Eubank, M. R. Patel, M. A. Barrufet, and B. Bozkurt. "Theoretical models of the electrical discharge machining process, the variable mass, cylindrical plasma model." *J. Appl. Phys.*, vol. 73, no. 11, pp. 7900–7909, 1993. DOI: 10.1063/1.353942.

45. P. Shankar, V. K. Jain, and T. Sundararajan. "Analysis of spark profiles during EDM process." *Mach. Sci. Technol.*, vol. 1, no. 2, pp. 195–217, 1997. DOI: 10.1080/10940 349708945647.

46. A. Singh and A. Ghosh. "A thermo-electric model of material removal during electric discharge machining." *Int. J. Mach. Tools Manuf.*, vol. 39, no. 4, pp. 669–682, 1999. DOI: 10.1016/S0890-6955(98)00047-9.

47. S. Dhanik and S. S. Joshi. "Modeling of a single resistance capacitance pulse discharge in micro-electro discharge machining." *J. Manuf. Sci. Eng. Trans. ASME*, vol. 127, no. 4, pp. 759–767, 2005. DOI: 10.1115/1.2034512.

48. X. Chu, K. Zhu, C. Wang, Z. Hu, and Y. Zhang. "A study on plasma channel expansion in micro-EDM." *Mater. Manuf. Process.*, vol. 31, no. 4, pp. 381–390, 2016. DOI: 10.1080/10426914.2015.1059445.

49. J. Roth. *Industrial Plasma Engineering*. Bristol, 1995.

50. S. H. Yeo, P. C. Tan, and W. Kurnia. "Effects of powder additives suspended in dielectric on crater characteristics for micro electrical discharge machining." *J. Micromech. Microeng.*, vol. 17, no. 11, 2007. DOI: 10.1088/0960-1317/17/11/N01.

51. S. H. Yeo, W. Kurnia, and P. C. Tan. "Critical assessment and numerical comparison of electro-thermal models in EDM." *J. Mater. Process. Technol.*, vol. 203, nos. 1–3, pp. 241–251, 2008. DOI: 10.1016/j.jmatprotec.2007.10.026.

52. M. R. Patel, M. A. Barrufet, P. T. Eubank, and D. D. DiBitonto. "Theoretical models of the electrical discharge machining process. II. The anode erosion model." *J. Appl. Phys.*, vol. 66, no. 9, pp. 4104–4111, 1989. DOI: 10.1063/1.343995.

53. R. Dijck, and F. V. Snoeys. "Investigations of EDM operations by means of thermo-mathematical models." *CIRP Ann.*, vol. 20, no. 1, pp. 35–36, 1971.

54. S. N. Joshi and S. S. Pande. "Thermo-physical modeling of die-sinking EDM process." *J. Manuf. Process.*, vol. 12, no. 1, pp. 45–56, 2010. DOI: 10.1016/j.jmapro.2010.02.001.

55. S. H. Yeo, W. Kurnia, and P. C. Tan. "Electro-thermal modelling of anode and cathode in micro-EDM." *J. Phys. D. Appl. Phys.*, vol. 40, no. 8, pp. 2513–2521, 2007. DOI: 10.1088/0022-3727/40/8/015.

56. D. K. Panda and R. K. Bhoi. "Developing transient three dimensional thermal models for electro discharge machining of semi infinite and infinite solid." *J. Mater. Process. Manuf. Sci.*, vol. 10, no. 2, pp. 71–89, 2001. DOI: 10.1177/1062065602010002604.

57. M. Williams. "Theory of electric spark machining." *Trans. Am. Institute,* vol. 71, no. II, 1952. Accessed April 30, 2021. https://ieeexplore.ieee.org/abstract/document/6371251/.

58. S. Tariq Jilani and P. C. Pandey. "Analysis and modelling of EDM parameters." *Precis. Eng.*, vol. 4, no. 4, pp. 215–221, 1982. DOI: 10.1016/0141-6359(82)90011-3.

59. P. Allen and X. Chen. "Process simulation of micro electro-discharge machining on molybdenum." *J. Mater. Process. Technol.*, vol. 186, nos. 1–3, pp. 346–355, 2007. DOI: 10.1016/j.jmatprotec.2007.01.009.

60. V. Yadav, V. K. Jain, and P. M. Dixit. "Thermal stresses due to electrical discharge machining." *Int. J. Mach. Tools Manuf.*, vol. 42, no. 8, pp. 877–888, 2002. DOI: 10.1016/S0890-6955(02)00029-9.

61. M. S. Murali and S. H. Yeo. "Process simulation and residual stress estimation of micro-electrodischarge Machining using finite element method." *Jpn. J. Appl. Phys., Part 1 Regul. Pap. Short Notes Rev. Pap.*, vol. 44, no. 7A, pp. 5254–5263, 2005. DOI: 10.1143/JJAP.44.5254.

62. R. Kumar and V. Yadava. "Finite element thermal analysis of micro-EDM." *Int. J. Nanoparticles*, vol. 1, no. 3, pp. 224–240, 2008. DOI: 10.1504/IJNP.2008.020898.

63. C. Liu, N. Duong, M. P. Jahan, J. Ma, and R. Kirwin. "Experimental investigation and numerical simulation of micro-EDM of bulk metallic glass with focus on crater sizes." *Procedia Manuf.*, vol. 34, pp. 275–286, 2019. DOI: 10.1016/j.promfg.2019.06.151.

64. S. S. Mujumdar, D. Curreli, S. G. Kapoor, and D. Ruzic. "A model of micro electro-discharge machining plasma discharge in deionized water." *J. Manuf. Sci. Eng. Trans. ASME*, vol. 136, no. 3, pp. 1–12, 2014. DOI: 10.1115/1.4026298.

65. S. S. Mujumdar, D. Curreli, S. G. Kapoor, and D. Ruzic. "Modeling of melt-pool formation and material removal in micro-electrodischarge machining." *J. Manuf. Sci. Eng. Trans. ASME*, vol. 137, no. 3, pp. 1–9, 2015. DOI: 10.1115/1.4029446.

66. B. Shao and K. P. Rajurkar. "Modelling of the crater formation in micro-EDM." *Procedia CIRP*, vol. 33, pp. 376–381, 2015. DOI: 10.1016/j.procir.2015.06.085.

67. B. Shao and K. P. Rajurkar. "Modelling of the crater formation in micro-EDM." *Procedia CIRP*, vol. 33, pp. 376–381, 2015. DOI: 10.1016/j.procir.2015.06.085.

68. D. G. Dilip, S. P. Ananthan, S. Panda, and J. Mathew. "Numerical simulation of the influence of fluid motion in mushy zone during micro-EDM on the crater surface profile of Inconel 718 alloy." *J. Brazilian Soc. Mech. Sci. Eng.*, vol. 41, no. 2, pp. 1–14, 2019. DOI: 10.1007/s40430-019-1595-0.

69. M. Kliuev, K. Florio, M. Akbari, and K. Wegener. "Influence of energy fraction in EDM drilling of Inconel 718 by statistical analysis and finite element crater-modelling." *J. Manuf. Process.*, vol. 40, pp. 84–93, 2019. DOI: 10.1016/j.jmapro.2019.03.002.

70. P. C. Tan and S. H. Yeo. "Modelling of overlapping craters in micro-electrical discharge machining." *J. Phys. D. Appl. Phys.*, vol. 41, no. 20, 2008. DOI: 10.1088/0022-3727/41/20/205302.

71. W. Kurnia, P. C. Tan, S. H. Yeo, and Q. P. Tan. "Surface roughness model for micro electrical discharge machining." *Proc. Inst. Mech. Eng. Part B J. Eng. Manuf.*, vol. 223, no. 3, pp. 279–287, 2009. DOI: 10.1243/09544054JEM1188.

72. M. P. S. K. Kiran and S. S. Joshi. "Modeling of surface roughness and the role of debris in micro-EDM." *J. Manuf. Sci. Eng. Trans. ASME*, vol. 129, no. 2, pp. 265–273, 2007. DOI: 10.1115/1.2540683.

73. K. P. Somashekhar, S. Panda, J. Mathew, and N. Ramachandran. "Numerical simulation of micro-EDM model with multi-spark." *Int. J. Adv. Manuf. Technol.*, vol. 76, no. 1–4, pp. 83–90, 2013. DOI: 10.1007/s00170-013-5319-9.

74. H. P. Jadhav, P. K. Mohanty, and S. Das. "Numerical simulation of multi-spark electric discharge machining analysis for Ti6Al4V alloy drilling." *Mater. Today Proc.*, vol. 5, no. 14, pp. 28337–28346, 2018. DOI: 10.1016/j.matpr.2018.10.118.

75. M. Singh, P. Saxena, J. Ramkumar, and R. V. Rao. "Multi-spark numerical simulation of the micro-EDM process: An extension of a single-spark numerical study." *Int. J. Adv. Manuf. Technol.*, vol. 108, nos. 9–10, pp. 2701–2715, 2020. DOI: 10.1007/s00170-020-05566-6.

5 Wire Electrical Discharge Machining Process

Mahavir Singh, J. Ramkumar, and V. K. Jain

CONTENTS

DOI: 10.1201/9780429160011-5

5.1 INTRODUCTION

The Wire Electrical Discharge Machining (Wire-EDM) process is an adaptation of the traditionally used Electrical Discharge Machining (EDM) process [1]. It is widely used for the processing of difficult-to-machine materials, namely titanium alloys, nickel alloys, carbides, stainless steel, etc. Typical applications of Wire-EDM include thin wafer machining for solar cells [2], machining of nickel alloys for turbine application [3], machining of delicate parts, namely miniature gears, microtools, etc. [4]. With the advancements in Wire-EDM machine tools, materials as thick as 300 mm are precisely sliced with a straightness accuracy of 7–8 μm [5]. Wire-EDM is probably the most resourceful and versatile variant of the EDM process. Notable advantages of a Wire-EDM technique for machining include the force-free system, as theoretically there is no mechanical contact of the wire with the workpiece, machinability independent of the workpiece's mechanical properties (hardness, yield-stress, toughness, etc.) and high relative accuracy as opposed to other machining techniques [1]. It utilizes a wire with a diameter of 30–300 μm as a tool electrode for cutting operations [1, 6]. It works on the same principle as the EDM process, i.e., spark generation between the tool (wire) and workpiece, and resultant melting and partial evaporation of the electrodes material. Enormously high-frequency electrical sparks are generated between the electrodes (cathode and anode) submerged in a suitable dielectric fluid during the Wire-EDM process. Two electrodes have a narrow gap of 20–100 μm submerged in the dielectric fluid. Among different dielectric fluids (hydrocarbon-based, water-based, and gases), deionized water possesses the characteristics of low viscosity and high heat-absorbing capacity, which are essential in the Wire-EDM process to prevent thermal damage to thin wire and its resulting breakdown [7, 8]. Moreover, its low viscosity facilitates effective flow in the small discharge gap due to which it is more regularly used in the Wire-EDM process. High frequency and comparatively low energy sparks result in the creation of a high-temperature plasma channel that expands with discharge time and generates high temperature (8000–10,000°C) for a very short duration [9]. Eventually, melting of the base material occurs, and intense heat flux may even evaporate a part of the workpiece material. An ultra-thin wire as small as 0.02 mm [10], traveling continuously to feed an unsullied wire, may be utilized as a tool electrode. However, the Wire-EDM machine used for industrial purposes (mass production) where dimensional accuracy is not very important, the wire is circulated and used again for a considerable duration. The choice of wire material depends on several factors that govern the proper wire tensioning, speed, and reduced vibration of the wire electrode. Commonly used wire materials are tungsten, molybdenum, copper, brass, coated wire, etc. [11].

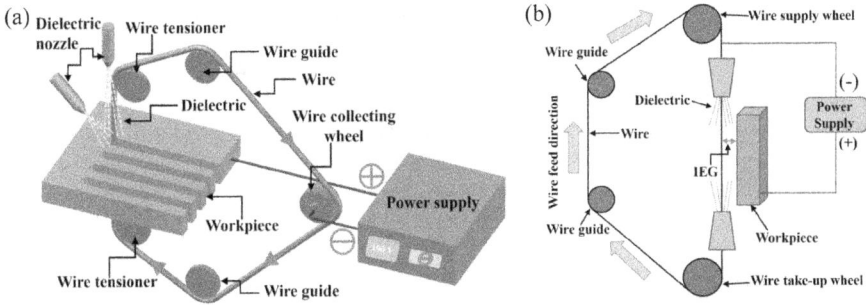

FIGURE 5.1 Schematic diagrams of a typical Wire-EDM system (a) for microslit cutting operation, (b) for billet slicing/thin wafer machining operation.

A typical Wire-EDM system for microslit cutting operation is shown in Figure 5.1(a). The machining system consists of a wire spool that supplies a fresh wire for machining, a series of intermediate rollers, a diamond guide-cum-tensioner in order to hold, guide, and maintain a stipulated tension of the wire electrode [12]. Sometimes the eroded wire is collected in a separate spool, as this wire cannot be reused. Figure 5.1(b) shows the identical Wire-EDM setup (as shown in Figure 5.1(a)) for the slicing of thin wafers. i.e., the slicing of thicker workpieces. Dielectric fluid is one of the critical elements of any EDM or Wire-EDM system. This fluid is generally supplied to the machining gap either axially (parallel to wire travel direction) or supplied as a jet flushing commonly used in the EDM system. Electrical power to the electrodes (cathode and anode) can be supplied through either a Resistance-Capacitance (R-C) based power generator or a transistor-based power generator. Both the power generators have their distinct advantages and limitations pertaining to particular machining conditions and desirability. The evolution of the electronics industry has driven EDM/Wire-EDM power systems to be replaced with fully reliable and versatile transistor-based systems. The R-C-based power generators find their unique capabilities in the Electrical Discharge Micromachining (EDMM) and Wire Electrical Discharge Micromachining (Wire-EDMM) processes. In the transistor-based pulse generator, a series of transistors act as a switch to allow the electrical power to be supplied for the pulse on-time [13]. The application of a transistor-based circuit makes it convenient for the user to arbitrarily select pulse on-time, pulse off-time, peak current, duty ratio (ratio of pulse on-time to the total pulse time), and supplied voltage. However, the switching action of the transistor itself takes a few microseconds. Therefore, this circuit is not suitable for micromachining or Wire-EDMM operations, wherein very low pulse energy and pulse on-time in nanoseconds is required [13]. R-C-based circuits have the distinct advantages of reducing the single pulse energy or pulse duration by merely reducing the capacitance of the employed capacitor [14]. The pulse on-time (discharging time), pulse off-time (charging time) and machining current are direct functions of capacitance. The higher the capacitance value, the higher is the charging and discharging time, and greater will be the machining current. Hence, higher or lower energy per pulse (E)

can be achieved by a suitable selection of capacitance $\left(E = \frac{1}{2}CV^2 \right)$, where C and V represent the capacitance and voltage, respectively. Conversely, for micromachining operations or Wire-EDMM, reducing the capacitance will substantially reduce the energy per pulse. The obtained single crater size depends on the energy per pulse, which determines the minimum feature dimension that can be accomplished [15].

5.2   WIRE-EDM VS WIRE-EDMM

The Wire-EDM process utilizes a wire electrode with a diameter in the range of somewhere between 100 and 300 μm [1], compared to an extremely small diameter wire (as small as 20–30 μm) [6, 16] in the Wire-EDMM process. Therefore, it is essential to control the process parameters for Wire-EDMM for the longevity of the wire electrode and prevent its frequent rupture during machining. Thus, the level of discharge energy of a pulse has to be lowered in the Wire-EDMM process. In contrast, in the Wire-EDM process, higher discharge energy levels are preferred for an enhanced machining rate and relatively rough profile of the cut. Depending on the nature of the cut required (kerf width and quality of kerf), the magnitude of the discharge energy can be obtained using two different types of power generators, namely a transistor-based power generator and R-C-based power generator. R-C-based power generators are most suitable for the Wire-EDMM process owing to their capability of reducing the discharge energy by reducing the capacitance of the capacitor and delivering the power for a smaller pulse duration, typically in the nanosecond range [11]. The machining rate is relatively low with R-C-based power generators compared with transistor-based systems. However, with a transistor-based power generator, there is a challenge in minimizing the pulse duration due to a delay in transistor switching action [13]. Thus, nanosecond pulses are relatively challenging to obtain

TABLE 5.1

Differences between Wire-EDM and Wire-EDMM processes [11, 13, 17]

S. No.	Criterion	Wire-EDM	Wire-EDMM
1.	Wire diameter	Greater than 50 μm, generally 100–300 μm, is used	Less than 50 μm wire diameter
2.	Power generator	Transistor-based power generator	Resistance-capacitance-based power generator
3.	Discharge energy and pulse duration	High discharge energy (greater than 10 μJ) and longer duration of a pulse	Significantly low discharge energy (0.1–10 μJ) and shorter duration of a pulse
4.	Wire feed	High feed	Low feed
5.	Wire circulation	Circulated continuously	Wire travels but not reused
6.	Wire material	Molybdenum, brass, tungsten	Tungsten wire is preferred
7.	Dielectric	Deionized water is preferred over hydrocarbon-based dielectrics	Hydrocarbon dielectrics can be used

through the transistor-based circuits. Nevertheless, these generators enable control over peak current and pulse on-time and pulse off-time, and deliver higher discharge energy per pulse. Therefore, it is possible to achieve a higher rate of machining with the Wire-EDM process than with the Wire-EDMM. Table 5.1 [11, 13, 17] summarizes the noticeable differences in the Wire-EDM and Wire-EDMM processes.

5.3 PROCESS PARAMETERS AND RESPONSES IN WIRE-EDM

The Wire-EDM process is a variant of the EDM process, and the majority of the input parameters are same as in the EDM process. However, the introduction of wire as a tool electrode brings a little intricacy to the system. Therefore, the wire electrode, its circulation, feed, and dielectric flushing in the narrow Interelectrode Gap (IEG) between the wire and workpiece are additional variables in a typical Wire-EDM system. The input process parameters and major output responses related to the Wire-EDM process are discussed as follows:

5.3.1 ELECTRICAL PARAMETERS

Electrical parameters are specific to the power generator being used for the delivery of input power to the electrodes. The following power generators are predominantly used in the Wire-EDM system:

5.3.1.1 Resistance-Capacitance-Based Power Generator

A Resistance-Capacitance (R-C) based power generator is characterized by two essential elements, i.e., open-circuit voltage and capacitance of the capacitor which largely determines the energy of a pulse. Generally, a R-C-based power generator is appropriate for smaller material removal per pulse and for higher dimensional accuracy of the feature by merely minimizing the capacitance value of the capacitor. Thus, based on the magnitude of the capacitor and charging resistance of the circuit, nanosecond pulse duration and low discharge energy are obtained. However, due to the reduction in discharge energy, the machining rate is relatively low. Typical input parameters pertaining to the R-C-based circuits are open-circuit voltage, resistance, capacitance of the capacitor, threshold value/sensitivity for short circuit detection and polarity.

5.3.1.2 Transistor-Based Circuit

A transistor-based power generator relies on the switching action of transistors for power delivery to the electrodes and its termination after a certain pulse duration. The input process parameters consist of open-circuit voltage, peak current, pulse on-time, pulse off-time, duty ratio (ratio of pulse on-time to that of the total pulse time), the threshold value, and polarity.

5.3.2 NON-ELECTRICAL PARAMETERS

Non-electrical parameters are common to both types of power generators. These parameters are wire diameter and material, wire traveling speed, and wire tension.

In addition, there are some miscellaneous input parameters, which consist of wire feed rate, IEG, dielectric fluid and it's pressure, etc.

5.3.3 TYPICAL OUTPUT RESPONSES

The typical output parameters or responses include linear/volumetric/mass material removal rate, kerf width/slit width, corner radius, average surface roughness, thickness of the recast layer, Heat-Affected Zone (HAZ), wire wear/breakage, etc. Among these parameters, the removal rate (mass or volumetric) represents the rate at which the material is removed during the operation. It considers the thickness of the workpiece being machined and the length of the cut and kerf width generated, whereas the linear machining rate/cutting rate merely considers the length of the feature to be cut in the given time duration. Therefore, the kerf width and thickness of the workpiece are ignored while calculating the linear machining/cutting rate. Kerf width is regarded as one of the critical outputs in the Wire-EDM process to assess the quality of machining operation. The input parameters and output responses prominent to the Wire-EDM process are displayed in Figure 5.2.

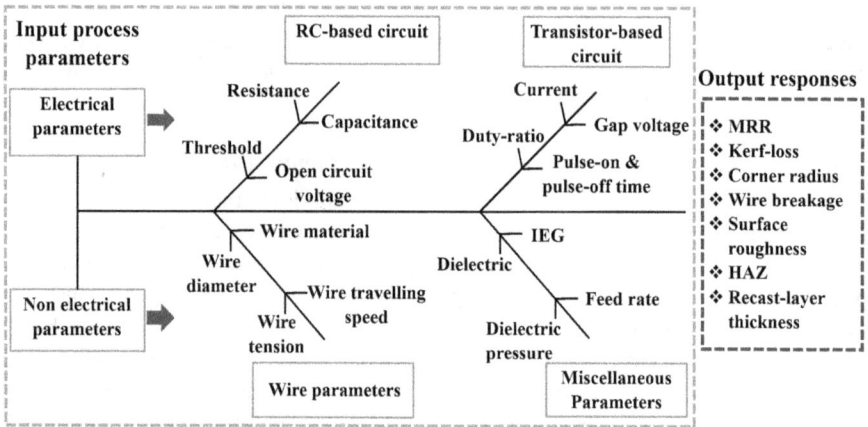

FIGURE 5.2 Fishbone diagram showing input parameters and output responses in the Wire-EDM process.

5.4 KERF WIDTH IN THE WIRE-EDM PROCESS

To appraise the performance characteristics of a typical Wire-EDM process, the kerf width is one of the essential criteria. The "*kerf width*" in Wire-EDM is the width of the material that is removed during the cutting process. It is the summation of the diameter of the wire, and twice the radial overcut between the wire and workpiece. The radial overcut is the consequence of secondary sparks, which occur in the gap in the radial direction between the wire and workpiece. The radial overcut also assists the evacuation of debris from the machining gap and allows the entry of fresh dielectric. However, enlarged overcutting deteriorates the dimensional accuracy of the machined features and sometimes may not serve the desired function. The kerf width in a typical Wire-EDM process is given as follows [18]:

$$y_k = D + 2a \qquad (5.1)$$

where y_k is the kerf width, D is the wire diameter, and a is the lateral discharge gap or radial overcut (the gap between the wire and workpiece on either side of the wire). The lateral gap is originated on both sides of the wire electrode due to secondary discharging. As the wire progresses in the cutting direction, similar to the discharge in the primary IEG, dielectric breakdown also occurs in the secondary IEG, which is also assisted by the presence of the machined debris.

The higher kerf width refers to the greater material loss and poor dimensional accuracy of the machined feature. Precise cutting operation of microgears, thin silicon wafers, bisecting of the microdrilled holes, etc., demand extremely narrow kerf width. The production cost of ultra-thin wafer commonly used in the semiconductor industry is reduced by minimizing kerf width and resultant thinning of the wafer [19]. Therefore, researchers across the globe are striving toward the minimization of kerf width, employing different techniques.

Figure 5.3 shows a typical wire slitting operation wherein a kerf width is generated after the completion of the machining operation. The remaining uncut material between two slits/kerfs is termed wafer/wall thickness. The thickness of the thin wafer depends upon the kerf width, wire vibration, effective dielectric flushing, etc. Therefore, prior knowledge of the kerf width generated with a particular set of input parameters and wire diameter is indispensable for accurate wafer/wall thickness prediction. For the two parallel successive passes of wire electrode keeping the separation between the passes invariant, higher wafer/wall thickness is achieved with lower kerf width and vice versa. Figure 5.4 shows the kerf width generation during

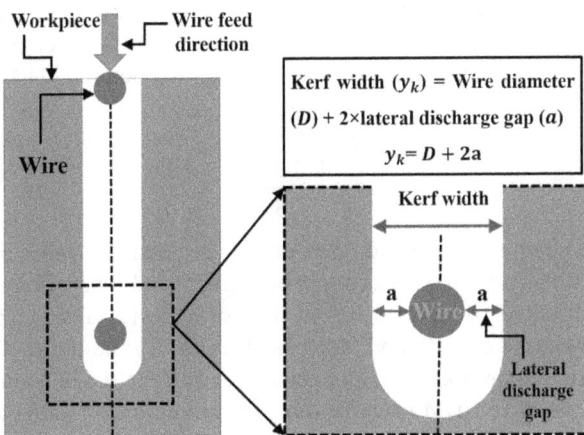

FIGURE 5.3 Schematic diagram showing wire electrode, kerf width, and radial overcut/lateral discharge gap in the Wire-EDM process.

FIGURE 5.4 Kerf quality representation of microslit machined at (a) high discharge energy, (b) low discharge energy. Reproduced with permission from [20].

the slit-cutting operation using two levels of discharge energy, i.e., low and high, at which the kerf width and quality of cut are observed to be different.

5.4.1 KERF WIDTH CONSIDERING WIRE VIBRATION

The kerf width expressed in Equation 5.1 is the ideal width of the kerf when there is no lateral vibration of the wire electrode. However, in reality there is a significant wire vibration in the direction normal to the wire feed, i.e., in the direction of the secondary discharge gap. Thus, the kerf width (y_k) in such situations is enlarged, which is given as follows [6]:

$$y_k = D + 2a + 2A \qquad (5.2)$$

where A is the amplitude of wire vibration in the lateral direction, and other symbols are the same as in Equation 5.1. The lateral discharge gap (a) is a function of the dielectric strength of the dielectric fluid. However, it varies significantly owing to the presence of impurities such as debris in the gap. There is an empirical relationship between lateral discharge gap (a) and breakdown potential (V), expressed as follows [6]:

$$a = \frac{V - 36.85}{28.5} \qquad (5.3)$$

This empirical relationship is valid only for the specified electrodes materials, and machining conditions. Such results cannot be extrapolated. The vibration of the wire electrode is attributed to the pressurized flow of dielectric fluid, exploding force exerted by the plasma column and vapor bubbles during the pulse off-period, and inadequate wire tension [8]. Thus, it is obvious to analyze the inherent vibration of wire to design the Wire-EDM machine tool for minimum wire vibration and higher dimensional accuracy of the process. During the Wire-EDM process, the discharging force is an impulse force acting on the wire, which is proportional to the frequency of discharge. This impulsive force (F_d) is given as follows [6, 21]:

$$F_d = f \times F' \times \delta(t) \tag{5.4}$$

where f is the frequency of discharge, F' is the force per discharge and $\delta(t)$ is the unit impulse function. The amplitude of wire lateral vibration is evaluated by solving the standard vibration equation of wire string, which accounts for wire tension, as follows [6]:

$$F \frac{\partial^2 y}{\partial z^2} - \mu \frac{\partial y}{\partial t} + F_d \delta(z - \xi) = \rho \frac{\partial^2 y}{\partial t^2} \tag{5.5}$$

where F is the wire tension, y is the wire lateral displacement, ρ is the density (linear) of wire material, μ is the damping coefficient, and ξ depicts the point in the wire. Equation 5.5 is generally solved by the variable separable method and contour integral. The initial and boundary conditions for solving the equation of wire vibration consists of the distance between the two wire guides (l), given as follows:

Initial conditions:

$$y(0,t) = 0, \quad \text{and} \quad y(l,t) = 0$$

Boundary conditions:

$$y(z,0) = 0, \quad \text{and} \quad \frac{\partial y}{\partial t}(z,0) = 0$$

Solving Equation 5.5 analytically, the required solution for wire displacement is represented as follows:

$$y(\xi, x, t) = \frac{2fF'}{l\rho} \sum_{n=1}^{n=\infty} \frac{e^{-\left(\frac{\mu}{2\rho}\right)t}}{\omega} \sin\left(\frac{n\pi\xi}{l}\right) \sin\left(\frac{n\pi x}{l}\right) \sin(\omega t) \tag{5.6}$$

$$\omega = \sqrt{\frac{n^2\pi^2 F^2}{\rho^2 l^2} - \frac{\mu^2}{4\rho^2}} \tag{5.7}$$

$$n = 1,2,3\ldots$$

Here, ω represents circular frequency of vibration of the wire. The wire vibrates with maximum amplitude when discharge happens at the center of the wire along the length, i.e., at $x = l/2$ and the middle point of the wire, i.e., $\xi = l/2$. Therefore, solving Equation 5.6, the maximum amplitude of wire vibration is deduced as follows:

$$y_{\max} = 4fF' \sum_{n=1}^{n=\infty} \frac{1}{\sqrt{4n^2\pi^2 F^2 - \mu^2 l^2}} e^{-\left(\frac{\mu}{2\rho}\right)t} \tag{5.8}$$

$$n = 1,2,3\ldots$$

As per Equation 5.8, the maximum amplitude of wire vibration is influenced by discharge force per pulse (F'), frequency of discharge (f), wire tension (F), specific damping coefficient (μ), and wire density (ρ). The determination of force per discharge is stochastic in nature and very difficult to determine. The vapor bubbles explode at the termination of power from the plasma column, which is greatly affected by the discharge energy per pulse and frequency of discharge [6]. Therefore, experimental investigations of Wire-EDMM can reveal the wire displacement in the lateral direction and kerf width with respect to various input parameters such as open voltage, capacitance, table feed rate, etc. Based on the experiments, low discharge energy level, higher wire tension, uninterrupted discharging, and lower separation between two wire guides are recommended for obtaining lower kerf width for a constant wire diameter [6].

In addition to the kerf width, corner radius also determines dimensional accuracy of the machined feature in the Wire-EDM process. Based on the analysis of wire vibration, a microwire of an average diameter 30 μm is used at the optimized parameters for microcutting of a 100 μm-thick stainless-steel sheet. The resultant kerf width and corner radius are measured to be 30.8 μm and 15.8 μm, respectively depicting a radial overcut of 1.33%, as shown in Figure 5.5 [6]. The variation in kerf width at the top (upper kerf width) and bottom (bottom kerf width) of the workpiece in the Wire-EDM process is found to be insignificant, and it is within 5% [22]. This is due to the continuous circulation of the wire electrode such that the same wire diameter is exposed for cutting the upper and bottom end of the workpiece. However, a slight variation in kerf width can be obtained depending on the thickness of the workpiece being processed.

The stability of wire in the Wire-EDM process is highly governed by the occurrence of locations of the sparks and their distribution along the wire length [23]. Uniform spark generation denotes stable wire operation compared to the concentration over an identical wire's segment. Experimental observation of sparks taking place during machining can reveal information regarding the distribution of discharges.

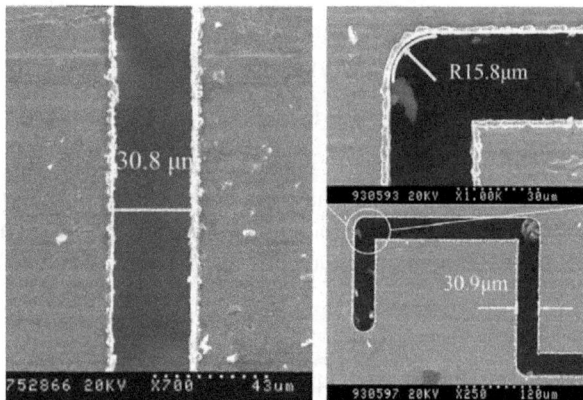

FIGURE 5.5 Kerf width generation using a microwire electrode of 30 μm diameter. Reproduced with permission from [6].

During the processing of a 1 mm-thick workpiece and segregating the thickness in small segments, the sparks' locations are evaluated with respect to one of the edges of the workpiece. The experimental results show a uniformity of sparks over the length of the wire without much concentration on any specific portion of the wire [23].

5.4.2 WIRE LAG IN THE WIRE-EDM PROCESS

"*Wire lag*" is defined as the deflection of the wire electrode in the direction opposite to the cutting direction in static conditions. It differs from wire vibration, which is the dynamic motion of the wire in the lateral direction. Both wire vibration and wire lag contribute significantly to the dimensional inaccuracies in the machined components. The effect of wire lag is more serious when processing a curved workpiece resulting in geometrical inaccuracies of the order of 100 μm [24]. Due to the number of influencing input parameters and stochastic nature of impact force due to discharge, dielectric fluid pressure, electrostatic and hydrodynamic forces and variation in wire tension, the accurate quantitative determination of wire lag is a difficult task [25]. However, qualitative analysis can be helpful in analyzing the influence of operating parameters on wire lag. Therefore, it is essential to analyze the wire lag to minimize its effects and assure higher dimensional accuracy of the Wire-EDM process. For the modeling of the wire-lag phenomenon of a wire electrode fixed between two wire guides, the following assumptions are made [25, 26]:

1. The wire tension remains constant throughout the discharge duration, and the wire is stationary.
2. The mass of the wire is uniformly distributed along the wire length.
3. Wire does not offer any resistance to bending, i.e., it is flexible.
4. The viscous force arising from the dielectric fluid is insignificant compared to the tension force, which can be neglected. The traverse force normal to the tension force is constant over the wire length supported between the wire guides.

Figure 5.6(a) illustrates the phenomenon of wire static deflection, i.e., wire lag while performing a cutting operation of a curved workpiece of height, h. x is the length of wire between the two wire guides, H depicts the workpiece distance from the wire guide, F is the axial wire tension, and q_0 is the force acting on the wire per unit length in the lateral direction. The wire's static deflection (y), i.e., wire lag, is evaluated by solving the following simplified differential equation [25, 26]:

$$F\frac{\partial^2 y}{\partial z^2} = q_0 \tag{5.9}$$

Solving Equation 5.9, the wire's static deflection comes out to be:

$$y = -\frac{q_0}{2F}z(x-z) \tag{5.10}$$

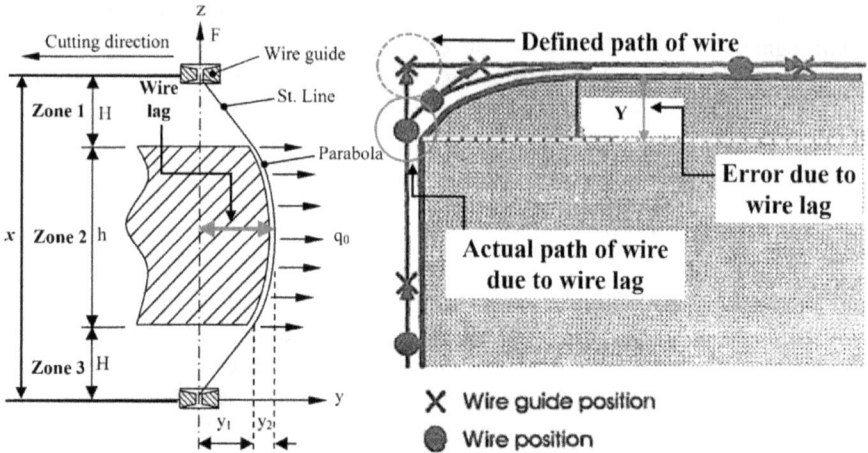

FIGURE 5.6 (a) Schematic diagram of Wire-EDM process showing wire lag. Reproduced with permission from [25], (b) geometrical inaccuracy caused by wire lag. Reproduced with permission from [25, 26].

The wire deflection equation (Equation 5.10) represents that the shape of the wire surrounding the workpiece is parabolic. The maximum deflection (y_{max}) of the wire is the sum of y_1 and y_2, as referred from Figure 5.6(a). Therefore, y_{max} is expressed as per the given equation:

$$y_{max} = y_1 + y_2 = -\frac{q_0}{2F} H(H+h) - \frac{q_0 H^2}{8F} \tag{5.11}$$

Moreover, the solution of Equation 5.9 for the wire outside the workpiece region, i.e., between the workpiece and wire guide, is given as follows:

$$y = -\frac{q_0}{2F} z(H+h) \tag{5.12}$$

This straight-line equation (Equation 5.12) depicts that the wire is straight outside the workpiece to the wire guide. A negative sign in the wire deflection (Equations 5.10 and 5.12) illustrates that the wire deflects in the opposite direction of machining. The geometric error caused by the wire lag is shown in Figure 5.6(b), wherein the wire trajectory deviates significantly from the position of the wire guide during the contour cutting. Owing to the wire lag the straight cut (sharp corner) becomes curved.

5.4.3 ROLE OF WIRE TENSION

To minimize wire vibration in the lateral direction, an appropriate wire tension is applied to the wire in the axial direction. Tension to the wire facilitates the reduced amplitude of wire vibration under the influence of electric discharge pressure and dielectric flushing pressure. Improper wire tensioning might hamper the proper functioning and sometimes instigate wire breakage. Therefore, the selection of appropriate wire tension must be

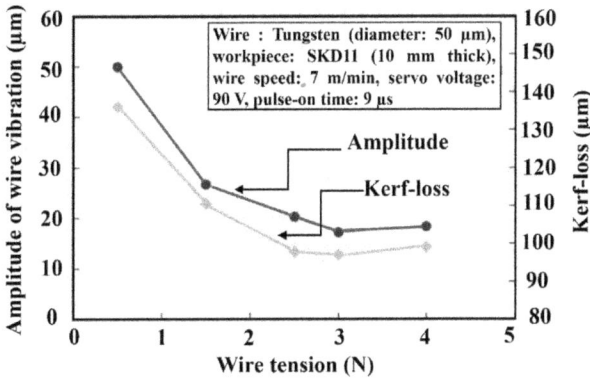

FIGURE 5.7 Influence of wire tension on the amplitude of wire vibration and kerf width. Reproduced with permission from [23].

based on the wire material, nature of cut (rough or finish cut), level of discharge energy, dielectric flushing conditions, etc. Figure 5.7 shows the variation in wire amplitude (in the lateral direction) and kerf width under increasing wire tension. At low wire tension, the wire vibrates severely, causing higher amplitude of vibration, which eventually enhances the kerf width. The kerf width is approximately 2.75 times the wire diameter, depicting a radial overcut of magnitude slightly lower than the wire diameter. However, as the wire is tensioned properly, the kerf width is reduced significantly (twice the wire diameter), and radial overcut is comparable to the wire radius only. A slight increase in vibration amplitude is observed around 3 N wire tension, which is attributed to the local extension of the wire, as the wire tension is comparable to or exceeds the elastic limit (3.2 N) of tungsten wire of 50 μm diameter. The kerf width is consistent with the amplitude of wire vibration [23]. Wire tension is influenced by the magnitude of discharge energy of the pulse, which determines the temperature generation and wire load during discharging. Thus, wire tension has to be adjusted according to the discharge energy by designing a wire tension control system [27].

5.5 TEMPERATURE FIELD ANALYSIS IN THE WIRE

A certain percentage of discharge energy going into the thin wire electrode raises its temperature. An upsurge in wire temperature beyond a specific limit would cause thermal damage to the wire, melting/evaporation, and reduction in yield strength. The combined effect of these adverse outcomes of wire heating can cause wire rupture during the machining operation. Therefore, prior understanding of temperature of the wire becomes of utmost importance for designing and selecting a wire material, its diameter, and wire tension. The amount of thermal load incident to the wire electrode is distributed as per the given equation [28]:

$$A_c k \frac{\Delta T_x(x,t)}{\Delta x} + h_a L \Delta x \left[T(x,t) - T_c \right] + A_c \Delta x \rho c \frac{\Delta T_t(x,t)}{\Delta t}$$

$$= A_c k \frac{\Delta T_x(x + \Delta x, t)}{\Delta x} + \eta \Delta \dot{Q}$$

(5.13)

where Δx is a segment of wire of total length x and cross-sectional area A_c. L is the wire circumference ($2\pi r$). k, ρ, and c are thermal conductivity, density, and specific heat of the wire material, T is temperature, T_c is dielectric temperature. $\Delta \dot{Q}$ is the power available in a small segment of wire ($\Delta x)\eta$ is the discharge energy proportion going to the wire. h_a is convective Heat Transfer Coefficient (HTC). The wire is traveling at a linear velocity v, which is given as follows:

$$v = \frac{\Delta x}{\Delta t}$$

$$\Delta t = \frac{\Delta x}{v}$$

(5.14)

The sparking is assumed to be uniform across the wire length engaged to the workpiece of thickness, h. Hence, power balance can be written as follows:

$$\frac{\Delta \dot{Q}}{\Delta x} = \frac{\dot{Q}}{h}$$

Finally, Equation 5.13 can be simplified by dividing each term by Δx, applying $\Delta x \to 0$ and putting Δt from Equation 5.14 and $\Delta T_t (x,t) = \Delta T_x (x,t)$ in Equation 5.13:

$$-\lambda_1 \frac{d^2 T(x)}{dx^2} + \lambda_2 \frac{dT(x)}{dx} + T(x) = \frac{\dot{Q}}{\Upsilon}$$

(5.15)

$$\text{Where, } \lambda_1 = \frac{kA_c}{h_a L}, \lambda_2 = \frac{\rho c A_c v}{h_a L} \text{ and } \Upsilon = \frac{Lhh_a}{\eta}$$

Equation 5.15 describes the temperature distribution in the wire electrode within the machining zone, i.e., corresponding to the workpiece thickness. Outside the machining zone (either side of the machining zone), there is no discharging. Therefore, energy transfer in these zones is due to conduction (along the wire axis) and convection (in the radial direction). Similar to Equation 5.15, the temperature distribution in the wire outside the machining zone is expressed as follows:

$$-\lambda_1' \frac{d^2 T(x)}{dx^2} + \lambda_2' \frac{dT(x)}{dx} + T(x) = 0$$

(5.16)

Temperature in zone 1 (outside), zone 2 (machining zone), and zone 3 (outside) are T_1, T_2, and T_3, which are evaluated by applying the following boundary conditions [29]:

Zone 1:

$$\frac{dT_1 (x = \infty)}{dx} = 0$$

$$T_1(x=0) = T_2(x=0)$$

$$\frac{dT_1(x=0)}{dx} = \frac{dT_2(x=0)}{dx}$$

Zone 2:

$$T_2(x=h) = T_3(x=h)$$

$$\frac{dT_2(x=h)}{dx} = \frac{dT_3(x=h)}{dx}$$

Zone 3:

$$\frac{dT_3(x=\infty)}{dx} = 0$$

Temperatures T_1, T_2, and T_3 in zones 1, 2, and 3, respectively (indicated in Figures 5.6(a) and 5.8) are evaluated as follows [28]:

Zone 1:

$$T_1(x) = C_1 e^{r_1 x}$$

Zone 2:

$$T_2(x) = C_2 e^{r_2 x} + C_3 e^{r_3 x} + \frac{\dot{Q}}{\Upsilon}$$

Zone 3:

$$T_3(x) = C_4 e^{r_4 x}$$

Characteristics roots r_1 and r_4 correspond to Equation 5.16, whereas r_2 and r_3 correspond to Equation 5.15.

$$r_1 = \frac{\lambda_2' + \sqrt{4\lambda_1' + \lambda_2'^2}}{2\lambda_1'}, r_2 = \frac{\lambda_2 - \sqrt{4\lambda_1 + \lambda_2^2}}{2\lambda_1}, r_3 = \frac{\lambda_2 + \sqrt{4\lambda_1 + \lambda_2^2}}{2\lambda_1}, r_4 = \frac{\lambda_2' - \sqrt{4\lambda_1' + \lambda_2'^2}}{2\lambda_1'}$$

The constants C_1 to C_4 are evaluated by the linear algebra method.

Figure 5.8 displays temperature distribution in the wire under the specified conditions. The maximum temperature of the wire electrode in the machining zone (zone 2) is around 160°C as opposed to the significantly lower temperature of the wire outside the machining zone (zones 1 and 3). Convective HTC (h_a) of the dielectric

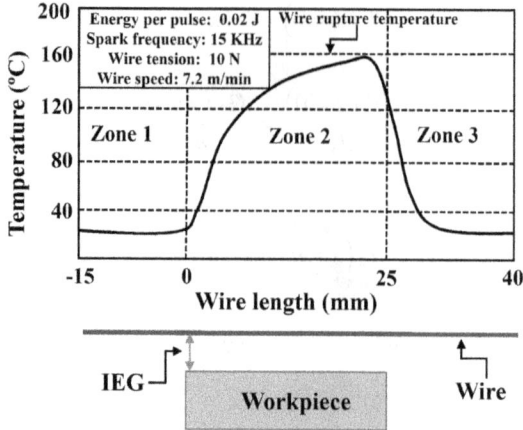

FIGURE 5.8 Variation in temperature in the wire at different zones. Reproduced with permission from [28].

(mixture of dielectric + reaction products + gas) in different zones has significant effect on the resultant wire temperature. Due to the narrower gap between the wire and workpiece in zone 2, HTC is relatively smaller than in zones 1 and 3, where flushing is adequate. Wire traveling speed (v) has less influence on temperature generation in the wire. However, higher wire velocity may affect the wire vibration [29].

5.6   WIRE BREAKAGE IN THE WIRE-EDM PROCESS

Owing to the thermal nature of the Wire-EDM process, intense heat is generated in the core of the plasma channel, which is shared by the wire, workpiece, and dielectric in different proportions. The amount of heat going into the wire electrode is likely to melt/vaporize the wire material. Therefore, if proper circulation of the dielectric is not provided, the concentration of sparks at the same location may result in wire rupture due to thermal loading. Wire breakage is an issue to be addressed in the Wire-EDM process, which hampers both the productivity of the process and dimensional accuracy of the machined features. In the Wire-EDM process, cutting is performed for a long duration without much intervention from the operator. In such a situation, wire breakage poses a serious concern. Automatic wire feed mechanisms are available in modern Wire-EDM machines, which automatically thread the wire as soon as it breaks, and continue with the machining operation from the position where the wire breakage took place [5]. However, broken wire leaves a permanent mark on the machined surface. Moreover, wire is vulnerable to re-breakage if the same machining parameters are used again. To avoid future wire breakage, the wire is circulated continuously from the machining zone to the wire-collecting roller, which ensures the availability of a fresh wire for cutting operations every time. Small diameter wires (usually wire diameter below 50 μm) are rarely used for rough cutting operations considering the tendency for wire breakage at high energy conditions. There are several inherent characteristics of the Wire-EDM process, which contribute to the frequent wire breakage. These include as follows [1, 30, 31]:

i. A sudden rise in the frequency of sparks such that there is a spark concentration at the same location on the wire surface. Substantial increase in spark frequency is observed at the onset of wire breakage while using higher flushing pressure.
ii. Surge in wire local temperature caused by spark localization.
iii. Increased number of short circuits caused by poor gap flushing and wire contacting the workpiece.

Sparking frequency can be controlled within an acceptable range to avoid any wire rupture through online monitoring/tuning of pulse off-time [32]. Besides thermal damage to the wire, mechanical properties of the wire electrode, such as yield strength and fracture toughness, also determine the wire breakage frequency. The yielding of wire material, crack formation and propagation reduce the wire strength and eventually ruptures the wire. The creation of small craters on the wire surface contributes to surface flaws, though these craters are smaller than the workpiece craters. These flaws on the wire surface propagate due to stress intensification at the crack. The stress field around the craters is defined in terms of Stress Intensity Factor (SIF) (K), which relates to the wire tension (F), wire diameter (D), and crack length (a'), as follows [33]:

$$K = \frac{4F\sqrt{\pi a'}}{\pi D^2} \tag{5.17}$$

Figure 5.9 shows the variation of K for different wire tensions and crater sizes for a wire diameter of 200 µm. SIF increases both with an increase in wire tension and crater size. Machining at a higher rate demands higher discharge energy of a pulse. It results in larger crater size, higher discharge pressure on the wire, and adaptation of

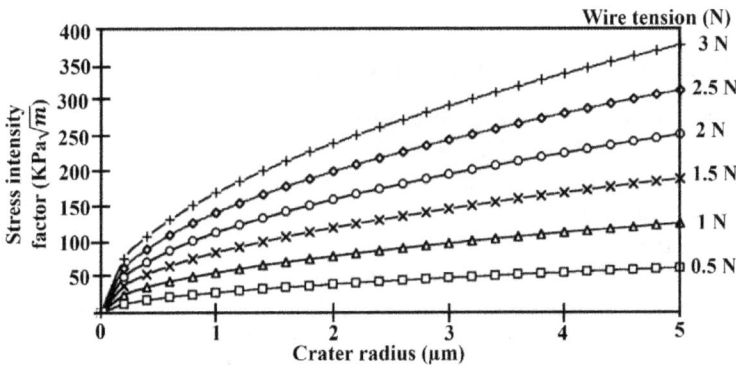

FIGURE 5.9 Effect of wire tension and crater size on stress intensity factor (wire diameter: 200 µm). Reproduced with permission from [33].

higher wire tension to prevent bowing of wire. A larger wire diameter decreases SIF but influences the machining performance as a higher workpiece material removal rate is obtained.

5.6.1 STRATIFIED WIRE

A stratified wire is used to avoid wire breakage and supply higher discharge energy. It consists of a core wire coated with a material such that it can handle a higher thermal load without breakage. A typical stratified wire consists of a copper core that is coated with zinc or its alloy [8]. Figure 5.10 depicts the cross-section of a typical stratified wire. The outer coated layer of material having relatively lower melting/vaporization temperature erodes due to the heat input to the wire, whereas the inner material or core wire remains protected, therefore allowing a higher cutting rate at higher discharge energy conditions. However, it is essential to keep the coating's thickness to a permissible limit; otherwise, the increased overall wire diameter will create larger kerf width.

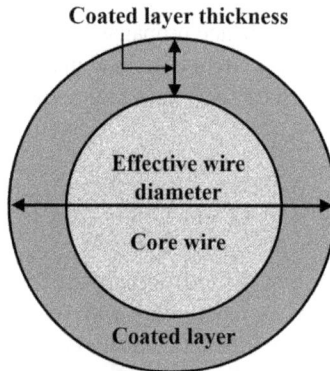

Coated layer thickness

Effective wire diameter

Core wire

Coated layer

FIGURE 5.10   Cross-sectional view of a typical stratified wire.

5.7   PARAMETRIC ANALYSIS OF WIRE-EDM PROCESS

The effect of some major input process variables (current, pulse on-time, sparks frequency) and secondary variables (wire speed and tension) [7, 34] of the Wire-EDM process on output responses is discussed. The trends of the results shown are based on a *one factor at a time approach*, wherein one parameter is varied at a time while keeping all other parameters constant.

Figure 5.11(a) depicts the variation in Material Removal Rate (MRR) and Surface Roughness (SR) with the current. Higher machining current refers to more significant ionization in the gap resulting in higher heat flux being applied to the workpiece (larger and deeper crater). Therefore, both MRR and SR are observed to be increasing with the current [28]. Pulse on-time denotes the duration for which energy is supplied to the workpiece. An increase in pulse on-time results in a higher amount

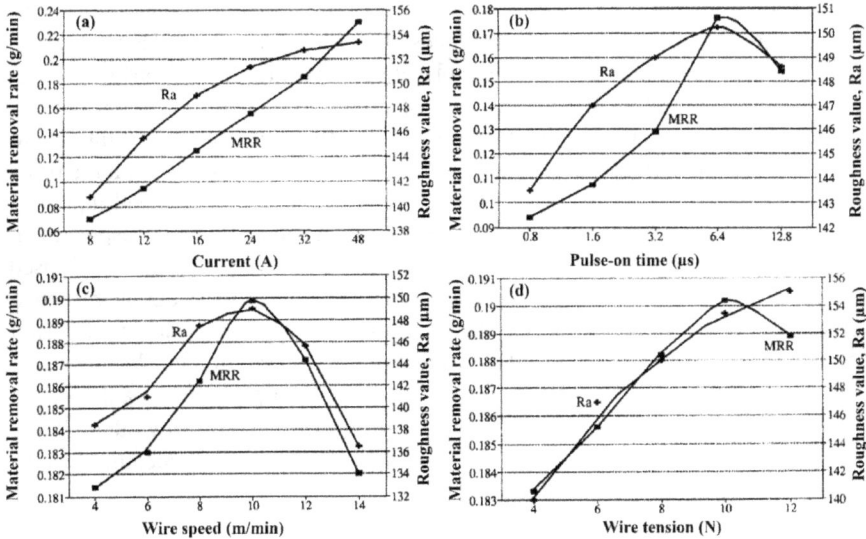

FIGURE 5.11 Variation of MRR and SR with Wire-EDM input parameters: (a) current, (b) pulse on-time, (c) wire speed, (d) wire tension. Reproduced with permission from [28].

of energy transfer and a greater volume of material removal per pulse (assuming that the pulse duration remains constant), therefore, both increasing MRR and SR value [35]. Beyond a certain pulse on-time, the plasma channel expands, which reduces heat flux causing a reduction in output responses (Figure 5.11(b)). Similar to the pulse on-time, with spark frequency, a similar variation in the outputs can be expected. Increase of wire speed enhances the evacuation of debris from the gap and reduces the abnormal discharges, thus improving the performance of the Wire-EDM process [35]. However, wire speed cannot exceed a certain limit, as it contributes to higher wire vibration and instability of sparks resulting in poor performance, as shown in Figure 5.11(c) [28]. Wire tension is probably one of the major influencing parameters, which also governs the vibration of the wire electrode. Higher wire tension reduces the wire vibration, which minimizes the detrimental impact on the machining performance. However, if it surpasses a limit, wire breakage may take place, hampering the machining rate, as given in Figure 5.11(d) [28]. It is important to note that as the MRR increases due to higher energy of discharge or reduction in abnormal discharges, the SR value also shows a similar trend, i.e., it generally increases (or it becomes rougher). Therefore, it is essential to optimize the input parameters for simultaneous maximization of MRR and minimization of SR.

5.8 OPTIMIZATION OF PROCESS PARAMETERS OF THE WIRE-EDM PROCESS

Multi-Objective (MO) optimization of input process parameters is carried out to determine a set of process parameters for simultaneous optimization of various output responses that have conflicting desirability. The machining rate (i.e., MRR), which defines the productivity of the process, must always be high. However, other outputs such as surface roughness, recast layer thickness, HAZ, and kerf width need to be minimized for better surface integrity and dimensional accuracy, respectively. The set of input parameters producing higher throughput (MRR) also result in deteriorated surface integrity and higher magnitude of kerf width. Therefore, it is a prime consideration in the Wire-EDM process to determine a set of optimized input process parameters at which a trade-off can be made between the output responses. To obtain the best set of input parameters, a number of optimization techniques is used. The results of experimental investigations are correlated to the output responses using prediction models/regression analysis. Then, an objective function is formulated for either maximization or minimization problems. Solving the objective function, the inferred input parameters are known to be the best input parameters. For simultaneous maximization of MRR (M) and minimization of kerf width $\left(y_k\right)$ for a set of input parameters, namely, voltage, wire speed, and pulse on-time, the objective function (Y) takes the following form [35, 36]:

$$Y_{\min} = \alpha_1 y_k * - \alpha_2 M *$$

where $y_k* = \dfrac{y_k}{y_{k_{\max}}}$ and $M* = \dfrac{M}{M_{\max}}$ are the normalized values of y_k and M, respectively. y_k and M are the regression equations (linear/higher-order or exponential) correlating the input parameters with the respective outputs (y_k or M). α_1 and α_2 are the weights assigned to the y_k* and $M*$, respectively. The objective function is subjected to the input parameters, whose numerical values vary over a range. Here, the objective function is the minimization function. Therefore, a negative sign appears before $M*$, as it is to be maximized. The optimized process parameters depend on the weightage (α_1 and α_2) assigned to an individual output response. This is decided by the criticality of the output response pertaining to the product requirements. Higher productivity would necessitate a higher weightage to the MRR than kerf width or surface roughness; however, precision machining requires higher weightage to the kerf width or surface roughness than the MRR. Therefore, the level of optimized input parameters varies according to the weightage given to the output responses. The objective function is solved using available techniques subjected to the range (minimum to maximum) of each input parameters to obtain a set of optimized parameters. The optimum process parameters for simultaneous optimization of MRR and kerf width for higher productivity (higher weightage to MRR than kerf width) require the highest possible value of input parameters (voltage, pulse on-time, and wire speed). However, while adopting higher weightage to kerf width than MRR, the same optimization technique requires minimum levels of input parameters. Assigning equal weightage to MRR and kerf width results in a different set of optimized input parameters [36]. Therefore, multi-objective (MO) optimization of the Wire-EDM process is instrumental in determining the value of input process parameters for conflicting outputs.

5.9 HYBRIDIZATION OF THE WIRE-EDM PROCESS

It is possible to combine Wire-EDM with other processes to develop a hybrid process. The hybrid process combines the advantages of the individual processes. The selected hybrid processes of the Wire-EDM are highlighted in the following sections:

5.9.1 WIRE-ELECTRIC DISCHARGE GRINDING

The concept of the Wire-EDM process has been successfully utilized for the generation of a new variant called the *"Wire-Electric Discharge Grinding"* (Wire-EDG) process. Owing to small material removal per pulse (similar to minute material removal in a grinding operation) for finer machined surface and utilizing a wire electrode for material removal, the combined process is termed Wire-EDG. Wire-EDG is especially useful in the on-site fabrication of microtools for Electric Discharge Drilling (ED-Drilling), and Electric Discharge Milling (ED-Milling) operations. Moreover, the fabricated microtools can also be used in the Electrochemical Machining (ECM) process, as the absence of tool wear and thermal load in the ECM process allow the use of a single microtool for an extended period. Similar to the Wire-EDM process, Wire-EDG also utilizes a thin wire as a tool electrode moving continuously around a disc through a wire-supplying spool and wire-collecting spool. Figure 5.12 shows the schematic of a typical Wire-EDG system for the precise machining of microtools. However, contrary to the Wire-EDM process where the cutting zone of the wire between two wire guides is extended for a wider span, in the Wire-EDG, the cutting zone is exposed through a V-shaped notch provided on the disc itself. This allows better control of the wire electrode and minimizes the wire deflection in the Wire-EDG process, making it appropriate for the fabrication of delicate thin tools which otherwise become deteriorated due to a slight wire vibration. Other than circular shaped microtools, Wire-EDG has the capabilities to fabricate complex-shaped microtools which are otherwise extremely difficult to machine. The complex cross-section of the electrodes includes tapered, wavy, flexure, etc. [37].

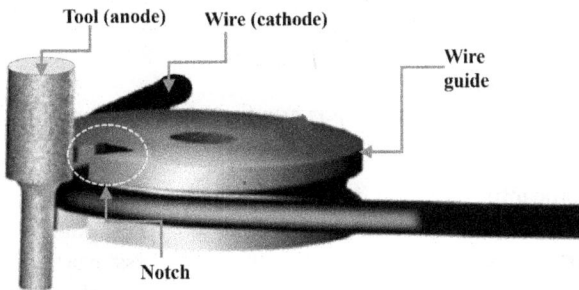

FIGURE 5.12 Schematic of the Wire-EDG process for microtool fabrication. Reproduced with permission from [38].

5.9.2   Ultrasonic Vibration and Magnetic Field-Assisted Wire-EDM

Mass transport from the narrow gap between the wire and workpiece is likely to augment through the integration of ultrasonic vibration to the Wire-EDM system. The primary objective of implementing ultrasonic vibration to the wire electrode or workpiece is to accelerate the dielectric circulation in the IEG, efficient ejection of debris, which eventually enhances the machining performance of the Wire-EDM process [39]. Similar to ultrasonic vibration, magnetic field assistance can be implemented in the Wire-EDM system to remove debris from the machining gap, provided the workpiece material is magnetic in nature. There is a scarcity of extensive research pertaining to the hybrid Wire-EDM (ultrasonic vibration, magnetic field assistance, and new hybrids). However, the authors feel that these techniques have the potential to enhance productivity significantly as mass transport is one of the critical issues to address.

5.10   WIRE-EDM: POTENTIAL APPLICATIONS

The Wire-EDM process is by far the most versatile and commonly used variant of the EDM process. The process can be employed for a plethora of cutting operations depending upon the nature of machining required. It can be used to slice thick difficult-to-machine billets; machining of complex 3-D and taper cutting can be performed by adjusting the upper and lower guides. Further, microtool fabrication, microgear cutting, etc., can also be done using this process. Some of the potential application regimes of the Wire-EDM or Wire-EDMM process are given as follows:

5.10.1   Machining of Fir Tree Shape for Turbines

Turbine blades' roots and the corresponding slots cut on the rotor disc have a complex shape resembling a fir tree. Typical jet engines consist of around 40 discs and approximately 10–12 blades per disc made of hard materials. Machining such a complex shape requiring a gradual change in the size and the tolerances in the range of 80–100 μm can be efficiently performed using the Wire-EDM process.

5.10.2   Thin Wall Micromachining for Slicing of Wafers

Thin wall micromachining is a domain of the Wire-EDM process to fabricate micro-wafers and microcantilever structures [20, 40]. Silicon wafers are primarily being used for solar cell applications. The slicing of silicon wafers is performed using the Wire-EDM process due to some advantages of this process over other slicing methods [41]. Silicon wafers with a minimum thickness of 130–150 μm are successfully sliced using the Wire-EDM process [42]. Figure 5.13 shows an array of thin walls machined on Ti-6Al-4V with a minimum wall thickness of around 10 μm with slight end deflection due to thermal effects.

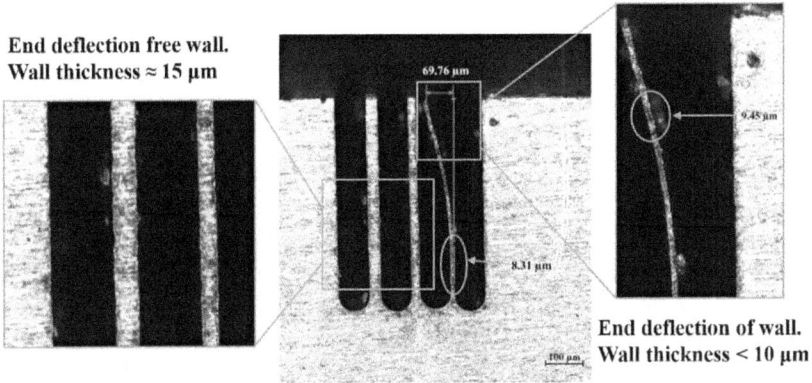

FIGURE 5.13 Fabrication of thin walls using a tungsten carbide wire of 70 µm. Reproduced with permission from [20].

5.10.3 Microgear Machining

Simultaneous machining of microgears with male and female parts is usually performed by the Wire-EDM process. The kerf width has to be maintained and controlled within a permissible limit to allow proper meshing of internal and external gears. The kerf width cannot be reduced below the wire diameter. Therefore, the minimum separation between the internal and external gears is the wire diameter, plus the radial overcut due to sparking in the secondary gap that contributes to the widening of the gap. Figure 5.14(a) shows a segment of microspur gear machined using the Wire-EDMM process. A tungsten carbide wire of an average diameter of 70 µm is used, whereas the gear is machined on a Ti-6Al-4V workpiece. An average kerf width of slightly smaller than 80 µm is generated. A series of miniaturized gears machined through the Wire-EDM process are shown in Figure 5.14(b). The

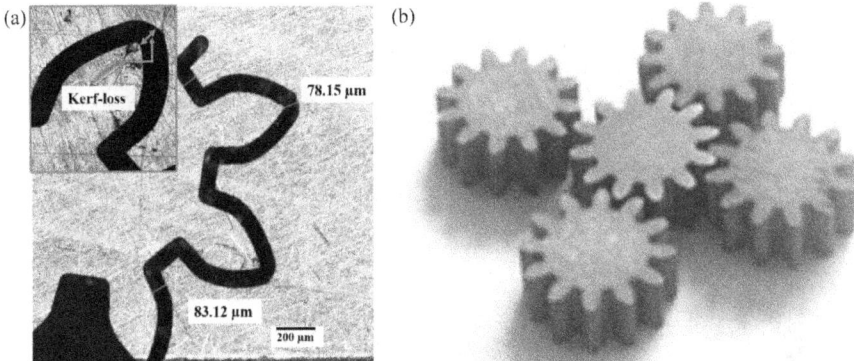

FIGURE 5.14 (a) Simultaneous machining of internal and external gear profile using the Wire-EDMM process. (b) Series of miniaturized gears machined using Wire-EDM. Reproduced with permission from [44].

typical applications of such miniaturized gears (meso or microgears) are in Micro-Electro-Mechanical System (MEMS) devices, automation, medical equipment, small motors, etc. [43, 44].

5.10.4 COMPOUND TOOL FABRICATION FOR EDM

The Wire-EDM process has the potential to fabricate a compound tool (consisting of a number of square/rectangular or irregular-shaped fins or tools) which is used for machining an array of circular or non-circular microholes, large area surface texturing on the metallic surfaces, etc. Figures 5.15(a) and (b) show the array of microelectrodes as tool fabricated by the Wire-EDM process, and an array of micro-holes machined using the same as a tool. Moreover, non-circular or prismatic cross-sectional tool electrodes, such as triangular, square, and hexagonal are machined using the Wire-EDM process. These tools are used in the die-sinking mode of the EDM process for replicating the tool shapes.

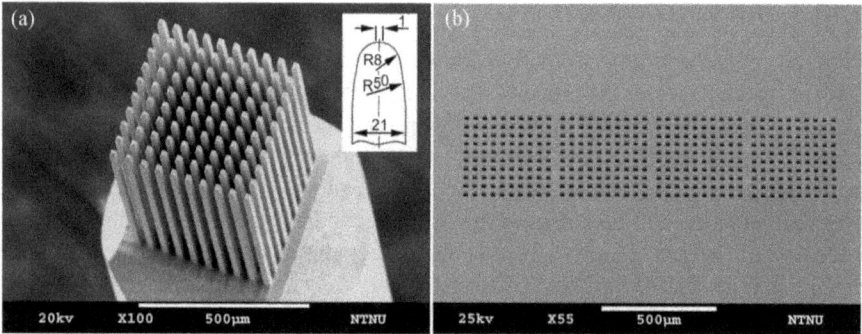

FIGURE 5.15 (a) An array of microelectrodes fabricated by the Wire-EDM process. (b) an array of microdimples machining using the fabricated tool. Reproduced with permission from [45].

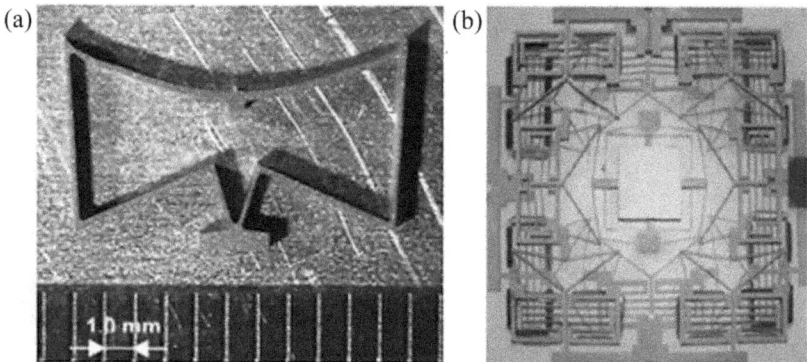

FIGURE 5.16 Machining of (a) a compliant mechanism. Reproduced with permission from [40], and (b) a precision linear stage machined using Wire-EDM process. Reproduced with permission from [46].

5.10.5 Fabrication of Compliant Mechanisms

Fabrication of compliant mechanisms for handling delicate and miniature components is one of the potential applications of the Wire-EDM process. Figure 5.16(a) shows the compliant gripper machined using the Wire-EDM process. With Wire-EDM, the flexibility of the grippers allows higher deformation at the output, and sharp corners are possible to machine. As shown in Figure 5.16(b), the miniature size stage is machined by the Wire-EDM process. The complexity of the shapes to be cut and larger cutting edges allow substantially longer time for complete fabrication.

5.11 RESEARCH OPPORTUNITIES

Some research opportunities in the Wire-EDM/Wire-EDMM domain can be explored for potential widespread utilization in advanced machining/micromachining regime.

5.11.1 Kerf Width Minimization

Though the Wire-EDM process is considered to be a more accurate cutting operation in the domain of Advanced Machining Processes (AMPs), kerf width minimization is still an area of investigation, especially when a higher rate of machining is desirable. Therefore, it is indispensable to optimize the process parameters for simultaneous minimization of kerf width, maximization of the machining rate, and an acceptable level of surface quality. Input pulse shapes can be changed from rectangular (or exponential waves in R-C generators) to triangular, sinusoidal etc., to achieve controlled material removal and greater dimensional accuracy of the process, similar to the Electrochemical Machining (ECM) or Wire Electrochemical Machining (Wire-ECM) processes [47, 48].

5.11.2 Utilizing Ultra-Thin Wire

One of the major challenges in the Wire-EDM process is to avoid wire breakage, especially for ultra-thin wires (typical wire diameter 10–50 μm). Reducing wire diameter significantly minimizes the kerf width, i.e., material loss. It also enhances machining productivity as the smaller size wire would cut at a faster rate due to the narrower width of the material to be cut. It necessitates a change in the machine tool itself, such as vibration-free linear axes, smooth wire circulation system, precise delivery of dielectric fluid to the cutting zone, optimized discharge parameters etc., to incorporate a wire of sizes below 50 μm.

5.11.3 Multiple Wire Electrodes for Enhanced Productivity

To enhance the productivity of the Wire-EDM process, a system of multiple wires performing simultaneous cutting can be used. These systems are especially useful for the mass production of thin wafers or the slicing of thick billets with ultra-precision [5]. The multiple wire system is suitable for microtexture fabrication or compound tool fabrication for higher productivity.

5.11.4   Mass Transport Methods

Mass transport from the minute IEG is of primary consideration, as poor flushing of the gap causes several abnormal discharges that may lead to short circuits, and wire retracting that can break the wire frequently and hamper the productivity of the process. Ultrasonic vibration can be augmented to the workpiece, a more advanced flushing system can be designed, the assistance of a magnetic field, wire reciprocating motion etc., can be the methods for efficient mass transport.

5.12   SUMMARY

The Wire-EDM process for microcutting operations for difficult-to-machine materials is discussed in this chapter. Analysis of kerf width and its variation with various process parameters and wire vibration is analyzed. The kerf width is one of the critical output responses of the Wire-EDM process, which reflects the effectiveness of the process. The effect of wire vibration, wire lag, and wire tension on the kerf width is discussed. Moreover, temperature analysis, wire breakage, and optimization of the Wire-EDM process parameters are also discussed. Potential applications, process limitations, and possible research opportunities in the Wire-EDM process are highlighted toward the end of the chapter.

ACRONYMS/ABBREVIATIONS

AMPs	Advanced Machining Processes
ECM	Electrochemical Machining
EDM	Electrical Discharge Machining
ED-Drilling	Electrical Discharge-Drilling
ED-Milling	Electrical Discharge-Milling
HAZ	Heat-Affected Zone
HTC	Heat Transfer Coefficient
IEG	Interelectrode Gap
MO	Multi-Objective
MRR	Material Removal Rate
R-C	Resistance-Capacitance
SIF	Stress Intensity Factor
SR	Surface Roughness
Wire-ECM	Wire Electrochemical Machining
Wire-EDG	Wire Electrical Discharge Grinding
Wire-EDM	Wire Electrical Discharge Machining
Wire-EDMM	Wire Electrical Discharge Micromachining

SYMBOLS

A	The amplitude of wire vibration
A_c	Wire cross-sectional area

a Lateral discharge gap
a' Crack length
C Capacitance
c Specific heat of the wire
D Wire diameter
E Discharge energy
F Wire tension
F' Force per discharge
F_d Impulse force
f Frequency of discharge
H Distance between wire guide and workpiece
h Workpiece thickness
h_a Convective heat transfer coefficient
K Stress intensity factor
k Thermal conductivity of the wire
L Circumference of the wire
l Distance between two wire guides
M MRR
$\dot{Q}$ Power
q_0 Force acting on wire per unit length
T Temperature
T_c Dielectric temperature
t Time
V Breakdown voltage
x Length of wire
y_k Kerf width
y_{max} Maximum wire deflection

GREEK LETTERS

α Weight assigned to the output response
$\delta(t)$ Unit impulse function
μ Damping coefficient
η Discharge energy fraction to wire
ρ Density of the wire material
ν Wire linear velocity

REFERENCES

1. K.H. Ho, S.T. Newman, S. Rahimifard, R.D. Allen. State of the art in wire electrical discharge machining (WEDM). *Int. J. Mach. Tools Manuf.* 44 (2004): 1247–1259. doi:10.1016/j.ijmachtools.2004.04.017.
2. Y.F. Luo, C.G. Chen, Z.F. Tong. Investigation of silicon wafering by wire EDM. *J. Mater. Sci.* 27 (1992): 5805–5810. doi:10.1007/BF01119742.
3. T.A. Spedding, Z.Q. Wang. Study on modeling of wire EDM process. *J. Mater. Process. Technol.* 69 (1997): 18–28. doi:10.1016/S0924-0136(96)00033-7.

4. J. Qu, A.J. Shih, R.O. Scattergood. Development of the cylindrical wire electrical discharge machining process, part 1: Concept, design, and material removal rate. *J. Manuf. Sci. Eng.* 124 (2002): 702. doi:10.1115/1.1475321.

5. Y. Takayama, Y. Makino, Y. Niu, H. Uchida. The latest technology of wire-cut EDM. *Procedia CIRP.* 42 (2016): 623–626. doi:10.1016/j.procir.2016.02.259.

6. S. Di, X. Chu, D. Wei, Z. Wang, G. Chi, Y. Liu. Analysis of kerf width in micro-WEDM. *Int. J. Mach. Tools Manuf.* 49 (2009): 788–792. doi:10.1016/j.ijmachtools.2009.04.006.

7. D.A.N. Scott, S. Boyina, K.P. Rajurkar. Analysis and optimization of parameter combinations in wire electrical discharge machining. *Int. J. Prod. Res.* 29 (1991): 2189–2207.

8. V.K. Jain. *Advanced Machining Processes*, 2nd Edition. Allied Publishers Private Limited, New Delhi, 2021.

9. S. Abulais. Current research trends in electric discharge machining (EDM): Review. *Int. J. Sci. Eng. Res.* 5 (2014): 100–118. http://www.ijser.org.

10. E. Uhlmann, S. Piltz, U. Doll. Machining of micro/miniature dies and moulds by electrical discharge machining – Recent development. *J. Mater. Process. Technol.* 167 (2005): 488–493. doi:10.1016/j.jmatprotec.2005.06.013.

11. T. Daniel, C. Liu, J. Mou, M.P. Jahan. Micro-wire-EDM. In: *Micro-Electrical Disch. Mach. Process. Mater. Forming, Mach. Tribol.*, Springer, Singapore, 2019, pp. 67–92. doi:10.1007/978-981-13-3074-2_4.

12. A.V. Shayan, R.A. Afza, R. Teimouri. Parametric study along with selection of optimal solutions in dry wire cut machining of cemented tungsten carbide (WC-Co). *J. Manuf. Process.* 15 (2013): 644–658. doi:10.1016/j.jmapro.2013.05.001.

13. M. Kunieda, B. Lauwers, K.P. Rajurkar, B.M. Schumacher. Advancing EDM through fundamental insight into the process. *CIRP Ann. – Manuf. Technol.* 54 (2005): 64–87. doi:10.1016/S0007-8506(07)60020-1.

14. V. Prakash, P. Kumar, P.K. Singh, M. Hussain, A.K. Das, S. Chattopadhyaya. Micro-electrical discharge machining of difficult-to-machine materials: A review. *Proc. Inst. Mech. Eng. Part B J. Eng. Manuf.* 233 (2019): 339–370. doi:10.1177/0954405417718591.

15. K. Egashira, K. Mizutani. EDM at low open-circuit voltage. *Japan Soc. Electr. Mach. Eng.* (2004).

16. F. Klocke, D. Lung, D. Thomaidis, G. Antonoglou. Using ultra thin electrodes to produce micro-parts with wire-EDM. *J. Mater. Process. Technol.* 149 (2004): 579–584. doi:10.1016/j.jmatprotec.2003.10.061.

17. A.B. Puri. Advancements in micro wire-cut electrical discharge machining. In: *Non-Traditional Micromach. Process. Mater.* Forming, Mach. Tribol., 2017. doi:10.1007/978-3-319-52009-4.

18. M. Singh, J. Ramkumar, R.V. Rao, J. Balic. Experimental investigation and multi-objective optimization of micro-wire electrical discharge machining of a titanium alloy using Jaya algorithm. *Adv. Prod. Eng. Manag.* 14 (2019): 251–263.

19. S. Sakamoto, K. Hayashi, Y. Kondo, K. Yamaguchi, T. Fujita, T. Yakou. Fundamental micro-grooving characteristics of hard and brittle materials with a fine wire tool. *Adv. Mater. Res.* 1136 (2016): 299–304. doi:10.4028/www.scientific.net/AMR.1136.299.

20. M. Singh, A. Singh, J. Ramkumar. Thinwall micromachining of Ti–6Al–4V using microwire electrical discharge machining process. *J. Brazilian Soc. Mech. Sci. Eng.* 41:338 (2019): 1–12.

21. N. Mohri, H. Yamada, K. Furutani, T. Narikiyo, T. Magara. System identification of wire electrical discharge machining. *Ann. CIRP.* 47 (1998): 173–176.

22. M.H.A.H. Ali, Mohammad Yeakub, Asfana Banu. Analysis of kerf accuracy in dry micro-wire EDM. *Int. J. Adv. Manuf. Technol.* 111 (2020): 597–608.

23. A. Okada, Y. Uno, M. Nakazawa, T. Yamauchi. Evaluations of spark distribution and wire vibration in wire EDM by high-speed observation. *CIRP Ann. – Manuf. Technol.* 59 (2010): 231–234. doi:10.1016/j.cirp.2010.03.073.

24. D.F. Dauw, I. Beltrami. High-precision wire-EDM by online wire positioning control. *Ann. CIRP.* 43 (1994): 193–197.
25. A.B. Puri, B. Bhattacharyya. An analysis and optimisation of the geometrical inaccuracy due to wire lag phenomenon in WEDM. *Int. J. Mach. Tools Manuf.* 43 (2003): 151–159. doi:10.1016/S0890-6955(02)00158-X.
26. I. Leuven, A. Bertholds, D. Dauw. A simplified post process for wire cut EDM. *J. Mater. Process. Technol.* 58 (1996): 385–389.
27. F. Han, G. Cheng, Z. Feng, S. Isago. Thermo-mechanical analysis and optimal tension control of micro wire electrode. *Int. J. Mach. Tools Manuf.* 48 (2008): 922–931. doi:10.1016/j.ijmachtools.2007.10.024.
28. K.P. Rajurkar, W.M. Wang. Thermal modeling and on-line monitoring of wire-EDM. *J. Mater. Process. Technol.* 38 (1993): 417–430.
29. W. Dekeyser, R. Snoeys, M. Jennes. A thermal model to investigate the wire rupture phenomenon for improving performance in EDM wire cutting. *J. Manuf. Syst.* 2 (1085): 179–190.
30. N. Kinoshita, M. Fukui, G. Gamo. Control of wire-EDM preventing electrode from breaking. *CIRP Ann. – Manuf. Technol.* 31 (1982): 111–114. doi:10.1016/S0007-8506(07) 63279-X.
31. Y.S. Liao, Y.Y. Chu, M.T. Yan. Study of wire breaking process and monitoring of WEDM. *Int. J. Mach. Tools Manuf.* 37 (1997): 555–567. doi:10.1016/S0890-6955(95)00049-6.
32. K.P. Rajurkar, W.M. Wang, R.P. Lindsay. On-line monitor and control for wire breakage in WEDM. *CIRP Ann. – Manuf. Technol.* 40 (1991): 219–222. doi:10.1016/ S0007-8506(07)61972-6.
33. Y.F. Luo. Rupture failure and mechanical strength of the electrode wire used in wire EDM. *J. Mater. Process. Technol.* 94 (1999): 208–215. doi:10.1016/S0924-0136(99)00107-7.
34. Y.S. Liao, J.T. Huang, H.C. Su. A study on the machining-parameters optimization of wire electrical discharge machining. *J. Mater. Process. Technol.* 71 (1997): 487–493.
35. N. Tosun. The effect of the cutting parameters on performance of WEDM. *KSME Int. J.* 17 (2003): 816–824. doi:10.1007/BF02983395.
36. N. Tosun, C. Cogun, G. Tosun. A study on kerf and material removal rate in wire electrical discharge machining based on Taguchi method. *J. Mater. Process. Technol.* 152 (2004): 316–322. doi:10.1016/j.jmatprotec.2004.04.373.
37. D. Rakwal, S. Heamawatanachai, P. Tathireddy, F. Solzbacher, E. Bamberg. Fabrication of compliant high aspect ratio silicon microelectrode arrays using micro-wire electrical discharge machining. *Microsyst. Technol.* 15 (2009): 789–797. doi:10.1007/ s00542-009-0792-7.
38. G.L. Chern, S. De Wang. Punching of noncircular micro-holes and development of microforming. *Precis. Eng.* 31 (2007): 210–217. doi:10.1016/j.precisioneng.2006.09.001.
39. Z.N. Guo, T.C. Lee, T.M. Yue, W.S. Lau. A study of ultrasonic-aided wire electrical discharge machining. *J. Mater. Process. Technol.* 63 (1997): 823–828.
40. S.F. Miller, C.C. Kao, A.J. Shih, J. Qu. Investigation of wire electrical discharge machining of thin cross-sections and compliant mechanisms. *Int. J. Mach. Tools Manuf.* 45 (2005): 1717–1725. doi:10.1016/j.ijmachtools.2005.03.003.
41. G. Dongre, S. Zaware, U. Dabade, S.S. Joshi. Multi-objective optimization for silicon wafer slicing using wire-EDM process. *Mater. Sci. Semicond. Process.* 39 (2015): 793–806. doi:10.1016/j.mssp.2015.06.050.
42. K. Joshi, A. Ananya, U. Bhandarkar, S.S. Joshi. Ultra thin silicon wafer slicing using wire-EDM for solar cell application. *Mater. Des.* 124 (2017): 158–170. doi:10.1016/j. matdes.2017.03.059.
43. K. Gupta, N.K. Jain, R.F. Laubscher. Spark erosion machining of miniature gears: A critical review. *Int. J. Adv. Manuf. Technol.* 80 (2015): 1863–1877. doi:10.1007/ s00170-015-7130-2.

44. K. Gupta, N.K. Jain. Analysis and optimization of micro-geometry of miniature spur gears manufactured by wire electric discharge machining. *Precis. Eng.* (2014).

45. S.T. Chen. Fabrication of high-density micro holes by upward batch micro EDM. *J. Micromech. Microeng.* 18 (2008): 1–9. doi:10.1088/0960-1317/18/8/085002.

46. M. Dinesh, G.K. Ananthasuresh. Micro-mechanical stages with enhanced range. *Int. J. Adv. Eng. Sci. Appl. Math.* 2 (2010): 35–43. doi:10.1007/s12572-010-0014-7.

47. V. Sharma, D.S. Patel, V. Agrawal, V.K. Jain, J. Ramkumar. Investigations into machining accuracy and quality in wire electrochemical micromachining under sinusoidal and triangular voltage pulse condition. *J. Manuf. Process.* 62 (2021): 348–367.

48. D.S. Patel, V. Sharma, V.K. Jain, J. Ramkumar. Reducing overcut in electrochemical micromachining process by altering the energy of voltage pulse using sinusoidal and triangular waveform. *Int. J. Mach. Tools Manuf.* 151 (2020). doi:10.1016/j.ijmachtools.2020.103526.

6 Electrical Discharge Diamond Grinding

Tribeni Roy and Satish Mullya

CONTENTS

6.1 INTRODUCTION

Developments of high strength, high hardness brittle materials are finding applications in various fields of engineering owing to their improved resistance against fatigue, wear, and corrosion. Machining such materials is difficult due to their brittle nature, leading to the formation of cracks. Thermal-assisted machining has become one of the ways by which conventional machining/finishing processes can be utilized to machine brittle materials [1]. This has led to a rapid development in hybrid machining processes wherein two or more machining processes are combined to utilize the advantages while overcoming the limitations of each process [2]. One such hybrid machining process extensively employed for machining of difficult-to-cut electrically conductive materials is the electrical discharge diamond grinding (EDDG) process [3]. It combines the electrical discharge machining (EDM) and the mechanical grinding processes. EDM is a well-established process with the ability to machine electrically conducting materials of any hardness. Two of the demerits of the EDM process are the formation of a recast layer and heat-affected zone (HAZ) which have material properties significantly different from the parent material [4]. Due to this, a post-processing technique is usually employed for its removal. On the other hand, mechanical grinding processes with diamond abrasives are extensively used for the finishing of brittle materials. A large force exerted on the abrasive grains leads to the formation of scratches on the finished surfaces as well as rapidly deteriorating the grinding wheel. EDDG uses a metal-bonded diamond grinding wheel that

DOI: 10.1201/9780429160011-6

removes material from the workpiece predominantly by abrasion of the electrically discharged thermally softened material [5]. The grinding forces are considerably reduced due to material softening by electrical discharges and it makes machining easy. On the other hand, the issues with EDM such as recast layer and HAZ are no longer present on the surfaces machined by EDDG. Another hybrid combining EDM with mechanical grinding is the electric discharge grinding (EDG) process where material removal takes place partly by melting and partly by abrasion whereas in EDDG, material removal is mostly by abrasion, while electric discharges help in softening the layer of material to be removed by grinding [6].

EDDG has major applications in the machining of high-strength electrically conducting materials such as superalloys, advanced ceramics, metal matrix composites, etc. [7, 8]. Important characteristics of the EDDG process include the following:

a) The tool electrode used in EDM is replaced with a metal-bonded grinding wheel containing diamond abrasives.
b) Material removal is achieved predominantly by grinding while EDM plays the role of thermally softening the workpiece material.
c) The material removal process is faster compared to the conventional grinding process.
d) No recast layer formation [9].
e) Surface roughness of less than 0.5 μm is achieved [10].
f) Exclusively used for machining brittle, hard, and high-strength electrically conducting materials.

6.2   DESCRIPTION OF EDDG PROCESS

A schematic view of a typical EDDG process is shown in Figure 6.1. A metal-bonded diamond grinding wheel is a cathode while the workpiece (brittle material to be machined) is an anode. The two electrodes are separated by a suitable liquid dielectric. The EDM machine's ram holds the grinding attachment, which allows the grinding wheel to rotate around its axis parallel to the horizontal machining table. The gap-sensing mechanism of the servo system prohibits direct physical contact

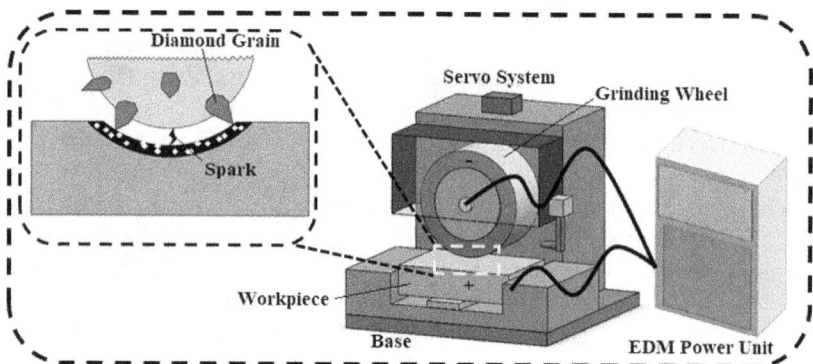

FIGURE 6.1   A schematic view of the EDDG process.

TABLE 6.1

Specification and Dressing Conditions of Grinding Wheel

Wheel Specifications		Wheel Dressing Conditions	
Wheel diameter	100 mm	Voltage	60 V
Abrasive material	Diamond	Current	10 A
Bond material	Bronze	Pulse-on time	100 μs
Bore	33 H7	Speed of wheel	1 m/s
Thickness	7 mm	Duty cycle	0.4–0.5
Impregnation	3 mm	Duration	1.8–2 min
Concentration	75%		
Grit size	80/100		

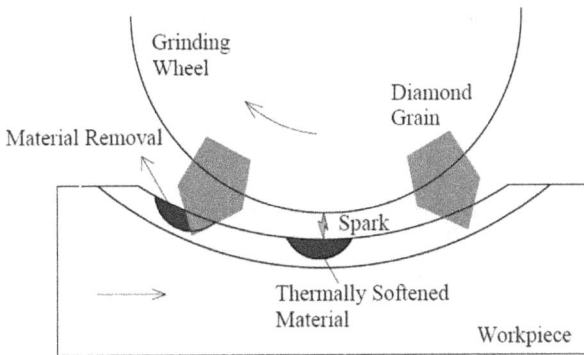

FIGURE 6.2   A schematic view of material removal process in EDDG.

between the electrodes. The rotational speed of the grinding wheel can be controlled by adjusting the motor's input voltage. Wheel dressing is carried out before machining in order to eliminate errors during machining and thereby maintaining standard wheel topography. Typical wheel specifications and dressing conditions are presented in Table 6.1 [11].

Figure 6.2 shows a schematic diagram of the material removal mechanism in EDDG. In this, due to breakdown of dielectric, a plasma channel is created that thermally softens the workpiece material. Due to this, localized ductility on the workpiece is achieved. Diamond abrasives remove the electrically discharged thermally softened material by abrasion. The force required for material removal is reduced drastically. There are certain advantages and limitations of the EDDG process. These are listed in Table 6.2.

6.3   MATHEMATICAL FORMULATION OF EDDG PROCESS

In order to have a scientific understanding of the EDDG process, it is important to understand the grinding forces acting in the EDDG compared to conventional grinding, as well as the mechanism of material removal.

TABLE 6.2

Advantages and Limitations of the EDDG Process

Advantages	Limitations
Compared to the EDM process, a better surface finish is achieved	Formation of high discharge craters due to high discharge energy is difficult to be removed by grinding
Grinding forces are substantially reduced as a result of material softening caused by electrical discharges	High discharge energy leads to premature grits pull out thereby affecting the material removal process by grinding
EDDG is environmentally friendly with the use of dielectric having low toxicity	There is a possibility of material removal by combined effect of melting and abrasion due to variation in discharge energy similar to EDG; this may lead to rough surfaces
With reduced grinding forces, material is removed at a rapid rate compared to conventional grinding	Cannot be used to machine electrically non-conducting high-strength materials

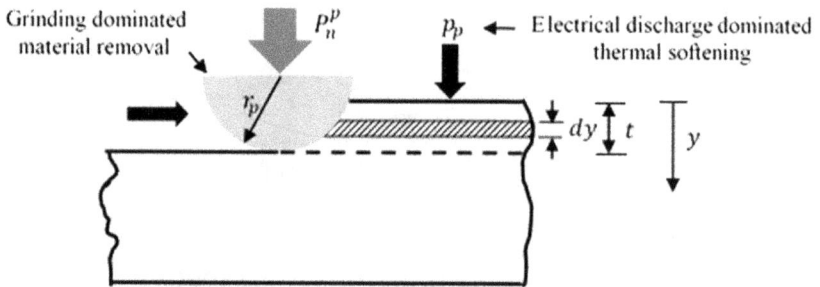

FIGURE 6.3 Model illustration to assess the normal force reduction in EDDG.

6.3.1 REDUCTION IN GRINDING FORCES DUE TO THERMAL SOFTENING OF MATERIAL

Electrical discharges in the EDDG process are responsible for thermally softening the work material. This in turn helps in reducing the grinding forces required for material removal. The component of normal force in grinding is due to the physical interaction between the work material and the diamond abrasives. This interaction is dependent on factors such as the grinding wheel topography and the kinematics of the process. This is purely probabilistic in nature as the diamond abrasives are randomly located on the grinding wheel. In order to understand the material removal mechanism in EDDG, it is first important to understand the forces responsible for material abrasion with respect to the hardness of the workpiece material (H_{wp}), power of the EDM pulse (p_p), and the depth of cut (t). A simple mathematical model is described here for comparing the forces in EDDG and conventional grinding. Figure 6.3 shows a schematic of a spherical abrasive grain of radius r_p plowing through the work material up to the depth of cut. The normal force (P_n) is given by Equation 6.1,

$$P_n = H_{wp}A_p \tag{6.1}$$

where $A_p = \dfrac{\pi}{2}\left(2r_p t - t^2\right)$ indicates the projected area of contact in a plane normal to the line of action of P_n [5] and H_{wp} is hardness of the workpiece material. Due to dielectric breakdown, an electrical discharge incident on the work material would lead to an increase in temperature. This will bring about a gradient in the hardness across the depth of cut and will ease material removal by the grain that is just behind the localized heated material. At depth, y, the hardness of the work material, $H_{wp}(y)$, depends on the temperature, $T(y)$. By considering small increments of thickness, dy, the normal force P_n^p can be evaluated by using Equation 6.2.

$$P_n^p = \int_0^t H_{wp}(y)A_p(y)dy \tag{6.2}$$

By adopting the anode erosion model of Patel et al. [12] for single discharge and considering the power flux incident on anode as $q_{flux} = \dfrac{fp_p}{\pi r_a^2(t)}$, the temperature distribution $T(y)$ along the axis of the discharge can be calculated as shown in Equation 6.3,

$$T(y) = \frac{2q_{flux}\sqrt{\alpha t}}{k}\left[\varphi\left(\frac{y}{2\sqrt{\alpha t}}\right) - \varphi\left(\frac{\sqrt{y^2 + r_a^2}}{2\sqrt{\alpha t}}\right)\right] \tag{6.3}$$

where f is the fraction of power of pulse expended at the anode ($\approx 8\%$), r_a is the radius of the spark assumed to vary with time as $r_a(t) = 0.8t_{on}^{0.75}$ µm, t_{on} is the pulse on time, k and α are the thermal conductivity and thermal diffusivity of the work material respectively, and φ represents the complementary integral work function [5].

In order to simplify the effect of temperature variation from room temperature to the melting temperature, average values of the thermophysical properties of the work material were assumed $\left(k = 25\ \mathrm{Wm^{-1}\,K^{-1}}, \alpha = 6.83\times10^{-6}\ \mathrm{m^2s^{-1}}\right)$. The maximum error due to the heat of fusion was estimated to be approximately 2% and therefore was ignored. Thermal softening due to frictional heating by abrasive sliding is ignored considering wet grinding conditions. Also, considering wet grinding conditions, frictional heating by abrasive sliding leading to thermal softening of work material is ignored.

High speed steels are high-strength materials that retain nearly 80% of their room temperature hardness up to 600°C. Above this temperature, the hardness sharply reduces followed by a gradual reduction in hardness and finally reducing to zero at melting point (1250°C) [5]. Piecewise polynomials were fitted to evaluate $H_{wp}(y)$ as a function of $T(y)$ that follow the trend for hardness as explained above. Figure 6.4

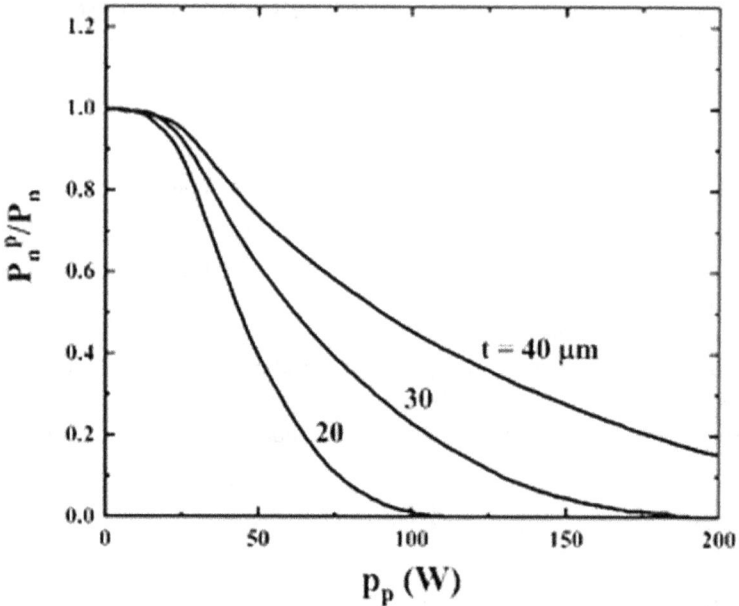

FIGURE 6.4 Simulated profile of the effect of p_p on $\dfrac{P_n^p}{P_n}$ for various depths of cut (t). Constant parameters used include $r_p = 80$ μm, $t_{on} = 100$ μs and discharge voltage$= 40$ V [5].

shows the simulated results for the effect of pulse power on the ratio of P_n^p and P_n for various depths of cut. With an increasing depth of cut, higher pulse power is required to reduce $\dfrac{P_n^p}{P_n}$. This is due to the fact that the amount of material to be removed increases with increasing depth of cut, therefore, the amount of power required for thermally softening the material also increases. This leads to a higher peak power requirement to reduce the forces at increasing depth of cut. This model is qualitatively validated by Koshy et al. [5] where they have experimentally shown that there is an increased current requirement for reducing the normal grinding force.

6.3.2 MECHANISM OF MATERIAL REMOVAL IN EDDG PROCESS

The phenomenon of material removal in the EDDG process is very complex owing to the superposition of thermal softening and mechanical energy for removal of material from the workpiece. Therefore, in order to model the behavior of material removal in EDDG, a few assumptions are considered, to simplify the model. They are as follows:

- Workpiece material is homogenous.
- The abrasive grains are similar in shape.

- EDM discharge leads to material softening and thereby achieves higher depth of cut by the abrasive grain during material removal.
- Only mechanical abrasion is responsible for material removal.
- The grains projecting from the wheel are of the same height, but their orientation on the wheel surface is different.

In the grinding process, Figure 6.5 shows the mechanics of cutting action caused by a grain [13]. The abrasive grain engaged in cutting generates a circular profile on the workpiece as a result of the wheel's rotational motion. The geometrical path of the previous cutting point is the same, but because the workpiece is moving, the abrasive grain's termination point is displaced by a distance A'A, as illustrated in Figure 6.5. As a result, in reference to the workpiece, the cutting path C'BA'AC' is practically trochoid rather than circular [13].

The workpiece's displacement distance A'A is caused by table movement and is equal to the feed per cutting point (f_c), which is defined (Equation 6.4) for inter-grain spacing (S) as S times the velocity ratio of workpiece to wheel [13]

$$f_c = \frac{SV_{wp}}{V_w} \tag{6.4}$$

where the workpiece speed is V_{wp}, and the wheel speed is V_w. The resulting chip resembles a triangular pyramid. As a result, the cross-sectional area (A_c) may be calculated as given by Equation 6.5.

$$A_c = \frac{1}{2} w_c h_c \tag{6.5}$$

The chip volume (C_V) may be estimated as given by Equation 6.6. Here, w_c, h_c, and l_c are width, height, and length of chip, respectively. The average effective cutting

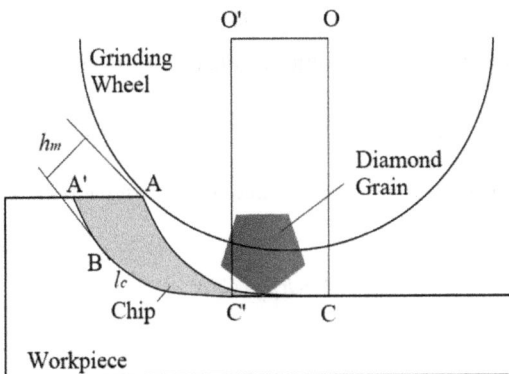

FIGURE 6.5 Schematic of cutting action by a grain.

width for each cutting point w_c may be estimated as given by Equation 6.7. Here, P_w is the number of cutting points on the wheel.

$$C_V = \frac{1}{3} A_c l_c$$
(6.6)

$$w_c = P_w h_c$$
(6.7)

The chip length l_c can be found by using Equation 6.8 where, t denotes depth of cut and d_w denotes wheel diameter.

$$l_c = \sqrt{t d_w}$$
(6.8)

The inter grain spacing S can be calculated as given by Equation 6.9. Here, P_a is the number of cutting points per unit area.

$$S = \frac{2}{P_a w_c} = \frac{2}{P_a P_w h_m}$$
(6.9)

The following relationship (Equation 6.10) can be used to calculate the chip's average height (h_m) [13],

$$h_m = 2S \left(\frac{V_{wp}}{V_w} \right) \left(\frac{t}{d_w} \right)^{\frac{1}{2}}$$
(6.10)

The chip volume C_V is simplified in terms of input process parameters by substituting Equations 6.4, 6.5, and 6.7–6.9 in Equation 6.6 and assuming that the chip height h_c is equal to average chip height h_m.

$$C_V = \frac{2t V_{wp}}{3 P_a V_w}$$
(6.11)

The number of chips produced per unit of time (N) is given by Equation 6.12.

$$N = P_a w V_{wp}$$
(6.12)

The rate of volumetric material removal due to grinding action is calculated as follows [13].

$$MRR_G = C_V N$$
(6.13)

$$MRR_G = \frac{2tw \left(V_{wp} \right)^2}{3 (V_w)}$$
(6.14)

The primary difference between material removal by grinding and EDDG is a higher depth of cut in EDDG caused by material softening by electrical discharges. Equation 6.14, therefore can be rewritten as Equation 6.15.

$$MRR_{EDDG} = MRR_G = \frac{2tw\left(V_{wp}\right)^2}{3(V_w)} \tag{6.15}$$

Figure 6.6 shows the effect of depth of cut on MRR. At a lower depth of cut, the primary action of material removal is by mechanical grinding wherein scratches will be observed on actual surfaces due to the brittle nature of the workpiece material. Due to electrical discharges, workpiece softening takes place and higher depths of cut are achieved thereby showing the transitioning toward the EDDG process. A higher amount of material removal can be obtained with smooth surfaces. It is worth noting that increasing the grinding wheel speed lowers the MRR. This could be caused due to the fact that the wheel rotating at a high speed does not provide sufficient time for contact between the grain and the thermally softened material, thereby removing less material. MRR, on the other hand, increases as the table speed increases (Figure 6.7). This could possibly be due to the constant exposure of a new workpiece surface to sparking and subsequent material removal due to grinding.

6.4   CONCLUSIONS

The EDDG process has proved to be one of the improved hybrid technologies for the machining of electrically conductive brittle materials. This chapter demonstrates mathematical models for reduced cutting forces in the EDDG process and the

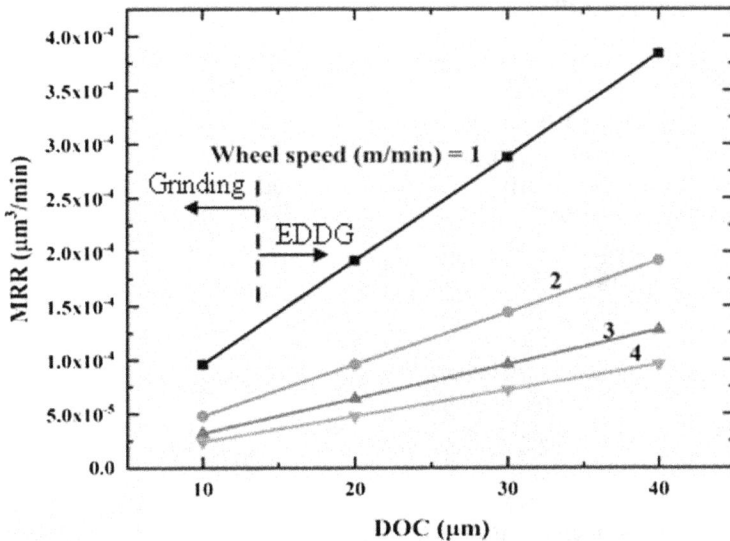

FIGURE 6.6   MRR vs depth of cut in EDDG.

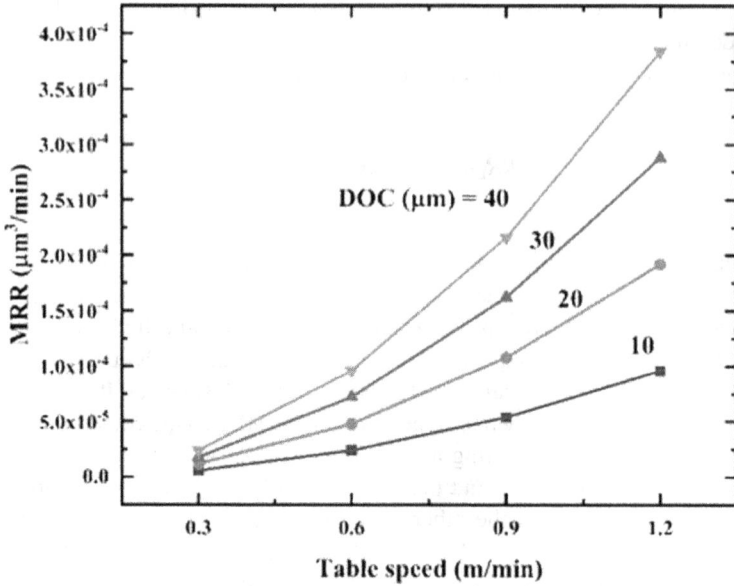

FIGURE 6.7   MRR vs table speed in EDDG.

mechanism of material removal. This provides readers with a basic guide for under-standing the process mechanics and for using the models as a building framework and adding complexity to the models for improved prediction accuracy.

6.5   FUTURE SCOPE

The EDDG process is a breakthrough in EDM and grinding technology for machin-ing brittle materials with good surface properties. By adding alternative abra-sives to the grinding wheels, advancements in the EDDG process can be made. Furthermore, integrating data driven models such as machine learning techniques will help further to identify the real-time process mechanics and improve the machining performance.

NOMENCLATURE

α	thermal diffusivity
ϕ	complementary integral work function
A_p	projected area of contact
A_c	cross-sectional area
C_V	chip volume
d_w	wheel diameter
dy	small increments of thickness
EDM	electrical discharge machining
EDG	electrical discharge grinding

EDDG	electrical discharge diamond grinding
f	fraction of power of pulse
f_c	feed per cutting point
h_c	chip height
h_m	average chip height
H_{wp}	hardness of workpiece material
HAZ	heat-affected zone
k	thermal conductivity
l_c	length of chip
MRR	material removal rate
MRR_G	volumetric material removal rate due to grinding
MRR_{EDDG}	volumetric material removal rate due to electric discharge diamond grinding
N	no. of chips produced per unit time
p_p	power of EDM pulse
p_n^p	normal force
P_a	no. of cutting points per unit area
P_w	no. of cutting points on the wheel
q_{flux}	power flux
r_a	radius of the spark
r_p	radius of abrasive grain
S	inter-grain spacing
t	depth of cut
t_{on}	pulse on time
$T(y)$	temperature
V_{wp}	workpiece speed
V_w	wheel speed
w_c	width of chip
Y	depth

REFERENCES

1. S. Lei, F. Pfefferkorn. A review of thermally assisted machining. In: *Proc. ASME Int. Manuf. Sci. Eng. Conf. 2007, MSEC2007*, 2007, 325–336 https://doi.org/10.1115/MSEC2007-31096.

2. K.K. Saxena, M. Bellotti, J. Qian, D. Reynaerts, B. Lauwers, X. Luo. Overview of Hybrid Machining Processes. In: *Hybrid Mach.*, 2018, 21–41. https://doi.org/10.1016/b978-0-12-813059-9.00002-6.

3. S.K. Choudhury, V.K. Jain, M. Gupta. Electrical discharge diamond grinding of high speed steel. *Mach. Sci. Technol.* 3 (1999): 91–105. https://doi.org/10.1080/10940349908945685.

4. T. Roy, A. Sharma, P. Ranjan, R. Balasubramaniam. Modeling the nano indentation behavior of recast layer and heat affected zone on reverse micro EDMed hemispherical feature. *J. Manuf. Sci. Eng.* 143 (2021): 101009. https://doi.org/10.1115/1.4050823.

5. P. Koshy, V.K. Jain, G.K. Lal. Mechanism of material removal in electrical discharge diamond grinding. *Int. J. Mach. Tools Manuf.* 36 (1996): 1173–1185. https://doi.org/10.1016/0890-6955(95)00103-4.

6. S.K. Choudhury, V.K. Jain, M. Gupta. Electrical discharge diamond grinding of high speed steel. *Mach. Sci. Technol.* 3 (1999): 91–105. https://doi.org/10.1080/10940349908945685.

7. D.R. Unune, H.S. Mali. Parametric modeling and optimization for abrasive mixed surface electro discharge diamond grinding of Inconel 718 using response surface methodology. *Int. J. Adv. Manuf. Technol.* 93 (2017): 3859–3872. https://doi.org/10.1007/s00170-017-0806-z.

8. P. Koshy, V.K. Jain, G.K. Lal. Grinding of cemented carbide with electrical spark assistance. *J. Mater. Process. Technol.* 72 (1997): 61–68. https://doi.org/10.1016/S0924-0136(97)00130-1.

9. D.R. Unune, C.K. Nirala, H.S. Mali. ANN-NSGA-II dual approach for modeling and optimization in abrasive mixed electro discharge diamond grinding of Monel K-500. *Eng. Sci. Technol. Int. J.* 21(2018): 322–329. https://doi.org/10.1016/j.jestch.2018.04.014.

10. S. Kumar, S.K. Choudhury. Prediction of wear and surface roughness in electro-discharge diamond grinding. *J. Mater. Process. Technol.* 191(2007): 206–209. https://doi.org/10.1016/j.jmatprotec.2007.03.032.

11. V.K. Jain, R.G. Mote. On the temperature and specific energy during electrodischarge diamond grinding (EDDG). *Int. J. Adv. Manuf. Technol.* 26 (2005): 56–67. https://doi.org/10.1007/s00170-003-1983-5.

12. M.R. Patel, M.A. Barrufet, P.T. Eubank, D.D. DiBitonto. Theoretical models of the electrical discharge machining process. II. The anode erosion model. *J. Appl. Phys.* 66(1989). https://doi.org/10.1063/1.343995.

13. M.K. Satyarthi, P.M. Pandey. Modeling of material removal rate in electric discharge grinding process. *Int. J. Mach. Tools Manuf.* 74 (2013): 65–73. https://doi.org/10.1016/j.ijmachtools.2013.07.008.

7 Laser Machining of Metals

Fundamentals & Applications

Deepak Marla

CONTENTS

DOI: 10.1201/9780429160011-7

7.1 INTRODUCTION

Lasers are capable of producing intense monochromatic light, which can be focused down to narrow spot sizes to generate high power densities that can rapidly heat any material, causing it to melt and vaporize. Unlike conventional tools, lasers can be used to process any material regardless of its hardness as long as the material is able to absorb the laser photons. Therefore, lasers are used to process various materials, including metals, ceramics, polymers, glass, and semiconductors. The development of lasers has revolutionized the machine tool industry, which has several applications – cutting, drilling, milling, texturing, cleaning, marking, and engraving. Laser cutting and laser drilling are some of the earliest applications of lasers that are still widely used in the industry today. For example, lasers are routinely used to cut sheet metals in the automotive industry and to drill holes on difficult-to-cut materials such as titanium alloys in the aerospace industry. Short and ultra-short pulsed lasers have proven to be effective in precision manufacturing. Ultra-short pulsed lasers with pulse durations in the range of femtoseconds (fs) can be used for precision cutting and drilling with excellent surface finish and no or minimum heat- affected zone. The ability of lasers to cut tiny and intricate parts with high precision and accuracy has made it a preferred choice for many applications in the electronics industry. Laser machining is also used in the biomedical industry. Laser cutting is the most preferred process for coronary stent manufacturing as stents require high precision and dimensional accuracy. Lasers are also used to create textured surfaces through controlled ablation with precise control over surface topography. Femtosecond lasers can be used to generate sub-micrometer features on metal surfaces, generally referred to as laser-induced periodic surface structures (LIPSS). Laser surface texturing to create antibacterial surfaces is a promising technique for texturing biomedical implants. Besides, engraving using lasers is a fast, precise, and clean process used to mark serial numbers and logos. It is prevalent in the jewelry industry to create artworks on gold and silver ornaments. Lasers are also used to clean surfaces contaminated with rust, paint, and grease without the use of any chemicals. Laser cleaning is a low-cost and environmentally friendly process that finds applications in the automotive and electronic industries.

Despite several advantages, laser machining processes are inherently complex and require appropriate processing conditions for their effective use. Therefore, the right choice of laser and its parameters are vital in producing parts with high accuracy and minimal thermal damage. The process mainly depends on the laser's interaction with the material and its subsequent response in the form of temperature rise, melting, and vaporization, governed by the material's optical and thermal properties. The optical properties, namely, reflectivity and absorption coefficient, determine the amount of laser radiation absorbed and the volume over which it is absorbed in the material. The optical properties of a given material change with the wavelength of the laser radiation incident on it [1]. Thus, the laser wavelength becomes an important factor in choosing the type of laser to be used for machining a given material, with the essential condition that the material must absorb at least some portion of the laser radiation.

In metals, the absorbed laser radiation is usually converted into heat, leading to a localized increase in the temperature of the material. Subsequently, the material may undergo melting and vaporization. These are the two primary mechanisms by which material removal occurs in laser beam machining. While the vaporized material gets removed automatically, the molten material is typically removed by blowing a high-pressure gas. Laser cutting using continuous-wave lasers is based on the principle of ejecting molten material using high-pressure gas. Besides, material removal due to explosive boiling or phase explosion may occur when the material is superheated beyond the normal boiling temperature and close to the liquid-vapor critical point. In that case, material removal occurs in the form of liquid droplets and vapor [2]. Under the application of femtosecond laser pulses, the material removal characteristics are quite different, with a very small heat-affected zone (HAZ) and the material removal is restricted only to the region of laser spot. The laser machining characteristics are primarily influenced by how the material removal happens and the extent of heat diffusion into the bulk material or the HAZ, which are influenced by laser intensity and thermal properties. Thus, the process is intricately dependent on the laser parameters and the optical and thermal properties of the material. Therefore, having a basic knowledge of the laser parameters is a prerequisite in understanding the laser beam machining process.

This chapter begins with a brief description of the laser system including the fundamental concepts of laser beam generation and laser optics, and a mathematical

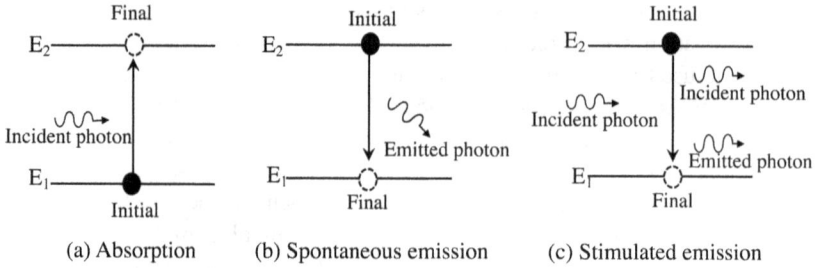

(a) Absorption   (b) Spontaneous emission   (c) Stimulated emission

FIGURE 7.1 Schematic showing absorption, spontaneous emission, and stimulated emission.

description of the laser beam parameters. This is followed by fundamentals of laser interaction with matter and material removal mechanisms to provide the reader with a basic understanding of the physical phenomena that occur in laser beam machining. Lastly, some of the important laser beam machining processes used in the industry are discussed.

7.2 LASER SYSTEM

Laser is the main component of the laser beam machining system. Besides, it is comprised of optics that are critical for beam delivery and beam modification, and a CNC-controlled machining center. There are several different lasers that are used in the machining processes, and they are characterized by their wavelength and the nature of beam output (pulsed or continuous). In this section, the basic concepts of lasers, the different lasers used in machining, and laser optics are described.

7.2.1 LASER BEAM GENERATION

7.2.1.1 Stimulated Emission

It is known that an atom can absorb radiation (photon), leading to an electron transition to an excited state, which is called absorption. Figure 7.1(a) schematically represents the absorption mechanism, where E_1 represents the lower energy level and E_2 represents the higher energy level or the excited state of the atom. The atom in its excited state is normally unstable, and subsequently, electron transition occurs between the excited state and the ground state, leading to photon emission (see Figure 7.1(b)). This is called spontaneous emission, and the emitted photon can be in any direction. However, in the presence of ambient radiation that is resonant with the bandgap of the excited atom, stimulated emission can occur. This leads to the emission of a photon that is identical in all aspects to the ambient photon, i.e., it has the same energy, direction, and phase as the ambient photon (see Figure 7.1(c)). Stimulated emission is the central idea of laser generation. The emitted photons are capable of further initiating stimulated emissions. Thus, photons generated through this process get amplified, producing an intense monochromatic light beam that is

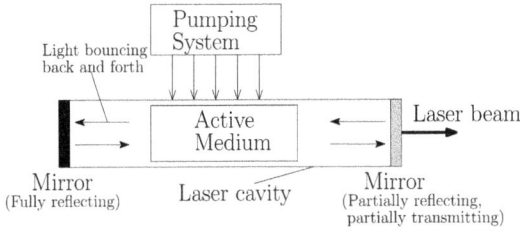

FIGURE 7.2 Schematic of laser cavity.

coherent and highly directional. Laser is an acronym for Light Amplification by Stimulated Emission of Radiation. In simple terms, the light emitted by the process of stimulated emission is amplified.

7.2.1.2 Laser Cavity

The material capable of sustaining stimulated emission and amplifying it is called gain medium. For stimulated emission to occur, the gain medium must be supplied with external energy to pump atoms from the ground state to an excited state, also called pumping. Pumping is essential in creating a population inversion, wherein the number of atoms in the excited state is more than that in the ground state. Pumping is typically done using electrical or optical energy. The most common pumping sources are a flash lamp or another laser source. Population inversion is necessary for the gain medium to amplify light produced by stimulated emission. Light amplification is achieved by providing optical feedback by placing a pair of mirrors on either side of the gain medium, as shown in Figure 7.2. The mirrors could be flat or curved. The setup is known as an optical cavity or a laser cavity. Light confined in the optical cavity bounces back and forth upon being reflected by the mirrors, producing standing waves. Every time the light passes through the gain medium, it gets amplified. One of the mirrors is fully reflective, while the other is partially transparent (output coupler), thus allowing some light to escape the optical cavity, generating a laser beam. The laser wavelength depends on the type of gain medium used, which could be a gas, liquid, or solid. Examples of gain medium include ruby crystal, CO_2 gas, helium-neon gas, Nd: YAG crystal, doped-fibers, etc.

7.2.1.3 Pulsing Techniques

The output from the laser cavity could be in pulsed or continuous mode. For example, a ruby crystal produces a pulsed output with a pulse duration of several milliseconds, whereas gain media such as CO_2 and Nd:YAG give a continuous-wave (CW) output. The power output from a CW laser is lower and limits its applications in processes such as machining where material vaporization is desirable. The output from CW lasers can be modulated to generate pulsed output using an external modulator that acts as a switch to allow light to transmit only for short periods. However, this method is very inefficient as most of the light is blocked by the modulator. Also, in such cases, the peak power of pulsed output will be the same as that of the power in

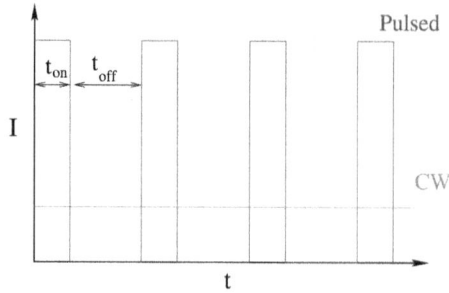

FIGURE 7.3   Schematic showing intensities of CW laser and pulsed laser.

CW mode. There are several techniques that use intra-cavity modulation to generate pulses with a specific repetition rate that have much higher peak powers, although the average power could be the same as that in CW mode (see Figure 7.3). These include cavity dumping, Q-switching, mode-locking, and pulsed pumping.

Cavity dumping involves the building up of intense energy in the laser cavity by reducing the losses and emitting the photons in short bursts, whereas Q-switching involves storing the energy in the gain medium by increasing the losses in the cavity using an optical switch. Since the losses in the cavity are high, optical feedback is inhibited, and no lasing action occurs. However, pumping of the gain medium occurs during this period to create a large population inversion. When a sufficiently high population inversion is achieved, the optical switch is used to minimize the losses in the cavity. This allows optical feedback in the cavity and all the excited atoms undergo stimulated emission, releasing a short and intense burst of photons. The quality factor (Q) is defined as the ratio of energy stored in a cavity to the energy lost per cycle. Since the Q value of the laser cavity is changed in this method, it is called Q-switching. Q-switching can be used to produce laser pulses of very high energy with pulse durations of a few nanoseconds and repetition rates between 1 Hz and 1 kHz.

Cavity dumping and Q-switching can produce laser pulses up to a few ns. However, in order to produce ultra-short pulses with pulse duration down to picoseconds (ps) or femtoseconds (fs), mode-locking is used. In general, the laser cavity can sustain a large number of longitudinal modes that travel back and forth to form standing waves. The principle of mode-locking is based on inducing a fixed phase relationship between the longitudinal modes sustained in a laser cavity to create a constructive interference that generates a train of extremely short pulses. The spectral bandwidth of the gain medium determines the pulse duration.

Therefore, different pulsing techniques can be chosen depending on the requirement of pulse duration, pulse repetition rate, and pulse energy.

7.2.2 Lasers in Machining

There are many different types of lasers available in the market depending on the type of lasing medium (active medium) used. These can be categorized as gas, liquid (dye), solid-state, fiber, and semiconductor lasers. For lasers to be applied in manufacturing, they must fulfill two criteria: (i) some fraction of the laser photons should be absorbed by the metal surface, and (ii) the laser must have sufficient power that can induce thermal effects on the workpiece material causing melting and vaporization. Although there are several types of lasers available, only a few lasers are suitable for applications in machining. Among the different lasers, only gas lasers, solid-state lasers, and fiber lasers have been used for machining applications. A description of the lasers used in machining is presented below.

7.2.2.1   Gas Lasers

In gas lasers, a glass or a ceramic tube is filled with a gas that acts as the active medium. Pumping is done by generating electric current by applying a voltage across the tube. The electrons collide with the gas atoms and excite them to higher levels. A variety of gases can be used to produce lasers. However, only excimer lasers and CO_2 lasers are used in machining applications. The CO_2 laser was one of the earliest gas lasers to be developed, a couple of years after the first laser was developed. It continues to be one of the widely used lasers in the industry mainly because of its high power conversion efficiency. The active medium in the CO_2 laser consists of CO_2, He, N_2, H_2, O_2, and Xe. CO_2 lasers typically emit laser radiation at a wavelength of 10.6 μm, but there can also be other emissions in the region of 9–11 μm. CO_2 lasers can be made to generate output both in CW form and in pulsed mode with a pulse duration of a few microseconds. Typical applications of CO_2 lasers in machining include cutting and marking. Excimer lasers typically use a combination of noble gas (Ar, Kr, Xe) and a reactive gas (F_2 or Cl_2) to create an excimer (or excited dimer) that is responsible for the lasing action. All the excimer lasers produce laser radiation in the UV regime with a pulsed output. Since the photon energies are higher, they are mostly preferred for machining polymers and semiconductors that have a wide bandgap. However, metals can also be machined using excimer lasers.

7.2.2.2   Solid-State Lasers

The active medium in solid-state lasers is optically transparent crystals doped with active ion species. The typically used dopant ions are transition-metal and lanthanide-metal (rare-earth) ions. The most commonly used ions are titanium (Ti), neodymium (Nd), ytterbium (Yb), and erbium (Er). These ions are held in a transparent crystal lattice and excited using an external source to cause lasing action. Crystals such as sapphire, yttrium aluminum garnet or YAG ($Y_3Al_5O_{12}$), yttrium vanadate (YVO_4), and glass are used as the host material. The crystals are used in the form of a cylindrical rod placed in the optical cavity between two mirrors. The dopant ions are dispersed in the crystal lattice with a concentration typically one part per hundred.

The ruby laser, which was the first laser developed, is a solid-state laser with synthetic sapphire as the active medium. Although it was initially used for machining applications, it is no longer used due to its low power efficiency. The most commonly used solid-state laser is the NdYAG laser, which contains Nd^{3+} ions doped in YAG crystal. NdYAG lasers emit light in the near-infrared region with a wavelength of 1064 nm. However, the laser output can be frequency doubled to produce a wavelength of 532 nm or higher harmonics with wavelengths of 355 nm, 266 nm, and 216 nm. NdYAG lasers are available in both continuous and pulsed form with a broad range of laser power. They are used for a variety of machining applications such as cutting, drilling, and engraving. NdYVO$_4$ and Nd-glass are the other lasers with Nd^{3+} as the dopant ion that is used in machining applications. A similar type of laser with Yb^{3+} as dopant ions such as YbYAG and Yb-glass are used in machining, both having a wavelength of 1030 nm. They generate laser output in both continuous-wave and pulsed form, with ultra-short pulses of ps or fs duration. These lasers generate very high power output and are used in micromachining applications. The other important laser in machining is the Ti-sapphire laser, containing Ti^{3+} ions doped in Al$_2$O$_3$. They are mainly known for their ability to produce ultra-short fs pulses with a tunable wavelength between 670 mm and 1100 nm. However, they can also be operated in CW mode. These lasers are commonly used in micromachining and texturing applications.

7.2.2.3   Fiber Lasers

In general, fiber lasers are also solid-state lasers. However, they are classified as different types because the technique involved is different from that of solid-state lasers. In fiber lasers, the gain medium is an optical fiber doped with Nd^{3+} ions or Yb^{3+} ions. Although the active medium in fiber lasers is similar to that of solid-state lasers, the properties of fiber lasers are entirely different due to the waveguide effects in the fiber. The active medium of the fiber laser can be coupled with another fiber for beam delivery. One of the key advantages of the fiber laser is its easier delivery to the target. The fibers can be bent and coiled, and the beam can be delivered over long distances. Besides, fiber lasers also have a high output power. There is a variety of fiber lasers with several different outputs, having a wide range of powers. Fiber lasers can produce both in CW mode and in pulsed form. Fiber lasers are also capable of producing ultra-short laser pulses with a pulse duration in femtoseconds. These unique advantages have made fiber lasers very attractive for the industry. Fiber lasers now have the highest market share among all the lasers in the industry. They are used for all kinds of machining applications.

7.2.3   Laser Optics

The laser beam generated from the cavity is modified before it is delivered to the workpiece for machining applications. Normally the beam diameter is reduced by using converging optics to generate high intensities, sufficient to cause melting and vaporization. Besides, the spatial variation of the beam intensity profile is also modified using certain optical elements. Laser optics includes all the elements that are used in modifying and transporting the beam from the source to the workpiece. In any laser beam machining application, the design of optics is crucial to the process performance. This section outlines the various elements used in laser optics.

7.2.3.1   Beam Delivery

The laser beam delivery basically means delivering the laser beam from the laser generator to the workpiece or point of application of laser. It can be done either by

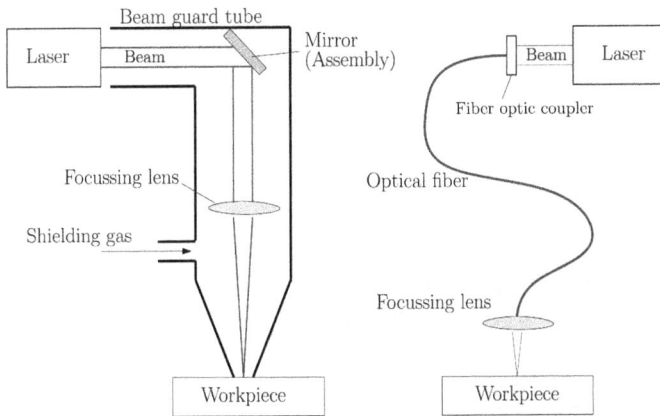

FIGURE 7.4 Schematic of (a) conventional and (b) fiber optic laser beam delivery systems.

using a conventional delivery system or a fiber optic system [3]. The delivery method depends on the laser wavelength and laser power.

Conventional beam delivery involves elements such as mirrors and lenses. As light travels in a straight line, mirrors are used to bend or change the direction of the laser beam by means of reflection. They are also used to precisely adjust the beam path. Lenses are used to focus or defocus the laser beam. The laser beam is often focused to very narrow spot sizes, which is achieved using a converging lens. A pair of lenses is also used to expand the beam, i.e., increase the beam's diameter and collimate the beam. In general, the lens assembly may consist of collimators, beam expanders, and focusing objectives. Besides, other optical elements may also be added to split the beam or change the profile of the beam. In order to have minimum losses, the optical elements must be chosen depending on the laser wavelength such that their absorption is minimum. Conventional beam delivery can be used for any type of laser. However, the mirrors and lenses have to be mounted with a high degree of precision. A schematic of conventional beam delivery is shown in Figure 7.4(a).

The laser beam can also be coupled with a fiber optic cable to transmit the beam through it to deliver it at the desired location. Fiber optic delivery offers great convenience as the flexible fiber cable can be routed through any orientation and delivered at distances far away from the laser source. Typically, the fiber can be between 5–10 m in length. However, fiber optics can only be used for specific wavelengths of the laser beam. It is mostly preferred when the wavelength of the laser beam is in the near-infrared region, where the absorption in the optical cable is minimum. Therefore, fiber optics cannot be used for CO_2 lasers and excimer lasers, as losses due to absorption are significant. Fortunately, most of the solid-state lasers and fiber lasers have wavelengths in the near-infrared region, and hence it is suitable for fiber optic delivery. Fiber optics cannot be used for high power lasers even in the near-infrared wavelength region as absorption in the optical fiber can occur at high intensities. A schematic of the fiber beam delivery is shown in Figure 7.4(b).

7.2.3.2 Beam Scanning

For machining applications, the laser beam has to be scanned over the workpiece. This can be achieved using:

    i. translation stages,
   ii. galvo-scanner, and
  iii. robotic arm.

The workpiece can be mounted on a translation stage (linear or rotary), and it can be moved using stepper and servo motors. This kind of setup is precise and can achieve adequate speeds. Galvo-scanners (or galvanometer scanners) can be used to achieve much higher scanning speeds than using CNC stages. A galvo-scanner consists of a set of rotating mirrors controlled using servomotors. When the laser beam falls on the rotating mirror, it gets steered along a straight path. Scanning along a plane is achieved using two mirrors that rotate simultaneously to achieve the desired scanning path. Galvo-scanners are typically combined with a focusing lens called an f-theta lens and are attached at the end of the beam delivery. They can be used for both conventional and fiber optics delivery. Galvo-scanners can be used to achieve incredibly high scanning speeds with good accuracy. However, they are limited by the scanning area, which is no more than 100×100 mm^2. A laser beam can also be scanned using a robotic arm. This can only be done for fiber optics delivery system. The delivery head consisting of focusing optics can be mounted on a servocontrolled, multi-axis mechanical arm. Although it can be used only for a certain type of laser, delivery using a robotic arm offers flexibility to scan the laser over large areas. Most importantly, it can also be used for non-planar workpieces. A sensor attached to the laser head can control the height and ensure that the beam is focused on the workpiece.

7.3 FUNDAMENTALS

7.3.1 LASER BEAM PARAMETERS

Laser is a complex tool and is characterized by several parameters. Apart from the material properties of the workpiece, laser parameters also play an important role in machining. One of the most fundamental characteristics of the laser is its wavelength. The wavelength of the photons generated from the laser cavity is solely dependent on the gain medium (or lasing source) used. Choosing an appropriate laser wavelength in machining is crucial, as it influences the optical absorption by the workpiece. Machining cannot occur if the workpiece is transparent or highly reflective to the laser radiation. Besides, other beam parameters such as power, intensity, and beam diameter are equally important. In the case of a pulsed laser, additional parameters such as pulse energy, fluence, pulse duration, and pulse repetition rate are used. This section presents the details of different laser parameters and their correlations.

7.3.1.1 Laser Beam Spatial Profile

The laser beam generated from the laser cavity has varying intensity along the plane perpendicular to the direction of propagation. This is described using transverse

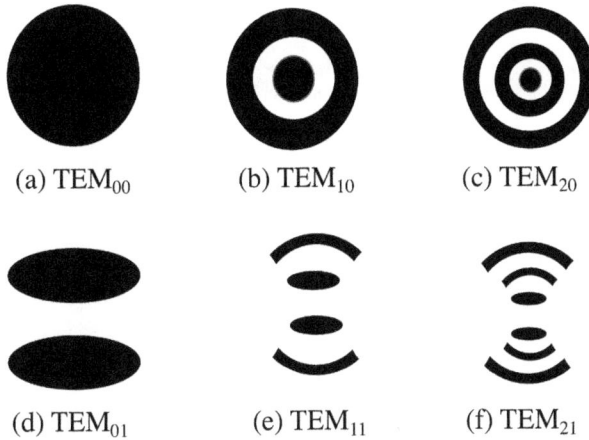

(a) TEM$_{00}$ (b) TEM$_{10}$ (c) TEM$_{20}$

(d) TEM$_{01}$ (e) TEM$_{11}$ (f) TEM$_{21}$

FIGURE 7.5   Patterns of different transverse electromagnetic modes for a cylindrical beam.

electromagnetic modes (TEM). It is characterized by two integers, m and n, that represent the number of transverse modes in two orthogonal directions in the plane of cross-section and is designated as TEM$_{m,n}$ [4]. The transverse modes are formed due to the interference of the propagating waves in the laser cavity. The number of transverse modes formed depends on the laser cavity's geometry, gain medium, and aperture at the output coupler. Figure 7.5 shows patterns of laser intensities along the beam cross-section for different transverse electromagnetic modes generated in a laser cavity. Here, the black color represents the regions of non-zero intensities, while the white color represents regions of zero intensity. For a cylindrical beam, m, n represents the number of modes along the radial and angular directions, respectively. TEM$_{00}$ is a zero-order mode, where "00" indicates that there are no modes. For the high-order mode lasers, there are dark bands (of intensity) in the profile. For example, TEM$_{21}$ has two modes along the radial direction and one mode along the angular direction.

The most common lasers used in machining are of TEM$_{00}$ type as they have high spectral purity and spatial coherence. The laser intensity distribution for TEM$_{00}$ is mathematically described using a Gaussian profile (see Figure 7.6).

For a beam with circular cross-section, the variation in laser intensity (I) along the radial direction (r) can be written as:

$$I(r) = I_0 e^{\frac{-2r^2}{w^2}}, \qquad (7.1)$$

where I_0 is the intensity along the beam axis ($r = 0$). Since the Gaussian profile is defined over an infinite range, it leaves a question about the beam diameter. There are several different ways to define the beam diameter for a Gaussian beam. One of the most widely used is the $1/e^2$ value. For the above equation (Equation 7.1), the beam diameter is equal to $2w$, where the intensity becomes $1/e^2$ of the peak value.

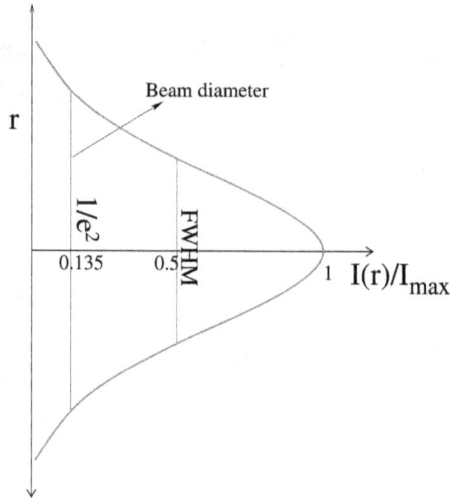

FIGURE 7.6 Gaussian beam intensity profile.

The power of the laser can be obtained by integrating the intensity over the entire area:

$$P = \int_0^\infty I \cdot 2\pi r dr$$

$$= \int_0^\infty I_0 e^{\frac{-2r^2}{w^2}} \cdot 2\pi r dr \tag{7.2}$$

$$= I_0 \frac{\pi w^2}{2} \left[e^{\frac{-2r^2}{w^2}} \right]_0^\infty .$$

This gives,

$$I_0 = \frac{2P}{\pi w^2}. \tag{7.3}$$

The above equation can be used to calculate the peak intensity for a given laser power, which can be used to get the intensity distribution for the TEM_{00} mode. In general, the power of the Gaussian beam as a function of radius r can be expressed as:

$$P(r) = \int_0^r I \cdot 2\pi r dr. \tag{7.4}$$

This can be simplified as:

$$P(r) = P\left[1 - e^{\frac{-2r^2}{w^2}}\right]. \tag{7.5}$$

7.3.1.2 Laser Pulse Profile

The laser pulse generated using various techniques described in Section 7.2 has a temporal nature. Similar to the spatial variation, the temporal variation of the pulse is also described using a Gaussian profile. Unlike the $1/e^2$ profile used for defining the beam diameter, the pulse width (t_p) is defined using the full width at half-maximum value (FWHM) [5]. Figure 7.7 shows the intensity variation of the Gaussian profile with time. The pulse width is equal to the width of the profile at half the peak intensity. The intensity variation with time can be written as:

$$I = I_{max}e^{-\beta\left(\frac{t}{t_p} - \frac{3}{2}\right)^2} \tag{7.6}$$

For a FWHM Gaussian pulse,

$$I(t_p) = I(2t_p) = 0.5I_{max} \tag{7.7}$$

Substituting in Eqn. 7.6 gives,

$$\beta = 4\ln 2 \tag{7.8}$$

Since the power varies with time for a pulsed laser, a laser pulse is also characterized using pulse energy (E) and laser fluence (F). Laser fluence represents the energy per unit area. The laser fluence can be written in terms of the intensity as:

$$F = \int_{-\infty}^{\infty} Idt = \int_{-\infty}^{\infty} I_{max}e^{-\beta\left(\frac{t}{t_p} - \frac{3}{2}\right)^2} dt. \tag{7.9}$$

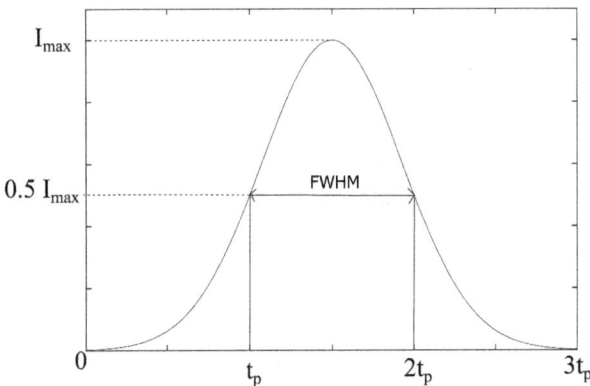

FIGURE 7.7 Temporal variation of laser intensity.

The above equation (Equation 7.9) can be evaluated using the Gaussian integral to obtain:

$$F = \sqrt{\frac{\pi}{\beta}} I_{max} t_p.$$ (7.10)

Therefore, for a given pulse energy or laser fluence, the peak intensity of the laser pulse can be obtained using the above equation (Equation 7.10), which can further be used to obtain the temporal variation of the laser intensity.

7.3.1.3 Beam Propagation

The nature of beam propagation is crucial in understanding the spot diameter when the laser is focused on a workpiece. The laser beam in TEM_{00} mode emerges from the laser cavity as a plane wave with the Gaussian intensity profile as shown in Figure 7.8. As the beam propagates, diffraction causes the beam to diverge laterally.

The spreading of the laser beam can be described using the diffraction theory. The following equations describe the variation of beam radius w and the radius of curvature of the wave front (r):

$$w(z) = w_0 \left[1 + \left(\frac{\lambda z}{\pi w_0^2} \right)^2 \right]^{1/2},$$ (7.11)

$$r(z) = z \left[1 + \left(\frac{\pi w_0^2}{\lambda z} \right)^2 \right],$$ (7.12)

where w_0 is the initial beam radius, and z is the distance propagated by the laser beam from the exit of the laser cavity.

Initially, the laser beam is a plane wave. Due to the divergence, it takes a curvature but eventually becomes a plane wave at infinity. Therefore

$$r(0) = r(\infty) = \infty.$$ (7.13)

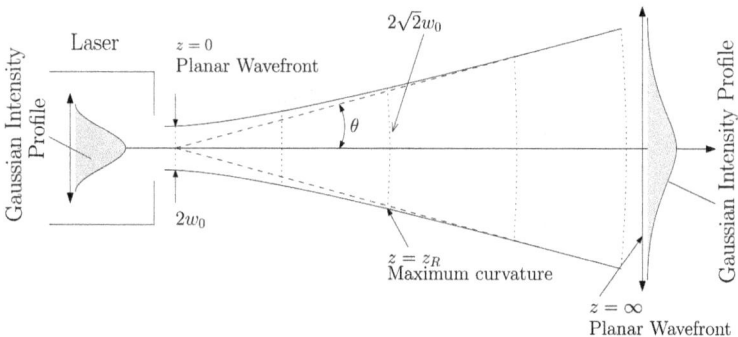

FIGURE 7.8   Schematic showing intensities of CW laser and pulsed laser.

This also implies that the curvature takes a minimum value at some arbitrary position, called the Rayleigh length, z_R. Therefore, at $z=z_R$, the differential of the radius of curvature becomes zero.

$$\frac{dr}{dz} = 0, \quad \text{for} \quad 0<x<\infty. \tag{7.14}$$

This gives,

$$z_R = \frac{\pi w_0^2}{\lambda}. \tag{7.15}$$

The beam divergence is defined using the far-field divergence angle, θ, shown in Figure 7.8. It is defined as:

$$\theta = \tan^{-1}\left(\frac{w(z)}{z}\right) \approx \left(\frac{w(z)}{z}\right). \tag{7.16}$$

The above approximation is valid since θ is very small. For large values of z, the beam radius (Equation 7.11) can be approximated as

$$w(z) \approx \frac{\lambda z}{\pi w_0}. \tag{7.17}$$

The far-field divergence angle can be obtained as:

$$\theta = \frac{w(z)}{z} = \frac{\lambda}{\pi w_0}. \tag{7.18}$$

7.3.1.4 Beam Focusing

In many machining applications, the laser beam is focused using a convex lens to reduce the beam diameter, which can result in very high fluences and intensities sufficient to cause material removal. Figure 7.9 shows a ray diagram of a laser beam

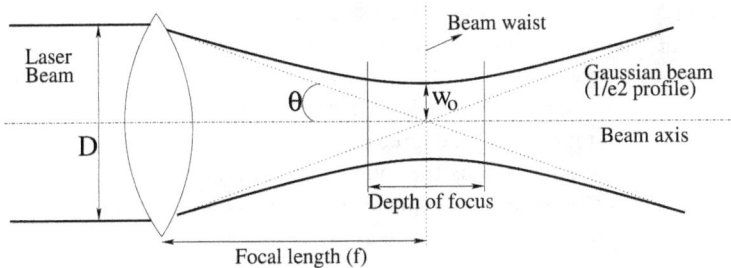

FIGURE 7.9   Schematic of beam focusing using a converging lens.

(TEM$_{00}$ mode) when focused using converging optics. The diameter of the laser beam is minimum at the focal point, after which it diverges.

An expression for the beam diameter at the focal point can be derived for the case of $f \gg D$, where f and D are the focal length and the initial beam diameter, respectively. For this case, the converging angle is nearly the same as the far-field divergence angle. This gives,

$$\theta = \frac{D/2}{f} = \frac{\lambda}{\pi w_0} \quad \left(\text{for } f/D \gg 1\right). \tag{7.19}$$

Therefore, the spot diameter $2w_0$ at the focal point can be obtained as:

$$2w_0 = \frac{4f\lambda}{\pi D}. \tag{7.20}$$

Apart from the spot diameter, depth of focus (z_f) is another important parameter for machining applications. Depth of focus is defined as the beam length with $\pm 5\%$ variation in beam diameter at the focal point. This can be obtained by substituting $w(z) = 1.05w_0$ in Equation 7.11 as

$$1.05w_0 = w_0\left[1+\left(\frac{z}{z_R}\right)^2\right]^{1/2}. \tag{7.21}$$

This gives,

$$z_f = 0.6402 z_R. \tag{7.22}$$

Depth of focus is crucial in applications such as drilling and cutting. Since the beam diverges beyond the focal point, the intensity reduces. In order to have a uniform hole diameter, it is recommended that the plate thickness be less than the depth of focus. Otherwise, the workpiece should be moved appropriately to ensure that the focal point falls on the instantaneous top surface.

7.3.1.5 Beam Quality (M^2)

Laser beams can be approximated as having a Gaussian intensity profile corresponding to the TEM$_{00}$ mode. This is a very close approximation for a single-mode laser. In general, the real beam produced from the laser cavity is not truly Gaussian, and it can exhibit any type of intensity profile. In order to characterize the quality of a real beam, the parameter, M^2 (called M-squared), is used. It is also called the beam quality factor, and it represents the degree of deviation of the real beam from an ideal Gaussian beam. It is expressed as the ratio of the product of beam waist diameter and far-field divergence of a real beam to that of a Gaussian beam with the same wavelength. Mathematically, it is written as [4]

$$M^2 = \frac{w_{0R}\theta_R}{w_0\theta}, \tag{7.23}$$

where w_{0R} and θ_R are the beam waist diameter and far-field divergence of the real beam. For an ideal Gaussian beam, the value of M^2 is equal to 1 and the product of beam waist diameter and beam divergence is given by,

$$w_0\theta = \lambda/\pi \quad (\text{See Eqn. 1.18}). \tag{7.24}$$

Therefore, the divergence of the real beam can be expressed as,

$$\theta_R = M^2 \frac{\lambda}{\pi w_{0R}}. \tag{7.25}$$

The above equation also indicates that the divergence of a real beam is the same as that of an ideal Gaussian beam having a wavelength of $M^2\lambda$ and the same waist diameter as that of the real beam. This allows modeling the real beam as a Gaussian beam using a similar set of equations described in Sections. 7.3.1.3 and 7.3.1.4 by replacing λ with $M^2\lambda$. The equations for beam radius and radius of curvature for a real beam can be written as:

$$w(z) = w_{0R}\left[1+\left(\frac{M^2\lambda z}{\pi w_{0R}^2}\right)^2\right]^{1/2}, \tag{7.26}$$

$$r(z) = z\left[1+\left(\frac{\pi w_{0R}^2}{M^2\lambda z}\right)^2\right]. \tag{7.27}$$

The Rayleigh length for the real beam can be expressed as,

$$z_R = \frac{\pi w_{0R}^2}{M^2\lambda}. \tag{7.28}$$

Similarly, the minimum spot diameter can be expressed as,

$$2w_{0R} = \frac{4fM^2\lambda}{\pi D}. \tag{7.29}$$

M^2 is a useful parameter in characterizing a real beam and theoretically evaluating the beam parameters such as spot diameter and divergence. For real beams, the value of M^2 is more than 1. For example, Nd:YAG lasers usually have an M^2 value of 1.2. For multimode lasers, the M^2 value could be very high.

7.3.2 LASER-METAL INTERACTION

7.3.2.1 Photon Absorption Mechanism

When the photons of the laser impinge on a metal surface, some of them get reflected while the rest are absorbed by the electrons in the conduction band. The fraction of radiation reflected by the material is represented by reflectivity (R), and the volume over which the photons are absorbed is represented by the absorption coefficient (α). The parameters R and α for a normal incidence of the laser can be obtained using the Fresnel equations as [2]:

$$R = \frac{(n-1)^2 + k^2}{(n+1)^2 + k^2}, \tag{7.30}$$

$$\alpha = \frac{4\pi k}{\lambda}, \tag{7.31}$$

where, n is the refractive index of the material and k is the extinction coefficient.

As the beam propagates, absorption of the laser photons by the material along the depth (z) is described using the Beer–Lambert law:

$$I(z) = I_0 \exp(-\alpha z), \tag{7.32}$$

where $I(z)$ represents the intensity variation inside the material along the beam propagation and I_0 is the laser intensity at the surface. The laser intensity at the surface of the workpiece can be written in terms of the laser beam intensity I_L as

$$I_0 = (1-R)I_L. \tag{7.33}$$

Laser intensity decreases along the depth, and the depth up to which the laser intensity falls to $1/e$ of the intensity at the top surface is referred to as the optical penetration depth or absorption depth (δ). The absorption depth is the inverse of the absorption coefficient ($\delta = 1/\alpha$). Typically, for most metals, the absorption depth is in the range of tens of nanometers, suggesting that a majority of the laser radiation is absorbed in a narrow region close to the top surface.

Upon absorption, the electrons get excited and increase in temperature. Consequently, the hot electrons interact among themselves and move freely in the metal lattice. In the process, the free electrons collide with the lattice phonons transferring energy to them. Further, the phonons interact among themselves until an equilibrium is established. Figure 7.10 schematically represents the microscopic interactions of photons, electrons, and phonons. The time required for the electrons and phonons to attain equilibrium is called "thermal relaxation time" (TRT), which

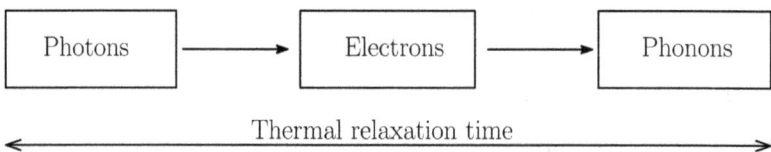

FIGURE 7.10   Schematic of microscopic interactions during the heat transfer process.

is of the order of several picoseconds for most metals [6]. It is the TRT that determines the nature of heating under the application of pulsed lasers.

7.3.2.2 Laser Heating

(i) *For CW and Short-Pulsed Lasers*

For CW lasers and pulsed lasers with pulse duration t_p much larger than the TRT, the microscopic interactions between photons–electrons–phonons can be ignored, and the heating can be assumed to occur instantaneously. Essentially, the laser heating process is considered to occur in a single step, i.e., radiation incident on the target leading to the rise in its temperature. The process is modeled using the Fourier heat conduction equation, with a source term to represent the radiation heating, given by [6],

$$C\frac{\partial T}{\partial t} = \nabla \cdot (K \nabla T) + S \tag{7.34}$$

where S is the laser heat source term, given by:

$$S = I_0 \exp(-\alpha z), \tag{7.35}$$

where α is the absorption coefficient and z represents the depth from the top surface. I_0 is the intensity of the laser at the surface (see Equation 7.1).

$$I_0 = (1 - R)I_L, \tag{7.36}$$

where, R is the reflectivity and I_L is the laser beam intensity.

Equation 7.34 is solved using the following initial and boundary conditions.

$$T(z, 0) = T_0, \tag{7.37}$$

$$\frac{\partial T}{\partial z}\bigg|_{z=0} = \frac{\partial T}{\partial z}\bigg|_{z=\infty} = 0. \tag{7.38}$$

The assumption here is that there is no heat loss at the top and bottom surfaces. While $z = 0$ indicates the top surface, $z = \infty$ indicates the bottom surface.

The above equation (Equation 7.34) is of the parabolic form, and hence, this model is also referred to as parabolic one-step (POS) model. Equation 7.34 is valid only under the condition that $t_p \gg$ TRT. Therefore, for all pulses up to nanosecond (ns) duration, the heating process can be modeled using Equation 7.34.

For CW lasers, the laser heat source is also assumed to be a surface heat flux. This is valid because the heat penetration depth is much larger than the laser absorption depth (δ). In such a case, S in Equation 7.34 becomes zero, and the top boundary condition is modified to account for the laser heating, as:

$$\left.\frac{\partial T}{\partial z}\right|_{z=0} = q'', \tag{7.39}$$

where, q'' is the surface heat flux due to laser radiation, and can be expressed in terms of the laser power (P) and the laser spot diameters (A) as

$$q'' = (1-R)\frac{P}{A}. \tag{7.40}$$

(ii) *For Ultra-Short-Pulsed Lasers*

For ultra-short pulses with pulse duration comparable to or less than TRT, Equation 7.34 becomes invalid. This is typically the case for ps and fs lasers. Under such cases, the electron temperature T_e and lattice temperature T_l are not the same during the pulse on-time. Therefore, the model has to consider the electron heating and phonon heating separately. Instead of the one-step model, a two-step model is used that considers the process occurring in two sub-stages [6]:

1. Electrons absorbing the photons, leading to an increase in their energy, and
2. The hot electron collisions with the lattice phonons, leading to heat exchange between the two.

Based on this, the heating process is modeled by characterizing the electrons and lattice subsystems separately by an electron temperature T_e and a lattice temperature (or phonon temperature) T_l, given by [6],

$$C_e \frac{\partial T_e}{\partial t} = \nabla \cdot \left(K \nabla T_e\right) - G\left(T_e - T_l\right) + S \tag{7.41a}$$

$$C_l \frac{\partial T_l}{\partial t} = G\left(T_e - T_l\right) \tag{7.41b}$$

The above governing equations are of a parabolic type and involve the two stages as described earlier. Hence, the model is referred to as the parabolic two-step (PTS) model. The model takes into account the electron-phonon thermalization time using electron-phonon coupling term $G(T_e - T_l)$, while it considers the electron thermalization and phonon thermalization to occur instantaneously.

The initial temperature conditions are given by:

$$T_e(x,0) = T_l(x,0) = T_0. \tag{7.42}$$

The electron temperature and the lattice temperature are equal to the initial temperature of the material, T_0.

The boundary conditions are given by:

$$\left.\frac{\partial T_e}{\partial x}\right|_{x=0} = \left.\frac{\partial T_e}{\partial x}\right|_{x=\infty} = \left.\frac{\partial T_l}{\partial x}\right|_{x=0} = \left.\frac{\partial T_l}{\partial x}\right|_{x=\infty} = 0. \tag{7.43}$$

7.3.3 MATERIAL REMOVAL MECHANISMS

Material removal in laser ablation may occur due to several different mechanisms depending on the pulse duration and laser fluence. Under the application of CW lasers or long-pulsed lasers (millisecond or μs pulses), the slow heating causes material removal to occur predominantly by melting [7]. However, for short and ultra-short pulses, vaporization is one of the dominant mechanisms by which material removal occurs. Besides, melt expulsion and phase explosion are also observed to be significant. The following sections discuss the various possible mechanisms by which material removal can occur in the laser machining of metals.

7.3.3.1 Vaporization

Vaporization refers to the passage from condensed phase (solid or liquid) to vapor by virtue of the emission of particles (atoms or molecules) from the top surface of the target. The rate of vaporization can be calculated using the Hertz-Knudsen equation. According to this, the flux (Φ) of particles (atoms or molecules) leaving the surface due to vaporization can be written as [8],

$$\Phi = \mu p_s (2\pi m k_B T)^{-1/2} \tag{7.44}$$

where μ is the condensation (or vaporization) coefficient, m is the mass of the particle (atom or molecule) and p_s is the saturated vapor pressure corresponding to temperature T. The above equation when multiplied by m/ρ gives the surface recession rate (u_s) in a 1D situation [8]:

$$u_s = \mu p_s \frac{m}{\rho} (2\pi m k_B T)^{-1/2}. \tag{7.45}$$

The vapor pressure (p_s) at a given temperature (T) can be evaluated using the Clausius–Clapeyron equation [2]:

$$p_s = p_0 \exp\left[\frac{\Delta H_v}{k_B}\left(\frac{1}{T_b} - \frac{1}{T}\right)\right], \tag{7.46}$$

where ΔH_v is the latent heat of vaporization (per atom), p_0 is the atmospheric pressure, and T_b is the normal boiling temperature.

According to the above equation (Equation 7.45), vaporization is mainly dependent on surface temperature and saturated vapor pressure. Since the vapor pressure is non-zero at all temperatures, vaporization occurs at all temperatures. Although vaporization is low at lower temperatures, it becomes very high at temperatures beyond the normal boiling point. Therefore, material removal due to vaporization should occur at all fluences in the nanosecond pulse regime as there is no threshold fluence. It has also been observed that vaporization accounts for a significant material loss in laser ablation with nanosecond pulses [9].

7.3.3.2 Normal Boiling

Normal boiling involves the removal of material due to heterogeneous nucleation in the form of vapor bubbles from the bulk liquid. The term normal boiling is used to distinguish it from explosive boiling, which is discussed later in the chapter. While vaporization is a surface phenomenon, normal boiling is a volume phenomenon. This phenomenon typically occurs when the surface reaches boiling point temperature, i.e., when the vapor pressure equals the external pressure. As a consequence, bubble nucleation may occur in the bulk of the liquid. The bubbles initiate heterogeneously from a variety of disturbances such as gas or solid impurities, or defects, or an underlying solid surface. Once the bubbles are formed, they move toward the surface and escape, enhancing vaporization. It has been deduced by Miotello and Kelly [10] that the formation and diffusion of a bubble to the outer surface take at least 100 ns, and the size of the critical (or stable) bubble is also very large in the range of 50–100 nm. For short and ultra-short pulses (ns, ps, and fs), as the rate at which a laser pulse deposits heat is much faster than the rate of formation and diffusion of bubbles, the surface will get superheated, and the temperature goes beyond the normal boiling temperature. Under such conditions, the velocity at which the surface recedes will be high enough that the nucleation cannot occur. Hence, nor mal boiling cannot be the mechanism of material removal in laser ablation for short and ultra-short pulses (ns, ps, and fs). Therefore, normal boiling can occur only for CW lasers or lasers with longer pulses (milliseconds).

7.3.3.3 Phase Explosion

Under high laser fluences, material removal also occurs by the ejection of microscopic droplets. Experimental evidence [11] has proved the existence of the explosive nature of material removal alongside vaporization that occurs only after a certain threshold fluence depending on the target material and pulse duration. The size of the particles varies from a few nanometers to a few micrometers. Normally, condensation in the expanding plume is used to explain the observation of small nanometer-size particles, typically consisting of thousands of atoms or molecules. However, the formation of large micron-sized particles cannot be explained based on condensation theory, as it would require an unrealistically large number of collisions to occur in the expanding plume [9]. Therefore, it was concluded that the larger particles of micrometer size are probably the result of direct ejection from the irradiated target.

This phenomenon of ejection of microscopic droplets has been explained to occur due to phase explosion, also called explosive boiling. At high laser fluences using nanosecond pulses, the rapid heating of the target occurs, and the target temperature goes well beyond the boiling temperature, and a portion of the target exists in a superheated state (metastable liquid). As the temperature of the superheated liquid reaches close to the thermodynamic liquid-vapor critical point, anomalies tend to occur, and the density rapidly falls. This typically occurs at around ~0.9 T_c. The fluctuations in the superheated matter give rise to the formation of vapor bubbles (or embryos) through out the superheated liquid. With an increase in temperature, the number of stable vapor bubbles also increases. The nucleation of these vapor

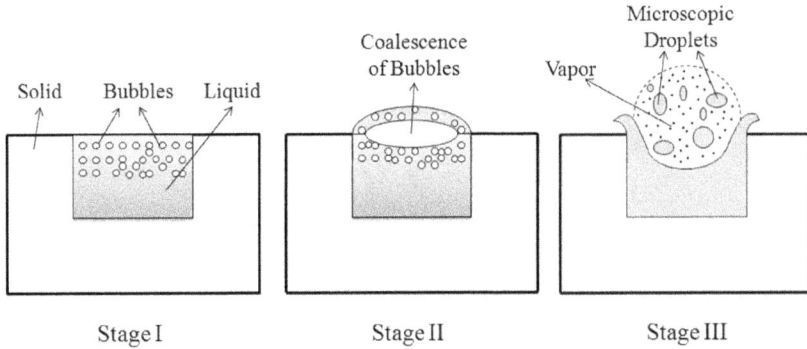

FIGURE 7.11   Schematic of phase explosion due to superheating.

bubbles prevents the liquid from approaching the critical temperature (T_c), and the highly superheated liquid breaks down into vapor and liquid droplets as shown in Figure 7.11.

As the vapor bubbles that are formed within the bulk of the liquid combine among themselves, this type of nucleation is called homogeneous nucleation. In the case of normal boiling, the nucleation arises with impurities or defects present in the liquid. The other difference is that normal boiling occurs at a temperature just above the boiling temperature, whereas phase explosion occurs when the temperature reaches the liquid-vapor critical point. Both normal boiling and phase explosion are volume processes, unlike vaporization, which is a surface phenomenon.

7.3.3.4   Melt Expulsion

Laser ablation usually involves intense vaporization of the target surface. Initially, the vaporizing atoms/molecules have their velocity vectors pointing away from the surface. Due to the collisions among the vapor particles, the velocity distribution in the vicinity of the vaporizing surface approaches equilibrium. This layer is called the Knudsen layer, which is of the order of several mean free paths and is often treated as a discontinuity. Due to this, the evaporating vapor plume exerts a recoil pressure on the irradiated surface. As a result of the plasma recoil pressure, the liquid portion in the target is pushed downward, as shown in Figure 7.12. This causes displacement of the molten metal into regions surrounding the ablation zone. This is also referred to as melt displacement or liquid splashing.

An approximate estimate of the plasma recoil pressure is given as [12]:

$$p_r = 0.56 p_s \tag{7.47}$$

where p_r and p_s represent the recoil pressure and saturation pressure of the vapor at T_s at the surface.

The plasma recoil pressure exerts a downward force on the melt pool causing it to flow outward. As the molten material cools, it resolidifies leading to the crater

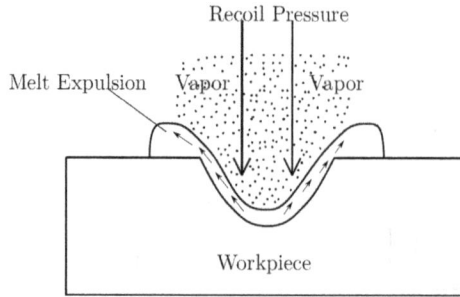

FIGURE 7.12 Schematic of melt expulsion due to plasma recoil pressure.

formation as shown in Figure 7.12. The liquid movement in the melt pool can be represented using the fluid flow equations [13]:

$$\frac{\partial \rho}{\partial t} + \nabla \cdot (\rho u) = 0, \tag{7.48}$$

$$\frac{\partial (\rho u)}{\partial t} + \nabla \cdot (\rho (u \times u)) = -\nabla p + \nabla \cdot (\tau) + \rho g, \tag{7.49}$$

where u is the flow velocity, p is the pressure, ρ is the density, τ is the stress tensor, and g is the acceleration due to gravity. The above equations must be solved with the heat transfer equation to obtain the final crater profile.

7.3.4 PLASMA SHIELDING

Vaporized material, as a result of laser-target interaction, forms a plasma above the target surface, as shown in Figure 7.13. The plasma consists of neutral atoms, ions, and free electrons. As the laser radiation propagates inside the plasma, a part of it is absorbed by the plasma constituents before it reaches the target surface. This causes a reduction in the laser intensity reaching the target surface. The phenomenon is also called plasma shielding. Consideration of the plasma shielding effect in the laser ablation modeling is important as it affects the laser-target interaction and material removal rate. The radiation absorption by plasma occurs by several mechanisms. A description of these mechanisms is presented in this section.

7.3.4.1 Inverse Bremsstrahlung

Inverse bremsstrahlung (IB) absorption involves the absorption of photons by free electrons. The two contributions to IB absorption are electron-neutral IB and electron-ion IB. The absorption of a photon by an electron moving through an electric field of a neutral atom is described by electron-neutral IB absorption, while that in an electric field of an ion is described by electron-ion IB. The absorption coefficient for the respective mechanisms is given by [14],

$$\alpha_{IB,e-n} = \sigma_p n_e n_0 \left[1 - \exp(-h\nu / k_b T) \right] \tag{7.50}$$

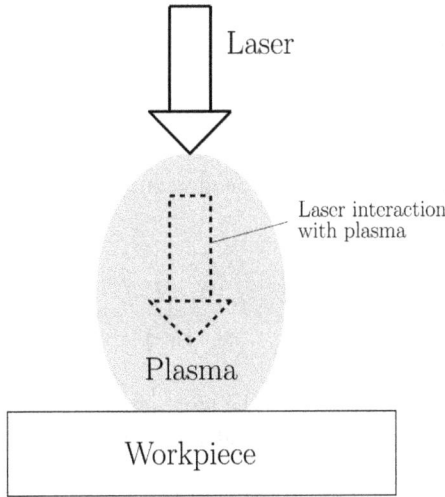

FIGURE 7.13 Schematic of plasma shielding during laser ablation.

and

$$\alpha_{IB,e-i} = \left(\frac{4e^6 n_e n_i}{3hcv^3 m_e} \right)^{1/2} \left[1 - \exp\left(-hv/k_b T \right) \right]$$ (7.51)

where v is the frequency of laser radiation, c is the velocity of light, k_B is the Boltzmann constant, h is Planck's constant, σ_p represents cross-section of photon absorption, and n_0, n_i, and n_e represent the number density of neutral atoms, ions and electrons respectively. The total absorption coefficient by IB can be written as,

$$\alpha_{IB} = \alpha_{IB,e-n} + \alpha_{IB,e-i}.$$ (7.52)

Based on the calculations of Fang et al., the probability of photons being absorbed by electron-neutral IB is very small, ~0.0034–0.02%. This is due to a very small value of σ_p. Therefore, the actual contribution to IB absorption is mainly through electron-ion IB. The IB absorption mechanism is dominant for laser radiation, corresponding to the IR radiation. At lower wavelengths, the IB absorption becomes important only at high plasma temperatures when the electron number density (n_e) is very high.

7.3.4.2 Photoionization

The absorption of a photon by a neutral atom is described by the photoionization (PI) mechanism. The photon is absorbed by a neutral atom that gets ionized. PI does not occur if the ionization energy is less than the energy of the photon. In such a case, ionization is possible only in the excited neutral atoms. For laser radiation corresponding to the visible and the UV range, the ionization of the excited atoms is mostly comparable to the photon energy. Therefore, PI is expected to occur, especially in the wavelength range corresponding to visible and UV radiation. A similar PI of ions can also occur, but it is negligible compared to that of excited neutral atoms. The radiation absorption coefficient by PI of excited neutral atoms is expressed as [14],

$$\alpha_{PI} = \sigma_{PI} n_0^*$$ (7.53)

where n^*0 is the number density of excited neutral atoms and σ_{PI} is the PI cross-section, expressed as:

$$\sigma_{PI} \approx 2.9 \times 10^{-17} \frac{\left(E_I - E^*\right)^{2.5}}{\left(hv\right)^3} \, \text{cm}^2$$ (7.54)

where E_I is the ionization potential of the atom and E^* is the energy of the excited state that can be photoionized. The typical values of PI absorption coefficient σ_{PI} is in the range of 10^{-21} m^{-2} [15].

7.3.4.3 Plasma Ionization

In laser ablation, the surface is heated to high temperatures. As a result, thermionic emission from a heated surface occurs that results in the ejection of electrons and ions. Therefore, the vaporized material consists of free electrons, neutral atoms, and ions.

In the above equations (Equations 7.50–7.53), the n_i and n_0 can be estimated by Langmuir–Saha equation, assuming the source of ionization to be thermionic, given by [16],

$$\frac{n_i}{n_0} = \frac{g_i}{g_0} \exp\left(-\frac{E_I - \phi}{k_B T}\right)$$ (7.55)

where g_i and g_0 are the degeneracy of states of ions and neutral atoms, respectively, and φ is the electronic work function.

Thermionic emission is the primary source of ionization. However, due to laser–plasma interaction, the plasma absorbs laser radiation and becomes further ionized. The electron and ion density can then be described by the Saha equation [1]:

$$\frac{n_i}{n_0} \approx 2.4 \times 10^{21} \frac{T^{3/2}}{n_i} \exp\left(\frac{-E_I}{k_B T}\right).$$ (7.56)

It is assumed that the plasma is in local thermal equilibrium, which means the temperature of free electrons, ions and neutral atoms are the same.

7.3.4.4 Intensity Drop Due to Shielding

The total absorption coefficient of plasma (α_{pl}) due to these various mechanisms can be expressed as:

$$\alpha_{pl} = \alpha_{IB} + \alpha_{PI}.$$ (7.57)

The drop in intensity due to shielding is estimated assuming that the attenuation of laser by plasma follows the Beer–Lambert law [5]. The laser intensity (I_0) at the surface of the workpiece after passing through the plasma can be expressed as:

$$I_0 = I_L \exp(-\alpha_{pl} H_{pl}),$$ (7.58)

where H_{pl} is the height of the plasma vapor above the target surface.

7.3.5 EMPIRICAL MODEL OF LASER ABLATION

In order to estimate the material removal during laser ablation, the heat transfer equation must be solved first to evaluate the temperature profile, followed by estimating the material removal by simultaneously accounting for the plasma effects. Obtaining a closed-form analytical solution for the above-described equations is quite challenging, and therefore the entire system of equations is solved using numerical techniques [1]. Thus, an empirical model of laser ablation is often used for its simplicity. This model is based on the experimental observations of the ablation depth vs fluence data. Experiments reveal that the ablation occurs only after a certain threshold fluence (F_{th}), below which no ablation occurs. For laser fluences more than the ablation threshold fluence, the ablation depth was found to show a logarithmic variation. Hence the ablation depth per pulse (h_{abl}) can be expressed as [17]:

$$h_{abl} = \frac{1}{\alpha'} \ln \frac{F}{F_{th}},$$ (7.59)

where α' refers to the effective absorption coefficient, which is found to be close to the absorption coefficient (α) obtained using Equation 7.31 only at lower laser fluences.

Similarly, the diameter of the hole or crater (D_{abl}) for a single pulse is expressed in terms of the focal spot radius (w_0) as [18]:

$$D_{abl}^2 = 2w_0^2 \ln \frac{F}{F_{th}}.$$ (7.60)

The above models are valid for a single pulse. However, when a train of pulses are used, the ablation threshold is found to decrease with the pulse number (N) due to accumulation of mechanical damage. This is also called the incubation effect. Accordingly, the threshold fluence for the Nth pulse ($F_{th}(N)$) is modified as [18]:

$$F_{th}(N) = F_{th}(1) N^{\zeta - 1},$$ (7.61)

where $F_{th}(1)$ is the threshold fluence for the first pulse or the single-shot ablation threshold, ζ is the incubation coefficient. For metals, the value of ζ is between 0 and 1.

These empirical models have been found to be suitable for all types of pulsed lasers. However, the fitting parameters such as α', F_{th}, and ζ are strongly dependent on the material and laser parameters.

7.4 APPLICATIONS

Laser beam machining has been used for a variety of machining applications. In this section, the most commonly used laser beam machining processes in the industry are discussed. These include cutting, drilling, texturing, cleaning, engraving, and micromachining.

7.4.1 LASER CUTTING

Cutting is one of the first applications of lasers in manufacturing. Laser cutting is a widely used process in the industry due to its simplicity and accuracy. It is generally used for cutting sheet metals. Laser cutting can be performed using both CW lasers and pulsed lasers. Laser cutting involves focusing the laser beam to narrow sizes on the workpiece to heat and remove the material through melting or vaporization. Cutting is achieved by moving the laser with a certain scanning speed along the workpiece by producing a slot, also called a kerf. It can be used for both straight and contour cutting. Laser cutting has several advantages. The process is not only fast but can also be used to cut intricate shapes, which are otherwise difficult to cut using conventional techniques. The kerf width in laser cutting is very narrow, and hence the material loss is minimal [3]. Based on the material removal mechanisms involved, laser cutting may be categorized into two different types – fusion cutting and sublimation cutting.

7.4.1.1 Fusion Cutting

Fusion cutting involves the use of CW lasers to melt and eject the material by using a high-pressure gas. Since CW lasers do not have high intensities, the material only gets melted upon laser irradiation. An assist gas at high pressure is then used to blow away the molten material resulting in a cut. Gas pressure is chosen depending on the thickness of the material to be cut. A schematic of the laser fusion cutting process is shown in Figure 7.14. The assist gas is applied using a nozzle that is coaxial with the laser beam.

The cutting process is largely influenced by the assist gas used, which may be active or inert. Oxygen, nitrogen, and air are some of the commonly used assist gases. If the assist gas used is inert, it is called laser fusion cutting, and when an active gas is used, it is called laser reactive fusion cutting. While the primary role of assist gas is to exert force on the molten material to eject it, reactive gases such as oxygen can induce oxidation reactions leading to energy release due to the exothermic reactions. This energy released due to the exothermic reactions improves the efficiency of the cutting process as it creates additional energy in the form of heat. If the temperature during the cutting process goes beyond the ignition temperature, the material may also get burned and vaporized. Oxygen is normally used as the assist

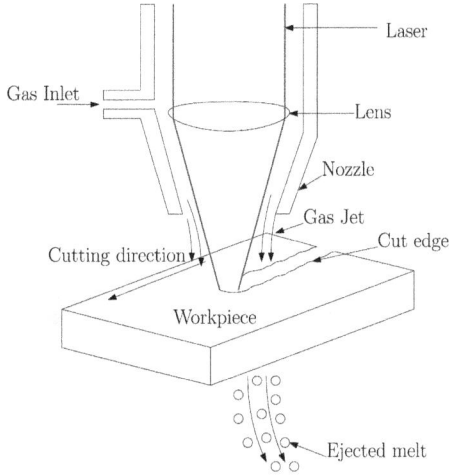

FIGURE 7.14   Schematic of fusion cutting process.

gas when cutting steel. Other metals such as aluminum are also cut with oxygen assist gas [3].

The fusion cutting process is extremely efficient compared to the other laser cutting processes as it consumes less energy per unit volume. The cutting speeds in fusion cutting are much higher than in other cutting processes. Besides, the ability of the laser to be tightly focused produces a relatively narrow kerf width, thereby reducing material wastage. However, the fusion cutting process invariably produces surfaces with striations and dross. Striations are the wavy patterns with peaks and valleys formed on the cut surface as a result of material ejection. Consequently, the surfaces produced by fusion cutting have a high roughness. Besides, during the ejection of molten material from the cutting region, some of the material clings onto the underside of the cut edge and solidifies to form a burr. This is called dross. Both striations and dross are inherent limitations of the fusion cutting process that invariably lead to the poor surface quality of the cut part. Another major disadvantage of the fusion cutting process is the generation of a HAZ. The heat generated in the cutting region propagates into the workpiece material due to conduction. Since only the molten material is removed during the cutting process, the cut surface invariably consists of a HAZ up to a certain depth, which depends on laser power, scanning speed, and the material's thermal diffusivity.

7.4.1.2 Sublimation Cutting

Cutting can also be achieved using pulsed lasers. Pulsed lasers have much higher intensities compared to CW lasers. Under pulsed laser irradiation, the material can be heated to very high temperatures beyond the normal boiling point, resulting in vaporization of the material. Laser cutting involving material removal predominantly by vaporization is called sublimation cutting (See Figure 7.15). The high-pressure vapor is easily removed from the cutting region. Although the process does

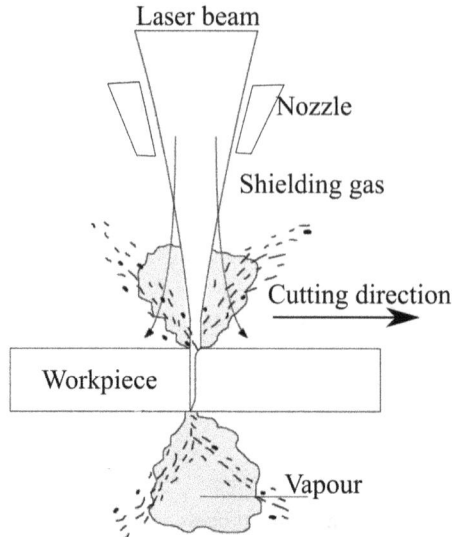

FIGURE 7.15   Schematic of sublimation cutting process.

not require an assist gas like in the case of fusion cutting, shielding gas such as nitrogen, helium, or argon are often used to protect the surface from oxidation. The assist gas also helps in protecting the focusing lens from getting damaged due to vapor deposition. Material removal in sublimation cutting occurs in small amounts as each pulse strikes the work surface. Hence, the cuts generated using sublimation cutting are precise and smoother in comparison to CW laser cutting. The kerf width is also much lower when pulsed lasers are used. Therefore, the surface quality obtained in sublimation cutting is much higher than in fusion cutting. Since material removal occurs by vaporization, sublimation cutting can be used only for thin components. However, the surface can have poor edges due to material deposition as a result of melt expulsion (discussed in Section 7.3.3.4.) by the high-pressure vapor generated during the process. Similar to that in the fusion cutting process, sublimation cutting also generates a HAZ on the cut surface. In comparison, a HAZ in sublimation cutting is much smaller than in fusion cutting due to the use of pulsed lasers, as the heating is localized. Hence, shorter pulses can generate a very thin HAZ [3].

74.2   LASER DRILLING

Laser drilling is usually performed using pulsed lasers with a pulse duration from several milliseconds to a few nanoseconds. Pulsed lasers with high intensities are focused on the workpiece material to cause sufficient heating to vaporize the material to generate holes. An assist gas jet that is coaxial with the laser beam is also used to assist material removal and to protect the focusing lens from ejected vapor or debris. Laser drilling has the unique capability of producing very high-aspect-ratio holes (~100:1) with hole diameters as small as a few micrometers, which is extremely difficult to produce using conventional drilling techniques. Besides, laser drilling can be used to produce holes with different shapes (non-circular) and sizes. Due to these advantages, it has wide application in many industries.

Material removal in laser drilling predominantly occurs by vaporization. However, at very high intensities, especially with short pulses, material removal can also occur by phase explosion (see Section 7.3.3), thereby ejecting material in the form of vapor and liquid droplets. While phase explosion can increase the material removal rate, it could also lead to the spattering of debris on the workpiece surface. Laser-produced holes also contain a recast layer and HAZ. Besides, the holes produced are tapered. This occurs mainly because the intensity of the Gaussian laser pulse decreases as the beam diverges along the depth of the workpiece. However, through an appropriate choice of laser parameters, it is possible to minimize the recast layer and taper and eliminate the spatter.

In general, there are four different types of laser drilling techniques. These include single-pulse drilling, percussion drilling, trepanning, and helical drilling. These are schematically represented in Figure 7.16.

7.4.2.1 Single-Pulse Drilling

As the name suggests, this technique involves the use of a single pulse to generate holes (see Figure 7.16(a)). The material removal, in this case, predominantly occurs by vaporization. This technique is used to produce through holes on thin plates or shallow blind holes on thick plates. A broad range of pulse duration from several milliseconds to a few nanoseconds has been used to drill holes on metals. In order to drill a hole using a single pulse, the pulse energy must be sufficiently high to cause melting and vaporization of the material. The pulse energies used are of several joules for millisecond pulses and a few millijoules for nanosecond pulses. The depth and diameter of the hole increase with an increase in pulse energy. The ability to drill a hole in a single pulse increases the production rate. On thin sheets of about 10–50 µm thickness, single-pulse laser drilling can be used to generate a few thousand holes per second having diameters of several hundred micrometers. However, single-pulse laser drilling has a poor tolerance owing to thick recast layer, spatter, and taper [19].

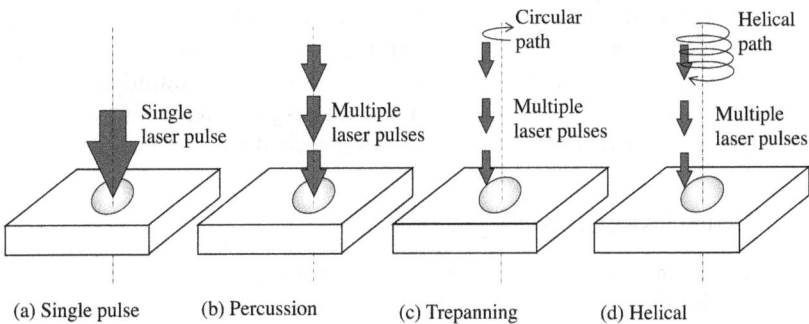

FIGURE 7.16 Schematic of different laser drilling processes.

7.4.2.2 Percussion Drilling

Percussion drilling involves the use of a series of pulses to drill a hole (see Figure 7.16(b)). As discussed earlier, lasers can produce pulses with very high pulse repetition rates up to several hundreds of kHz. The pulses are identical, and each pulse removes material in small quantities. Therefore, the pulse energies used in percussion drilling are much lower than those used in single-pulse drilling. Since a small amount of material is removed by each pulse, the effects of plasma shielding are very low. Thus, it is more efficient than single-pulse drilling, where plasma shielding is quite significant due to higher material removal per pulse. The spatter associated with this type of drilling is minimal. In addition, the hole taper is very small. This technique is particularly used to drill deep holes [19]. However, the use of a series of pulses reduces the production rate.

7.4.2.3 Trepanning

Trepanning is used to cut large diameter holes. The size of the holes is much larger than the laser beam's spot diameter. In this technique, a focused laser beam is used to cut along the perimeter of the hole through a relative motion between the laser beam and the workpiece (see Figure 7.16(c)). Material removal happens only along the perimeter of the hole. It is a form of contour cutting and can be used to produce holes of different shapes (circular and non-circular) and shapes with high repetition rates [19]. Trepanning is done using both CW lasers and pulsed lasers.

In the case of thick plates, first a hole is drilled using percussion drilling. Once a through hole is formed, the laser moves outward along a spiral path with increasing diameter until the required hole is created. While some of the material vaporizes from the top surface, the molten material gets ejected through the bottom of the hole that has been created.

7.4.2.4 Helical Drilling

Helical drilling is very similar to trepanning but involves the movement of the laser along the depth of the hole to be generated in addition to the movement along its perimeter. Overall, the laser moves along a helical path (see Figure 7.16(d)). The movement of the laser downward (along the depth of the hole) is done in order to focus the laser beam on the bottom surface of the hole. This is extremely important while drilling deep holes because the laser gets defocused as the hole depth increases, which leads to a decrease in laser intensity as the spot diameter increases. The molten material gets ejected from the top surface (in the upward direction) due to the recoil pressure created by the vapor and the gas jet. After cutting out the hole, rough edges are smoothened by moving the laser several times around the hole [19].

7.4.3 Ultra-Fast Laser Micromachining

Ultra-fast laser micromachining refers to micromachining performed using ultra-fast lasers. Ultra-fast lasers are the lasers with ultra-short pulse duration, in the range of picoseconds to femtoseconds. Ultra-fast lasers have emerged as potential tools for precision micromachining of different materials [20]. Owing to their extremely short

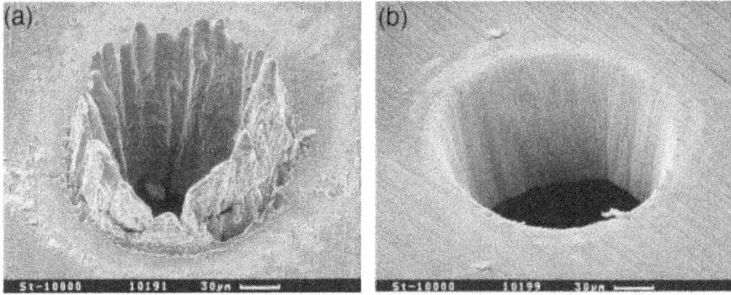

FIGURE 7.17 SEM images of holes drilled on a steel sheet using (a) a femtosecond laser and (a) a nanosecond laser. (Reprinted from [17], copyright (1997), with permission from Elsevier.)

pulse durations, ultra-fast lasers cause localized heating effects as the heat does not diffuse into the surrounding areas, which is quite significant for short (ns) and long pulses or CW. The diffusion of heat into the surrounding regions causes heating of the material in the regions beyond the focal spot, leading to unwanted material removal and HAZ. Besides this, melting ejection causes material redeposition at the cut edges, leading to poor machining characteristics. When ultra-fast lasers are used, the material vaporizes even before the heat is diffused into the surrounding regions. Also, since the melting effects are minimal, the redeposition of the material at the cut edges is almost negligible. Figure 7.17 shows the images of features generated using a nanosecond laser and a femtosecond laser. It is clearly evident that the machined surface generated using a femtosecond laser has superior surface quality than that produced using a nanosecond laser. Due to these advantages, ultra-fast laser micromachining is now used in many industries to produce small parts, such as coronary stents, microcantilevers, and microfluidic devices [21].

7.4.4 WATER JET GUIDED LASER MACHINING

The conventional focused beam has a limited working distance due to beam divergence as discussed in Section 7.3.1.4, defined by the depth of focus. This is typically about a few millimeters or less. As the beam diverges, the laser intensity decreases. Therefore, the cuts produced are tapered, which become prominent when the plate thickness increases. This poses a challenge for efficient cutting or drilling.

The technique of water jet guided laser machining offers an excellent solution for the machining of thick plates. This process uses a high-pressure water jet with a very small diameter to guide the laser beam, similar to an optical fiber cable through internal reflection as shown in Figure 7.18. This ensures that the beam spot diameter does not change as it propagates. The beam can be guided over sufficiently long distances up to 10 cm, thus allowing thicker plates to be machined with very low taper or parallel kerfs. Besides, the water jet acts as a coolant and eliminates thermal damage. The parts machined using this process are free of thermal defects and burrs. This makes it a potential technique for high-aspect-ratio drilling.

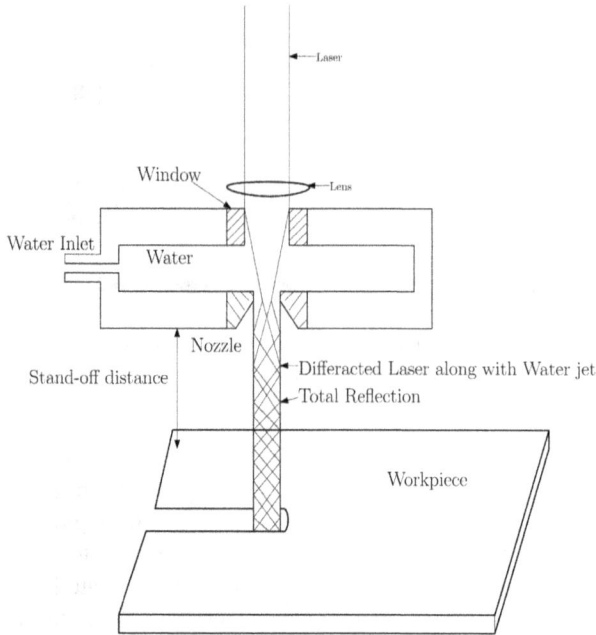

FIGURE 7.18   Schematic of water jet guided laser machining.

One of the key challenges in this process is the ability to generate a thin water jet and couple it with the laser beam. This is done by focusing the laser beam into a nozzle, which is placed beneath a chamber filled with water. Water is pumped into the nozzle at high pressure (50–600 bar) to generate the water jet. The nozzle diameters are typically between 20–150 μm [22]. The most commonly used lasers in the process are the Nd:YAG lasers at 1064 nm and 532 nm. Other lasers such as CO_2, fiber lasers, excimer lasers, and high power diode lasers have also been used, both in pulsed and CW mode.

7.4.5   LASER TEXTURING

Laser surface texturing (LST) involves surface modification by generating very fine periodic structures. The structures are typically in the range of several micrometers or nanometers, with shapes such as bumps, dimples, cones, grooves, etc. LST is usually used to impart functional characteristics on a surface for various industrial applications. The earliest commercial application of laser surface texturing was in honing IC engine cylinders. It was later used for the texturing of piston rings to reduce friction. The process is routinely applied to improve tribological performance on several mechanical components. LST can also be used to modify surface adhesion and stiction, which finds application in texturing surfaces of modern magnetic storage devices mainly to reduce adhesion and stiction. Fabrication of hydrophobic surfaces has also been achieved by modifying the surface wettability using LST.

More recently, LST has been used to enhance the optical properties of materials and in the fabrication of antibacterial surfaces for biomedical applications.

Compared to conventional techniques such as sandblasting and chemical etching, laser surface texturing is faster and more accurate. Besides, it can be used for texturing both 2D and 3D parts. The process is also easily controllable by varying the laser parameters, and it can be applied to a variety of materials.

Several methods can be used to achieve the generation of fine periodic structures required in LST, and these include (i) direct laser ablation and (ii) laser interference.

7.4.5.1 Direct Laser Ablation

This method generates periodic structures by selective ablation of the surface using a pulsed laser. When a pulsed laser irradiates a metal surface, material ablation can occur due to melting and vaporization, leading to the formation of a crater with a width nearly equal to the beam spot size and depth varying from several nanometers to micrometers. Several such craters are produced on the surface by scanning the laser beam. A schematic of laser scanning used in LST is shown in Figure 7.19(a). The pulses could be overlapping or non-overlapping. Depending on the application, various structures can be obtained by controlling the laser parameters and the spacing between the pulses represented by a and b in Figure 7.19(a). For example, an array of micro dimples can be generated by equispaced non-overlapping pulses (see Figure 7.19(b)) and microgrooves produced by overlapping pulses along the scanning direction (see Figure 7.19(c)).

7.4.5.2 Laser Interference

Periodic structures on surfaces can also be produced through the interference of two or more coherent laser beams. This method is also called direct laser interference patterning (DLIP). The multiple laser beams required for the interference are usually achieved by splitting the primary laser beam using beam splitters. Further, all the beams are superimposed on the work surface to create interference. The method of beam splitting and superposition is illustrated in Figure 7.20(a). Due to

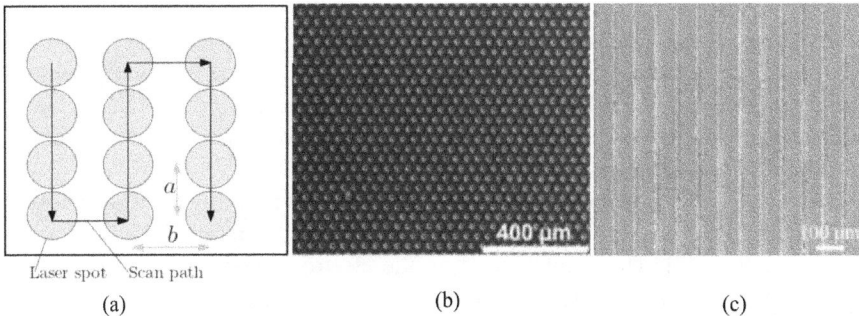

(a) (b) (c)

FIGURE 7.19 (a) Schematic of laser scanning, (b) laser textured micro dimples (Reprinted with permissions from [23], copyright (2014) American Chemical Society), (c) laser textured microgrooves. (Reprinted from [24], copyright (2016), with permission from Elsevier.)

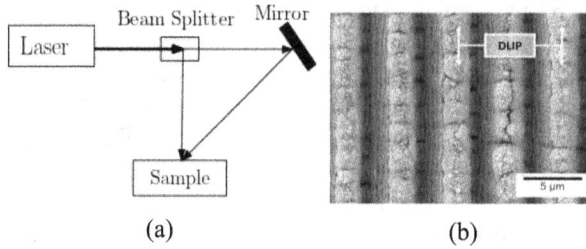

(a)   (b)

FIGURE 7.20 (a) Schematic of DILP, (b) sample of textured surface. (Reprinted from [25], copyright (2022), with permission from Elsevier.)

the interference, the laser intensity in the spot area varies spatially, leading to non-uniform heating. The material is locally melted in the regions of constructive interference with the highest laser intensity, while no melting happens in the regions of destructive interference. Subsequently, material movement occurs due to Marangoni convection from the region of high temperature to low temperature, thus generating surface structures. Periodic surface structures such as lines, dots, and cross-like patterns can be generated using this technique by varying the laser wavelength, laser-target distance, and the distance between the laser beams. Figure 7.20(b) shows a sample of surface texture obtained using this process. Compared to the direct laser ablation technique, this method has higher texture resolution and a smaller HAZ.

7.4.5.3 Laser-Induced Periodic Surface Structures (LIPSS)

When a linearly polarized laser beam is irradiated on a surface, it leads to the generation of periodic ripples. These ripples are also called laser-induced periodic surface structures (LIPSS). These periodic ripples (or structures) are produced within the focal spot and have dimensions in the range of several micrometers to nanometers. LIPSS are surface modification processes, and they can be generated on all types of materials, including metals, semiconductors, and insulators. The shape and size of the structures are largely influenced by the laser parameters such as wavelength, fluence, number of pulses, and pulse overlap. LIPSS generated using femtosecond lasers can produce structures with dimensions much smaller than the laser wavelength. Based on their spatial period (Λ), LIPSS is categorized into two groups: (i) low-spatial frequency LIPSS (LSFL) with ($\Lambda > \lambda/2$) and (ii) high-spatial frequency LIPSS (HFSL) with ($\Lambda > \lambda/2$) [26]. Figure 7.21 shows the two different types of LIPSS formation obtained at different laser parameters.

There are several theories that explain the formation of LIPSS. Broadly, they can be classified as: (i) based on electromagnetic effects and (ii) based on matter reorganization [26]. The theory based on the electromagnetic effects argues that the structures are formed due to the interference of the incident laser beam with the surface plasmons. This leads to alternative dark and bright fringes causing spatial modulation of the local energy distribution. Further the material absorbs this inhomogeneous energy, resulting in material removal only in the regions of high energy, thus forming periodic ripples or structures. The other group of theories based on the

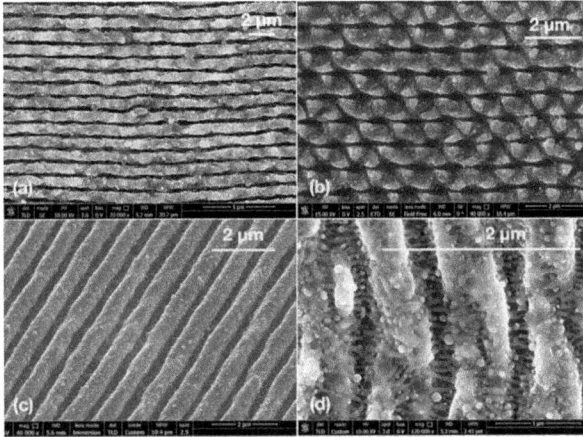

FIGURE 7.21 SEM images of LIPSS obtained used a 1030 nm femtosecond laser of pulse 230 fs (a) LSFL on silicon, (b) LSFL on AISI 316, (c) regular LIPSS on Molybdenum, (d) HFSL in the valleys of LSFL on AISI 316. (Reprinted from [27], copyright (2021), with permission from Elsevier.)

matter reorganization explains that the structures are generated due to the hydrodynamic effects of the melted surface and material instabilities, which leads to reshaping of the surface topography. It has been accepted that one or both of the phenomena could be responsible for the LIPSS formation depending on the material and the processing conditions. Electromagnetic effects could be dominant under the application of ultra-short pulses, and matter reorganization could dominate either for long pulses or under the application of multiple pulses.

7.4.6 Laser Cleaning

Lasers are also used to clean metal surfaces contaminated with rust, paints, grease, or dust [28]. The contaminants are removed by using a focused laser beam with sufficient intensity to cause ablation. The ablation mechanisms could vary depending on the type of contaminant. The key idea is to vaporize the contaminant layer rapidly, and therefore, pulsed lasers are suitable for this application as they have high intensities. The vaporized material is mostly removed using vacuum suction. The process can be carried out with different kinds of lasers depending on the nature of the contaminant. Ruby lasers that produce pulsed output were the first to be used for cleaning. The other lasers that are used in cleaning include NdYAG lasers, excimer lasers, and fiber lasers, all of which typically have pulse durations in the order of nanoseconds. Excimer lasers are found to be a better choice when the contaminants are paints or polymers, as the photon energy is sufficient to cause chemical degradation. Laser cleaning is now an industrial process with many applications, such as cleaning molds, rusted metallic surfaces, artworks, and artifacts. Figure 7.22 shows a rusted steel sample before and after cleaning. The top layer, consisting of rust, is

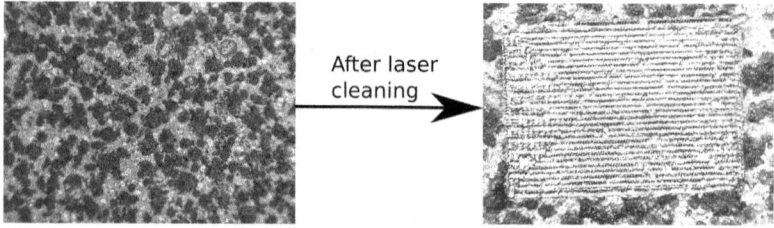

After laser cleaning →

FIGURE 7.22 A rusted iron sample before and after cleaning.

removed by ablation, leaving a cleaner surface. The damage to the surface due to temperature build-up is also very little.

Laser cleaning has several advantages compared to traditional cleaning methods, such as chemical purging or mechanical abrasion. Firstly, it does not require any chemicals or abrasive particles; thus, no secondary wastage is produced, making it environmentally friendly. Secondly, the process can be precisely controlled and localized, making it suitable for the cleaning of sensitive parts. The ability to localize the process to narrow areas also gives the advantage of cleaning small holes blocked with dust particles. Lastly, the process is very fast, roughly about 10–15 times faster than the traditional cleaning methods. Although the initial cost of the laser cleaning setup is high, the running costs are meager, and the life of the laser is very long, which makes it an affordable technology at the industrial level.

7.4.7 LASER MARKING AND ENGRAVING

Focused laser beams can be used to create marks on any metal surface. Laser marking broadly refers to the technique of generating marks on a surface through focused laser irradiation. Marks are produced either due to material ablation or coloration due to chemical change (such as oxide formation) upon laser irradiation. The ejected material is blown off or removed using vacuum suction. In general, marks could be just on the surface or can contain deeper grooves. Laser engraving is a term that is specifically used when the mark needs to penetrate to a certain depth (typically 200–300 µm). This is done by scanning the laser multiple times until the desired depth of the groove (or mark) is achieved by ablation. Marks produced by laser engraving have greater permanency and can withstand harsh conditions. They remain visible even after some coatings are applied. In comparison, marks produced by the coloration of a surface due to chemical changes may disappear upon surface damage (such as scratches) or after applying coatings.

Laser marking and engraving can be used to produce any kind of mark on all kinds of metals. The laser is attached to a galvo-scanner, and marking can be achieved by selectively scanning the laser beam in the desired regions. The process is extremely fast and can generate marks with precision. Unlike the conventional techniques, it does not use inks or tool bits that need to be replaced regularly. Laser marking is the most preferred technique in the industry to mark text, serial numbers, identification codes, bar codes, 2D data matrix codes, and logos. Figure 7.23(a) shows marking

FIGURE 7.23 Laser marking on steel samples: barcode, QR code and logo. (Reprinted from [29], copyright (2021), with permission from Elsevier.)

of a workpiece surface with numbers, and Figure 7.23(b) shows marking on a shaft with numbers and a data matrix code. Laser engraving is also used in the jewelry industry to generate a variety of features such as customization, personalization, and to enhance sheer beauty. Laser marking and engraving are also used to create fine art on metal surfaces. Software is used that can read digital drawings or images and produce the image on a metal surface by selective scanning of the laser beam. Although any type of laser can be used that can generate a mark, fiber lasers are now the most preferred choice as they have the ideal wavelength that can be absorbed by any metal, and they have very high pulse repetition rates due to which the beam can be scanned at very high speeds.

7.5 SUMMARY AND OUTLOOK

Laser beam machining is a well-established manufacturing process that has a plethora of applications ranging from micro to meso to macro scales. The ability to machine any metal with great precision and speed makes the laser a very important tool in the machine tool industry. Several different types of lasers exist, both in CW and in pulsed mode, each having its own unique applications. Lasers can be operated in a pulsed mode by using pulsing techniques that include cavity dumping, Q-switching, and mode-locking. The pulse duration can be varied from a few milliseconds to as short as a few femtoseconds. The use of short and ultra-short pulses helps in localizing the heating effects. Besides, the lasers can be focused to narrow spot sizes of a few micrometers using a converging lens. Thus, lasers can generate very high intensities, sufficient to melt and vaporize the material. Pulsed lasers have very high intensities, which can be used to rapidly heat the material even beyond its normal boiling temperature. Material removal in laser beam machining is very complex and is still a subject of intense research. Material removal may occur due to melt expulsion, vaporization, and phase explosion. Phase explosion occurs only when the pulse duration is very short, and the intensities are high, whereas CW lasers do not have sufficient intensity to vaporize the material. Hence, material removal with CW lasers invariably involves melt ejection. Machining processes such as cutting and drilling can easily be performed using both CW lasers and pulsed lasers. CW lasers are usually used along with a high-pressure assist gas to eject the molten material. Lasers are currently routinely used in industry in the cutting of sheet metals, drilling high-aspect-ratio holes for automobile, aerospace, and biomedical applications. Apart from cutting and drilling, lasers are also used for texturing, cleaning, marking, and engraving. Laser texturing can produce sub-micron-sized features for application in tribology and medicine. Lasers are used in industry to clean rust, paints, and grease on metal samples. Lasers have also made marking and engraving very easy to perform with high speed and great accuracy.

It has been six decades since the invention of lasers. The machine tool industry has slowly replaced conventional machining processes with laser-based processes. The last decade has seen a dramatic rise in the use of lasers in machining, driven by the development of fiber lasers and their ability to deliver a laser beam conveniently using fiber optics. However, the effective application of lasers in machining requires a great deal of understanding of its several process parameters. Laser interaction with matter is very complex and depends on the pulse duration and laser intensities. Material removal is usually followed by the formation of an intense plasma, which can further interact with the laser and makes the process phenomena even more complex. The use of physics-based models is seen as an approach to optimize the process mainly because of the small length scales and short time scales at which it happens. Laser interaction with matter is highly non-linear with picosecond and femtosecond pulses. This still continues to attract the attention of many researchers to understand the physics of the process. Due to the complex nature of the laser beam machining process, future research in laser beam machining could be driven by the use of data-driven models for process optimization. The breakthrough in laser machining could be in the use of machine-learning algorithms. This could pave the way for creating smart laser beam machining systems.

NOMENCLATURE

ABBREVIATIONS

ns	Nanoseconds
ps	Picoseconds
fs	Femtoseconds
CW	Continuous wave
TEM	Transverse electromagnetic mode
FWHM	Full width at half-maximum
TRT	Thermal relaxation time
IB	Inverse Bremsstrahlung
PI	Photoionization

SYMBOLS

A	Area (m^2)
C	Heat capacity ($Jm^{-3} K^{-1}$)
D	Beam diameter (m)
D_{abl}	Diameter of ablated hole or crater (m)

E	Energy (J)
E_I	Ionization potential (eV)
f	Focal length (m)
F	Fluence (J/m^2)
g	Acceleration due to gravity (m/s^2)
g_0	Degree of degeneracy of neutral atoms
g_i	Degree of degeneracy of ions
G	Electron-phonon coupling constant (W K^{-1} m^{-3})
h_{abl}	Ablation depth per pulse (m)
H_{pl}	Plasma height (m)
I	Intensity (W/m^2)
I_L	Laser beam intensity (W/m^2)
I_0	Laser intensity at the target surface (W/m^2)
I_m	Laser intensity inside the target (W/m^2)
I_{pl}	Radiation intensity emitted by plasma (W/m^2)
k	Extinction coefficient
K	Thermal conductivity (Wm^{-1} K^{-1})
M	Beam quality factor
m	Mass (kg)
m_e	Electron rest mass (kg)
n	Refractive index
n_e	Electron number density (m^{-3})
n_i	Ion number density (m^{-3})
n_a	Neutral atom number density (m^{-3})
P	Laser power (W)
p	Pressure (N m^{-2})
p_s	Saturated vapor pressure (N m^{-2})
p_r	Plasma recoil pressure (N m^{-2})
q''	heat flux (W/m^2)
Q	Quality factor
r	Radius (m)
R	Reflectivity
S	Source term (W/m^3)
t	Time (s)
t_{on}	Pulse on-time (s)
t_{off}	Pulse off-time (s)
t_p	Pulse duration (s)
T	Temperature (K)
T_b	Normal boiling temperature (K)
T_c	Critical point temperature (K)
T_e	Electron temperature (K)
T_l	Lattice temperature (K)
u	Velocity (m/s)
u_s	Surface recession rate (m/s)
w	Beam radius (m)

z_R	Rayleigh length (m)
Z	valency of the ion
α	Absorption coefficient (m^{-1})
α_{IB}	Absorption coefficient due to IB (m^{-1})
α_{PI}	Absorption coefficient due to PI (m^{-1})
α_{pl}	Absorption coefficient of plasma (m^{-1})
δ	Absorption depth (m)
θ	Far-field divergence angle (radians)
λ	Wavelength of laser light (m)
μ	Condensation coefficient
ν	Frequency of the laser light (s^{-1})
ζ	Incubation coefficient
ρ	Density (kg m^{-3})
σ_p	Spectral absorption cross-section (m^2)
τ	Viscous stress tensor (N/m^2)
φ	Electron work function (eV)
Φ	flux of particles ($m^{-2}\,s^{-1}$)

CONSTANTS

k_b	Boltzmann constant (1.38065×10^{-23} J/K)
h	Plank's constant (6.62606×10^{-34} Js)
e	Elementary charge (1.602×10^{-19} C)

REFERENCES

1. D. Marla, U.V. Bhandarkar, and S.S. Joshi. A model of laser ablation with temperature dependent material properties, vaporization, phase explosion and plasma shielding. *Applied Physics A* 116(1): 273–285, 2013.
2. D. Marla, U.V. Bhandarkar, and S.S. Joshi. Models for predicting temperature dependence of material properties of aluminum. *Journal of Physics D: Applied Physics* 47(10): 105306, 2014.
3. E. Kannatey-Asibu Jr . *Principles of Laser Materials Processing.* John Wiley & Sons Inc., New Jersey 2009.
4. W.M. Steen and J. Mazumder. *Laser Material Processing (4th edition).* Springer Verlag London Limited, 2010.
5. D. Marla, U. V. Bhandarkar, and S. S. Joshi. Transient analysis of laser ablation process with plasma shielding: One dimensional model using finite volume method. *Journal of Micro and Nano-Manufacturing* 1: 011007, 2013.
6. D. Marla, U. V. Bhandarkar, and S. S. Joshi. Critical assessment of the issues in the modeling of ablation and plasma expansion processes in the pulsed laser deposition of metals. *Journal of Applied Physics* 109: 021101, 2011.
7. D.J. Lim, H. Ki, and J. Mazumder. Mass removal modes in the laser ablation of silicon by a Q-switched diode-pumped solid-state laser (DPSSL). *Journal of Physics D: Applied Physics* 39(12): 2634, 2006.
8. R. Kelly and A. Miotello. Contribution of vaporization and boiling to thermal-spike sputtering by ions or laser pulses. *Physical Review E: Statistical Physics, Plasmas, Fluids, and Related Interdisciplinary Topics* 60(3): 2616–2625, 1999.

9. A. Bogaerts, Z. Chen, R. Gijbels, and A. Vertes. Laser ablation for analytical sampling: What can we learn from modeling? *Spectrochemica Acta Part B* 58: 1867–1893, 2003.

10. A. Miotello and R. Kelly. Laser-induced phase explosion: New physical problems when a condensed phase approaches the thermodynamic critical temperature. *Applied Physics A: Materials Science and Processing* 69: S67–S73, 1999.

11. K.H. Song and X. Xu. Explosive phase transformation in excimer laser ablation. *Applied Surface Science* 127–129: 111–116, 1998.

12. N.B. Dahotre and S.P. Harimkar. *Laser Fabrication and Machining of Materials.* Springer Science, New York, 2008.

13. V. Narayanan, R. Singh, and D. Marla. A computational model to predict surface roughness in laser surface processing of mild steel using nanosecond pulses. *Journal of Manufacturing Processes* 68: 1880–1889, 2021.

14. S. Amoruso. Modeling of UV pulsed-laser ablation of metallic targets. *Applied Physics A* 69: 323–332, 1999.

15. Z. Chen and A. Bogaerts. Laser ablation of Cu and plume expansion into 1 atm ambient gas. *Journal of Applied Physics* 97: 1–12, 2005.

16. R. K. Singh and J. Narayan. Pulsed laser evaporation technique for depositing thin films: Physics and theoretical model. *Physical Review B* 41: 8843–8859, 1990.

17. C. Momma, S. Nolte, B. N. Chichkov, F. Alvensleben, and A. Tünnermann. Precise laser ablation with ultra short pulses. *Applied Surface Science* 109–110: 15–19, 1997.

18. N. Lasemi, U. Pacher, L. V. Zhigilei, O. Bomat-Miguel, R. Lahoz, and W. Kautek. Pulsed laser ablation and incubation of nickel, iron and tungsten in liquids and air. *Applied Surface Science* 433: 772–779, 2018.

19. A.K. Nath. Laser drilling of metallic and nonmetallic substrates. In *Comprehensive Materials Processing*, edited by Saleem Hashmi, Gilmar Ferreira Batalha, Chester J. Van Tyne, and Bekir Yilbas, pp. 115–175. Oxford: Elsevier, 2014.

20. S. Lei, X. Zhao, X. Yu, A. Hu, S. Vukelic, M.B.G. Jun, H.E. Joe, Y.L. Yao, and Y.C. Shin. Ultrafast laser applications in manufacturing processes: A state-of-the-art review. *Journal of Manufacturing Science and Engineering* 142(3): 2020.

21. R.E. Samad, L.M. Machado, N.D. Vieira Jr, and W. de Rossi. Ultrashort Laser Pulses Machining. In *Laser Pulses: Theory, Technology, and Applications*, edited by Igor Peshko. IntechOpen, 2012.

22. V.M. Tabie, M.O. Koranteng, A. Yunus, and F. Kuuyine. Water-jet guided laser cutting technology: An overview. *Lasers in Manufacturing and Materials Processing* 6: 189–203, 2019.

23. C. Greiner, M. Schäfer, U. Popp, and P. Gumbsch. Contact splitting and the effect of dimple depth on static friction of textured surfaces. *ACS Applied Materials & Interfaces* 6(11): 7986–7990, 2014.

24. Z. Wang, Y.B. Li, F. Bai, C.W. Wang, and Q.Z. Zhao. Angle-dependent lubricated tribological properties of stainless steel by femtosecond laser surface texturing. *Optics & Laser Technology* 81: 60–66, 2016.

25. F. Schell, S. Alamri, A. Hariharan, A. Gebert, A.F. Lasagni, and T. Kunze. Fabrication of four-level hierarchical topographies through the combination of lips and direct laser interference pattering on near-beta titanium alloy. *Materials Letters* 306: 130920, 2022.

26. J. Bonse and S. Gräf. Maxwell meets marangoni—A review of theories on laser-induced periodic surface structures. *Laser Photonics Reviews* 14: 2000215, 2020.

27. L. Orazi, L. Romoli, M. Schmidt, and L. Li. Ultra fast laser manufacturing: From physics to industrial applications. *CIRP Annals* 70(2): 543–566, 2021.

28. V. Narayanan, R. Singh, and D. Marla. Laser cleaning for rust removalon mild steel: An experimental study on surface characteristics. *MATEC Web Conference* 221: 01007, 2018.

29. L. Lazov, E. Teirumnieks, T. Karadzhov, and N. Angelov. Influence of power density and frequency of the process of laser marking of steel products. *Infrared Physics & Technology* 116: 103783, 2021.

8 Fundamentals of Plasma Polishing

Hari Narayan Singh Yadav,
Manjesh Kumar, and Manas Das

CONTENTS

8.1 INTRODUCTION

In recent years, nanofinishing through non-conventional processes has become the key technology for the processing of optical materials [1]. Nanofinished products are widely used in space applications, and the electronics, aviation, biomedical devices, optics, and communication industries [2, 3]. In the past few decades, there has been continual growth in the awareness of the significance of the surface quality of components produced in manufacturing sectors, because improving the surface quality frequently leads to better performance and greater service lifespan [4, 5]. Enhancing the surface topography of the finished product is one of the excellent and effective ways to establish superior properties [6, 7]. Over the years, several researchers have made consistent efforts to achieve a nanolevel surface finish [8]. Engineers face

DOI: 10.1201/9780429160011-8

difficulty when it comes to the machining of quartz and glass materials with great dimensional accuracy [9]. In the past few decades, copper and silicon polishing in the nanometer range has been of key interest in the manufacturing of super-finished integrated circuit chips [10]. The innovative processing technique promises to increase the functionality and quality of products [11]. The performance and durability of the parts rely largely on the material removal mechanism during manufacturing. The surface or surface texture quality increases the machine's functional efficiency and its lubricating, wear, and bearing resilience. Surface roughness is a critical parameter for predicting a component's performance in the natural environment [12]. For different purposes, a diverse surface roughness range is required. Wear and friction occur more in the case of a rough surface rather than a smooth surface. Also, components with uneven surfaces suffer more from corrosion and component failure [13]. Surface roughness affects the operational efficiency of the component [14].

Generally, any three-dimensional complex geometric component without a rotational axis is considered a freeform surface [15]. Regular surfaces like planes, conics, cylinders, etc., are different from freeform surfaces. It does not have a rigid dimension. This type of surface consists of non-planar surfaces. In product design, not only functionality, but also aesthetics are considered. The use of freeform surfaces is increasing in many industrial applications. Many components of the freeform surface are utilized extensively in the aircraft, automotive, electronics, die and molding, and medical sectors, etc. Ultra-precision optical components find applications in microelectromechanical systems (MEMS), semiconductors for electronic equipment, telescopes, and a multitude of defense equipment [16, 17]. Optical parts are often produced using conventional mechanical techniques, for example, lapping, grinding, and buffing [18]. However, such mechanical processes produce plastic deformation and brittle ruptures on the surfaces resulting in deterioration of optical properties [19]. To overcome these problems, a plasma polishing process has been developed [20]. A recently developed finishing method called *plasma etching* has also been recognized to surpass the constraints of conventional polishing processes [21]. Surface roughness ($R_a < 1$ nm) and machining accuracy between 0.1 and 100 nm are achieved using the ultra-precision machining method [22]. The definition of "*ultra-precision*" is ever-evolving as new technological advances are made. There are numerous applications of freeform optics in a reflective, refractive, and diffractive optical system that demand finishing multiple surfaces on a single microstructure or inside surfaces that cannot be accessed by any tool. Nanofinishing of such freeform surfaces in optics and sensor microstructure realized from glass or glass-like materials is thought to be not achievable. However, newer and innovative processes are being developed for the nanofinishing of such components. Presently, researchers are concentrating on improving every aspect of industrial systems that can have an influence on the performance of a given material [23]. Figure 8.1 shows the transformation process of manufacturing systems.

In this chapter, the principle of the material removal mechanism and applications of plasma polishing methods for optical materials are discussed in detail. A comprehensive introduction of the evolution of plasma polishing processes is also

FIGURE 8.1 Transformation process in manufacturing [23].

discussed. This literature survey presents the basic background of plasma physics, characteristics, and the processing of optical components and it also reports valuable knowledge on the impact of various process parameters on the plasma polishing of glass or glass-ceramic material used in the optical industry.

8.2 BACKGROUND OF PLASMA PHYSICS

Plasma is the word used to describe a gathering of partially or fully ionized gases [24] consisting of charged particles: Electrons, ions, neutral atoms, and possibly molecules that exhibit effects such as conducting electrical currents and generating magnetic fields. The electrical conductivity of plasma is the leading property that differentiates it from neutral gases, which are an electric insulator. As the temperature rises, the molecule becomes highly energetic and changes nature in the following order: Solid, liquid, gas, and finally plasma, thereby justifying the term "*fourth state of matter*" [25]. Plasma is electrically conductive, interactive, and strictly sensitive to the electromagnetic field due to the presence of free electrons, electrical charge, and ions. Electrically neutral ionized gases are frequently called plasma (i.e., positive ions are balanced with electron density) and hold a substantial number of electrically charged particles, enough to influence their electrical characteristics and nature. Irving Langmuir, who initiated the technical analysis of ionized gases [26], provided this novel aspect of the matter with the name *plasma* in 1928 when studying oscillations in ionized gases. Langmuir discovered that, except at the electrodes, where sheaths store relatively fewer electrons, the ionized gases contain similar amounts of electrons and ions, resulting in very little leftover space. This leftover area carrying coordinated ion and electron charges is referred to as plasma.

Plasma naturally exists and can also be produced in industry or a laboratory, which offers openings for different uses, featuring electronics, thermonuclear chemistry, fluorescent lamps, and lasers, etc. To be more precise, plasma technologies are used in the manufacture of the majority of computer and mobile phone parts. Plasma has three main characteristics that make it appealing for use in chemistry and related fields: (1) The temperature of at least some plasma components and energy density can significantly exceed those in conventional chemical technology, (2) plasma have the ability to generate extremely large quantities of energetically and chemically active entities (e.g., ions, electron, atom, radicals, and various wavelength photons), and (3) plasma systems allow exceptionally large quantities of chemically active species while maintaining the bulk temperature as low as normal room temperature. These plasma properties allow substantial escalation of conventional chemical procedures and improvement in their performance. The plasma polishing process has made huge advances in the field of surface modification [27].

Since the early research of Irving Langmuir and his colleague Lewi Tonks, the field of plasma has increased in size and variety. Some other engineers and scientists that contributed immensely to the area of plasma physics include:

1. Edward Appleton and his colleague K. G. Budden – credited with developing the theory of electromagnetic wave propagation in non-uniformly magnetized plasmas.
2. Hannes Alfven – who generated the magnetohydrodynamics (MHD) theory, which treats plasma as a conducting fluid.
3. James Van Allen – credited with pioneering the investigation of the Earth's magnetosphere and developed the Van Allen radiation belts surrounding the Earth.

Laboratory and naturally occurring plasmas (produced in gas discharge) obtained over inclusive ranges of pressure, electron densities, and electron temperatures are shown in Figure 8.2. Industrially produced plasmas have temperatures ranging from just above room temperature comparable to the interior of the stars and electron densities span over 15 orders of magnitude.

In the gas phase, electrical conductivity is very low; however, generally, electrical conductivity is very high in the plasma phase. A comparison of the gas and plasma phases is presented in Table 8.1.

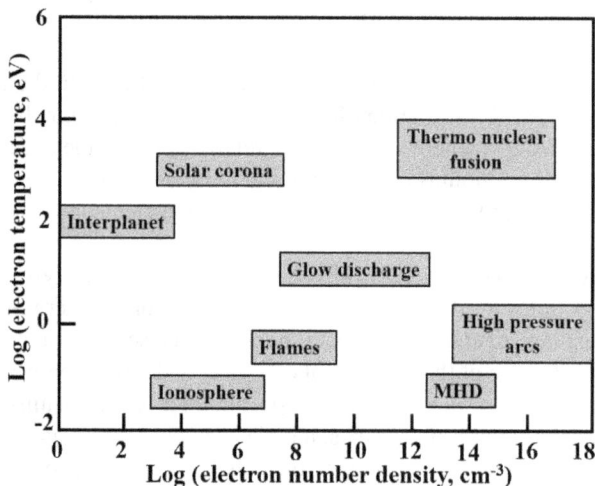

FIGURE 8.2 Plasma temperatures and densities [25].

TABLE 8.1
Comparison between Gas and Plasma Phase

Property	Gas	Plasma
Electrical conductivity	Very low	Generally, very high.
Independently acting species	One: All gas particles behave in a similar way.	Two or three: Protons ions, electrons, and neutrons behave independently in many circumstances.
Interactions	Binary: Two-particle collisions are the standard, and three-body collisions are exceptionally rare.	Collective: Organized motion of plasma is very crucial because particles can interact over long distances through magnetic and electric forces.
Velocity distribution	Binary: Normally, two-particle collisions are there, three-body collisions extremely rare.	Collective: Waves, or organized motion of plasma is very essential as particles can interact at long ranges through the electric and magnetic forces.

8.3 PLASMA POLISHING PROCESS

The plasma polishing process is a new "green" method that is capable of polishing optical components and produce a high-quality ultra-smooth surface [28]. In this process, the optimal ratio of process gas is combined and then introduced into the vacuum plasma chamber. Further, a radio frequency (RF) power source is used to ionize the gases. The reaction gases are stimulated in the plasma to produce higher energy and higher density reactive radicals. The radical produced then reacts with the component's surface atom, resulting in the removal of materials from the component. Plasma is basically the fourth state of matter; it is a fully ionized gas with an equal number of negative and positive particles. Ions consist of positive or negative charges which come from the removal of an electron from substances. Radicals are neutral particles that exist in a condition of partial chemical bonding, which is chemically highly reactive. It is formed when gas molecules are split by the high energy of electron collisions. A different combination of plasma and reactive gas is used. Like for silicon, He is used as plasma, and carbon tetrafluoride (CF_4) is used as reactive gas. This gives a radical F^* and SiF_4, both of which are volatile, and the machined surface is left with no contamination. The formation of oxygen radicals is shown in Equation 8.1 (Table 8.2).

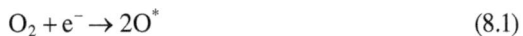

$$O_2 + e^- \rightarrow 2O^* \tag{8.1}$$

Low-pressure plasma is known for its high material removal rate (MRR). Anisotropic plasma machining has been used effectively for high surface finish. As the pressure of the plasma is increased, isotropic property is gained, and the energy level keeps coming down. Atmospheric pressure plasma is a low-energy isotropic plasma process. Inside surfaces or where the plasma torch cannot reach, atmospheric plasma

TABLE 8.2
Plasma Species Density

Species	Density
Electrons	$10 * 10^8/cm^3$
Radicals	$10 * 10^{14}/cm^3$
Positive ions	$10 * 10^8/cm^3$

TABLE 8.3
Mechanism Spectrum for Plasma Polishing

Pressure	Plasma Process	Energy
1 bar	Atmospheric pressure plasma (small aperture).	Very low energy with more surface integrity.
1–100 mbar	Bulk isotropic polishing due to cold chemical plasma.	Low energy cold plasma model assumes lower plasma temperature.
<1 mbar	Low-pressure plasma.	Surface damage is high due to higher energy.

fails to be useful. When the plasma pressure is at 1 bar, the plasma is regarded as atmospheric pressure plasma with a small aperture. The plasma material removal mechanism spectrum is shown in Table 8.3.

A comparison of different pressure plasma with its application is shown in Table 8.4. Table 8.4 demonstrates that different plasma pressures (low, medium, and atmospheric) have different applications and give more productivity with maximum use.

A comparison between the conventional and plasma-polishing processes in respect of industrial/research use is shown in Table 8.5.

Chemical interaction between the surface atoms and plasma is formed. This plasma-polishing method enables atom-by-atom material removal, as illustrated in Figure 8.3. Because this technique is purely chemical in nature, subsurface flaws that are common in traditional polishing methods are absent in the atmospheric pressure plasma polishing (APPP) process [29]. Uniform low-temperature plasma may be created across a larger area at a lower cost and for a wider variety of applications. Wet chemical processes generate more active and diverse species, which rapidly interact with the surfaces of the component. The reactive and processing gases with the best ratio are combined and sent into the plasma chamber. High energy reactive radicals are generated by ionization through RF excitation, which initiates a chemical interaction with the atoms on the surface. This process can achieve atomic-scale material removal [30].

The process parameter is crucial during the polishing of optical materials. The chemical qualities of the materials influence the choice of process gas. When the RF power is increased, the electron density and radicals rise near the

TABLE 8.4
Comparison of Different Pressure Plasma and Their Applications

Application	Low		Medium		Atmospheric	
	Advantages	Disadvantages	Advantages	Disadvantages	Advantages	Disadvantages
Generation of plasma	Plasma is evenly distributed	Complex vacuum technology	Plasma can be made evenly distributed inside the chamber	Complex vacuum and flow control	Vacuum setup is not required	Treatable area is limited
Treatment of metal	Plasma may be used to treat oxidation-sensitive materials	Microwave plasma may convey energy to overheat	Cleaning can be done	Microwave plasma may convey energy to object	Very thin layer generated on aluminum	Oxidation-sensitive objects are limited
3D Objects	All objects inside the vacuum chamber are treated	Under observation	All surfaces of an object can be polished uniformly	Isotropic chemically polishing is a challenge	Local surfaces treatment is possible	Complex robotics technologies are necessary for the treatment of deep grooves
Surface finish of optics	Isotropic polishing is possible	Limited application	Surface integrity is very good due to non-contact chemical reactive plasma	Only incremental surface finish can be achieved	Surface integrity and surface finish improves	Only locally very small area can be covered
Electronic/semiconductor	Plasma treatment of state of art electronic device, printed circuit boards, and semiconductors can be done	Not known	To be explored	To be studied	Plasma finishing of metal is possible directly	Can reduce the ability to the polished surface with deep grooves

TABLE 8.5
Conventional Process vs Plasma Polishing Process

Parameter	Conventional Process	Plasma Polishing Process
Contact between tool and workpiece	Contact	Non-contact
Surface finish (nm)	Less	More
Dimensional accuracy	Less	More
Induced defects	More	Less
Sub-surface damage	More	Less
MRR	More	Less

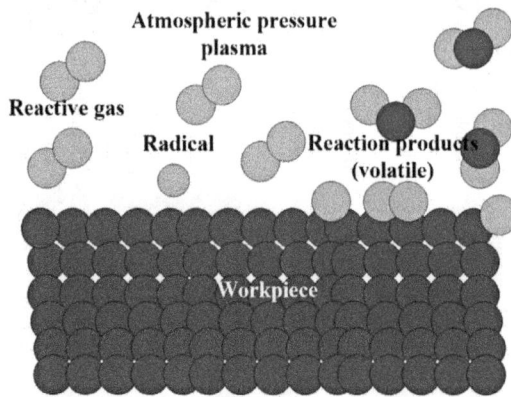

FIGURE 8.3   Principle of atmospheric pressure plasma polishing method [31].

substrate. The surface quality improves within a certain limit as the polishing time increases; however, as the processing time grows, the surface quality deteriorates due to undesirable removals of material from the workpiece, resulting in extra peaks and valleys. An optimal pressure setting in the plasma chamber is essential to sustain the static plasma for improved polishing. It is important to consider raising the gas flow ratio, which results in a lower surface roughness value to some extent. The fish diagram of the plasma polishing process parameters is displayed in Figure 8.4.

A schematic of a plasma polishing machine is displayed in Figure 8.5. A high vacuum discharge chamber made up of Zerodur (a lithium-aluminosilicate glass-ceramic)/sital (a crystalline glass-ceramic) material with external and internal diameters of 55.5 mm and 51 mm, respectively, are used. The setup consists of three process gas cylinders (He/Ar, SF_6, and O_2), and a mass flow meter which is used to control the flow rate of gases. A matching box is used to perform impedance matching between the RF generator and the discharge chamber. A vacuum pump system is used to create a vacuum inside the chamber and also to remove the exhaust gas from the vacuum chamber.

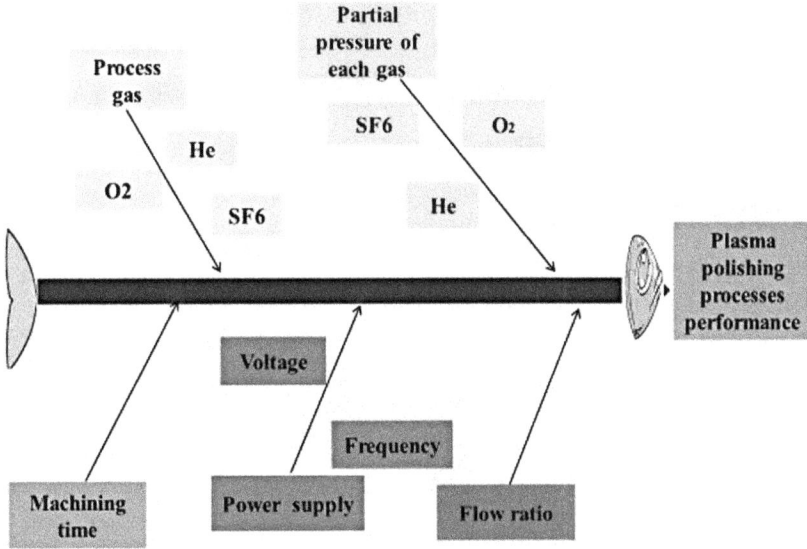

FIGURE 8.4   Fish diagram representing the input process parameters of plasma polishing process.

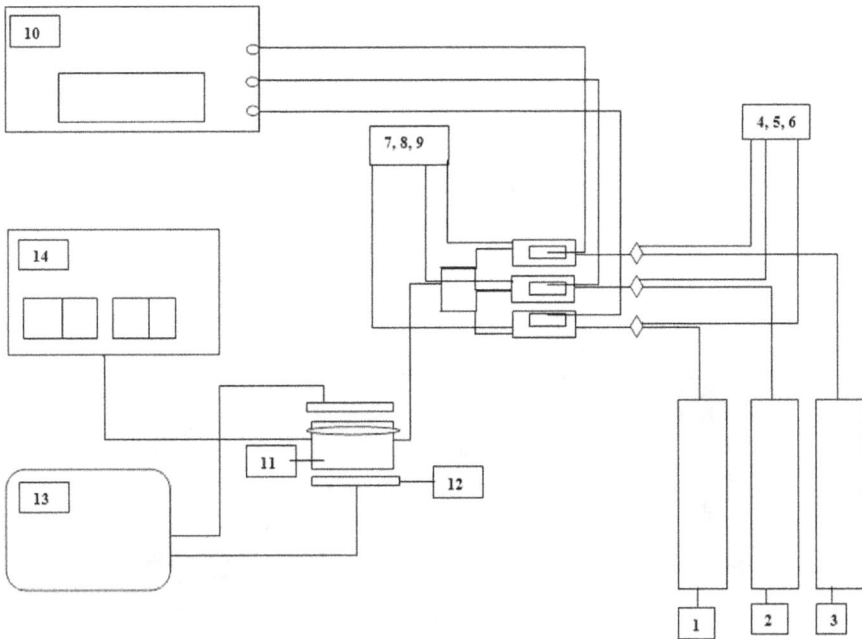

FIGURE 8.5   Schematic diagram of plasma polishing setup; 1,2,3 – Process gas cylinders; 4, 5, 6 – entry valves of mass flow meter; 7, 8, 9 – mass flow meters; 10 – mass flow controller; 11 – vacuum chamber; 12 – electrode; 13 – RF power source; 14 – vacuum pump.

The working flow chart of the plasma polishing experiment is demonstrated in Figure 8.6. There are total of six primary processes that occur during a plasma etching inside the chamber. The sequence is as follows:

1. Reactive species formation.
2. Surface diffusion.
3. Adsorption.
4. Physical and chemical reaction.
5. Desorption.
6. Diffusion into a large volume of gas.

If any one of these does not occur, the entire process will stop.

8.3.1 COMPONENTS OF PLASMA POLISHING SETUP

Various instruments are used in the setup of the plasma polishing process, which is discussed in the following section.

8.3.1.1 Thermocouple

A thermocouple is used to measure the temperature inside the vacuum chamber during the experiment. The knob of the thermocouple is fixed inside the chamber and connected to the display device. The display device reads the temperature of the plasma chamber, which is depicted in Figure 8.7.

8.3.1.2 Optical Emission Spectroscopy

Optical emission spectroscopy is used to measure plasma species density in excited states. It is very useful to interpret the mechanism of material processing under the plasma. It is demonstrated in Figure 8.8.

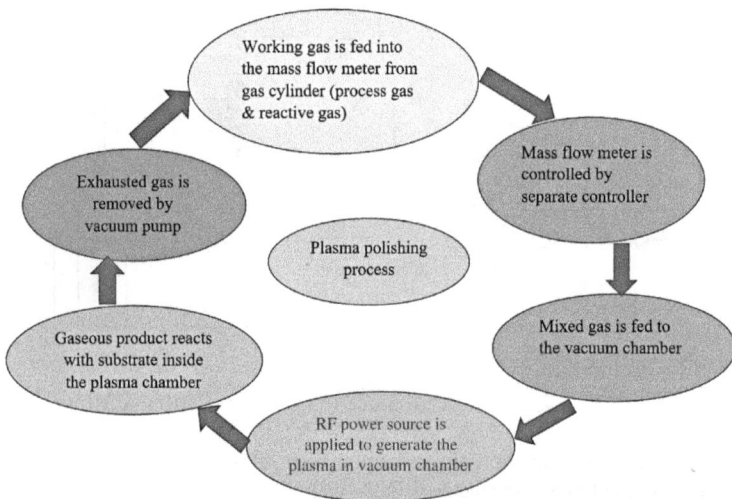

FIGURE 8.6 Working flow chart for plasma polishing.

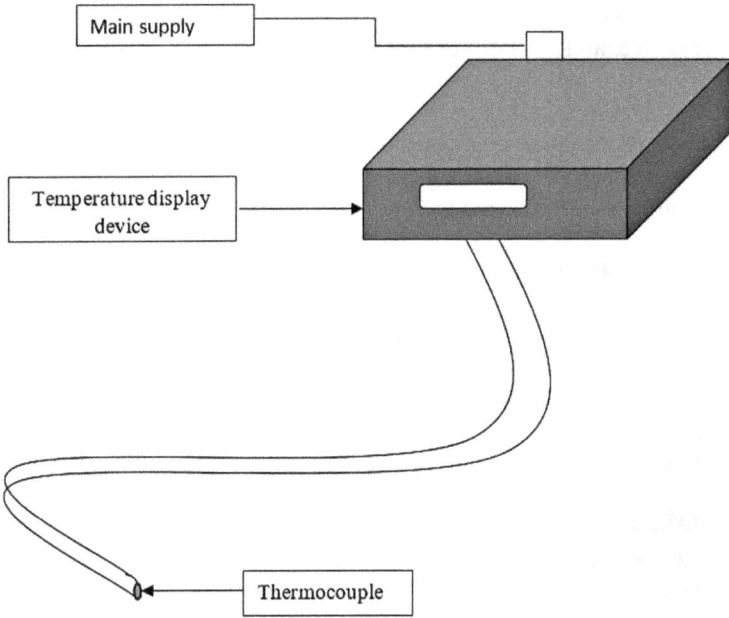

FIGURE 8.7 Schematic diagram of thermocouple.

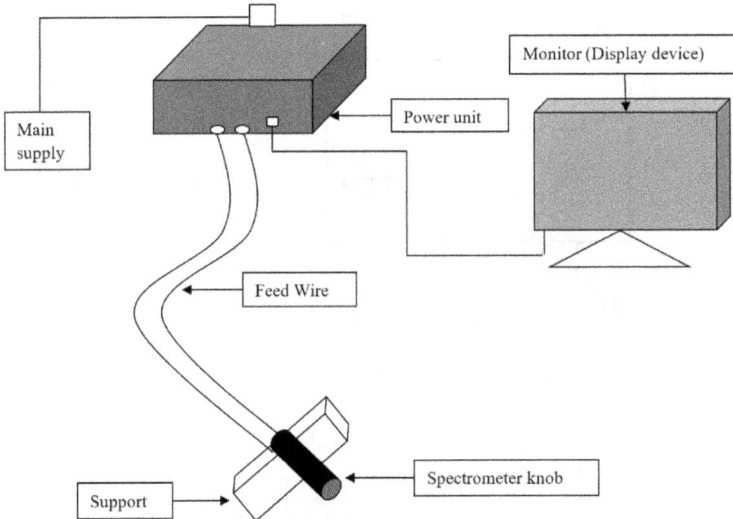

FIGURE 8.8 Schematic diagram of optical emission spectroscopy.

8.3.1.3   RF Generator

The power source, an RF generator with a frequency of 40.68 MHz and a maximum power of 85 W, is used. The discharge voltage is provided by an RF generator and a matching box. During excellent matching, most of the driving power dissipates during the discharge, and the reflected power should have a negligible value. The circular copper electrode is attached to the RF generator and placed at the top and bottom of the plasma chamber. A schematic block diagram of an RF generator is shown in Figure 8.9.

8.3.1.4   Mass Flow Meter

Mass flow meter instruments are used to control the flow of gases (SF_6, O_2, He, etc.), which comes from the gas cylinder and provides the required flow rate for plasma processing. The mass flow meters and controllers are basically linear and designed to properly detect and manage the mass flow from 2.5 to 250 standard cubic centimeters per minute (SCCM). A schematic view of the mass flow meter is demonstrated in Figure 8.10.

8.3.1.5   Exhaust Pumping System

The exhaust pumping system is used to create a vacuum inside the plasma chamber just before the start of the experiment and also to remove the waste gases which generate during plasma polishing. A schematic diagram is demonstrated in Figure 8.11.

8.3.1.6   Process Gas Cylinder (O_2, He, SF_6)

Plasma is generated using three different gases: He/Ar, O_2, and SF_6. In the initial stage of the experiment, the gases in the chamber are filled one by one at different

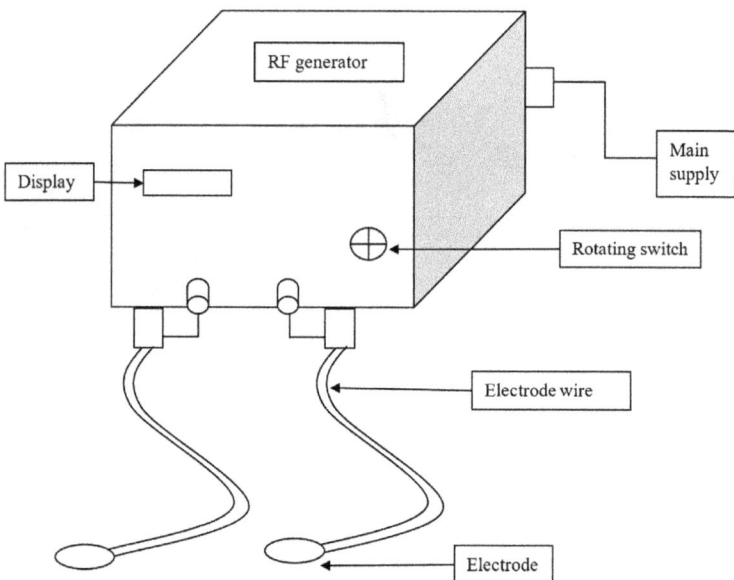

FIGURE 8.9   Schematic diagram of RF generator.

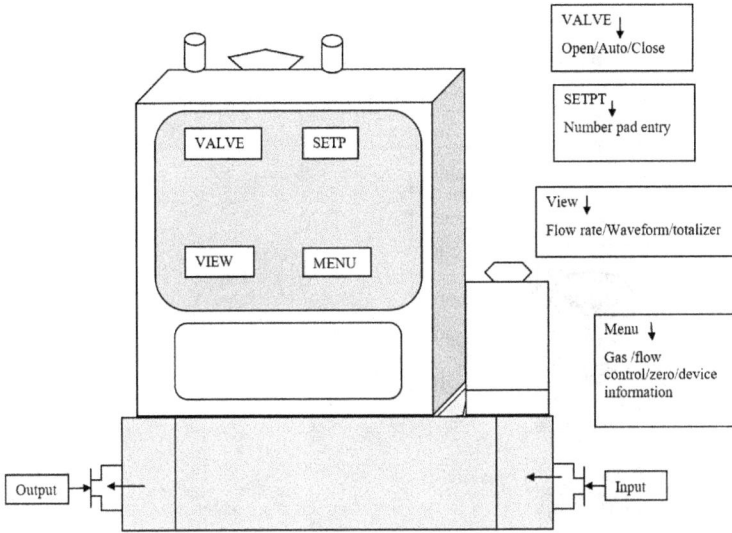

FIGURE 8.10 Schematic diagram of mass flow meter.

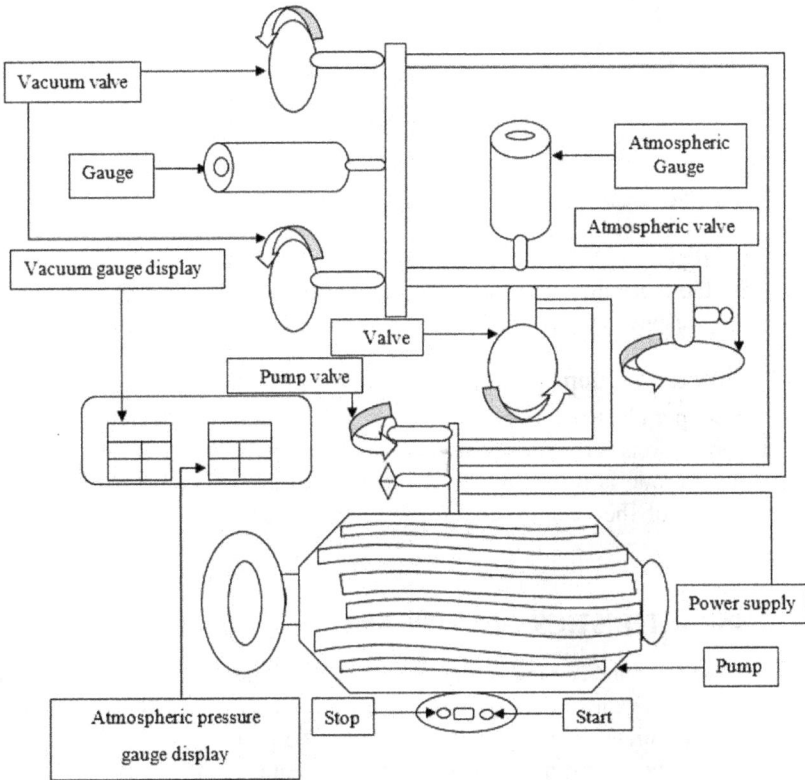

FIGURE 8.11 Schematic diagram of vacuum pump.

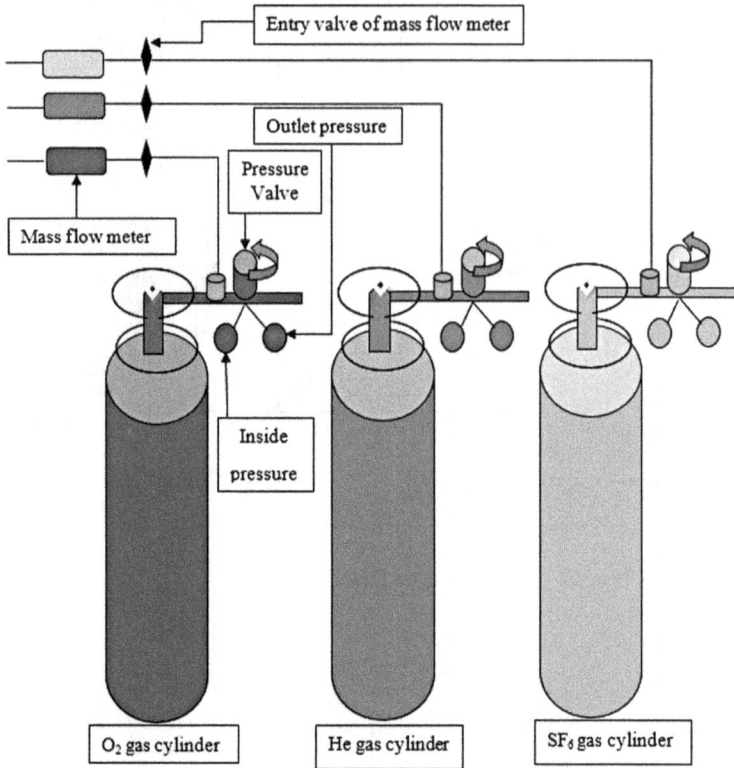

FIGURE 8.12 Schematic diagram of process gas cylinder (O_2, He, SF_6).

flow rates and pressure. The different gases pass through these separate mass flow meters, and the mass flow controller device controls the flow rate. Schematic diagrams of the various cylinders with flow diagrams are shown in Figure 8.12.

8.3.1.7   Process Vacuum Chamber with Two Electrodes

A circular copper electrode is placed on top and bottom of the vacuum chamber, with the gap between electrode and vacuum chamber maintained between 1 and 2 mm. An electrode is linked to the RF power source during the experiment. A schematic view of the vacuum chamber with two electrodes is demonstrated in Figure 8.13.

8.4   CHARACTERISTICS OF OPTICAL COMPONENTS

Optical components need a high degree of form accuracy as well as a smooth surface [32]. The qualities of the optical component have a considerable influence on propagation direction, intensity, and polarization [33]. The properties of optical components are primarily determined by their chemical composition and atomic structure [34]. As a result, it is vital to use a non-contact approach. Contact between the work

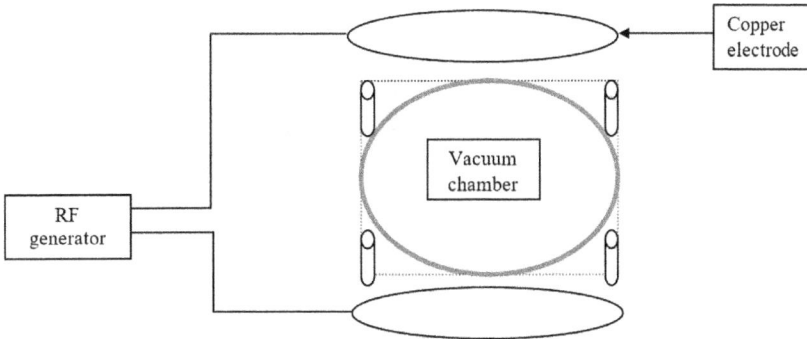

FIGURE 8.13 Schematic diagram of vacuum chamber with two electrodes.

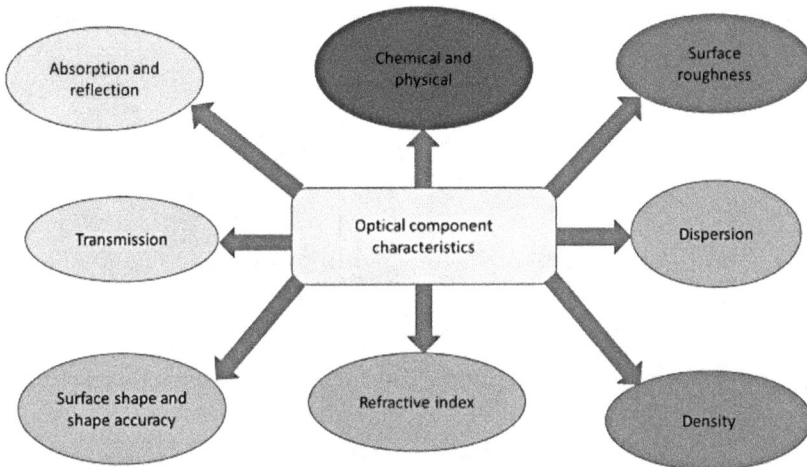

FIGURE 8.14 Optical component characteristic.

surfaces and the tool is avoided in this technique, resulting in a good surface quality without any surface or subsurface damage. Figure 8.14 shows the surface characteristics of optical components [35].

8.5 PROCESSING OF VARIOUS MATERIALS BY PLASMA POLISHING PROCESS

Knizikevicius [36] determined the chemical etching of Si and SiO_2 in $SF_6 + O_2$ plasma. By this etching rate analysis of SiO_2 and Si in $SF_6 + O_2$ plasma, it is found that the Si etching rate is dependent on the O_2 contents in the feed. As O_2 content in the feed increases, initially, F atoms increase because of reactions of SF_5 radicals with O atoms. Later, the concentration begins to fall in proportion to the quantity of injected SF_6 molecules, as shown in Figures 8.15(a) and (b), respectively.

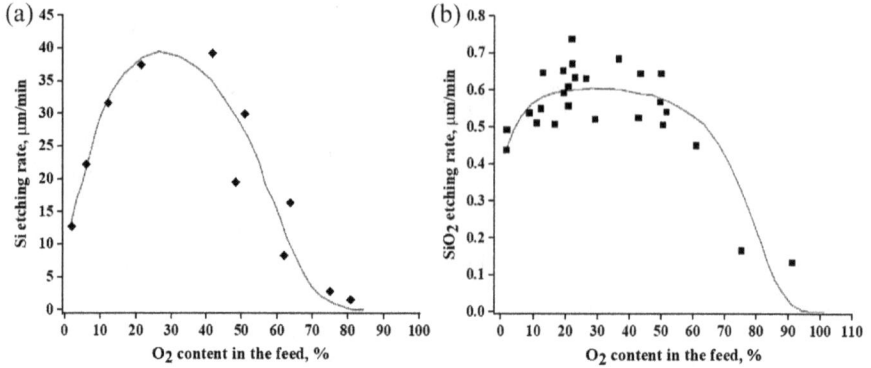

FIGURE 8.15 (a) Effect of O_2 content on Si etching rate. (b) Effect of O_2 content on SiO_2 etching rate [36]. With kind permission from Elsevier, License No. 5172541337231.

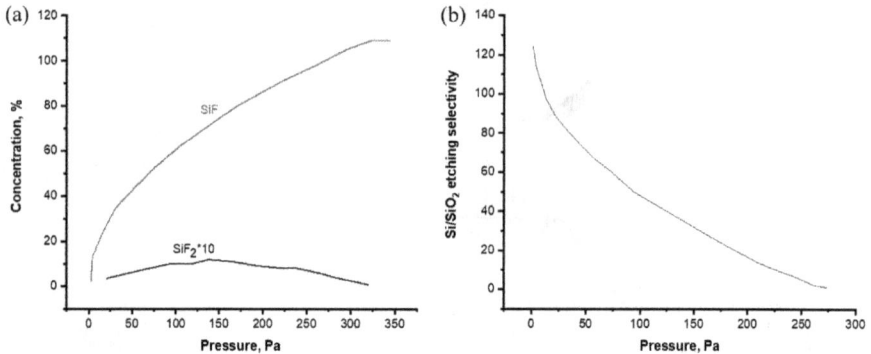

FIGURE 8.16 Effect of pressure on (a) concentration % SiF molecules and (b) Si/SiO2 etching selectivity [38]. With kind permission from Elsevier, License No. 5172550641162.

Yao et al. [37], through atomic emission spectroscopy, showed that the material removal rate in the APPP process for the Zerodur component is strongly dependent on working distance, RF powder, SF_6 and O_2 ratio, and the type of work material. Knizikevicius [38] found that pressure has no linear effect on the reactive ion etching (RIE) rates of SiO_2 and Si in CF_4 plasma. The concentration of SiF molecules grows monotonically as pressure increases until it reaches 100%. Because of minimum surface coverage, the higher concentration of SiF molecules in the adsorbed layer stops the etching process. With increasing pressure, the selectivity of Si/SiO_2 etching reduces monotonically until the RIE of Si is stopped. The RIE rates start to decrease due to the increased surface coverage, as shown in Figures 8.16(a) and (b), respectively.

Vana et al. [39] used plasma polishing for a stainless steel X10CrNi specimen and observed the change in surface roughness and gloss level values as a function of time, as shown in Figure 8.17(a). Experimentally, they found that surface roughness

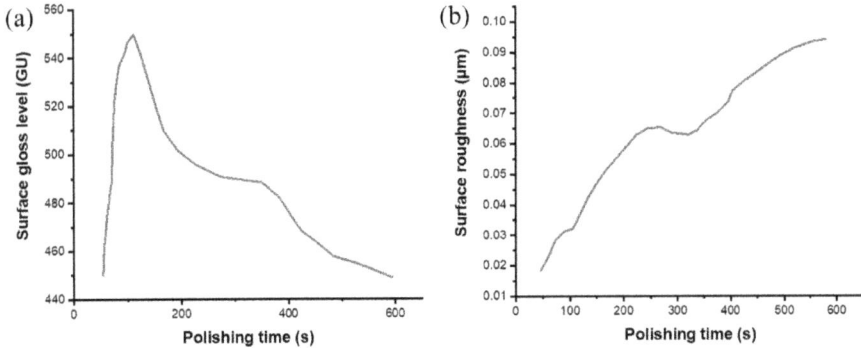

FIGURE 8.17   Effect of polishing time on (a) surface gloss level, and (b) surface roughness [39].

improved after plasma polishing with varying gloss levels, and its value depended on finishing time, as shown in Figure 8.17(b). After 60 seconds, the gloss level was maximum.

A new non-contact type plasma-assisted freeform surface finishing method is investigated on fused silica. This process is used for the finishing of ultrafine surfaces without any surface and sub-surface damage. He and Ne work as plasma processing gases, whereas SF_6 and O_2 are selected as reactive gases. The pressure is maintained up to 30 mbar by using a dielectric barrier capacitive coupled RF discharge. The relative density of excited species in the plasma is investigated by atomic emission spectroscopy to study the material removal mechanism. Surface roughness was enhanced up to 68% by using He–O_2 plasma, whereas surface waviness was enhanced up to 85%. Furthermore, the maximal MRR obtained with the He-SF_6–O_2 gas combination was 0.008 mm³/min [40]. Dasen Wang et al. [41], using plasma, produced with the help of a capacitive coupled hollow cathode (CCHC) RF discharge method, polished the fused silica up to 1.4 nm by a highly stable SF_6 and Ar/O_2. In this study, the process parameters are selected as the rate of gas flow, discharge power, and pressure. The rate of material removal is affected by the gas flow rate. It is mostly affected by the O_2 and SF_6 gas flow rate ratio. When the flow rate ratio of the gases O_2 and SF_6 were set to 1:1, the greatest MRR was attained. Furthermore, raising the input plasma power increases the removal rate. However, when the power is more than 100 W, the value of surface roughness increases. The influence of gas flow rate on MRR, and the influence of RF power on surface roughness are shown in Figures 8.18(a) and (b), respectively.

Wang et al. [41] investigated the hardness and surface modulus of the substrate, which were changed by using the APPP process, and surface mechanical properties were improved. They found that residual surface stresses decrease when the distorted layer is removed from the surface atom by atom. Liu et al. [42] concluded that low-pressure plasma is capable of removing subsurface damage on complex and freeform surfaces in which the pressure range varies from 10^{-2}–0.5 mbar. Though surface finish does not improve significantly, surface integrity has been improved in terms of residual surface stresses. Zhang et al. [43] machined the silicon wafer to an

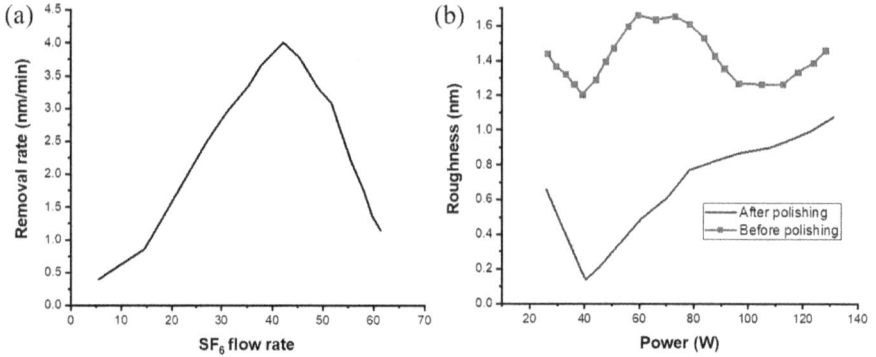

FIGURE 8.18   (a) Effect of gas flow rate on the material removal rate. (b) Effect of RF power on surface roughness [41]. With kind permission from Elsevier, License No. 5172561191602.

ultra-smooth surface. The spatial gas and temperature distributions on the workpiece surface were analyzed using the finite element method. They found a peak temperature of about 90°C at the center. The surface roughness, R_a of 0.6 nm, was measured through an atomic force microscope (AFM). The finished surface elemental composition was investigated by X-ray photoelectron spectroscopy (XPS). The minimum achieved surface roughness was 0.63 nm. Gerhard et al. [44] reported that the fused silica could easily be finished by using the APPP process on the nanoscale. By using this method, the transmission of the treated glass sample was dramatically reduced. Further, ellipsometry instrument measurements revealed a 3.66% reduction in the surface index of refraction at a wavelength of 636.7 nm and also a reduction in surface polarity by 30.23% (measured by using a surface energy measurements instrument). The absolute change in reflection and transmission of the fused silica was obtained after 12 min of plasma processing in the wavelength range of 190–400 nm as shown in Figure 8.19. It is also found that surface energy (γ_s), strength (σ), and polarity (P) of fused silica are reduced after plasma treatment, as displayed in Table 8.6.

Dev et al. [45] have carried out plasma processing at a pressure of 20 mbar and a power of 40 W with uniform plasma distribution. Experiments are conducted to investigate the influence of the SF_6/O_2 ratio on the MRR of the shell (Figure 8.20). The result depicts that the highest MRR was obtained at $(SF_6/O_2 = 2)$.

Non-contact plasma technology was used to eliminate the atomic-level defects in fine finished total internal reflecting (TIR) optics. Raman microscopy was used to understand the chemical structure of fused silica [20, 46]. A novel process is introduced to investigate the cracks on the optics surface through in situ laser illumination. Fused silica observes 80% reduction in a higher spatial wavelength using power spectral density after plasma polishing.

8.6   PLASMA PROCESS SIMULATION

The plasma processing chamber is fabricated using a Zerodur/sital material. The top portion of the chamber is enclosed with a circular disc that is optically transparent

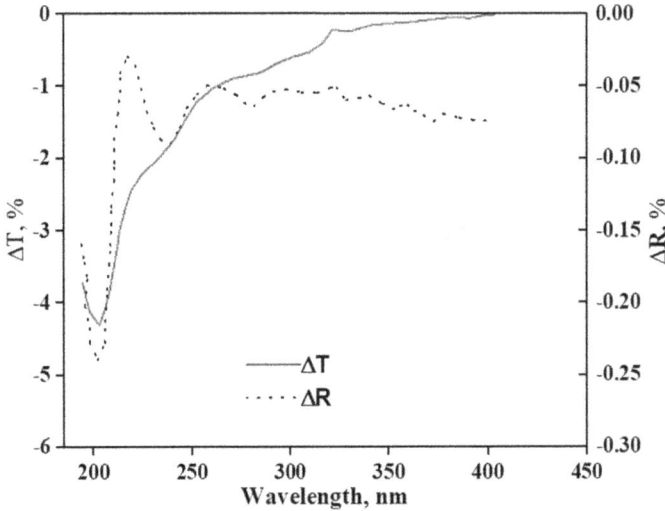

FIGURE 8.19 Change in reflection (ΔR) and transmission (ΔT) at different wavelengths [44].

TABLE 8.6

Surface Energy, Strength, and Polarity of Fused Silica Before and After Polishing

	γ_s(mJ/m^2)	P	σ (N/m^2)
Before processing	67.31	0.57	534.48
After processing	60.88	0.42	508.16

for the wavelength of 300 to 1200 nm. The body of the plasma processing chamber works as a dielectric barrier for exciting the electrodes. The excitation frequency of the RF power source is used as 40.68 MHz, and it is chosen against 13.56 MHz to reduce ion bombardment [45, 47]. The chamber is connected to distinct gas feeding lines to admit the gases from gas cylinders, and also the other side is connected to the vacuum pump to evacuate the exhaust gases and create the vacuum after keeping the substrate inside the chamber. The fiber optic probe head of optical emission is used to measure the plasma species density in the excited electronic state, which is very useful to understand the mechanism of material processing under plasma. This vacuum plasma chamber is developed for uniform polishing of optical components up to the size of 40 mm as shown in Figure 8.21.

A two-dimensional transient study is carried out using the microwave plasma module of Comsol® Multiphysics software. The thickness of the electrode in the microwave plasma module was assumed to be equal to the thickness of the chamber. The simulation was performed by using oxygen, helium, and species (a) active

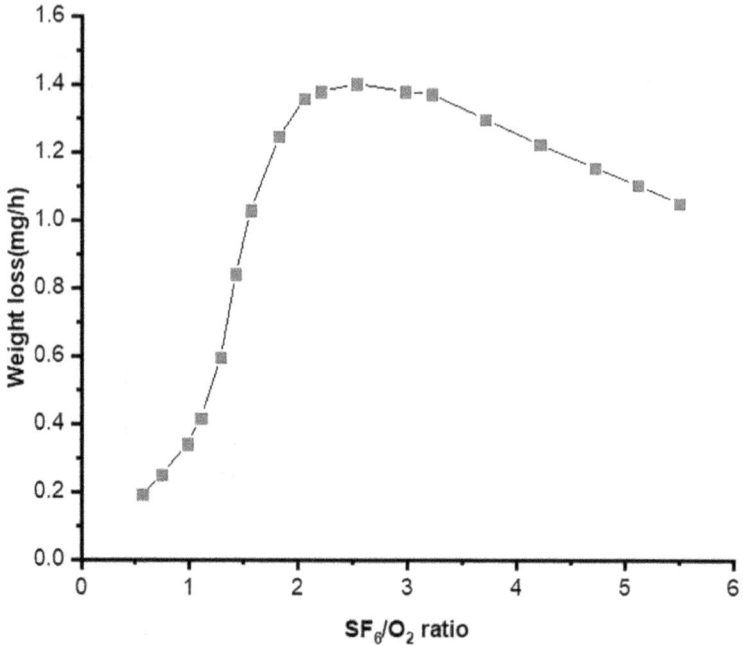

FIGURE 8.20   Weight loss versus ratio of SF_6/O_2 [45].

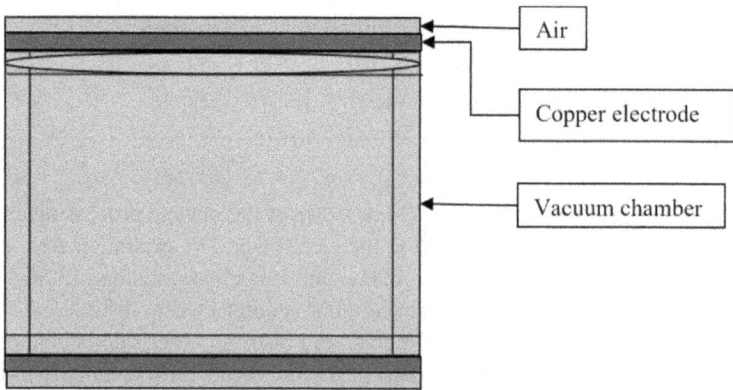

FIGURE 8.21   Two-dimensional domain of vacuum plasma chamber.

radicals as He, O1d, O_2 a1d, O_2b1s and O1s, (b) neutral atoms/molecules as O_2 and He, and (c) reactive ions as O^{2+}, O, e, $O^{+,}$ and He^+. The following results were achieved at 1 s. The electron density 10^{13} $1/m^3$ are obtained without the specimen inside the plasma chamber, as shown in Figure 8.22.

The maximum electron temperature (V) inside the plasma obtained using simulation analysis is 0.45 V (Figure 8.23), and the maximum electric potential

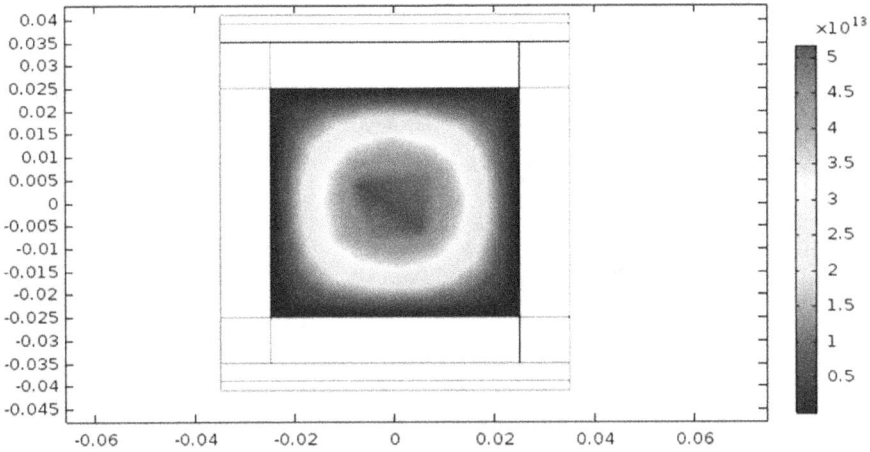

FIGURE 8.22 Electron density distribution within the plasma chamber.

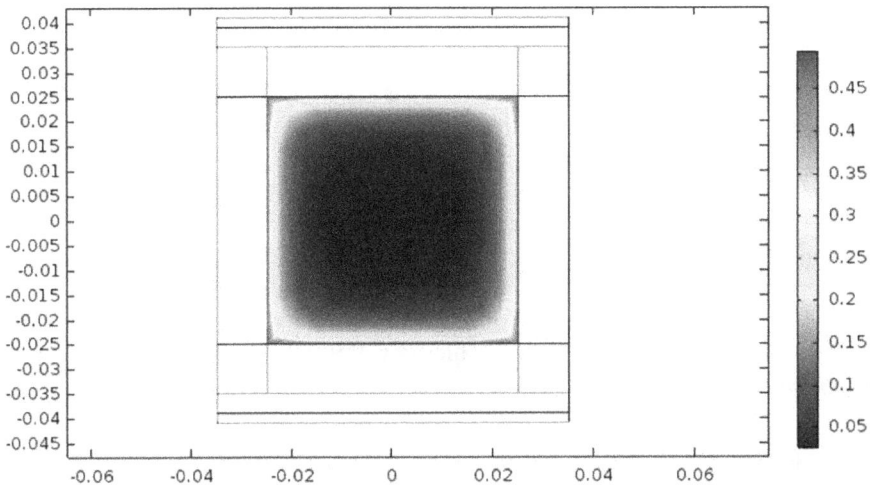

FIGURE 8.23 Electron temperature distribution within the plasma chamber.

obtained inside the plasma chamber is 0.2 V using microwave plasma simulation (Figure 8.24) [20].

8.7   RESEARCH CHALLENGES

The most challenging task in optical component fabrication is achieving high surface accuracy and surface finish. The modern optics sector needs a high level of surface quality with zero surface defects, which is a challenge to optics machining technology. In traditional contact machining methods, there are usually some

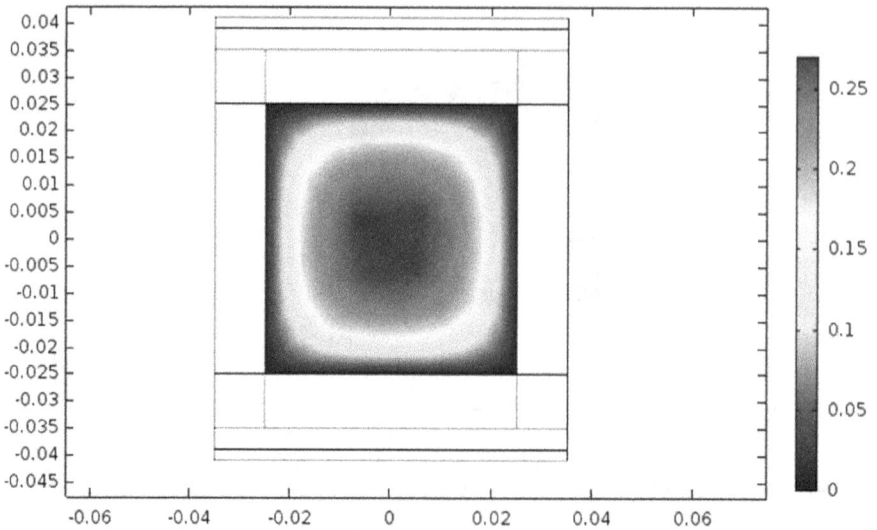

FIGURE 8.24 Electric potential distribution within the plasma chamber.

defects like lattice disturbances, micro-cracks, plastic deformation, dimensional inaccuracy, etc., on the final machined surfaces of the components. Because of the unique properties of the plasma polishing process, it is easy to polish the hard-brittle materials (crystals, glass, and ceramics, etc.). While most studies use simple plain substrate to polish with SF_6, He, Ar, CF_4, and O_2 gases, a few researchers attempted the polishing of complex and freeform metallic surfaces. The challenging work for this plasma process is to find the uniform inlet and outlet flow rate of the gases during processing. To date, there have been very few research papers reported in which volumetric MRR achieved within 5–32 mm^3/min during plasma polishing on optical surfaces.

8.8   FUTURE SCOPE

Plasma polishing requires appropriate composition of processing gases to be filled inside the plasma chamber at a specified pressure for a given component. The composition of the gas mixture in the plasma chamber and the pressure within the chamber vary with time because of volatile products such as SiF_4 and SiO_2 gases.

- Failure to perform regular cleaning of both the plasma chamber and the component surfaces might lead to a reduction in the surface quality of the polished components. When there is a consistent replenishment and waste disposal of processing and reactive gases, the polishing process will be more effective and durable. Investigations into the continuous replenishment of plasma gases are critical in establishing this procedure across several sectors.

- The majority of plasma polishing investigations have only been conducted on optical components. However, by employing alternative processing gases under optimal process parameter settings, the metallic component may also be viewed as a substrate. Furthermore, when finishing intricate freeform substrates, the MRR may be enhanced.
- The inlet and outlet conditions of reacting gas flow in the simulation model can also be added for a better understating of plasma processing.

8.9 CONCLUSIONS

Ultra-smooth surface finish and excellent dimensional precision with no surface or subsurface imperfections are the key criteria of ultra-precision optics. This chapter mainly focuses on the literature survey related to the plasma polishing method applied on optical materials. The block diagram of the plasma polishing process setup is systematically presented in this chapter. The mechanism of the plasma polishing process, along with the basic background, is also discussed in detail. A simulation study of the plasma processing chamber is studied. Non-conventional and non-contact plasma processes can easily polish complex shaped glass or fused silica components and can achieve surface roughness up to 0.6 nm without surface and sub-surface damage. Various experiments are conducted on the polishing of planer, freeform, and complex fused silica components using plasma processing by different researchers. The plasma polishing process can polish different materials such as hard-brittle materials (ceramics, glass, crystal, fused silica, Safire, etc.), non-metals at atmospheric pressure achieving surface roughness in the nanometer range. Plasma polishing gives a uniform MRR at various processing conditions. The polishing process combines the merits of anisotropic etching with low-pressure plasma and chemical vaporization with atmospheric pressure plasma.

ACKNOWLEDGMENT

We acknowledge the Science and Engineering Research Board, New Delhi, India, for its financial support for project No. ECR/2018/002801 entitled "Design and development of a novel plasma processing set up for uniform nanopolishing of a prism and any freeform surfaces of fused silica".

REFERENCES

1. X. Liu, R. DeVor, S. Kapoor, and K. Ehmann. "The mechanics of machining at the microscale: Assessment of the current state of the science." *J. Manuf. Sci. Eng.*, vol. 4, pp. 666–678, 2004. doi: 10.1115/1.1813469.
2. X. Luo, K. Cheng, D. Webb, and F. Wardle. "Design of ultraprecision machine tools with applications to manufacture of miniature and micro components." *J. Mater. Process. Technol.*, vol. 167, nos. 2–3, pp. 515–528, 2005. doi: 10.1016/j.jmatprotec.2005.05.050.
3. K. Ehmann, D. Bourell, M. Culpepper, R. E. DeVor, T. Hodgson, T. Kurfess, M. Madou, and K. Rajurkar. "An international assessment of micro-manufacturing research technology." *Mech. Mater. Eng.*, vol. 155, no. 9, pp. 211–224, 2005.

4. M. Kumar, H. N. S. Yadav, A. Kumar, and M. Das. "An overview of magnetorheological polishing fluid applied in nano-finishing of components." *J. Micromanuf.*, vol. 12, no. 5, pp. 1–19, 2021. doi: 10.1177/25165984211008173.

5. M. Kumar, A. Kumar, A. Alok, and M. Das. "Magnetorheological method applied to optics polishing: A review." *IOP Conf. Ser. Mater. Sci. Eng.*, vol. 804, no. 1, pp. 321–328, 2020. doi: 10.1088/1757-899X/804/1/012012.

6. H. N. S. Yadav, H. Bishwakarma, N. Kumar, S. Kumar, P. K. Singh, S. Mohanty, and A. K. Das. "Production of tungsten carbide nanoparticles through micro-EDM and its characterization." *Mater. Today Proc.*, vol. 18, pp. 1192–1197, 2019. doi: 10.1016/j.matpr.2019.06.580.

7. M. Kumar, S. Ahmad, and M. Das. "Magnetorheological-finishing of miniature gear teeth profiles using uniform flow restrictor." *Mater. Manuf. Process.*, vol. 36, no. 13, pp. 17–25, 2021. doi: 10.1080/10426914.2021.1954193.

8. A. Barman and M. Das. *Generation of Nano-Level Surface Finish By Advanced Nano-Finishing Processes.* Springer, Singapore, 2020, pp. 199–214.

9. N. Kumar, N. Mandal, and A. K. Das. "Micro-machining through electrochemical discharge processes: A review." *Mater. Manuf. Process.*, vol. 35, no. 4, pp. 363–404, 2020. doi: 10.1080/10426914.2020.1711922.

10. D. Manolakos and A. P. Markopoulos. "Micro and nanoprocessing techniques and applications." *J. Mater. Process. Technol.*, vol. 1, no. 1, pp. 31–52, 2005.

11. M. Kumar and M. Das. "Improvement in surface characteristics of SS316L tiny gear profiles by magnetorheological-polishing fluid using flow restrictor." *Trans. Indian Inst. Met. 2021*, vol. 15, no. 14, pp. 1–10, 2021. doi: 10.1007/S12666-021-02339-X.

12. M. Kumar, V. Kumar, A. Kumar, H. N. S. Yadav, and M. Das. "CFD analysis of MR fluid applied for finishing of gear in MRAFF process." *Mater. Today Proc.*, vol. 8, no. 6, pp. 245–252, 2021. doi: 10.1016/j.matpr.2021.01.116.

13. A. Alok, M. S. Niranjan, A. Kumar, M. Kumar, and M. Das. "Synthesis and characterization of sintered magnetic abrasive particles having alumina and carbonyl iron powder." *IOP Conf. Ser. Mater. Sci. Eng.*, vol. 804, no. 1, pp. 231–237, 2020. doi: 10.1088/1757-899X/804/1/012002.

14. N. R. Dhar, M. Kamruzzaman, and M. Ahmed. "Effect of minimum quantity lubrication (MQL) on tool wear and surface roughness in turning AISI-4340 steel." *J. Mater. Process. Technol.*, vol. 172, no. 2, pp. 299–304, 2006. doi: 10.1016/j.jmatprotec.2005.09.022.

15. X. Jiang, P. Scott, and D. Whitehouse. "Freeform surface characterisation – A fresh strategy." *CIRP Ann. – Manuf. Technol.*, vol. 56, no. 1, pp. 553–556, 2007. doi: 10.1016/j.cirp.2007.05.132.

16. Y. Li, Y. Wu, J. Wang, W. Yang, Y. Guo, and Q. Xu. "Tentative investigation towards precision polishing of optical components with ultrasonically vibrating bound-abrasive pellets." *Opt. Express*, vol. 20, no. 1, p. 568, 2012. doi: 10.1364/oe.20.000568.

17. W. Peng, C. Guan, and S. Li. "Ultrasmooth surface polishing based on the hydrodynamic effect." *Appl. Opt.*, vol. 52, no. 25, pp. 6411–6416, 2013. doi: 10.1364/AO.52.006411.

18. G. Ghosh, A. Sidpara, and P. P. Bandyopadhyay. "Review of several precision finishing processes for optics manufacturing." *J. Micromanuf.*, vol. 1, no. 2, pp. 170–188, 2018. doi: 10.1177/2516598418777315.

19. H. Takino. "Plasma chemical vaporization machining with a pipe electrode for optical fabrication: A review." *Int. J. Electr. Mach.*, no. 17, pp. 1–6, 2012.

20. H. N. S. Yadav, M. Kumar, A. Kumar, and M. Das. "COMSOL simulation of microwave plasma polishing on different surfaces." *Mater. Today Proc.*, vol. 256, no. 8, pp. 1–7, 2021. doi: 10.1016/j.matpr.2021.01.266.

21. A. F. Bastawros, A. Chandra, and P. A. Poosarla. "Atmospheric pressure plasma enabled polishing of single crystal sapphire." *CIRP Ann. – Manuf. Technol.*, vol. 64, no. 1, pp. 515–518, 2015. doi: 10.1016/j.cirp.2015.04.037.

22. W. M. Chiu. "Development of ultra-precision machining technology." *Technol. Exploit. Process*, vol. 1997, no. 435, pp. 486–490, 1997. doi: 10.1049/cp:19970187.

23. R. S. Rai and V. Bajpai. *Optimization in Manufacturing Systems Using Evolutionary Techniques*, 2020. doi: 10.1007/978-3-030-19638-7_9.

24. K. T. A. L. Burm. "Plasma: The fourth state of matter." *Plasma Chem. Plasma Process.*, vol. 32, no. 2, pp. 401–407, 2012. doi: 10.1007/s11090-012-9356-1.

25. A. Fridman. *Plasma Chemistry*, Cambridge University Press, 2008.

26. L. Tonks and I. Langmuir. "A general theory of the plasma of an arc." *Phys. Rev.*, vol. 34, no. 6, pp. 876–922, 1929. doi: 10.1103/PhysRev.34.876.

27. E. I. Meletis. "Intensified plasma-assisted processing: Science and engineering." *Surf. Coat. Technol.*, vol. 149, nos. 2–3, pp. 95–113, 2002. doi: 10.1016/S0257-8972(01)01 441-4.

28. J. Wang, L. Suo, L. Guan, and Y. Fu. "Analytical study on mechanism of electrolysis and plasma polishing." *Adv. Mater. Res.*, 472–475, pp. 350–353, 2012. doi: 10.4028/.472-475.350.

29. B. G. Beutel, N. R. Danna, R. Gangolli, R. Granato, L. Manne, N. Tovar, and P. G. Coelho. "Evaluation of bone response to synthetic bone grafting material treated with argon-based atmospheric pressure plasma." *Mater. Sci. Eng. C*, vol. 45, pp. 484–490, 2015. doi: 10.1016/j.msec.2014.09.039.

30. R. R. Borude, H. Sugiura, K. Ishikawa, T. Tsutsumi, H. Kondo, J. G. Han, and M. Hori. "Modifications of surface and bulk properties of magnetron-sputtered carbon films employing a post-treatment of atmospheric pressure plasma." *Jpn. J. Appl. Phys.*, vol. 58, no. 21, pp. 189–198, 2019. doi: 10.7567/1347-4065/aaec87.

31. J. Zhang, B. Wang, and S. Dong. "Application of atmospheric pressure plasma polishing method in machining of silicon ultra-smooth surfaces." *Front. Electr. Electron. Eng. China*, vol. 3, no. 4, pp. 480–487, 2008. doi: 10.1007/s11460-008-0072-9.

32. J. Yuan, B. Lyu, W. Hang, and Q. Deng. "Review on the progress of ultra-precision machining technologies." *Front. Mech. Eng.*, vol. 12, no. 2, pp. 158–180, 2017. doi: 10.1007/s11465-017-0455-9.

33. O. W. Faehnle and H. H. van Brug. "Novel approaches to generate aspherical optical surfaces." *Opt. Manuf. Test. III*, vol. 3782, no. 11, pp. 170–180, 1999. doi: 10.1117/12. 369181.

34. D. Lee, J. Gwak, T. Badloe, S. Palomba, and J. Rho. "Metasurfaces-based imaging and applications: From miniaturized optical components to functional imaging platforms." *Nanoscale Adv.*, vol. 2, no. 2, pp. 605–625, 2020. doi: 10.1039/c9na00751b.

35. G. S.-A. O. and O. Engineering and U. 1987. "The precision machining of optics." *Appl. Opt. Opt. Eng.*, vol. 6, no. 12, pp. 505–515, 1987.

36. R. Knizikevičius. "Simulations of Si and SiO_2 etching in SF6 + O_2 plasma." *Vacuum*, vol. 83, no. 6, pp. 953–957, 2009. doi: 10.1016/j.vacuum.2008.11.002.

37. Y. Yao, B. Wang, J. H. Wang, H. L. Jin, Y. F. Zhang, and S. Dong. "Chemical machining of Zerodur material with atmospheric pressure plasma jet." *CIRP Ann.*, vol. 59, no. 1, pp. 337–340, 2010.

38. R. Knizikevičius. "Simulation of Si and SiO_2 etching in CF4 plasma." *Vacuum*, vol. 82, no. 11, pp. 1191–1193, 2008. doi: 10.1016/j.vacuum.2008.01.047.

39. D. Vana, R. Suba, and M. Hurajt. "The change of surface properties on tested smooth stainless steel surfaces after plasma polishing." *Int. J. Eng. Sci. Invent.*, vol. 2, no. 6, pp. 7–11, 2013.

40. D. S. D. Dev, E. Krishna, and M. Das. "A novel plasma-assisted atomistic surface finishing on freeform surfaces of fused silica." *Int. J. Precis. Technol.*, vol. 6, nos. 3–4, pp. 262–276, 2016.

41. D. Wang, W. Liu, Y. Wu, L. Hang, H. Yu, and N. Jin. "Material removal function of the capacitive coupled hollow cathode plasma source for plasma polishing." *Phys. Procedia*, vol. 19, pp. 408–411, 2011. doi: 10.1016/j.phpro.2011.06.183.

42. W. Liu, D. Wang, M. Hu, Y. Wang, H. Liang, and L. Hang. "Roughness evolution of fused silica during plasma polishing processes." *Adv. Opt. Manuf. Technol.*, vol. 7282, no. 21, p. 72822T, 2009. doi: 10.1117/12.831002.

43. S. Zhang, Jufan Wang, and Bo Dong. "Application of atmospheric pressure plasma polishing method in machining of silicon ultra-smooth surfaces." *Front. Electr. Electron. Eng*, vol. 245, no. 6, pp. 480–487, 2008.

44. C. Gerhard, T. Weihs, D. Tasche, S. Brückner, S. Wieneke, and W. Viöl. "Atmospheric pressure plasma treatment of fused silica, related surface and near-surface effects and applications." *Plasma Chem. Plasma Process.*, vol. 33, no. 5, pp. 895–905, 2013. doi: 10.1007/s11090-013-9471-7.

45. D. S. D. Dev, K. Enni, and M. Das. Novel Finishing Process Development for Precision Complex-Shaped Hemispherical Shell by Bulk Plasma Processing, in *Precision Product-Process Design and Optimization,* Springer, 2016.

46. D. S. D. Dev, E. Krishna, and M. Das. "Development of a non-contact plasma processing technique to mitigate chemical network defects of fused silica with life enhancement of He-Ne laser device." *Opt. Laser Technol.*, vol. 113, pp. 289–302, 2019. doi: 10.1016/j.optlastec.2018.12.028.

47. H. N. S. Yadav, M. Kumar, A. Kumar, and M. Das. "Plasma polishing processes applied on optical materials: A review." *J. Micromanuf.*, p. 251659842110388, 2021. doi: 10.1177/25165984211038882.

9 Nanofabrication Using Focused Ion Beam

Bhaveshkumar Kamaliya and Rakesh G. Mote

CONTENTS

9.1   INTRODUCTION

The focused ion beam (FIB) system has been utilized mainly in the semiconductor industry to repair, modify, and debug integrated circuits (ICs) and perform failure analysis of ICs. In principle, using the FIB system, a highly energetic (typically with acceleration voltage in the range of 5–50 keV) narrow beam of ions is produced and focused on the target sample. And with this, direct writing by micro/nanoscale machining, and imaging processes are carried out. The FIB machining process exhibits ultra-high accuracy, and it is highly reproducible on the range of materials. Hence, it has also been widely utilized as a nano/microfabrication and imaging tool [1, 2]. Usually, for advanced material characterization, very thin (less than 100 nm) samples need to be prepared for transmission electron microscopy (TEM), known as TEM lamella. Due to the machining capabilities of FIB with nanometer precision,

DOI: 10.1201/9780429160011-9

it has also been used widely for TEM lamella sample preparations [3]. The maskless nano/microscale machining capability of the FIB system is advantageous for the fabrication of complex two-dimensional (2D) and three-dimensional (3D) nanostructures. However, direct machining using FIB shows limitations for the feature sizes of less than the beam size. To overcome such limitations, attempts are made to explore FIB-induced surface self-organization for a few nanometer-scaled features. Also, the operational parameters of FIB need to be optimized in order to achieve reduced redeposition while machining.

This chapter provides an overview of the FIB system, its mechanism for micro/nanomachining, modeling of the sputtering processes, and a fundamental understanding of FIB parameters. Also, examples of nanostructure fabrication using milling, deposition, and self-organization processes are discussed in this chapter.

9.2 FOCUSED ION BEAM SYSTEM

The FIB system is similar to the scanning electron microscope (SEM), where the electron source is replaced by an ion source and the lens system is modified accordingly. In FIB, the electrostatic lens system and gun are different than those in the SEM. A schematic of the FIB is given in Figure 9.1(a). At the top of the FIB column, the liquid metal ion source (LMIS) produces a stream of ions on the application of

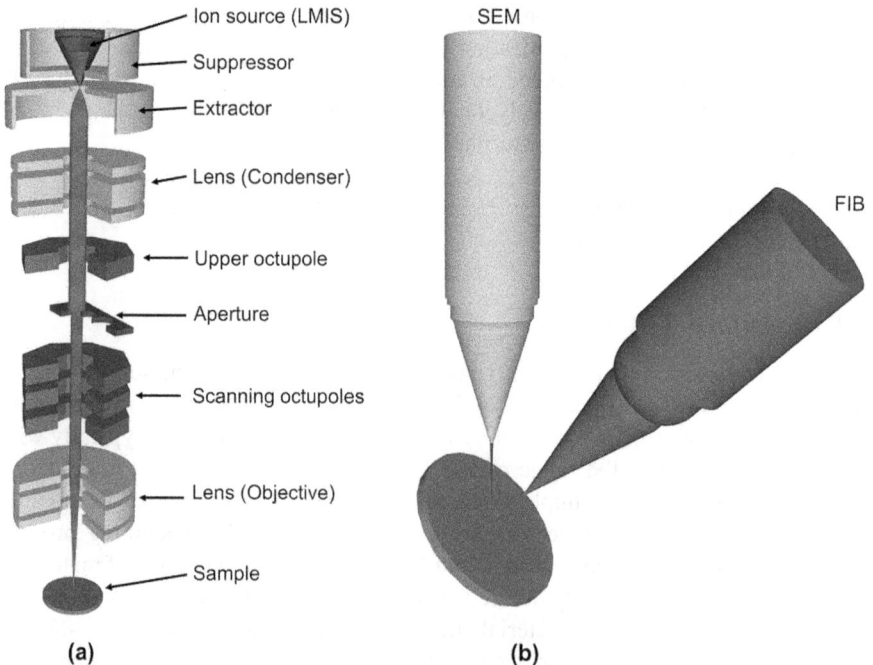

(a)   (b)

FIGURE 9.1   Schematics of (a) FIB column and (b) dual-beam FIB/SEM system.

electric potential. Gallium is commonly used as an ion source due to its low melting point (32°C), high mass, and low volatility. However, argon, indium, neon, etc., can also be used as an ion source. By applying a high electric field, ions are evaporated from LMIS and are focused within a narrow beam by the electrostatic lens system. Passing through the deflector plate and apertures, the ion beam reaches the sample surface. Upon the impact of focused ions on the surface, elastic and inelastic collisions happen. The elastic ion–atom collision removes surface atoms, which is called the sputtering process, while inelastic ion–atom collision results in the production of secondary electrons and x-rays. These secondary electrons are then detected, amplified, and analyzed to produce images of the sputtered surface. Thus, in-situ imaging is also possible by FIB. In general, for simultaneous imaging, the column of an electron beam is attached along with the ion beam. Therefore, it is called a dual-beam FIB/SEM system, as seen in Figure 9.1(b).

9.2.1 FIB: PRIMARY FUNCTIONS

The bombardment of a FIB on the surface of the substrate mainly results in physical effects like (i) sputtering of surface atoms (either in neutral or ionized form), (ii) emission of electrons, (iii) displacement of atoms in the solids (damage, defect creation), (iv) photon emission, and (v) ion beam-induced chemical reaction at the selected site. These effects can lead to four basic operations using the focused ion beam (see Figure 9.2), namely,

1. Milling (by sputtering of materials).
2. Deposition (by deposition of solid product on the materials).
3. Implantation (ion implantation within the bulk of the material).
4. Imaging (by detecting the emitted secondary electrons).

Selective area milling of the surface atoms happens by sputtering of the atoms due to high energy FIB bombardment (Figure 9.2(a)), and 2D as well as 3D nano/microstructures can be fabricated [4]. Deposition can be achieved by inserting some precursor molecules (mainly gases) in the chamber near the ion beam spot (Figure 9.2(b)). The interaction of ion beam, surface atom, and precursor lead to a chemical reaction, and the resultant product gets deposited on the surface while volatile byproducts are removed through a vacuum pump system. This process can also be called FIB-chemical vapor deposition (FIB-CVD) [4]. The FIB-CVD process is highly suitable for the fabrication of 3D functional nanostructures [4]. When the ion beam is bombarded with much higher kinetic energy such that ions penetrate up to a certain depth on the surface, they get implanted on the surface as an impurity atom and make the crystalline surface amorphous. This is called the implantation process (Figure 9.2(c)). The implantation process is mainly used in the semiconductor industry for doping intrinsic semiconductors with impurity atoms. Using the electron detector (such as a photomultiplier tube), the emitted secondary electrons can be detected, and topographical imaging is possible (Figure 9.2(d)). This process is called "*ion beam microscopy*".

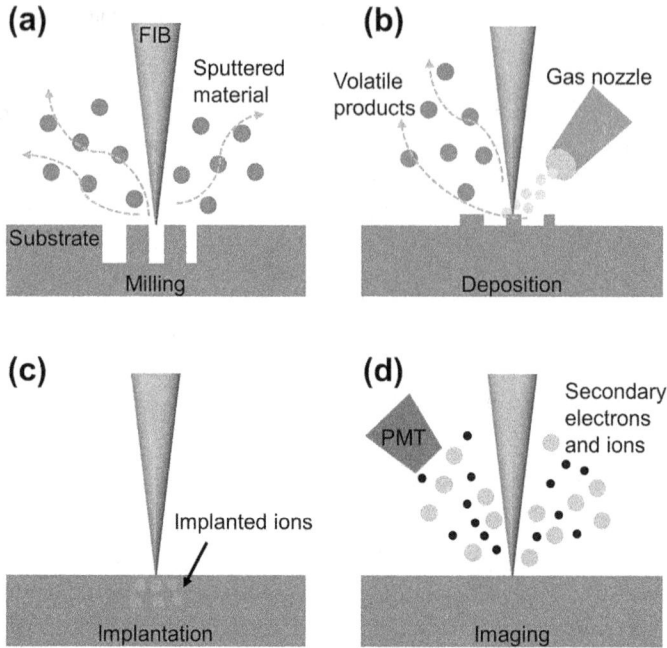

FIGURE 9.2 Basic operations using the FIB system: (a) Material removal by milling, (b) material addition by deposition, (c) ion implantation, and (d) imaging by capturing emitted secondary electrons.

9.2.2 SPUTTERING DUE TO ION–SOLID INTERACTIONS

During ion beam irradiation on the material, elastic ion–atom collision results in different mechanisms such as ion–solid collision cascades, sputtering of the atoms, and redeposition of sputtered atoms. These processes take place through the bulk of the solid, on the surface, and above the surface of the target. The ion–atom collision cascade due to ion–solid interactions in the bulk region of the substrate can be estimated by the binary collision approximation (BCA) model. According to the BCA model, ions undergo elastic collision with the target atoms in a manner which could be considered as hard-sphere collisions with an assumption that the ions and target atoms are classical perfect spheres colliding with each other. Interaction of irradiated ions and atomic nucleus in the material is considered to be an elastic collision in the BCA model, and the model has been widely used toward underlying kinetics during ion–atom collisions. Generally, based on the BCA model, the stopping and range of ions in matter (SRIM) can be estimated using the Monte Carlo simulations known as transport of ions in matter (TRIM) [5, 6]. In these simulations, every incident ion and recoiled atom are traced, and their trajectories are stored up until their kinetic energy reduces to the cut-off value (i.e., stopping of ion or atom) or until they have escaped from the simulation region.

Ion–silicon interactions are calculated using Monte Carlo TRIM calculations and presented for He and Ga ions in Figure 9.3. It can be observed from the trajectories of

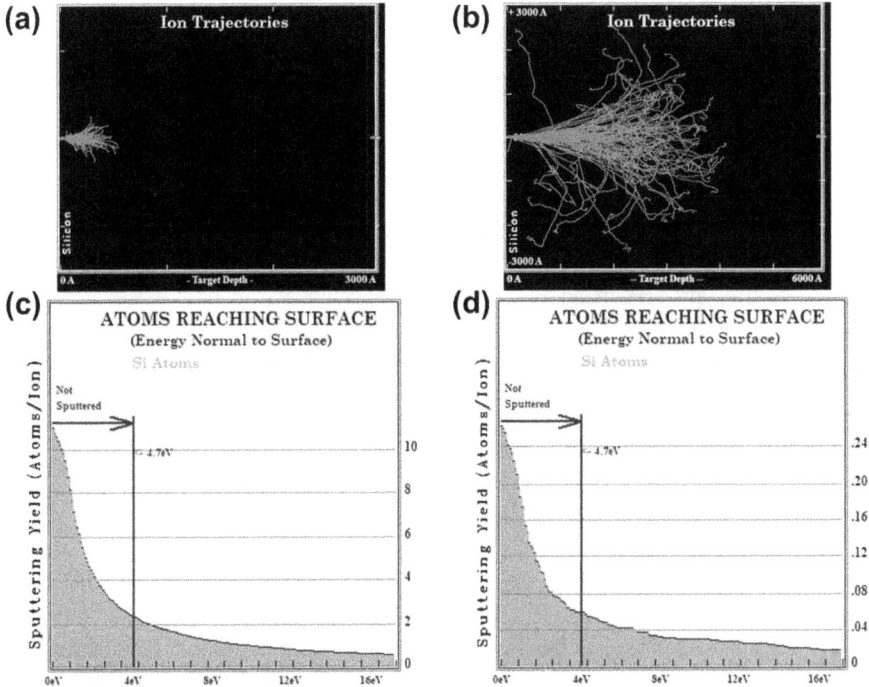

FIGURE 9.3 Monte Carlo TRIM calculations for ion–silicon interactions. Ion trajectories and sputtered yield for the Si crystal due to 30 keV irradiation of different ion species, gallium ions (a, c) and helium ions (b, d).

He and Ga ions in silicon that He ions exhibit a larger penetration range (Figure 9.3(b)) compared to that of Ga ions (Figure 9.3(a)). The penetration depth of ions into a solid is inversely proportional to the atomic weight of the ions. The nuclear stopping is more prominent for the heavier ions than the lighter ions. Hence, the He ions being lighter (atomic weight 4 amu), exhibit a larger stopping range of around 300 nm, whereas Ga ions (atomic weight 69 amu) show a stopping range of about 30 nm. It should also be noted that the recoil atoms on the surface can result in sputtering, which leads to the milling of the target surface. Sputtering is the essential ion–solid interaction, which defines the use of the FIB system for the milling process. To undergo the sputtering process, the recoil atom needs kinetic energy greater than the surface binding energy through the ion-induced collision cascade. The extent of sputtering is measured by sputtering yield, which is estimated from the number of atoms sputtered away from the material surface per incident ion. The distribution of energy for the recoil silicon atoms is shown in Figure 9.3(c) and Figure 9.3(d) for He and Ga ion bombardment, respectively. The recoil atoms having kinetic energy less than the binding energy of Si (i.e., 4.7 eV) cannot get knocked out from the surface. Hence, the sputtering yield (~0.06 atoms/ion) is far less for the He irradiation than the sputtering yield (~2.35 atoms/ion) from Ga ion bombardment. Thus, the Ga^+ FIB is a widely used system for the micro/nanomachining of metals and semiconductors.

When some atoms are sputtered out of the target surface, they possess finite kinetic energy, and they undergo a projectile motion out of the surface. During the projectile motion, they may stick to the walls of the feature being milled or on the edge of the features, which is known as the redeposition process. The extent of redeposition depends on the sticking factor of the material. It is helpful to optimize the FIB system parameters such that minimal redeposition occurs during the milling process.

9.3 BASIC OPERATIONAL PARAMETERS OF THE FIB PROCESS

The machining process using FIB involves important parameters associated with the operation such as sputter yield, ion beam energy (i.e., acceleration voltage), beam current, dwell time, ion dose (i.e., ion flux or ion fluence), incident angle, beam overlap, etc. Optimization of these parameters depending on the substrate material and desired FIB operation is essential during the FIB fabrication process. These crucial FIB parameters are explained as follows.

9.3.1 Sputter Yield

As described previously, sputter yield is used to measure the material removal characteristics during the ion beam irradiation process. The average number of sputtered atoms per ion bombarded is known as the sputter yield. The fundamental sputtering model was developed by Sigmund in 1969, which is known as the "Theory of Sputtering" [7]. Sigmund reported the relation of the total sputtering yield $Y(E)$ with the surface binding energy and transfer of momentum due to the bombarded ion as,

$$Y(E) = \frac{3\alpha T_m}{4\pi^2 U_s} \tag{9.1}$$

where α is described as the mass-dependent factor which provides a function of energy transfer, T_m is the momentum transferred due to the ion bombardment and U_s is the surface binding energy for the material.

Sigmund [7] also estimated the angular sputter yield y for the polar angle $\varnothing$ above the surface of the target material. The angular sputter yield is considered diffuse and axisymmetric in nature, relating to the total sputter yield as,

$$y = \frac{Y\cos(\varnothing)}{\pi} \tag{9.2}$$

Though these equations are very important in the modeling of the ion-induced sputtering, they are based on the assumptions that the ion incidence is normal to the target surface and velocities of the sputtered atoms are uniform in all directions. The incident angle-dependent total sputter yield, using the linear collision cascade provided by Sigmund, can be given as follows,

$$Y(E,\theta) = \frac{0.042\alpha}{U_s} S_n(E)\cos^{-f_{mr}}(\theta) \tag{9.3}$$

where θ is the incident angle for the bombarded ions, $S_n(E)$ is the stopping power for the elastic collision and f_{mr} is the function depending on the ratio of atomic masses for the incident ion to the atom. Equation 9.3 provides a total sputter yield that is a function of incident ion energy, angle of incidence for bombarded ions, and material-specific milling characteristics. To estimate the incident angle-dependent sputter yield, Yamamura suggested the analytical expression [8], which has been widely recognized for the non-normal incidence conditions (Equation 9.4).

$$Y(\theta) = Y(0)\cos^{-f}(\theta)e^{-c\left(\cos^{-1}(\theta)-1\right)} \tag{9.4}$$

where $Y(0)$ is total sputter yield for the normal incidence angle of the bombarded ions and f and c are the parameters, which are adjustable depending on the ion-solid combinations.

9.3.2 BEAM ENERGY

The applied acceleration voltage ensures the extraction and acceleration of the ions produced from the LMIS. The applied acceleration voltage defines the energy and velocity for the ions passing through the ion beam column. The unit for the beam energy (acceleration voltage) is stated as kV or keV. The penetration range inside the material will be higher for the ions accelerated with higher acceleration voltage. Hence, the ion implantation and the milling depth can be controlled by beam energy. In general, an acceleration voltage of 30 keV is used [4, 9].

9.3.3 FIB INTENSITY PROFILE – BEAM SIZE, BEAM CURRENT, AND DOSE

Ion beam irradiation-induced machining on the materials depends on crucial parameters such as beam size, beam current, dwell time, incidence angle, and ion dose. The key parameters are the function of the intensity profile of the FIB and are defined as follow,

- The aperture size and condenser lens control the beam current through the ion beam column. For a higher beam current, the aperture will allow an increased number of ions to pass through the column, and thus, the diameter of the beam (i.e., beam size) will be higher. Hence, a smaller beam current is required for the higher resolution, but the smaller beam current will significantly decrease the throughput during the milling procedure. The unit of beam current is nA or pA.
- The number of ions bombarded per unit area is known as the ion dose. The product of total dwell time and beam current is proportional to the number of ions bombarded on the surface. Ion dose is expressed in units of ions/cm^2.
- The duration of beam irradiation at each position (pixel) is known as dwell time. Generally, the dwell time is in the range from μs to ms. A higher dwell time value leads to an increased depth of the milling features.

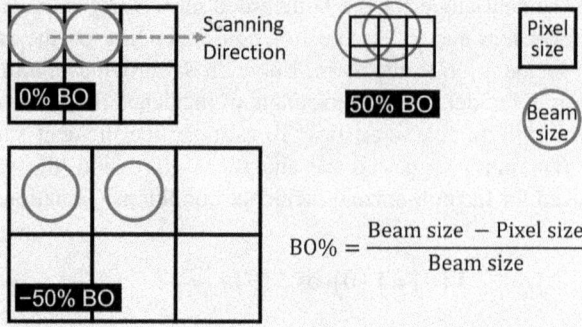

FIGURE 9.4 Illustration of beam overlap (BO) with different pixel sizes. The pixel size is lesser than the beam size for positive beam overlap, whereas it is greater than the beam size for negative beam overlap.

- The ion impingement angle onto the target sample is known as an incident angle or the angle of incidence. The angle of incidence is defined with respect to the surface normal of the sample substrate.
- As shown in Figure 9.4, during FIB scanning on the surface, after the set dwell time, the beam position is moved to the next pixel position (depending on the set pixel size) toward the scanning direction. When the FIB is moved, the beam is blanked until it is positioned at the next pixel position. The beam overlap (BO) is defined by the step size (pixel size), as illustrated in Figure 9.4.

9.3.3.1 Intensity Distribution

Usually, FIB intensity distribution across the beam cross-section is considered to be Gaussian distribution [10]. According to the experimental observations, the intensity distribution $D(r, \sigma)$ of ions for the FIB can be expressed as,

$$D(r,\sigma) = \frac{D_0}{\sigma\sqrt{2\pi}} e^{-\left(\frac{r}{\sigma\sqrt{2}}\right)^2} \tag{9.5}$$

where D_0 is the constant dose depending on the incident ion current and dwell time, the radial distance is r from the beam center ($r = 0$). The standard deviation is σ for the Gaussian distribution function. The beam diameter is given by full width at half maximum (FWHM) of the Gaussian distribution, which is given by

$$d_f = 2.35\sigma \tag{9.6}$$

where d_f is also known as beam size. If beam current I is irradiated at a spot for a dwell time of t_d with energy E then the total number of bombarded ions is $It_d/(nC_e)$, where n is a charge of the ion (i.e., 1 for singly charged and 2 for doubly charged ions) and C_e is a charge of a single-charged ion equivalent to one electron charge (i.e., 1.6×10^{-19} C). Hence, assuming the conservation of kinetic energy, for the beam energy E the dose constant can be given by

$$D_0 = \frac{EIt_d}{nC_e\sqrt{2\pi\sigma}} = \frac{0.94EIt_d}{nC_ed_f} \tag{9.7}$$

Researchers have also reported through experimental observations that, in practice, the actual beam profile for a FIB is different from the Gaussian distribution [11–13]. With detailed investigations, it was found that for the central section (axial region) of the ion beam, the Gaussian distribution can be well fitted. However, for the edge of the beam, the same Gaussian distribution function does not fit well. Hence, Assayag et al. [11] proposed a bi-Gaussian function representing different Gaussian distributions for the central region of the beam and for the edge part of the beam. This bi-Gaussian model has been widely accepted [14, 15], and the intensity distribution profile for a distance d from the beam center can be expressed as

$$D(d) = \frac{D_0}{\eta}\left[we^{-\left(\frac{d}{\sigma_1\sqrt{2}}\right)^2} + (1-w)e^{-\left(\frac{d}{\sigma_2\sqrt{2}}\right)^2} \right] \tag{9.8}$$

where σ_1 and σ_2 are standard deviation for the central and edge region of the beam, respectively, w is the weight factor for the intensity distribution, and η is atomic density. Han et al. [15] have used Equation 9.8 for prediction and experimental validation of the milling geometries by modeling the bi-Gaussian intensity distribution function.

9.4 3D NANOFABRICATION USING FIB

The study of nanostructures and sub-nanometer-sized structures is of great interest today because of their novel applications and efficient performance. The FIB is widely used in the areas of fabrication, surface modification, and surface analysis. Nano/microscale photonic crystals, MEMS/NEMS devices, TEM sample preparation, 2D/3D micro/nanostructure fabrication, mechanical machining tools, biological tools, microfluidics, etc., are the applications where the fabrication capabilities of FIB have been utilized. Through the material removal and material deposition processes, milling (see Figure 9.2(a) and Figure 9.5(a)) and deposition (see Figure 9.2(b) and Figure 9.7(a)) are the two main direct writing functionalities of FIB for fabricating micro/nanostructures using material removal and material deposition processes [16]. In addition to these direct methods, ion beam-induced surface self-organization has also emerged as a promising nanostructuring strategy toward creating features of sub-nanometer to few nanometers in size.

9.4.1 MATERIAL REMOVAL: FIB MILLING

Using the milling feature of FIB, 2D as well as complex (3D) nano/microstructures can be fabricated on the variety of substrates. During the milling process, pixel-by-pixel material removal is achieved in a manner depicted in Figure 9.5(a) which

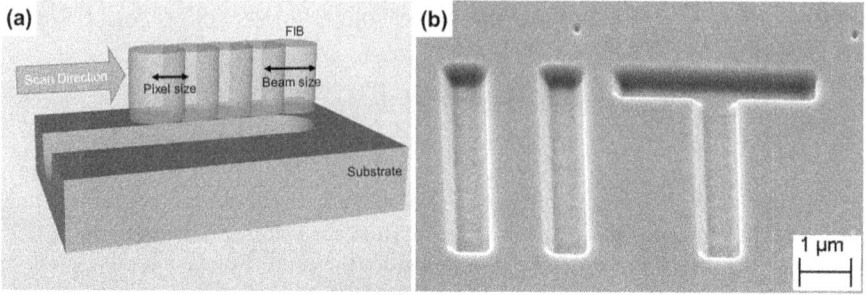

FIGURE 9.5 (a) Schematic for 3D fabrication process by material removal by FIB milling process and (b) SEM image of the letters "IIT" on silicon created by FIB milling.

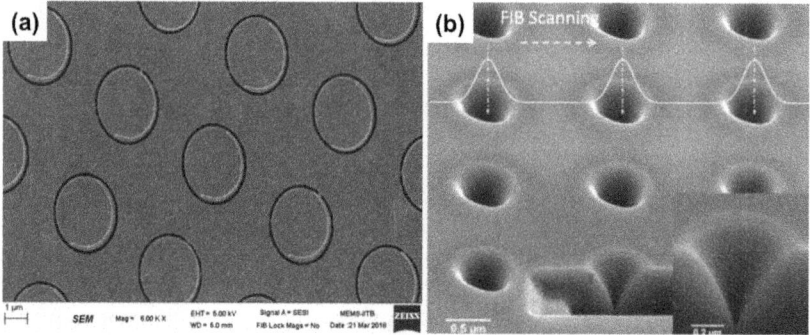

FIGURE 9.6 FIB milling of nanostructures. SEM images for (a) array of ring structures (plasmonic lens) in gold thin film [17], and (b) nanoholes fabricated on silicon with negative beam overlap [18].

enables us to write nano/microscaled features on the substrates (Figure 9.5(b)). The FIB-presented milling was performed using 30 keV Ga+ FIB with a dual-beam FIB/SEM system (Auriga® compact, Carl Zeiss) at IIT Bombay.

Nano/microscale fabrication can be carried out using FIB milling on a variety of materials. Examples of nanofabrication in gold thin film and silicon are presented in Figure 9.6. In both cases, 30 keV Ga+ FIB was used. The circular ring structures in Figure 9.6(a) are fabrication on gold thin film using positive beam overlap conditions during FIB milling [17]. In comparison, the array of tapered nanoholes in Figure 9.6(b) were created with negative beam overlap so that each FIB spot generates single nanohole morphology [18]. These results demonstrate that FIB milling can be an ideal method to create complex nanostructures which have continuous features (e.g., Figure 9.6(a)) and isolated nanoholes (e.g., Figure 9.6(b)) simultaneously without any mask required. The capability of manipulating the beam positions at the nanoscale has proven FIB milling to be highly suitable for nanoscale machining on tiny optical components such as optical fibers [19].

9.4.2 Material Deposition: FIB-CVD Method

In the FIB-CVD processes, the reactive gas is injected while pixel-by-pixel scanning of FIB, as shown in Figure 9.7(a). Due to the reaction activation energy provided by the energetic ions, the injected gases undergo a chemical reaction, and the desired solid product is deposited at the site of ion bombardment on the substrate. The reaction precursors are designed in such a way that the byproduct is always volatile, and it should be removed through the vacuum system. An example of platinum deposition to produce embossed letters "IIT" is demonstrated in (Figure 9.7(b)) on the silicon surface using 30 keV Ga$^+$ FIB with a dual-beam FIB/SEM system (Auriga® compact, Carl Zeiss). Generally, organometallic precursor gases are used for metal deposition using FIB-CVD.

A continuous supply of precursor gas is required during FIB scanning so that the deposition rate is always higher than the milling rate. To ensure less milling, the ion beam current is generally kept lower for the FIB-CVD mode. Such conditions provide good confirmation of the deposited structures, but the throughput is less, so large area fabrication is challenging. With a combination of beam scanning capability in the lateral directions and stage rotation capabilities, it is also possible to fabricate complex 3D structures [21, 22], as shown in Figure 9.8. This is the most exciting fabrication strategy using FIB as it provides a plethora of possibilities in the complexities of the fabricated structures.

9.4.3 Surface Self-Organization

Recently, FIB has attracted much research into its capability of forming well-ordered nanostructures via self-organization. Lithography by FIB milling has its limitations of beam diameter (in practice, the lowest possible beam diameter is ~10 nm depending on ion source), which limits direct writing of few nanometer and sub-nanometer features on the various substrates [23]. Important structural features of ripples like

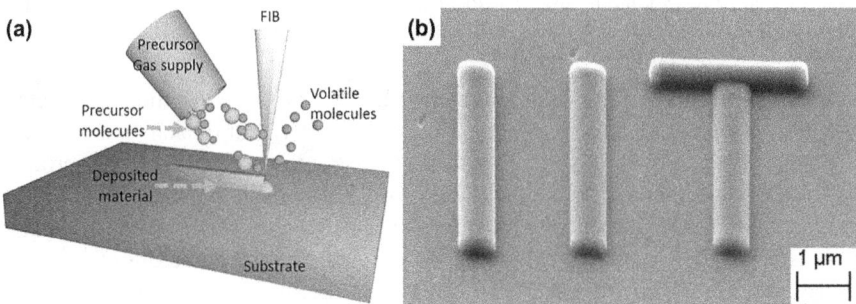

FIGURE 9.7 Schematic for 3D fabrication process by material deposition. (a) Material deposition by FIB-CVD process [20]. (b) The letters "IIT" on silicon by Pt deposition using FIB-CVD method.

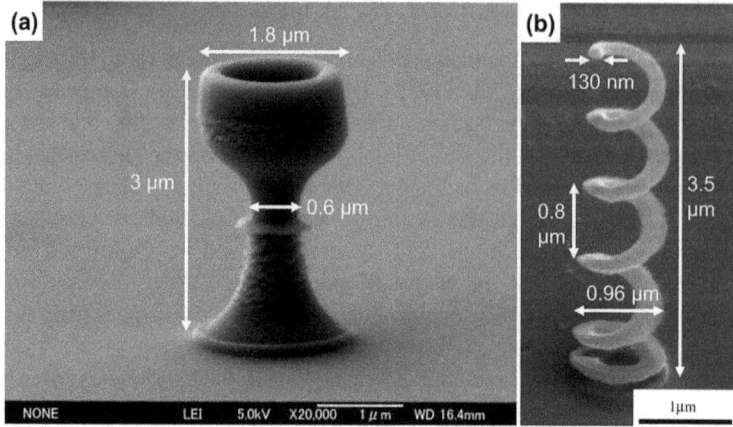

FIGURE 9.8 SEM images of FIB-CVD-fabricated 3D structures. (a) Wineglass structure [21] and (b) nanospring (helical) structure [22].

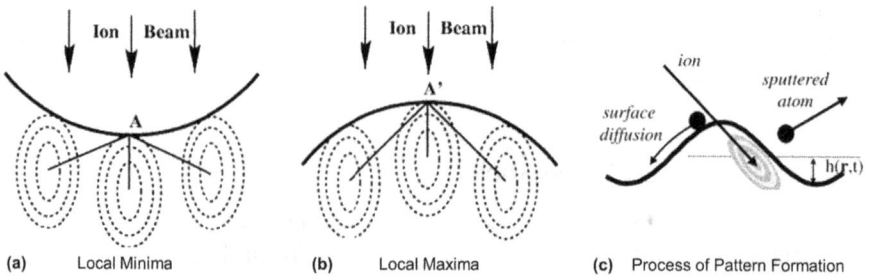

FIGURE 9.9 Schematic diagrams to understand surface imbalance during ion beam erosion of surfaces. (a) The surface with local minima (convex geometry) erodes faster than (b) the local maxima (concave geometry) because the energy has to travel shorter distances (solid lines) to sputter nearby atoms than the local maxima. (See points A and A') [24]. (c) The process of pattern formation [25].

wavelength, height, period, and orientation depend on the irradiation parameters such as type of ion, ion doses, angle of incident, the energy of the ion, and duration.

Sigmund conducted the first theoretical understanding of the sputtering process of energetic ions on the substrate and the formation of nanostructures in 1973 [24, 26, 27]. Sigmund suggested that on sputtering of the ions on the off-normal angle to the surface, the sputtering rate at the site depends on the curvature of the local site (Figure 9.9(a) and (b)). Once the curvature of the surface is formed due to erosion by ion sputtering, further sputtering depends on the local surface rate, which is now different at different surface curvatures. At the surface minima, further erosion will be faster than the surface maxima, creating one surface minimum adjacent to the surface maxima; this leads to the creation of wave-like ripples and increases roughness of the surface (Figure 9.9(c)). These kinds of unbalanced surfaces are the cause of nanostructure formations [24–28]. An apparent mechanism manifesting local diffusion originates from selective sputtering, and as discussed, an actual mass movement is not involved.

9.4.3.1 Nanorippled Structures

Controlled perturbation on the surface due to ion beam irritation with optimized ion beam parameters, leads to the development of ordered and aligned self-organized nanostructures. As discussed in the previous section, the self-organized nanostructures are the nanoscaled pattern formation attributed to the reorganization of the surface atoms leading to periodic or quasi-periodic nanostructures, which occur during the ion irradiation on the surface. It has been known that nanoripples can be fabricated by the non-normal bombardment of FIB on the surfaces [29, 30]. Nanoripples can be achieved with high-order and uniform amplitude on optimizing the parameters of FIB suitable to the surface.

In 2003 Aziz et al. reported rare and spontaneous ripple formation on normal FIB bombardment attributed to linear propagation of FIB-induced microexplosions in the FIB scanning direction [31]. The nanoripples (Figure 9.10) were fabricated on germanium with 30 keV Ga⁺ FIB, and the raster scanning was performed with 50% beam overlap and the ion dose was 1.04×10^{18} ions/cm^2. The observed orientation of ripples in the direction of fast scanning was further confirmed by changing the direction of the raster scan. It can be seen in Figure 9.10 that the nanoripples are oriented in the raster scanning direction. And for the raster scanning by 90°, then ripples are aligned with the new orientation (Figure 9.10(b)). Also, for the spiral scanning from center to outward direction, the ripples were spirally oriented (Figure 9.10(c)). These results are noteworthy in producing the preferred design by varying ripple orientation, but the approach is limited to quasi-periodicity and not suitable for well-ordered nanoripples.

In order to accomplish well-ordered nanostructures on a germanium surface, Mote et al. have demonstrated the nanoscale-controlled progression of various nanoscale structures by FIB-induced self-organization, such as aligned ripples [32, 33] and periodic polygons [33, 34]. The orientation of the nanoripples was found to be strongly dependent on the beam overlap conditions [32]. During this experiment, raster scanning was performed by 30 KeV Ga⁺ FIB with beam current on

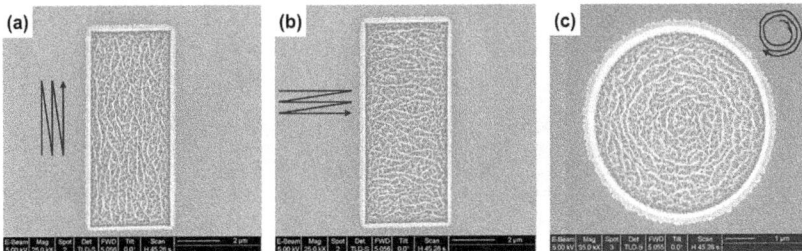

FIGURE 9.10 SEM images showing the ripple orientation toward the fast scan direction as indicated by arrows [31].

the germanium (100) surface. As a result of varying the beam overlap from zero to negative during the FIB irradiation, differently oriented nanoripples were realized with respect to the scanning direction. This orientation manipulation of nanoripples was realized with the FIB incidence angle normal to the surface. The SEM images in Figure 9.11(a) and (b) show the alignment modification for the nanoripples merely by switching the BO from 0% (perpendicular to the scanning direction) to −250% (parallel to the scanning direction) [32]. Moreover, these nanoripples demonstrate 95% light absorptance in the wideband of the visible-near infrared (Vis-NIR) spectrum (Figure 9.11(d)). This enhanced light absorption is attributed to the multiple reflections and improved light trapping through rippled morphologies, as depicted in Figure 9.11(c) [32].

9.4.3.2 Periodic Polygonal Structures

The strategic operation of Ga^+ FIB (FIB/SEM dual-beam system FEI Quanta 3D) scanning with negative beam overlap of −600% using 30 keV energy and 50 pA beam current, the nanoscale self-organization was achieved on a germanium surface to produce periodic polygonal nanostructures. It can be seen from Figure 9.12(a) that the nanoholes are initially circular at the lower dose of ion irradiation (with a dwell time of 0.1 ms, 10-pass scan); however, for the higher dose of irradiation (with a dwell time of 1.8 ms, 10-pass scan) polygonal morphologies such as squares and hexagons are obtained [33, 34]. The researchers have demonstrated that such a polygonal formation mechanism is due to nanoscale-controlled viscous-fingering driven self-organization instead of direct milling.

Concentric scanning with an outer diameter of 5 μm was performed to obtain hexagon-shaped nanoholes and spiraling zigzag geometries [33], as shown in Figure 9.12(b–d). The same beam overlap was set radially for concentric scanning, giving a pitch of −600% BO between adjacent spots. In order to get different doses over the 5 × 5 μm² area, the dwell time was varied from 0.001 to 0.025 s. Figure 9.12 shows a comparison of the nanostructures obtained from low-, intermediate-, and high-dose concentric scans. It is observed from Figure 9.12(b) that the low dose gives circular nanoholes similar to the serpentine scan, and on an increase in the dose,

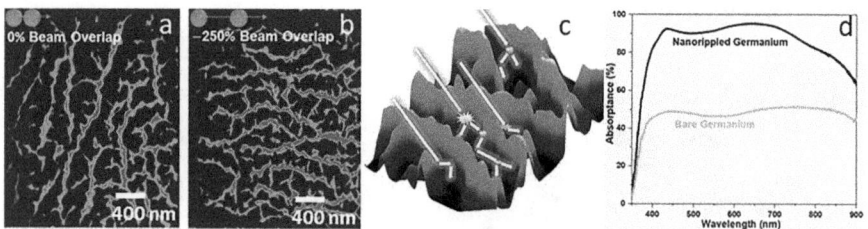

FIGURE 9.11 Self-organized germanium nanoripples via FIB irradiation: The false-colored SEM images of nanoripples on Ge for (a) perpendicular- and (b) parallel-oriented with respect to the scanning direction, (c) schematic for interaction of light with nanoripples, (d) light absorption for nanorippled and bare Ge surface [32].

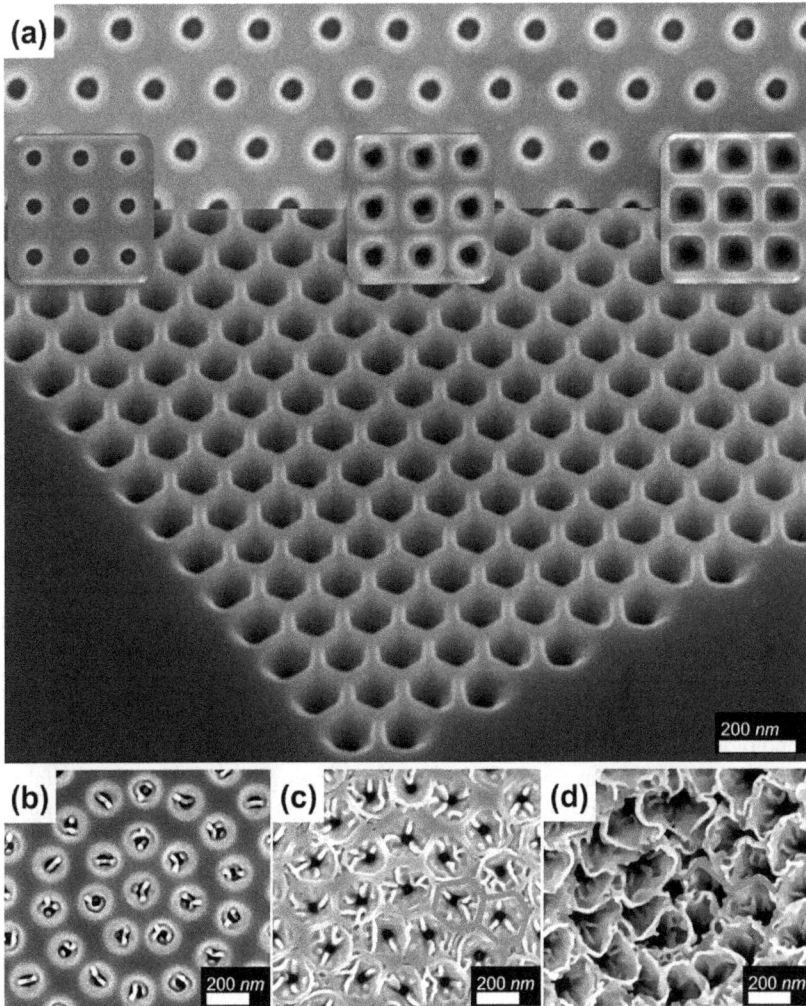

FIGURE 9.12 Evolution of polygonal nanoholes. (a) An array of hexagonal and square nanoholes transformation from circular (at low dose) to polygonal morphologies (at high dose) [34]. (b–d) Concentric FIB scanning for hexagon- and V-shaped nanostructures with (b) low dose, (c) intermediate dose, and (d) high dose (red lines in (c) and (d) represent geometric shapes) [33].

hexagonal geometries are evolved (Figure 9.12(c)). For the higher dose, the walls form a V-shape in the direction opposite to the scan propagation (Figure 9.12(d)). It is observed from Figure 9.12(b) that six beam spots surround any typical spot on the concentrically scanned area; interaction of boundaries of these adjacent spots leads to the formation of hexagon-shaped nanowalled structures for intermediate beam doses, similar to the results in Figure 9.12(a). However, for the higher beam doses, the wall forms zigzag nanostructures which are concentrically spiraled pointing opposite to the scan propagation.

9.4.3.3 Complex Morphologies

In order to fabricate complicated morphologies of periodic nanostructures on the Ge, 30 keV Ga$^+$ FIB was employed with 9.7 pA beam current [33, 35]. The self-organized nanostructures obtained with complex morphologies and high periodicity are shown in Figure 9.13. A range of complicated morphologies for the 3D nanostructures is attained with strategically optimized beam current and beam overlap conditions, as shown in Figure 9.13. These complex morphologies are comprised of interesting structures like corrugations in Figure 9.13(a), nanomushrooms in Figure 9.13(b), and nanoscaled needle-hole-type dual structures in Figure 9.13(c).

The properly focused FIB scanning of a single pass, with a dwell time of 10 ms and −400% BO, produced highly periodic corrugated nanostructures. In contrast, with −600% BO, a little elongated FIB overlap produced nanomushrooms (Figure 9.13(b)) and dual-structured nanofeatures (Figure 9.13(c)). In the case of nanomushrooms, the 10-pass scanning with a dwell time of 3.8 ms was used and 20-pass with a dwell time of 3.2 ms created dual structures. It is important to note that variation in the number of passes and dwell time causes a significant influence on the morphology of structures formed by the elongated FIB due to corresponding self-organization in a preferred manner.

The demonstrated methodologies for varied morphologies provide manipulation of the individual FIB milling spots with nanoscale-control to realize the complex morphologies of the nanostructures. The control accomplished on the transformation of nanoholes can have an advantage in developing relatively effortless fabrication techniques for desired polygonal and complex geometries by engineering each focused ion beam spot.

9.5 SUMMARY

An overview of the fundamentals on the FIB system has been given in this chapter. Examples of FIB-based nanofabrication by material removal and deposition are discussed. Formation and manipulation of self-organized nanostructures by FIB have

FIGURE 9.13 SEM images of complex-shaped self-organization (a) corrugated, (b) nanoscaled mushrooms (inset, tilted at 30°), and (c) nanoneedle-hole type dual structures (inset, tilted at 30°). Scale bars in (a–c) correspond to 200 nm [35].

been reported. Self-organized nanoripples, nanoscale polygons and complex nano-structures have been reported to be formed by FIB irradiation on the various surfaces. The morphologies and spacing of the self-organized nanostructures are found to be strongly influenced by the surface type and operating conditions of the FIB system. FIB has been found to be an ideal instrument for fabricating, modifying, and imaging innovative nanostructured materials and devices. Thus, it can be considered a multi-purpose tool for milling, deposition, characterizing, and tailoring the surface of nano-materials. In a dual-beam FIB/SEM system, by using an SEM column, in-situ imaging and monitoring can also be performed simultaneously while the FIB milling/deposi-tion process is running.

LIST OF SYMBOLS AND ABBREVIATIONS

α	Mass-dependent factor
η	Atomic density
θ	Incident angle for the bombarded ions
μm	Micrometer
μs	Microsecond
σ	Standard deviation
σ_2	Standard deviation for the edge region of the beam for bi-Gaussian model
σ_1	Standard deviation for the central region of the beam for bi-Gaussian model
$\varnothing$	Polar angle
2D, 3D	Two-dimensional and three-dimensional
amu	Atomic mass unit
BCA	Binary collision approximation
BO	Beam overlap
c	The adjustable parameter depending on the ion-solid combinations
C	Coulomb
°C	Degree Celsius
C_e	Charge of a single-charged ion
cm	Centimeter
CVD	Chemical vapor deposition
D_0	Constant dose
$D(d)$	Intensity distribution at a distance d from the beam center
d_f	Beam size
$D(r, \sigma)$	Intensity distribution
eV	Electronvolt
E	Energy
f	The adjustable parameter depending on the ion-solid combinations
FIB	Focused ion beam
f_{mr}	Function depending on the ratio of atomic masses for ion to atom
FWHM	Full width at half maximum
Ga	Gallium
Ge	Germanium

$h(r, t)$	Height at a position $r = (x, y)$ and time t
He	Helium
I	Beam current
IC	Integrated circuit
keV	Kiloelectronvolt
kV	Kilovolt
LMIS	Liquid metal ion source
MEMS	Microelectromechanical systems
ms	Millisecond
nA	Nanoampere
NEMS	Nanoelectromechanical Systems
nm	Nanometer
pA	Picoampere
PMT	Photomultiplier tube
Pt	Platinum
r	Distance
s	Second
SEM	Scanning electron microscopy
Si	Silicon
$S_n(E)$	Stopping power for the elastic collision
SRIM	Stopping and range of ions in matter
t_d	Dwell time
TEM	Transmission electron microscopy
T_m	Momentum transferred due to the ion bombardment
TRIM	Transport of ions in matter
U_s	Surface binding energy
Vis-NIR	Visible-near infrared
w	Weight factor
y	Sputter yield
$Y(0)$	Total sputter yield for the normal incidence angle
$Y(E, \theta)$	Angle and energy-dependent total sputter yield
$Y(E)$	Total sputtering yield

REFERENCES

1. R. M. Langford. Focused Ion Beam Systems: Application to Micro- and Nanofabrication. In *Encyclopedia of Materials: Science and Technology*, edited by K. H. J. Buschow, R. W. Cahn, M. C. Flemings, B. Ilschner, E. J. Kramer, S. Mahajan, and P. Veyssière (Elsevier, Oxford, UK, 2010), pp. 1–13.

2. N. Yao and A. Epstein. Surface Nanofabrication Using Focused Ion Beam. In *Microscopy: Science, Technology, Applications and Education*, edited by A. Mendez-Vilas and J. Díaz, Vol. 3 (Formatex, Badajoz, Spain, 2010).

3. R. G. Mote and L. Xiaomin. Focused Ion Beam (FIB) Nanofinishing for Ultra-Thin TEM Sample Preparation. In *Nanofinishing Science and Technology*, edited by V. K. Jain (CRC Press, Boca Raton, USA, 2016).

4. C.-S. Kim, S.-H. Ahn, and D.-Y. Jang. Review: Developments in Micro/Nanoscale Fabrication by Focused Ion Beams. *Vacuum* 86: 1014 (2012).

5. J. F. Ziegler, M. D. Ziegler, and J. P. Biersack. SRIM – The Stopping and Range of Ions in Matter (2010). *Nucl. Instrum. Methods Phys. Res. B* 268: 1818 (2010).

6. J. Ziegler. *SRIM & TRIM*. http://www.srim.org/.

7. P. Sigmund. Theory of Sputtering. I. Sputtering Yield of Amorphous and Polycrystalline Targets. *Phys. Rev.* 184: 383 (1969).

8. Y. Yamamura, Y. Itikawa, and N. Itoh, *Angular Dependence of Sputtering Yields of Monatomic Solids*, Vol. IPPJ-AM-26 (Institute of Plasma Physics, Nagoya University, Nagoya, Japan, 1983).

9. K. Gamo. Nanofabrication by FIB. *Microelectron. Eng.* 32: 159 (1996).

10. A. A. Tseng, I. A. Insua, J. S. Park, B. Li, and G. P. Vakanas. Milling of Submicron Channels on Gold Layer Using Double Charged Arsenic Ion Beam. *J. Vac. Sci. Technol. B* 22: 82 (2004).

11. G. B. Assayag. New Characterization Method of Ion Current-Density Profile Based on Damage Distribution of Ga^+ Focused-Ion Beam Implantation in GaAs. *J. Vac. Sci. Technol. B* 11: 2420 (1993).

12. S. Tan, R. Livengood, Y. Greenzweig, Y. Drezner, and D. Shima. Probe Current Distribution Characterization Technique for Focused Ion Beam. *J. Vac. Sci. Technol. B* 30: 06F606 (2012).

13. E. Chang, K. Toula, and V. Ray. Reconstructing Focused Ion Beam Current Density Profile by Iterative Simulation Methodology. *J. Vac. Sci. Technol. B* 34: 06KO01 (2016).

14. Y. Drezner, Y. Greenzweig, S. Tan, R. H. Livengood, and A. Raveh. High Resolution TEM Analysis of Focused Ion Beam Amorphized Regions in Single Crystal Silicon—A Complementary Materials Analysis of the Teardrop Method. *J. Vac. Sci. Technol. B* 35: 011801 (2017).

15. J. Han, H. Lee, B.-K. Min, and S. J. Lee. Prediction of Nanopattern Topography Using Two-Dimensional Focused Ion Beam Milling With Beam Irradiation Intervals. *Microelectron. Eng.* 87: 1 (2010).

16. K. Kant and D. Losic. Focused Ion Beam (FIB) Technology for Micro- and Nanoscale Fabrications. In *FIB Nanostructures*, edited by Z. M. Wang (Springer International Publishing, Cham, Switzerland, 2013), pp. 1–22.

17. M. S. Darak, R. G. Mote, and S. Shukla. Single-Spot Focusing With Plasmonic Phase Manipulation. *Ann. Phys.* 530: 1800193 (2018).

18. V. Garg, R. G. Mote, and J. Fu. Focused Ion Beam Direct Fabrication of Subwavelength Nanostructures on Silicon for Multicolor Generation. *Adv. Mater. Technol.* 3: 1800100 (2018).

19. P. Romagnoli, M. Maeda, J. M. Ward, V. G. Truong, and S. NicChormaic. Fabrication of Optical Nanofibre-Based Cavities Using Focussed Ion-Beam Milling: A Review. *Appl. Phys. B* 126: 111 (2020).

20. V. K. Jain, R. Balasubramaniam, R. G. Mote, M. Das, A. Sharma, A. Kumar, V. Garg, and B. Kamaliya. Micromachining: An Overview (Part I). *J. Micromanufact.* 3: 142 (2020).

21. Y. Kang, S. Omoto, Y. Nakai, M. Okada, K. Kanda, Y. Haruyama, and S. Matsui. Nanoimprint Replication of Nonplanar Nanostructure Fabricated by Focused-Ion-Beam Chemical Vapor Deposition. *J. Vac. Sci. Technol. B* 29: 011005 (2011).

22. J. Igaki, K. Kanda, Y. Haruyama, M. Ishida, Y. Ochiai, J. Fujita, T. Kaito, and S. Matsui. Comparison of FIB-CVD and EB-CVD Growth Characteristics. *Microelectron. Eng.* 83: 1225 (2006).

23. W. J. Moberly Chan. Dual-Beam Focused Ion Beam/Electron Microscopy Processing and Metrology of Redeposition During Ion-Surface 3D Interactions, From Micromachining to Self-Organized Picostructures. *J. Phys. Condens. Matter.* 21: 224013 (2009).

24. M. A. Makeev, R. Cuerno, and A.-L. Barabási. Morphology of Ion-Sputtered Surfaces. *Nucl. Instrum. Methods Phys. Res. B* 197: 185 (2002).

25. E. Chason, J. Erlebacher, M. J. Aziz, J. A. Floro, and M. B. Sinclair. Dynamics of Pattern Formation During Low-Energy Ion Bombardment of Si(0 0 1). *Nucl. Instrum. Methods Phys. Res. B* 178: 55 (2001).

26. P. Sigmund. A Mechanism of Surface Micro-Roughening by Ion Bombardment. *J. Mater. Sci.* 8: 1545 (1973).

27. J. Munoz-Garcia, L. Vazquez, R. Cuerno, J. A. Sanchez-Garcia, M. Castro, and R. Gago. Self-Organized Surface Nanopatterning by Ion Beam Sputtering. In *Toward Functional Nanomaterials*, edited by Z. M. Wang (Springer, New York, US, 2009), pp. 323–398.

28. R. M. Bradley and H. Hofsäss. A Modification to the Sigmund Model of Ion Sputtering. *J. Appl. Phys.* 116: 234304 (2014).

29. A. Datta, Y.-R. Wu, and Y. L. Wang. Real-Time Observation of Ripple Structure Formation on a Diamond Surface Under Focused Ion-Beam Bombardment. *Phys. Rev. B* 63: 125407 (2001).

30. Y. Zhang and G. Ran. Self-Assembly of Well-Ordered and Highly Uniform Nanoripples Induced by Focused Ion Beam. *Physica E* 41: 1848 (2009).

31. W. Zhou, A. Cuenat, and M. J. Aziz. Formation of Self-Organized Nanostructures on Ge During Focused Ion Beam Sputtering. In Conference *Series-Institute of Physics*, edited by A. G. Cullis and P. A. Midgley, Vol. 180 (CRC Press, Boca Raton, USA, 2003), pp. 625–628.

32. B. Kamaliya, R. G. Mote, M. Aslam, and J. Fu. Enhanced Light Trapping by Focused Ion Beam (FIB) Induced Self-Organized Nanoripples on Germanium (100) Surface. *APL Mater.* 6: 036106 (2018).

33. B. Kamaliya. *Study of Ion Beam Interaction With Materials and Nanostructure Fabrication*. Thesis, Monash University, 2021.

34. B. Kamaliya, V. Garg, A. C. Y. Liu, Y. (Emily) Chen, M. Aslam, J. Fu, and R. G. Mote. Tailoring Surface Self-Organization for Nanoscale Polygonal Morphology on Germanium. *Adv. Mater.* 33: 2008668 (2020).

35. B. Kamaliya, V. Garg, R. Mote, M. Aslam, and J. Fu. Controlled Self-Organization on Germanium Using Focused Ion Beam (FIB): From Quasi-Periodic Nanoripples to Well-Ordered Periodic Nanostructures. *Microsc. Microanal.* 26: 1684 (2020).

10 Electrochemical Machining

Vyom Sharma and Divyansh Patel

CONTENTS

DOI: 10.1201/9780429160011-10

10.1   INTRODUCTION

In electrochemical machining (ECM), the volume of material dissolved from the work-piece is directly proportional to the machining current and atomic weight of the anode material, and it is inversely proportional to its dissolution valency. Since the mechanism of material removal in ECM does not appreciate any degree of hardness in the work-piece, materials with high strength and hardness, which are difficult to process using other machining techniques, can be easily machined using ECM. During the process of electrolysis, hydrogen gas evolves at the cathode and no material is dissolved from it. Therefore, the shape of the tool is retained after machining, and this makes ECM one of the most suited options for micromachining operations. The surface generated after ECM is smoother compared to other advanced machining processes (AMPs). However, the machining of alloys still remains a challenge. Since it is a non-thermal and non-contact process, there are no thermal or mechanical stresses induced into the work-piece after machining. Unlike in electrical discharge machining (EDM) and laser beam machining (LBM), there is no heat-affected zone or burr formation in this process [1]. However, it is equally important to mention that controlling the machining tolerance in this process is a challenge and requires a sound knowledge of the effect of different parameters influencing process performance and tool design. Another problem which restricts its industrial applications is the need to design a profiled tool before perform-ing actual machining. This is because the profile of the workpiece after machining is an approximate negative replica of the tool profile. Over a period of time, research-ers have developed different hybridizations and variants of ECM to perform various operations such as grinding, milling, drilling, profiling, turning, parting, and slitting. In this chapter, different fundamental aspects of ECM are presented. The fundamental principle of material dissolution in ECM is presented using the different chemical reac-tions involved. The mechanics of machining in the ECM process are explained using a detailed mathematical formulation for the material removal rate (MRR). Various parameters which influence the process performance in ECM are introduced and a detailed classification of different types of electrolytes used in the process is also pre-sented. Finally, different variants of ECM such as electrochemical drilling, shaped-tube electrochemical machining, electrolytic stream drilling, electrochemical jet drilling, and wire electrochemical machining are also discussed in detail.

10.2   PRINCIPLE OF ELECTROLYSIS

In electrolysis, two electrodes are connected to the opposite terminals of a battery, and they are dipped into an electrically conducting solution called an electrolyte. On passing electrical current through them, a physical change occurs at the elec-trode which is connected to the positive terminal (anode) of the battery as shown in Figure 10.1. The flow of current in the electrochemical (EC) cell thus formed is due to the movement of ionic species in the electrolyte. These ionic species are gener-ated due to the dissociation of the salt present in the solution into its constituents. Ions of the anode material leave the interface, enter the bulk solution, and move towards the cathode. Electrons leave the tool (cathode) and move to the opposite end of the battery through the external circuit. In presence of the applied electric field,

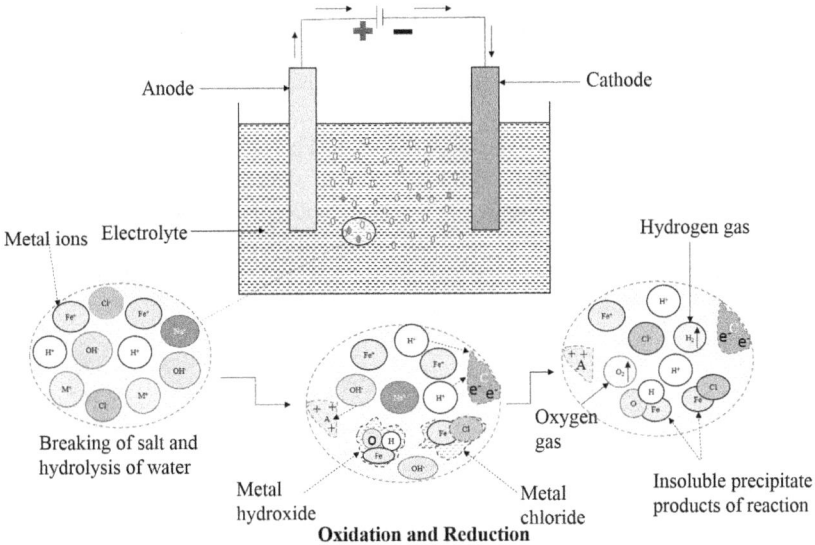

FIGURE 10.1 Schematic showing the dissociation of salts, reduction of metal ions, and precipitation of electrolysis products during ECM.

hydrolysis of water (present in the electrolyte solution) takes place. H^+ ions, thus generated, migrate towards the cathode and are reduced to generate hydrogen gas. The chemical reactions which take place at the two electrodes with metal M as anode are expressed in Equations 10.1 and 10.2 as follows [2, 3]:

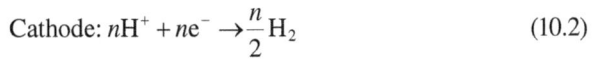

$$\text{Anode: } M \rightarrow M^{n+} + ne^- \tag{10.1}$$

$$\text{Cathode: } nH^+ + ne^- \rightarrow \frac{n}{2}H_2 \tag{10.2}$$

where n is the number of valence electrons.

Metal ions present in the solution combine with the hydroxyl ions generated due to the hydrolysis to form metal hydroxides. The metal ions may also combine with the anions generated due to the splitting of the salt into its constituents to form metal halides. For example, when sodium chloride (NaCl) is used as a salt in an electrolyte, metal hydroxide and metal chloride generally form and are precipitated in the solution. These precipitates are electrically non-conducting in nature, and they may lead to the clogging of the gap between the two electrodes. Therefore, it is essential to flush out these electrolysis products to ensure smooth motion of the ionic species in the solution.

10.3 THEORY OF ECM

10.3.1 MECHANICS OF MACHINING

In 1834 Michael Faraday gave two laws which quantify the process of electrolysis. These two laws form the basis for estimating the amount of material which gets dissolved from the anode in a given time span. The first law states that the mass of material dissolved from the anode is directly proportional to the quantity of charge (measured in coulombs) passed between the two electrodes. The second law states that the mass of material dissolved from the anode is also directly proportional to the gram equivalent weight of that anode material. The gram equivalent weight of an element is a ratio of atomic mass and the valency of dissolution of that element. The valency of dissolution depends upon the reaction undergone by the element. The reaction undergone by the element depends upon its Gibbs free energy. Quantitatively, the relationship between the charge passed and the mass of material dissolved can be expressed by combining the aforementioned two laws as shown in Equation 10.3.

$$M = \left(\frac{Q}{F}\right)\left(\frac{A}{Z}\right) g \qquad (10.3)$$

where Q is the total electric charge passed, F is Faraday's constant ($\approx$96,500 C/mol), A is the atomic mass of anode (workpiece) material, and Z is the valence number of ions of anode material. $\left(\dfrac{A}{Z}\right)$ is the equivalent weight of the altered substance (anode). For a case when current density is constant, M can be expressed in grams as shown in Equation 10.4 [3].

$$M = \left(\frac{It}{F}\right)\left(\frac{A}{Z}\right) g \qquad (10.4)$$

Here, I is the current passed in the electrochemical cell and t is the dissolution time. Let the current efficiency of the system be η, then MRR can be expressed as follows (Equation 10.5).

$$MRR = \left(\frac{\eta IA}{FZ}\right) g/s \qquad (10.5)$$

Let the density of workpiece material be ρ, the volumetric material removal rate (MRR_V) can be expressed as shown in Equation 10.6.

$$MRR_V = \frac{\eta IA}{FZ\rho} \ m^3/s \qquad (10.6)$$

Also, let the common area of tool and workpiece through which the current is flowing be A_C, then linear material removal rate MRR_l can be expressed by Equation 10.7.

$$MRR_l = \frac{\eta IA}{FZ\rho A_C} \ m/s \qquad (10.7)$$

Let, $\left(J = \dfrac{I}{A_C}\right)$ be the current density over the common area A_C. Using Ohm's law, the current I can be expressed in terms of applied potential V, over-potential ΔV, and gap resistance R as shown in Equation 10.8.

$$MRR_l = \frac{\eta IA}{FZ\rho A_c} = \frac{\eta JA}{FZ\rho} = \frac{\eta(V-\Delta V)A}{FZ\rho R} = \frac{\eta(V-\Delta V)A\kappa}{FZ\rho y} \text{ m/s} \quad (10.8)$$

where y is the interelectrode gap (IEG) and κ is the electrical conductivity of the electrolyte between the two electrodes (Figure 10.2).

Further, for the case of non-zero tool/workpiece feed rate f, the time rate of change of IEG $\left(\dfrac{dy}{dt}\right)$ can be expressed as shown in Equation 10.9.

$$\frac{dy}{dt} = MRR_l - f \quad (10.9)$$

On substituting the expression of MRR_l from Equation 10.6 in Equation 10.7, the expression for $\left(\dfrac{dy}{dt}\right)$ becomes as follows.

$$\frac{dy}{dt} = \frac{\eta(V-\Delta V)A\kappa}{FZ\rho y} - f \quad (10.10)$$

For the condition of equilibrium to exist $\left(\dfrac{dy}{dt} = 0\right)$, the expression for f can be written as expressed in Equation 10.11.

$$f = \frac{\eta(V-\Delta V)A\kappa}{FZ\rho y} \quad (10.11)$$

On rearranging Equation 10.11, the expression for equilibrium IEG (y_e) can be written as expressed in Equation 10.12.

$$y_e = \frac{\eta(V-\Delta V)A\kappa}{FZ\rho f} \quad (10.12)$$

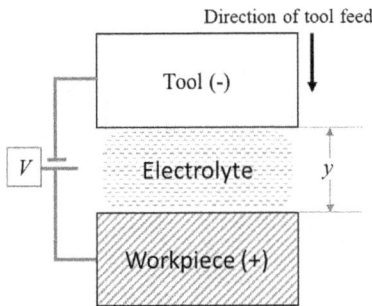

Direction of tool feed

Tool (-)

V

Electrolyte y

Workpiece (+)

FIGURE 10.2 Schematic showing the configuration of tool and workpiece in the ECM process.

Now, consider Equation 10.10, let $\dfrac{\eta(V-\Delta V)A\kappa}{FZ\rho} = C$ (any constant). For the case when $f = 0$, the rate of change of IEG can be expressed as shown in Equation 10.13.

$$\frac{dy}{dt} = \frac{C}{y} \tag{10.13}$$

On integrating Equation 10.13, the expression for y can be written as expressed in Equation 10.14.

$$y = y(t) = \sqrt{\left(2Ct + y_o^2\right)} \tag{10.14}$$

Here, y_o is the IEG at time $t = 0$ s. This expression suggests that IEG increases in proportion to the square root of the machining time (Figure 10.3(a)). For the case when $f \neq 0$, the relationship between y and other process parameters can be expressed as shown in Equation 10.15.

$$y + \frac{\eta E(V-\Delta V)\kappa}{Fpf} \ln\left(\frac{\eta E(V-\Delta V)\kappa}{F\rho} - fy\right)$$
$$= y_0 + \frac{\eta E(V-\Delta V)\kappa}{Fpf} \ln\left(\frac{\eta E(V-\Delta V)\kappa}{F\rho} - fy_0\right) - ft \tag{10.15}$$

When $f = MRR_l$, the IEG remains constant, and machining becomes stable (that is equilibrium condition of machining). For the case when $f < MRR_l$, y increases initially and attains a value greater than y_e. Due to this, the gap resistance increases and the current density over the anode surface decreases. This leads to the reduction in MRR_l and y starts decreasing and tends to attain the equilibrium value y_e. Conversely, if $f > MRR_l$, the value of y initially starts decreasing. This leads to the

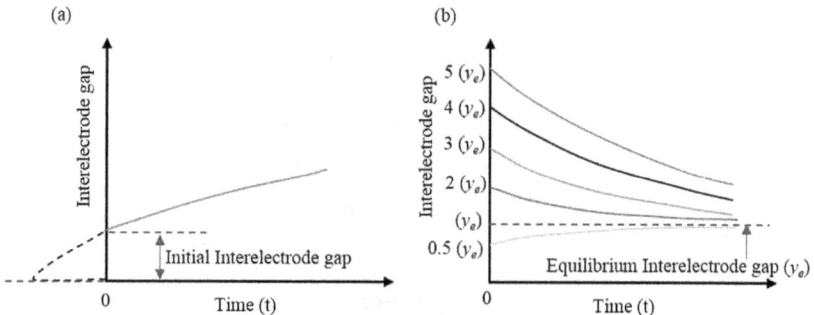

FIGURE 10.3 Graph showing (a) the variation of IEG with time and (b) the IEG attaining the equilibrium value with time for different values of initial IEG.

reduction in the gap resistance, and consequently, the current density on the anode surface increases. Due to this, the value of MRR_t increases and y starts increasing again to attain the equilibrium value (Figure 10.3(b)). This feature of ECM to attain the equilibrium IEG during machining is called the "*self-regulating*" feature.

10.3.2 Electrochemical Equivalent of Alloys

Shown in Equation 10.3 is the weight of metal M which gets dissolved from the anode consisting of a single element. However, the calculation of M for alloys is not straightforward as the electrochemical equivalent $\left(\dfrac{A}{Z}\right)_{alloy}$ of the alloy is not known. This is calculated using the "percentage by weight" method, wherein the electrochemical equivalents of individual elements are multiplied by their weight fractions and then they are summed up. Let $\left(\dfrac{A_i}{Z_i}\right)$ be the electrochemical equivalent of element i in the alloy and has a percentage by weight fraction of X_i. Then, the electrochemical equivalent of an alloy can be expressed as in Equation 10.16.

$$\left(\frac{A}{Z}\right)_{alloy} = \frac{1}{100}\sum_{i=1}^{p}\left(\frac{A_i}{Z_i}\right)X_i \qquad (10.16)$$

where p is the number of elements present in the alloy. In Equation 10.16, $X_1 + X_2 + X_3 + \ldots X_n = 1$.

10.3.3 Equivalent Density of Alloys

Let us consider one gram of an alloy with a density of $\left(\rho_a\right)$ and volume of V_a. Then, it is convenient to write:

$$\left(\rho_a\right)\times\left(V_a\right) = 1 \qquad (10.17)$$

Alternatively, we can also write:

$$\left(\rho_a\right) = \frac{1}{\left(V_a\right)} = \frac{1}{V_1 + V_2 + V_3 + V_4 + \ldots + V_p} \qquad (10.18)$$

In Equation 10.18,

$$V_1 = \frac{X_1}{100\rho_1}, V_2 = \frac{X_2}{100\rho_2}, \ldots, V_p = \frac{X_p}{100\rho_p}.$$

Therefore, the expression for $\left(\rho_a\right)$ can be written as shown in Equation 10.19.

$$\left(\rho_a\right) = \frac{100}{\sum\nolimits_{i=1}^{p}\left(\dfrac{X_i}{\rho_i}\right)} \tag{10.19}$$

By using Equation 10.19, the equivalent density of an alloy can be calculated.

10.4 PROCESS PARAMETERS

A fishbone diagram showing the classification of parameters which influence ECM process performance is given in Figure 10.4. The criteria for evaluating the process performance are MRR, accuracy of machining which is estimated by measuring the overcut generated during the machining process, and quality of the surface generated after machining. Quality of surface is estimated by measuring the surface roughness, surface integrity, and the presence of microcracks. Process parameters are those parameters which are independent and are controlled by the operator or by the sensors in the automatic/computer numeric control (CNC) machine tools. In ECM, the common parameters are applied potential (constant or pulsed DC), electrolyte conductivity, temperature, and flowrate; tool/workpiece feed rate and its geometrical parameters; and the IEG. Other parameters which also have a significant influence on the performance in ECM are classified under power supply, tool, feed rate, IEG, electrolyte, and mechanical parameters.

10.5 ECM EQUIPMENT

The system mainly consists of electrodes, electrolyte, supplier, triaxial pulse motor, piezoelectric (PZT) power supply, and personal computer [4]. The electrode can

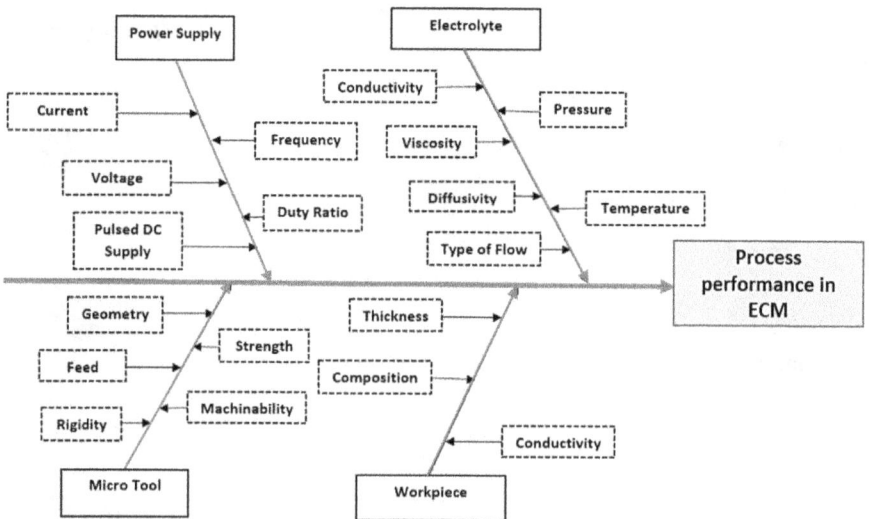

FIGURE 10.4 Fishbone diagram showing different process parameters in ECM.

move in Z direction by using the pulse motor. The PZT actuator enables high-speed X, Y, and Z-direction movements of the electrode. Electrolyte at a high flow rate can be supplied in the machining region through a flexible tube to flush the electrolysis products out of the narrow IEG and the two electrodes can be kept submerged in the electrolyte tank (see Figure 10.5). The movement of triaxial drive is controlled with the help of a microcontroller which is connected to the computer. Commands are sent to the controller in the form of electrical signals with the help of a graphic user interface. Power supply in ECM can be DC or pulsed DC. Different waveforms of pulsed voltage can be used in ECM such as half-wave rectified rectangular, sinusoidal, or triangular. A function generator can be used to generate such types of waveforms. In ECM, online monitoring of the machining current and tool/workpiece position is very important. This helps in controlling the movement of the X, Y, and Z stages accordingly. IEG can be predefined and maintained constantly during the machining of microfeatures. Machining of 3-D shapes can be conveniently carried out by utilizing the scanning movement of a prismatic electrode under the optimum parametric conditions. Pulsed DC voltage is supplied to the electrodes which helps in localizing the dissolution.

10.6   ELECTROLYTES IN ECM

10.6.1   ROLE OF ELECTROLYTES

The electrolyte in ECM provides a passage for the movement of ions between the two electrodes. The flow of electrolyte also ensures the removal of electrolysis products out of the machining zone. Electrical conductivity of the electrolyte plays an important role as it is responsible for the exchange of ions (charge transport) between the electrodes. Even a small change in its value may significantly alter the machining conditions. A change in its temperature and the presence of electrolysis products alters the electrolyte conductivity. As machining starts, there is an increase in concentration of the ions present in the electrolyte. This corresponds to more

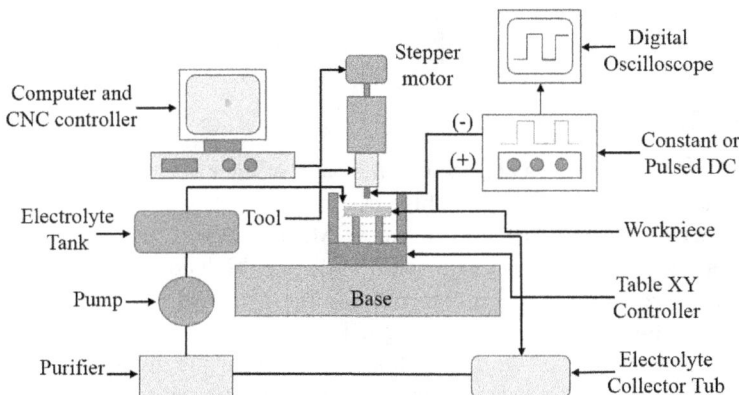

FIGURE 10.5   Schematic showing the different components of an ECM setup.

charge carriers for a unit volume of electrolyte, hence, an increase in concentration increases electrolyte conductivity. Metal ions released into the electrolyte combine with negative ions and form precipitates, reducing the effective number of charge carriers. Also, the formation of hydrogen replaces the electrolyte from the cathode–electrolyte interface by a significant amount. Hence, the formation of precipitates and hydrogen bubbles reduces the electrolyte conductivity. Initial conductivity κ_0 of an electrolyte can be expressed as given by Equation 10.20:

$$\kappa_0 = F \sum_{i=1}^{m} z_i u_i c_i \qquad (10.20)$$

where z_i is valency, u_i is mobility, and c_i is concentration of ions. To account for the influence of hydrogen gas bubbles generated at the cathode on the electrical conductivity of electrolyte, Hopenfield et al. proposed an expression for the effective electrolyte conductivity (κ_{eff}) as shown in Equation 10.21 [5]:

$$\kappa_{eff} = \kappa_0 \left(1 + \alpha \Delta T\right)\left(1 - \alpha_v\right)^n \qquad (10.21)$$

where κ_0 is initial conductivity of the electrolyte, α is temperature coefficient of specific conductance, ΔT is change in temperature, α_v is void fraction in electrolyte and n is exponent (value ranges from 1.5 to 2). However, Equation 10.21 is silent about the influence of sludge particles on the electrolyte conductivity that are generated due to the formation of metal oxides, halides, and hydroxides as reaction products in the machining zone. This makes electrolyte in the machining zone act as a three-phase medium. Jain et al. [6] proposed an improved equation for estimating more realistic electrical conductivity of the electrolyte, and it is given by Equation 10.22:

$$\kappa_{eff} = \kappa_0 \left(1 + \alpha \Delta T\right)\left(1 - \alpha_v\right)^n \left(1 + \alpha_m\right)^m \qquad (10.22)$$

where α_m is the solid fraction in the electrolyte and m is an exponent. This is a generalized equation and the + or – sign and exponent value will depend upon the properties of the reaction products formed.

10.6.2 Types of Electrolytes

Electrolytes in ECM are categorized on the basis of their pH value as acidic, basic, and neutral. A broad classification of the electrolytes is shown in Figure 10.6 [7].

When acidic electrolytes are used in ECM, the electrolysis products are dissolved into it and do not clog the machining gap. A high concentration of hydrogen ions in the solution aids in neutralizing the hydroxyl ions generated at the cathode [8]. However, while machining using hydrochloric acid, there is a problem of the blackening of the tool. This is because of precipitation of iron due to the migration of iron ions towards the cathode [9]. Basic electrolytes such as sodium hydroxide are usually avoided in ECM because high hydroxide ion

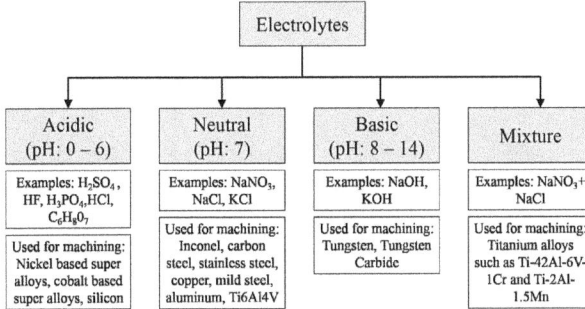

FIGURE 10.6   Schematic showing a broad classification of electrolytes used in ECM.

concentration increases the precipitation of metal hydroxides [9]. This may clog the narrow IEG and even lead to the termination of machining. However, potassium hydroxide is often used for the machining of pure tungsten as the hydroxyl ions assist in rupturing the oxide film which forms during machining [10]. In ECM, neutral electrolytes are the most widely used. Sodium chloride offers high machining efficiency and stability, but results in low machining accuracy and a high corrosion rate. Sodium nitrate offers high machining accuracy, but the machining efficiency is compromised significantly [11]. Another neutral electrolyte, potassium chloride, is non-passivating in nature and is widely used for machining aluminum [12]. Apart from this, electrolytes prepared by mixing two distinct salts have also been tested for the machining of alloys [13]. Although preliminary results are encouraging, more research is needed in this area.

10.6.3   Different Electrolyte Flow Configurations

In ECM, it is necessary to maintain sufficient electrolyte flow between the tool and workpiece in order to carry away the heat and the electrolysis products out of the narrow IEG. This is necessary to conduct machining at the designed machining rate. In the absence of electrolysis products in the machining region, the current density distribution over the anode surface remains stable and results in a good surface finish. Cavitation, stagnation, and vortex formation should be avoided in the path of electrolyte flow. All the corners in the flow path should have some corner radius as shown in Figure 10.7.

Although a tool with electrolyte supply slot is simple to manufacture, such a slot often leaves small ridges on the workpiece. The distance between the tip of the slot and the corners should be at least 1.5 mm, whereas a slot with a width 0.7–0.8 mm is recommended. When a workpiece corner is rounded, the slot end should be made larger. The shape and location of the slot should be such that every portion of the surface is supplied with electrolyte flow and no passive area exists (see Figure 10.8).

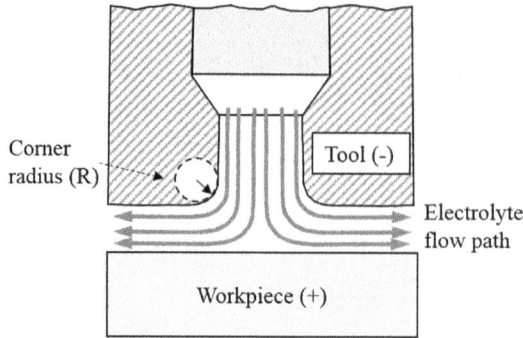

FIGURE 10.7   Schematic showing rounded corners in the electrolyte flow path in ECM.

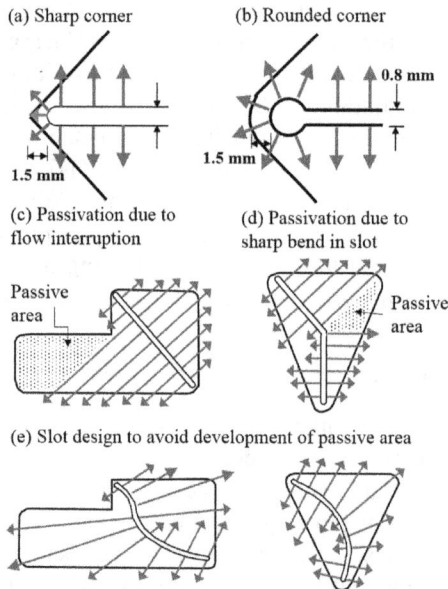

FIGURE 10.8   Schematic showing electrolyte flow scheme for the workpiece with (a) a sharp corner, (b) rounded corner, (c) passivation due to flow interruption, (d) passivation due to sharp bend in slot, and (e) slot design to avoid development of passive area.

10.7   POWER SUPPLIES IN ECM

Material removed from the workpiece surface largely depends upon the amount of electric current flowing through the electrolyte, which makes the power supply a most important part of the process. Power supply during machining can be of direct current (DC) or pulsed DC. When power supply in the form of DC is applied, material removal from the workpiece takes place continuously and production of a large quantity of reaction products takes place. The reaction products are to be removed continuously from the machining chamber in order to continue the machining process. A major portion of these reaction products in the narrow IEG are removed by the high velocity electrolyte. The remaining product starts to damage the tool material by becoming accumulated over its surface. It results in the slowing down of the workpiece dissolution rate. These accumulated reaction products may also lead to a change in the electrolyte temperature which adversely affects the machining accuracy and performance. These problems, along with other problems such as non-localized electric field and boiling of electrolyte in the narrow IEG, can be solved by using pulse power supply. With such a power input, machining takes place only during the pulse on-time. The reaction products are removed during the pulse off-time. This minimizes the problem of the boiling of the electrolyte and improves the process performance. Generally, rectangular waveform of a pulsed voltage is used for machining. Since this waveform requires high frequency pulses for the localization of the dissolution zone, the machining rate is usually compromised. To overcome this problem, researchers have also proposed the use of sinusoidal and triangular waveforms of pulsed voltage for machining. It is already shown using the experimental results that low frequency triangular voltage pulses can provide better machining accuracy and surface quality compared to high frequency rectangular voltage pulses [14]. For ECM, half-wave rectified and amplified voltage pulses can also be used [15]. A schematic diagram showing the process of rectification and amplification of the signal obtained from a function generator is shown in Figure 10.9.

10.8 APPLICATIONS

As the mode of material removal in ECM is by anodic dissolution at an ionic level, the process performance is not dependent on the thermal and mechanical properties of the material to be machined. Due to this, ECM is widely used for the machining of turbine blades made of high-strength temperature-resistant (HSTR) alloys. Different operations that are possible using ECM are drilling, etching, turning, grooving, deburring, die sinking, parting, grinding [16], profiling, etc. [7]. Also, since there is, theoretically, no limitation on the minimum dimension of the tool, the process is one of the most suited options for micromachining operations. ECM is also used for generating textures on a large, conducting, flat or a freeform surface [17]. Parts and features made using ECM find extensive applications in industries related to biomedical devices, aerospace, maritime, automobile, energy generation, and defense.

10.9 ADVANTAGES AND LIMITATIONS OF ECM

ECM offers a few unique advantages over other advanced machining processes which makes it an industrially viable option for machining parts and features at both the macro and micro scales. However, there are also some limitations in the process which limit its scope for commercialization. These advantages and limitations are listed below.

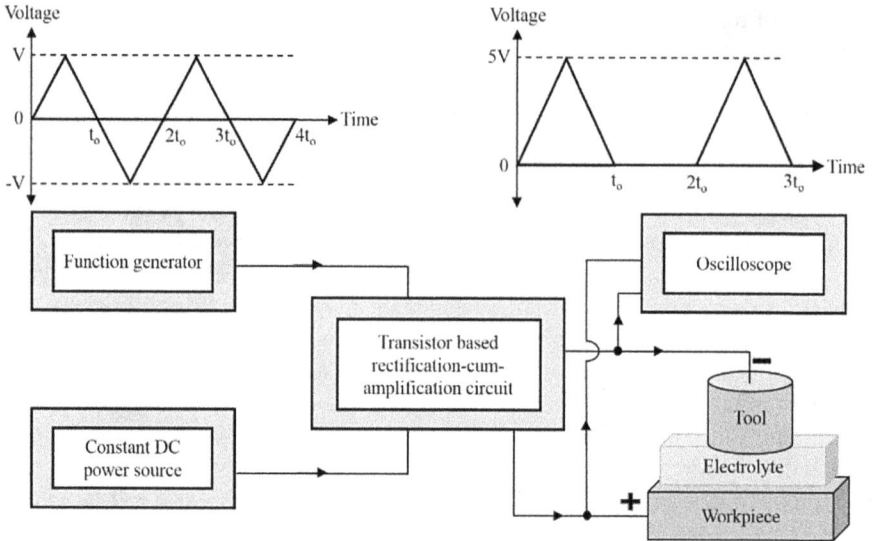

FIGURE 10.9 Rectification cum amplification electronic circuit developed for obtaining rectangular, sinusoidal, and triangular voltage pulses.

10.9.1 ADVANTAGES

1. Due to the nature of material dissolution in ECM process, its machining rate depends only upon the atomic weight and valency of ions, the magnitude of electric charge passed, and the duration of machining. In ECM, the thermal and mechanical properties of the anode material do not influence the machining rate.
2. The machined surface is burr-free and is free from thermal and mechanical stress, a heat-affected zone, and microcracks.
3. Complex 3-D macro/microfeatures and end-use products can be conveniently produced in a single step by designing a single tool (cathode) for die sinking-type ECM.
4. The ECM process offers a high degree of scalability and repeatability. In comparison to other advanced machining processes, ECM is highly cost effective. Its setup requires simple components, and the operation cost is also comparatively very low.
5. In ECM, the machined surface is burr-free and possesses a high degree of surface integrity. Thermal or mechanical stresses are not induced, and the properties of the parent material are retained after machining.

10.9.2 LIMITATIONS

1. As the depth of the machined feature increases, the availability of adequate electrolyte at the bottom of the structure where anodic dissolution has to

take place is restricted. Consequently, high aspect ratio features are comparatively difficult to machine using ECM.

2. In ECM, there is a presence of undesirable current flux that is not taking part in the actual machining direction according to the requirement based on the designed features of the workpiece. This current is termed as "stray current" and it cannot be eliminated completely during the machining process. Due to this, material is dissolved from the undesired regions of the workpiece, and it results in poor machining accuracy.

3. Since, in ECM, the shape of the machined feature is an approximate negative replica of the tool, there is a need to follow a tool design procedure, according to the required anode shape, before performing actual machining.

10.10 METHODS OF MINIMIZING STRAY CURRENT EFFECT IN ECM

10.10.1 TOOL INSULATION

The areas on a tool through which ECM is not desirable should be insulated. In die sinking, the tool should be properly insulated to minimize stray machining (see Figure 10.10). The insulation must be tough and securely bonded to the tool surface. It can be provided by securing the reinforced solid plastic material to the tool with epoxy resin cement and plastic screws. Sometimes, the insulation can also be done by applying a synthetic rubber coating on the artificially oxidized copper tool surface. For this, a hot chemical oxidizing solution is used. The boundaries of the insulation layer should not be exposed to a high velocity electrolyte flow as this may tend to tear up the glued layer.

10.10.2 WORKPIECE INSULATION USING A MASK

Another way of minimizing the stray current effect in ECM is by using a mask. A mask made of a photoresist material is prepared by creating the openings into it

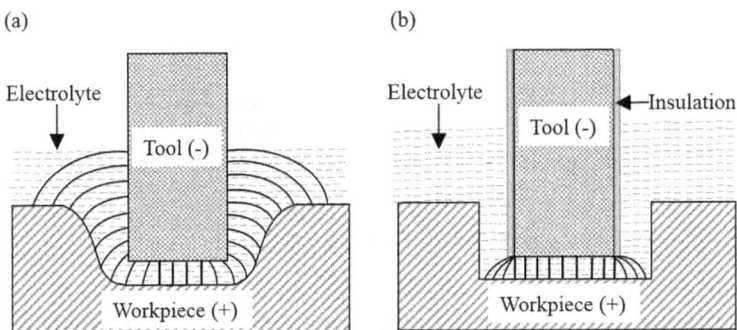

FIGURE 10.10 Schematic showing ECM with (a) uninsulated tool, (b) insulated tool.

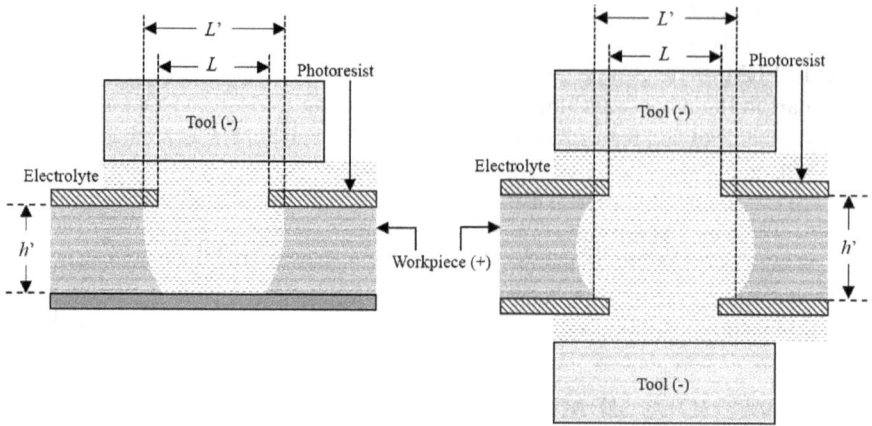

FIGURE 10.11 Schematic showing through mask ECM process from (a) one side of the workpiece, (b) both sides of the workpiece.

according to the desired anode geometry. This mask is then adhered on the surface of the workpiece. Since the photoresist material is insulating in nature, material gets dissolved from the regions where the anode material without a mask is exposed to the electrolyte. A schematic diagram of through-mask ECM from one side and both the sides of the workpiece is shown in Figure 10.11. It is important to note here that since the material removal in ECM is isotropic in nature, some parts of the workpiece beneath the mask also get dissolved in the process. This is called "undercutting". A factor, termed an etch factor (EF), is used to determine the machining quality in through-mask ECM. It is defined as the ratio of the length of straight through cut h' to the length of the undercut $(L'-L)/2$. The mathematical expression for EF for one sided and two-sided etching are given by Equations 10.23 and 10.24, respectively [4].

$$\text{For one-sided etching: } (EF) = \frac{h'}{(L'-L)/2} \tag{10.23}$$

$$\text{For two-sided etching: } (EF) = \frac{h'}{(L'-L)} \tag{10.24}$$

In through-mask ECM, the undercut and shape of the evolved surface depends upon the feature-aspect ratio, spacing-to-opening ratio, and the ratio of film thickness and the anode thickness. A higher value of EF indicates a smaller undercut.

10.11 ANODE SHAPE PREDICTION IN ECM

To predict the evolution of the anode profile during ECM, the finite element method (FEM) is used. This method was first proposed by Jain and Pandey [18], wherein, the simulation domain is discretized into small parts using elements such as triangle, rectangle, polygon (in 2-D), tetrahedron, cuboid, etc. (in 3-D). At each node, the value of electric potential is obtained by solving the Laplace equation for the potential function φ in a Cartesian coordinate system as expressed in Equation 10.25.

$$\frac{\partial^2 \left[\varphi(x,y) \right]}{\partial x^2} + \frac{\partial^2 \left[\varphi(x,y) \right]}{\partial y^2} = 0 \tag{10.25}$$

Figure 10.12 shows a schematic diagram of the discretized simulation domain with different boundaries marked therein. n_x and n_y represent the unit vectors along x and y directions, respectively.

Equation 10.25 is solved numerically for the value of φ at different nodes in the simulation domain with the help of the boundary conditions mentioned in Table 10.1.

Over a period of time, researchers have simulated this process to predict the shape of the final anode profile which is specific to a given tool geometry. However, it is important to mention here that some of the key aspects of anodic dissolution, such as a potential drop due to charging and discharging of the electric double layer (EDL), reduction in the material dissolution rate due to the formation of passive oxide film, and variation in the instantaneous current density distribution in the machining gap due to the formation of hydrogen gas bubbles and electrolysis products are yet to be included in numerical models.

10.12 ENVIRONMENTAL AND HEALTH HAZARDS

In ECM, the total machining cost also includes the cost incurred in the disposal of waste electrolyte after machining. The waste electrolyte may contain acids, alkalis, nitrates, sulphates, chromates, and other solid precipitates [19]. Compounds containing chromium-(IV) are considered toxic in nature and their overground or underground disposal is not permitted in most countries. Consequently, detoxification methods are adopted to convert chromium-(IV) compounds into less hazardous

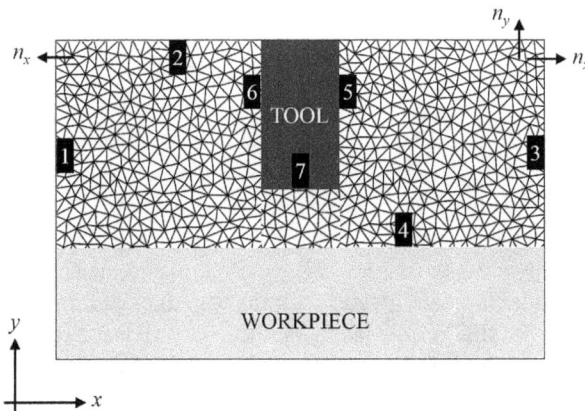

FIGURE 10.12 Schematic of the discretized simulation domain in ECM with a triangular element.

TABLE 10.1

List of Different Boundary Conditions Used for Solving Equation 10.24

Boundary Number	Condition	Implication
1 and 3	$\dfrac{\partial \varphi}{\partial n_x} = 0$	These are insulated boundaries. No electric flux escapes the simulation domain bounded by these boundaries.
2	$\dfrac{\partial \varphi}{\partial n_y} = 0$	This is an insulated boundary. No electric flux escapes the simulation domain bounded by this boundary.
4	Surface potential = φ	This is an isopotential surface. The value of potential on this surface is equal to the applied potential in ECM.
5, 6, and 7	Surface potential = 0	This is an isopotential surface. The value of potential on this surface is equal to the zero.

chromium-(III) compounds and also for the detoxification of nitrides present in the waste electrolyte. This adds to the total machining cost of the process [20]. Also, precautions must be taken by the workers to ensure that they are not exposed to such toxic wastes in the electrolyte stream. The hydrogen gas which is generated due to the hydrolysis of water is flammable in nature, and therefore it must be vented out of the machining area and breathing zone of the workers.

10.13 SOME VARIANTS OF ECM

Over a period of time, researchers across the globe have developed different variants of ECM for performing operations such as milling, drilling, slitting, parting, turning, etc. A variant of ECM is a process which has the same mechanism of material removal as that of ECM, but the geometry of the tool and kinematics of the tool–workpiece interaction (non-physical) are different. The different variants of this process are discussed in the following section.

10.13.1 ELECTROCHEMICAL DRILLING

In electrochemical drilling (ECD), the electrolyte is supplied from the center of the tool and comes out through the side-machining gap which forms between the walls of the tool and the drilled hole walls [18]. The dissolution rate is highest where the magnitude of current density on the anode surface is highest. Side EC dissolution occurs laterally between the sidewalls of the tool and the component. The hole diameter produced is therefore greater than the tool diameter. In ECD, the passage of current through the IEG results in the anodic dissolution at a rate that is governed by Faraday's laws of electrolysis. Sparking takes place at a critical feed rate where the speed of tool advance toward the workpiece is greater than the anodic dissolution rate. Under these circumstances, the gap in front of the tool decreases to a critical value at which sparking occurs and may damage both tool and workpiece. It is also important to note that ECD is not limited to circular holes since a tool with any cross-section can machine the corresponding shape in the workpiece. A schematic diagram of ECD configuration is shown in Figure 10.13.

10.13.2 Shaped Tube Electrolytic Machining

Shaped tube electrolytic machining (STEM) is a variant of ECM which is used to machine high aspect ratio turbulated holes in the workpiece. Such turbulated holes are used in turbine blades through which water is supplied to carry away the heat. The mechanism of material removal in this process is same as that in ECM, except that the tool is specifically shaped/selectively insulated according to the requirement. Since the gap between the tool and the workpiece is very narrow, the flushing of electrolysis products from the deeper regions of the workpiece is very difficult. To overcome this, acidic electrolytes are generally used in this process as the electrolysis products get dissolved in the acidic medium. After the metal ions are dissolved in the solution, they are removed by the electrolyte flow. Machining setup in STEM is similar to that of the ECM process, except that it must be resistant to acids, and it should also have a periodically reverse polarity power supply. A periodic reversal of polarity is used to prevent the accumulation of the undissolved machining products on the tool surface. STEM does not leave a heat-affected layer, which may lead to the development of microcracks. a schematic diagram of STEM configuration is shown in Figure 10.14.

10.13.3 Electrostream (Capillary) Drilling

Electrostream drilling (ESD) is a technique used for the machining of fine holes that are very deep to machine by EDM and very small to drill by STEM. The tool used in ESD is made of a glass nozzle, the diameter of which is smaller than the required hole diameter. To conduct ESD, the current through the acid electrolyte that fills the IEG is passed through a platinum wire electrode that is fitted inside the glass nozzle, as shown in Figure 10.15. Acidic electrolytes are

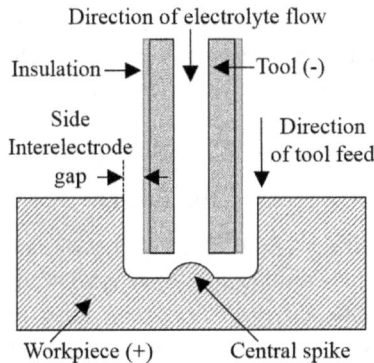

FIGURE 10.13 Schematic showing the tool and workpiece configuration in ECD.

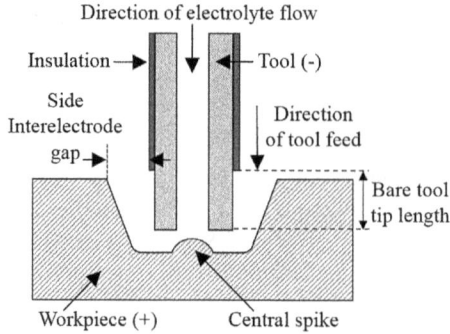

FIGURE 10.14   Schematic showing the tool and workpiece configuration in STEM.

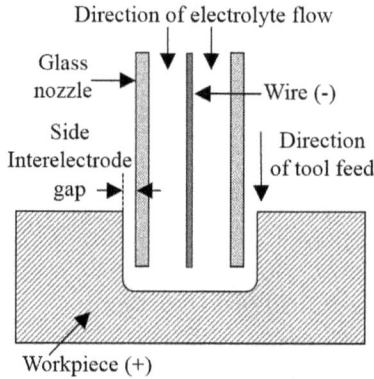

FIGURE 10.15   Schematic showing the tool and workpiece configuration in ESD.

commonly used in this process as the electrolysis products get readily dissolved in these electrolytes and do not clog the narrow IEG. Using this process, many high aspect ratio holes can be machined simultaneously without causing any metallurgical defect in the workpiece. A unique application of this process is the ability to machine blind and intersecting holes with high aspect ratios. However, to drill a single hole, the process is slow, and the handling of acidic electrolyte is a challenge.

10.13.4   ELECTROCHEMICAL JET DRILLING (ECJD)

In electrochemical jet drilling (ECJD), the tool does not enter the workpiece in which a hole is to be drilled. Consequently, there is no need for a tool in this process. The nozzle acts as a cathode and machining takes place through an electrolyte jet. Diluted acidic electrolyte is used for machining and enough room is required for the electrolyte to exit, preferably in the form of a spray. The holes machined using

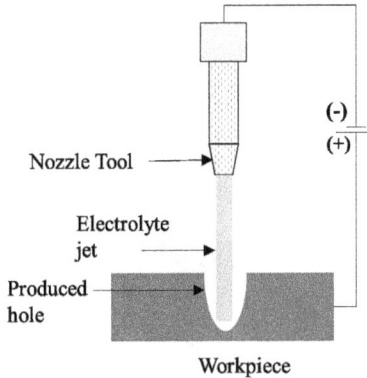

Workpiece

FIGURE 10.16 Schematic showing the tool and workpiece configuration in ECJD.

this process have an aspect ratio lower than that in ESD. In ECJD, the lower limit of the hole size is determined by the smallest hole that can be drilled in the nozzle, the pressure required to pump the electrolyte in the form of a jet from the nozzle, and the amount of permitted overcut. A schematic diagram of ECJD configuration is shown in Figure 10.16.

10.13.5 WIRE ELECTROCHEMICAL MACHINING

Wire electrochemical machining (wire-ECM) is a variant of ECM where, a wire is used as a tool to cut a desired profile in the workpiece. Wire should be made of a material with high tensile strength, electrical conductivity, chemical stability, and etchability. Tungsten is the most commonly used tool material due to its high tensile strength. Although, copper is a better conductor of electricity, its wire is not commonly used due to its poor tensile strength. Recently, researchers have also tested wires made of carbon nanotube fibers for machining in wire-ECM. The wire diameter in this process varies from as low as 2 μm and up to 3 mm. A schematic diagram of wire-ECM is shown in Figure 10.17. In the machining zone, the wire is subjected to forces generated due to the collapse of gas bubbles. These forces may result in radial swinging of the wire which hampers the machining accuracy. To prevent this, the wire is kept in a tensed position and the tension generally varies between 2 and 50 N.

A typical kerf machined by this process is shown in Figure 10.18(a). The width of this kerf can be calculated using Equation 10.26.

$$w = 2\Delta_s + D_w \tag{10.26}$$

here Δ_s is the side-IEG and D_w is the wire diameter. The relationship between IEG (y) at any angle (θ) from the direction of tool feed (as shown in Figure 10.18(b)) and other process parameters can be written as expressed in Equation 10.27:

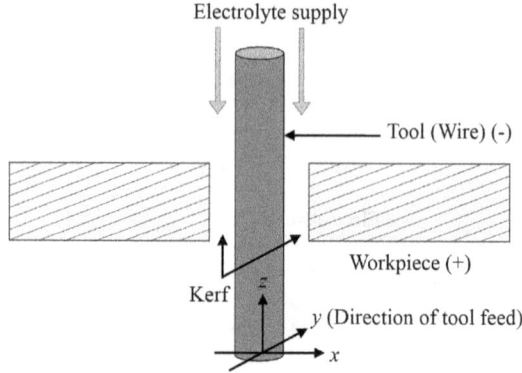

FIGURE 10.17 Schematic showing the tool and workpiece configuration in wire-ECM.

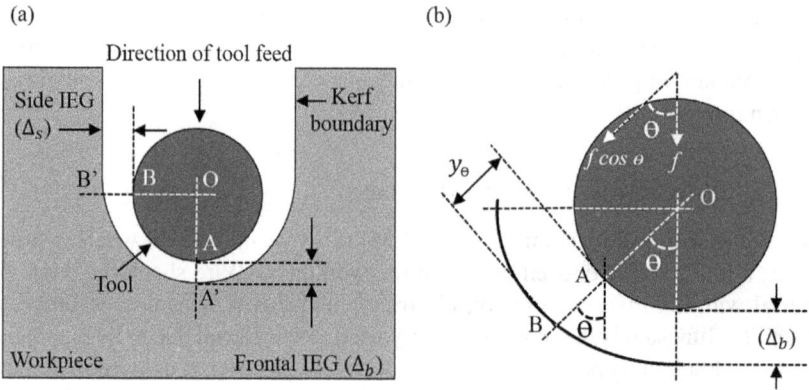

FIGURE 10.18 Schematic showing (a) a typical kerf profile and (b) variation of the machining gap along the periphery of the wire in the wire-ECM process.

$$
y + \frac{\eta E V k}{F \rho f \cos(\theta)} \ln \left| \frac{\eta E V k}{F \rho} - f \cos(\theta) y \right|
$$

$$
= y_0 + \frac{\eta E V k}{F \rho f \cos(\theta)} \ln \left| \frac{\eta E V k}{F \rho} - f \cos(\theta) y_0 \right| - f \cos(\theta) t
$$

(10.27)

By substituting the values of different process parameters, Equation 10.27 can be solved numerically for y.

Apart from profiling, wire-ECM can also be used for operations such as slitting, turning, grooving, indexed cutting, and texturing. Since a very small diameter wire can be used, wire-ECM has become more popular in micromachining operations. Over a period of time, researchers have demonstrated the capabilities of this process in machining complex microfluidic circuits, microcantilever beam balance, selective thinning of very thin-walled tubes, microgears, cams, and microsprings for microassemblies.

10.14 RESEARCH OPPORTUNITIES

The machining of alloys with ECM is a challenge because of the difference in the dissolution rate of its constituents in the presence of a specific electrolyte type and applied potential. An electrolyte can be prepared by mixing more than one type of salt, in the presence of which, different constituents of a specific alloy can dissolve at an approximately similar rate. Research is due in this area. Another major problem with ECM is the formation of a passive film over the anode surface, although a passive film is non-conducting in nature and may be useful in protecting the machined parts from further corrosion. Formation of such a film during machining may be detrimental. In ECM, a passive film generally consists of metal oxides and hydroxides. The presence of oxygen near the machining region is prominently due to the hydrolysis of water. To prevent this, research can be conducted to perform ECM in the presence of non-aqueous electrolytes. Such electrolytes can be prepared by using oil or ethylene glycol as a solvent.

10.15 SUMMARY

ECM is an AMP where the material is removed from the workpiece through anodic dissolution at an ionic level in the presence of an electrolyte and an applied electric field. This process is primarily used in industries for the shaping of large and electrically conducting materials which require a profiled tool. Since there is theoretically no tool wear/dissolution in this process, a tool of very small dimensions can be used for machining. This makes ECM one of the most suited processes for micromachining operations. A major challenge in ECM is in controlling stray cutting, which refers to the dissolution of material from an undesired region of the workpiece in the presence of an electric current field which is not localized in the intended machining region. To minimize the stray current affect, researchers have proposed the use of insulation over the surface of the tool and/or the workpiece. ECM can be performed either with a constant DC or a pulsed DC voltage depending upon the required machining rate and tolerance. Over a period of time, different variants of ECM have been developed. Using these variants, the application of ECM is extended for processes such as drilling, milling, turning, slitting, and surface texturing.

ACRONYMS

AMP	Advanced Machining Process
CNC	Computer Numeric Control
DC	Direct Current
EC	Electrochemical
ECD	Electrochemical Drilling

ECM	Electrochemical Machining
EDL	Electrical Double Layer
EDM	Electric Discharge Machining
ESD	Electrostream Drilling
ECJD	Electrochemical Jet Drilling
EF	Etch Factor
FEM	Finite Element Method
HSTR	High Strength Temperature Resistant
IEG	Interelectrode Gap
LBM	Laser Beam Machining
MRR	Material Removal Rate
MRR_v	Volumetric Material Removal Rate
MRR_l	Linear Material Removal Rate
PZT	Piezoelectric
STEM	Shaped Tube Electrolytic Machining
Wire-ECM	Wire-Electrochemical Machining

SYMBOLS

A	Atomic mass (g)
a	Transfer coefficient
D_w	Wire diameter (m)
E	Gram equivalent weight of anode material (g)
F	Faraday's constant (96,000 C/kmol)
f	Tool feed rate (m/s)
I	Circuit current (A)
M	Mass of material dissolved (g)
n	Number of valence electrons
p	Number of elements in an alloy
Q	Charge (C)
R_{el}	Resistance of the electrolyte band (Ω)
t	Time interval (s)
V	Magnitude of applied DC potential (V)
w	Width of the kerf (m)
y_o	Interelectrode gap at time t = 0 (m)
Z	Electrochemical valency

GREEK LETTERS

Δ_s	Side interelectrode gap (m)
κ	Electrical conductivity of electrolyte (S/m)
κ_o	Initial electrical conductivity of electrolyte (S/m)
κ_{eff}	Effective electrical conductivity of electrolyte (S/m)
κ_d	Electrical conductivity of the dispersed medium (S/m)
Δ_b	Frontal interelectrode gap (m)

ΔT Change in temperature (°C)
ρ Density of anode material (kg/m^3)
ϕ Potential distribution function
η Current efficiency
a Transfer coefficient
α Temperature coefficient of resistance of electrolyte
α_s Void fraction of sludge
α_v Void fraction of bubble
β,γ Constants

REFERENCES

1. A. Ghosh, A.K. Mallik. *Manufacturing Science*, 2nd ed. New Delhi: East-West Press Private Limited, 2016.
2. M.M. Lohrengel, K.P. Rataj, T. Münninghoff. Electrochemical machining—Mechanisms of anodic dissolution. *Electrochimica Acta*. 201 (2016): 348–353.
3. V.K. Jain. *Advanced Machining Processes*, 2nd ed. New Delhi: Allied Publisher, 2002.
4. B. Bhattacharyya. *Electrochemical Micromachining for Nanofabrication, MEMS and Nanotechnology*, 1st ed. William Andrew, 2015. doi: 10.1016/C2014-0-00027-5.
5. J. Hopenfeld, R.R. Cole. Prediction of the one-dimensional equilibrium cutting gap in electrochemical machining. *Journal of Engineering for Industry* (1969): 755–763.
6. V.K. Jain, A.K. Chouksey. A comprehensive analysis of three-phase electrolyte conductivity during electrochemical macromachining/micromachining. *Proceedings of the Institution of Mechanical Engineers, Part B: Journal of Engineering Manufacture* 232 (2018): 2449–2461. doi: 10.1177/0954405417690558.
7. V. Sharma, D.S. Patel, V.K. Jain, J. Ramkumar. Wire electrochemical micromachining : An overview. *International Journal of Machine Tools and Manufacture* 155 (2020): 103579. doi: 10.1016/j.ijmachtools.2020.103579.
8. R.J. Leese, A. Ivanov. Electrochemical micromachining: An introduction. *Advances in Mechanical Engineering* 8 (2016). doi: 10.1177/1687814015626860.
9. K. Xu, Y. Zeng, P. Li, D. Zhu. Study of surface roughness in wire electrochemical micro machining. *Journal of Materials Processing Technology* 222 (2015): 103–109. doi: 10.1016/j.jmatprotec.2015.03.007.
10. X. Qi, X. Fang, D. Zhu. Investigation of electrochemical micromachining of tungsten microtools. *International Journal of Refractory Metals & Hard Materials* 71 (2018): 307–314. doi: 10.1016/j.ijrmhm.2017.11.045.
11. H. Wu, S. Xiao, Q. Li, Y. Zhang. Research of high efficiency combined wire electrochemical machining technology and its application. *Advanced Materials Research* 316 (2011): 1015–1019. doi: 10.4028/www.scientific.net/AMR.314-316.1015.
12. A. Tyagi, V. Sharma, V.K. Jain, J. Ramkumar. Investigations into side gap in wire electrochemical micromachining (wire-ECMM). *The International Journal of Advanced Manufacturing Technology* 94 (2017): 4469–4478. doi: 10.1007/s00170-017-1150-z.
13. Q. Ningsong, F. Xiaolong, L. Wei, Z. Yongbin, Z. Di. Wire electrochemical machining with axial electrolyte flushing for titanium alloy. *Chinese Journal of Aeronautics* 26 (2013): 224–229. doi: 10.1016/j.cja.2012.12.026.
14. V. Sharma, D.S. Patel, V. Agrawal, V.K. Jain, J. Ramkumar. Investigations into machining accuracy and quality in wire electrochemical micromachining under sinusoidal and triangular voltage pulse condition. *Journal of Manufacturing Processes* 62 (2021): 348–367. doi: 10.1016/j.jmapro.2020.12.010.

15. D.S. Patel, V. Sharma, V.K. Jain, J. Ramkumar. Reducing overcut in electrochemical micromachining process by altering the energy of voltage pulse using sinusoidal and triangular waveform. *International Journal of Machine Tools and Manufacture* 151 (2020): 103526. doi: 10.1016/j.ijmachtools.2020.103526.

16. D.S. Patel, V.K. Jain, J. Ramkumar. Electrochemical Grinding. In *Nanofinishing Science and Technology*, edited by V.K. Jain, 1st ed., pp. 321–352. CRC Press, Boca Raton, 2016.

17. D.S. Patel, V. Agrawal, J. Ramkumar, V.K. Jain, G. Singh. Micro-texturing on free-form surfaces using flexible-electrode through-mask electrochemical micromachining. *Journal of Materials Processing Technology* 282 (2020): 116644. doi: 10.1016/j.jmatprotec.2020.116644.

18. V.K. Jain, P.C. Pandey. Tooling design for ECM. *Precision Engineering* 2 (1980): 195–206. doi: 10.1016/0141-6359(80)90012-4.

19. H. El-hofy, H. Youssef. Environmental Hazards of Nontraditional Machining. In *ASME/WSEAS International Conference on Energy & Environment (EE'09)*, 2009, pp. 140–145.

20. H. Tönshoff, R. Eggerl, F. Klocke. Environmental and Safety Aspects of Electrophysical and Electrochemical Processes. *CIRP Annals – Manufacturing Technology* 42 (1996): 553–568.

11 Wire Electrochemical Micromachining

Vyom Sharma and Divyansh Patel

CONTENTS

DOI: 10.1201/9780429160011-11

11.1   INTRODUCTION

The introduction of new engineering materials such as alloys of titanium, tungsten, nickel, and metallic glasses, has applied a significant thrust on the manufacturing industry to develop new machining processes. Conventional options of micromachining such as micromilling, microdrilling, microturning, and microgrinding are all contact-type processes, and they are affected by the hardness of the workpiece material. Issues of rapid tool wear, chatter due to undamped vibrations, induction of mechanical stresses in the workpiece, and inferior machined surface quality often arise during machining. To overcome these problems, researchers have developed different non-contact-type machining processes such as Laser and Electron Beam Machining (LBM/EBM), Electric Discharge Machining (EDM), Ultrasonic Machining (USM), Abrasive Jet Machining (AJM), Abrasive Water Jet Machining (AWJM), and Electrochemical Micromachining (ECM). Amongst these, ECM does not alter the surface morphology of the machined component as there are no heat and mechanical forces involved [1].

Over a period of time, researchers have developed various hybridizations and variants of ECM. Hybridization which combines kinematics and material removal mechanisms of two or more than two machining processes [2] such as Electrochemical Grinding (ECG) [3, 4], Electrochemical Honing [5], and Electrochemical Discharge Machining (ECDM) [6, 7] process. These variants have the same mechanism of material removal as that of ECM, but the tool geometry and kinematics of machining are modified according to the application [8, 9]. These include Shaped Tube Electrochemical Drilling (STED), Electrochemical Turning (ECTrg) [10], Jet

Electrochemical Machining [11], and the Wire-ECMM. Process developments in Wire-ECMM have now reached a significant level of maturity and a decent volume of literature has been reported to date.

11.2 FUNDAMENTALS OF WIRE-ECMM

11.2.1 DEFINITION

In Wire-ECMM, the tool is a wire which is connected to the negative terminal (cathode) of a DC power source (Figure 11.1). This wire traces a defined trajectory in a workpiece which is connected to the positive terminal (anode). Electrically conducting fluid (electrolyte) is supplied between the two electrodes and a feature of the desired shape is machined. When the tool is made of a material which is chemically stable toward the adjacent electrolyte, no material gets dissolved from it. Consequently, the same wire can be used repeatedly for machining provided it does not erode beyond a certain limit. For example, tungsten metal resists attack by oxygen, acids, and alkalis, and therefore, it is the most suitable tool material in Wire-ECMM. Machining capabilities of this process such as accuracy, efficiency, and surface quality are unaffected by the mechanical properties (hardness, toughness) of the workpiece.

11.2.2 MECHANISM OF MATERIAL REMOVAL

When the electrodes are placed in the electrolyte and electric current is passed through them, ions are transferred between the electrodes and electrolyte (Figure 11.2). This leads to the occurrence of physical change at the anode. A quantitative relationship

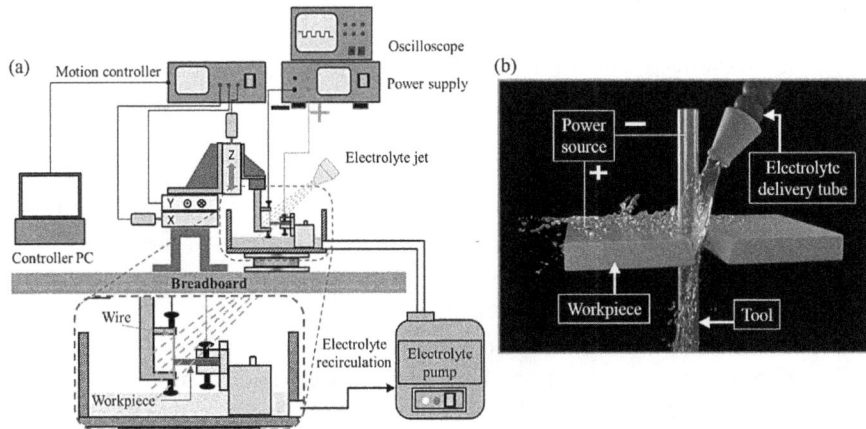

FIGURE 11.1 Schematic showing (a) the experimental setup of Wire-ECMM process and (b) direction of wire feed with respect to the workpiece and the generated kerf (reproduced with permission from [12]).

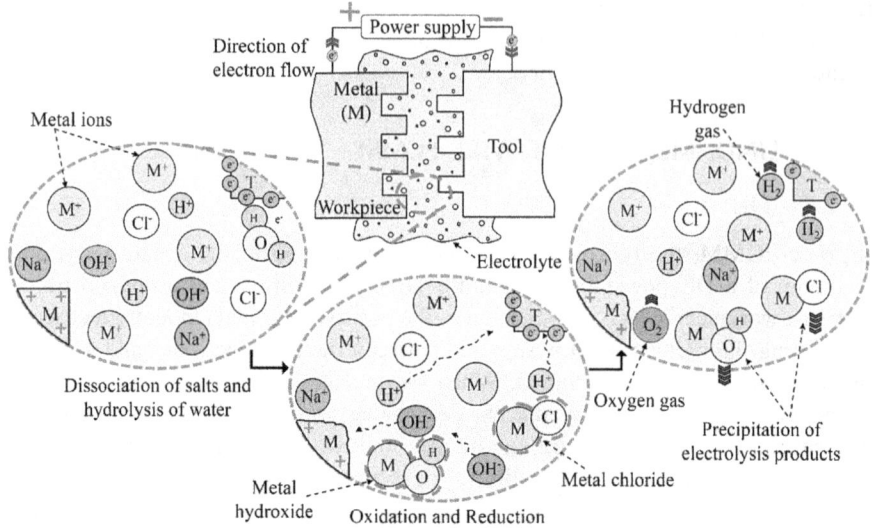

FIGURE 11.2 Schematic showing the movement of electrons and ionic species during electrolysis and dissolution of anode material. Reproduced with permission from [15].

between the mass of material removed (M) and the charge passed is given by Faraday's laws of electrolysis (Equation 11.1) [13, 14].

$$M = \left(\frac{Q}{F}\right)\left(\frac{A}{Z}\right) \text{grams} \tag{11.1}$$

where Q is total electric charge passed, F is Faraday's constant ($\approx 96,500$ C/mol), A is atomic mass of anode material, and Z is valence number of ions of anode material.

During electrolysis, positive metal ions leave the workpiece and electrons leave the tool to reach their respective opposite ends. Reactions taking place at the anode and cathode are given in Equations 11.2 and 11.3, respectively [13].

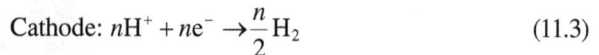

$$\text{Anode: } M \rightarrow M^{n+} + ne^- \tag{11.2}$$

$$\text{Cathode: } nH^+ + ne^- \rightarrow \frac{n}{2}H_2 \tag{11.3}$$

where n is the number of valence electrons.

The anodic reaction leads to the formation of metal anion compounds (heavier than other ions) and the loss of H^+ ions (which reduces the electrolyte conductivity) but reaction at the cathode leads to electron interaction. Hydrolysis of water generates H^+ and OH^-. As machining progresses, the conductivity continues to decrease due to the formation of electrically non-conducting electrolysis products and pH increases due to an increase in the hydrogen ion concentration. In practice, metal hydroxide and metal chloride would form and get precipitated as sludge when aqueous NaCl is used as an electrolyte for machining steel.

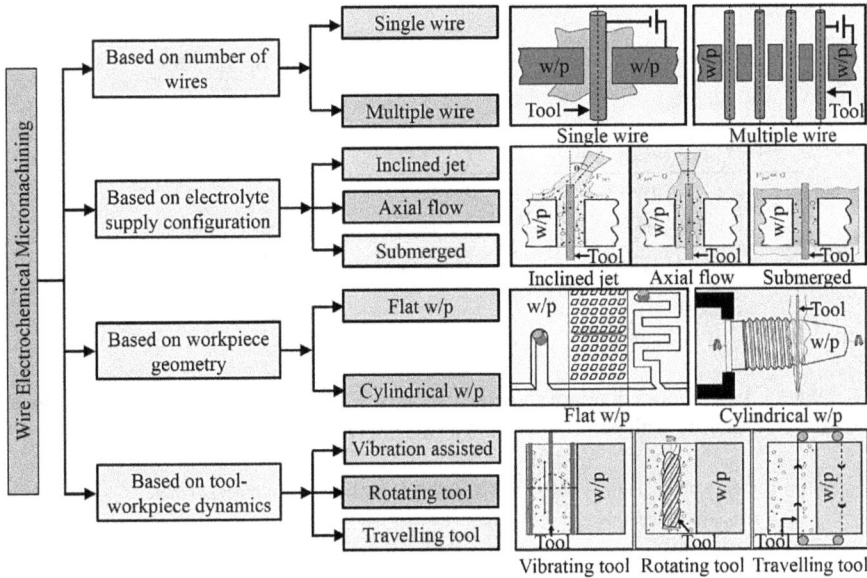

FIGURE 11.3 Broad classification of the Wire-ECMM process. Reproduced with permission from [15].

11.2.3 CLASSIFICATION

Wire-ECMM is classified on the basis of the number of tools, workpiece geometry, electrolyte configuration, and tool–workpiece dynamics. A broad classification of this process is shown in Figure 11.3.

11.2.4 PROCESS PARAMETERS

A classification of parameters which are directly associated with the process performance in Wire-ECMM, is shown in Figure 11.4. Process performance can be evaluated on the basis of machining efficiency, accuracy, and the quality of the machined component. Machining efficiency is quantified in terms of maximum possible tool/workpiece feed rate (f_m) which is achievable only under the given set of working parameters. Machining accuracy is measured in terms of side Interelectrode Gap (IEG) generated during machining. Machining quality is quantified in terms of roughness of the surface produced after machining, kerf width, material removal rate, and kerf depth.

11.2.5 SOME OTHER INFLUENCING FACTORS

There are some factors in Wire-ECMM which are not directly controlled by the operator, and they are dependent upon the combination of other input parameters and the electrochemistry of the process. These factors significantly influence the machining stability and quality. To understand these, it is important to understand

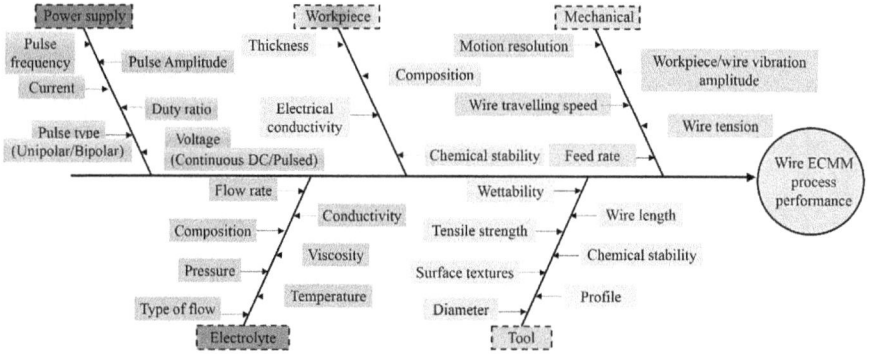

FIGURE 11.4 Ishikawa diagram showing the classification of process parameters in Wire-ECMM. Reproduced with permission from [15].

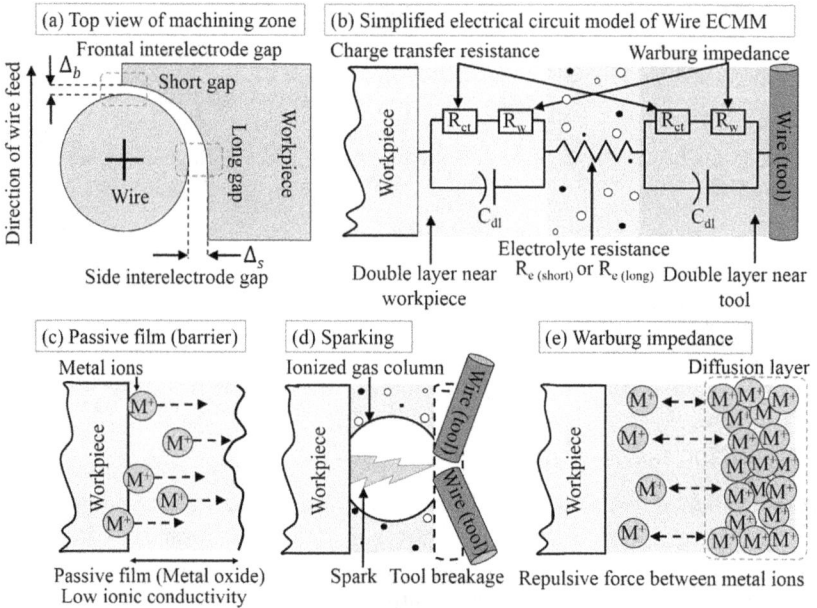

FIGURE 11.5 Schematic showing (a) the machining zone in Wire-ECMM, (b) different resistances and capacitances in the IEG which represent a simplified electrical circuit model of the process, (c) passive film formation which acts as a barrier for ion mobility, (d) sparking leading to wire breakage, and (e) formation of diffusion layer giving rise to Warburg impedance. Reproduced with permission from [15].

the profile of a kerf machined by Wire-ECMM (Figure 11.5(a)). The machining region is typically characterized by frontal IEG, Δ_b, and side IEG, Δ_s. During electrolysis, there is a formation of a layer of ionic species near the two electrodes, called Electric Double Layer (EDL), which acts as a capacitor with a capacitance (C_{dl}) (Figure 11.5(b)). Before an external potential is applied between the two electrodes, a state of equilibrium exists as the rate of metal ions leaving the metal surface and entering the electrolyte is equal to the rate of metal ions getting reduced and diffusing back to the metal electrode. To break the equilibrium of charge transfer during electrolysis and force the metal ions to leave the metal surface and diffuse into the electrolyte, a resistance is experienced which is known as charge transfer resistance (R_{ct}). Also, due to insufficient transport of mass from one electrode to another, a diffusion layer is created which offers resistance known as Warburg impedance (R_w) [16] (Figure 11.5(e)). EDL discharges through these two resistances (Figure 11.5(b)). Some portion of the total machining current in pulse duration is consumed in charging of this capacitor [17]. Consequently, total machining current and effective machining time are reduced [18].

In a case of constant DC power supply, the effect of EDL is not pronounced because once charged, it does not discharge until the power is turned off. Depending on the metal and electrolyte combination, there is a formation of an oxide film over the workpiece surface which acts as a barrier to the electric current flow between the anode surface and electrolyte. This oxide film is of low ionic conductivity and reduces the anodic dissolution rate. The phenomenon is called "passivation" and may also lead to the termination of machining [13, 19, 20] (Figure 11.5(c)). Hydrogen and oxygen gas liberated during electrolysis may also get ionized in the presence of an electric field, leading to the generation of sparks [13] (Figure 11.5(d)). In Wire-ECMM, sometimes even microsparks can damage the tool and destabilize machining [21].

11.3 MATERIALS AND DESIGN OF WIRE IN WIRE-ECMM

11.3.1 WIRE MATERIALS

Wire material in Wire-ECMM should possess good chemical stability, good etchability and wettability, low electrical resistivity, and good tensile strength. Wire made of tungsten [22], stainless steel (SS 304) [23], molybdenum [24], copper [19], carbon nanotubes [25], platinum [26], brass [27], and tungsten carbide [28] have been used to date. A tool made of chemically stable material does not form any precipitate during machining, which may restrict the mobility of ionic species in the IEG. Although, copper has low electrical resistivity, its poor chemical stability limits its use as a tool (wire) material. Carbon Nanotubes (CNTs) offer high tensile strength and wettability, but very limited work has been reported on their use as a tool material. This is possibly due to their poor electrical conductivity [29]. Although, CNTs have five times more electrical conductivity compared to copper at the nanoscale [30], when thousands of CNTs are woven together to make a wire, its conductivity decreases drastically. A wire made of CNTs has to be electroplated with conducting materials such as aluminum to improve its conductivity [25]. The diameter of tool used in Wire-ECMM varies from 2 μm (tungsten) [31] (not commercially available) to 2.8 mm (brass) [32].

To prevent the wire from swinging radially due to the force exerted by the gas bubbles and momentum of electrolyte jet, it is kept in a tensed position. This tensile force usually varies from 2 N [29] to 50 N [33]. Tool material must have sufficient tensile strength to withstand this tensile force even when the wire diameter is very small (≤ 10 μm). Platinum is a chemically stable metal and its wires of a diameter as small as 0.6 μm are commercially available, but their use in Wire-ECMM is limited due to their poor tensile strength. Tungsten is the most widely used tool material due to its high tensile strength and good chemical stability. A tungsten wire can be used with acidic and basic, as well as neutral electrolytes. Its wires of a diameter up to 4 μm are commercially available. It is important to note here that the handling of a wire with a very small diameter is difficult. Such wires easily get deformed and they are difficult to mount on a machining fixture [34].

11.3.2 Wire Design

A cylindrical wire with a smooth surface is not very efficient in pulling the electrolysis products out of the IEG when axial vibrations are imparted to it or the workpiece. This is because of its limited wettability. It has been shown that on imparting some degree of roughness on the surface of a wire, it becomes more hydrophilic in nature [29]. When an etched wire (Figure 11.6(a)) is used for machining, slits have better homogeneity and lower surface roughness as compared to those machined with a smooth wire [35]. A wire with dimples on its surface (Figure 11.6(b)) is effective in promoting the detachment of gas bubbles from its surface by increasing hydrophilicity and reducing the bubble-surface contact area [36]. The adhesion energy between solid and liquid surface increases by increasing the hydrophilicity of the solid surface (as per the Young–Dupre equation [37]). Therefore, a wire made of CNT with a series of microstructures is more wettable than a smooth wire (Figure 11.6(c)). When such a wire vibrates axially during machining, the momentum is more efficiently

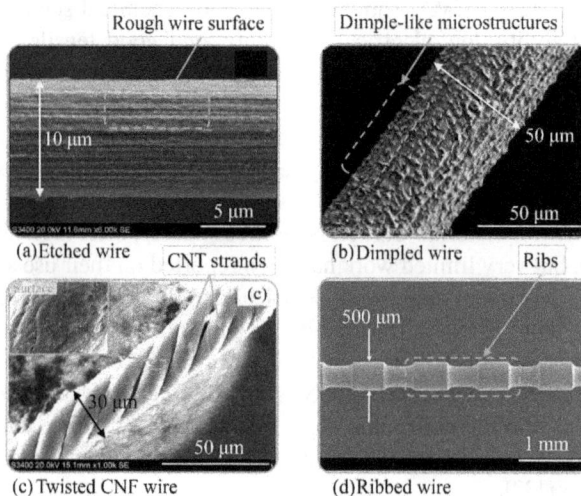

FIGURE 11.6 SEM image of (a) etched wire, (b) dimpled wire, (c) twisted CNT fiber, and (d) ribbed wire. Reproduced with permission from [15].

transferred from the tool to the electrolyte in its vicinity. Due to this momentum transfer, the flow of electrolyte gets accelerated. Along with the electrolyte, the mass transport of electrolysis products also increases and the deposition of sludge precipitates on the tool surface becomes difficult. This provides stability to the machining process and under identical machining conditions, the feed rate (f_m) is observed to double in comparison with a smooth wire [29].

The use of ribbed wire with sequential grooves in Wire-ECMM is also proposed (Figure 11.6(d)). Ribs in the wire assist in squeezing and dragging the gas bubbles liberated at the cathode out of the machining zone [38], and hence electrolyte conductivity remains largely unaltered. An improvement of 22% in machining efficiency is reported [23]. However, the existing literature is silent about the technique used for fabricating such a wire. Experimental study suggests that machining efficiency improved by 2.5 times and variation in kerf width is reduced to half. Although machining efficiency and quality improve significantly by the use of textured wire, fabrication of such wires is difficult. Microtexturing of a wire is done either by chemical etching or by laser machining. This makes Wire-ECMM a two-step process. Also, the mounting of such thin wires on the setup for processing is not very convenient. To overcome these problems, researchers have proposed the use of a microdrill, cutting-edge electrode, and fluted metal tube in place of a flexible wire for improving flushing efficiency in the process. A microdrill (Figure 11.7(a)) and cutting-edge electrode (Figure 11.7(b)) are easy to mount on a spindle and rotate at high rpm (up to 50,000 rpm). The rotation of the microdrill imparts axial velocity (along the thickness of the workpiece) to the electrolyte in its vicinity and accelerates the machining process [28, 39]. When a cutting-edge tool is used, the average current density in the machining gap is reduced due to an increase in IEG in the zone facing the flat surface of the tool [40]. This leads to the reduction in the side IEG. In case of a fluted tube cathode (Figure 11.7(c)), electrolyte is ejected from its microholes. The workpiece is vibrated in the direction parallel to the axis of the tube to ensure uniform spread of the electrolyte [41], and the amplitude of vibration is kept more than the hole spacing on the tube. This method is especially useful in the machining of thick workpieces [42] and also in preventing the electrolyte from flowing to the non-machining region which results in stray corrosion [10].

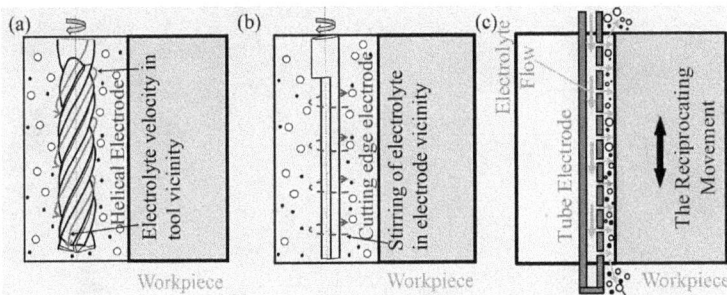

FIGURE 11.7 Schematic of (a) helical tool, (b) cutting-edge tool, and (c) tube with microhole array. Reproduced with permission from [15].

11.4 ELECTROLYTE TYPES AND METHODS OF SUPPLY IN WIRE-ECMM

11.4.1 ELECTROLYTE TYPES

Electrolyte's electrical conductivity is a crucial factor in Wire-ECMM as it indicates the ease of charge transport between the electrodes. Any variation in its value affects the machining conditions. Electrolytes used in Wire-ECMM can be categorized on the basis of their pH value as acidic, neutral, and basic as shown in Figure 11.8.

11.4.1.1 Neutral Electrolytes

Among neutral electrolytes, aqueous solution of NaCl (a non-passivating electrolyte) has the advantage of high processing efficiency, better stabilization, and low cost but has a disadvantage of low machining precision and causes greater corrosion of equipment. Materials such as copper, carbon structural steel, and stainless steel 304 have been machined using this. In Wire-ECMM, $NaNO_3$ (a passivating electrolyte) is the most widely used as it offers high precision and low corrosion of equipment, but it gives low processing efficiency [43]. When machining stainless steel 304 with $NaNO_3$ (usually 0.05–0.5 M), reactions occurring at the anode often lead to the formation of $Fe(OH)_2$ which adheres on the surface and slows down machining [44]. To prevent this, B_4C particles are added (concentration of 1–8 g/l) in the electrolyte. These particles assist in dispersing hydrogen bubbles from the IEG and aid in stabilizing the machining. Machining tungsten carbide (WC) in the presence of $NaNO_3$ (0.7 M) results in the formation of NaOH. This dissolves passive tungsten oxide film and stabilizes machining. Aqueous solution of KCl (0.25 M) is non-passivating in nature and it is used for machining aluminum. It is the most suitable electrolyte for achieving uniform and stable machining without being affected by the passive oxide film [19].

11.4.1.2 Basic Electrolytes

Basic electrolytes such as NaOH are generally avoided in ECM because high hydroxide ion concentration increases the precipitation of metal hydroxides. This may

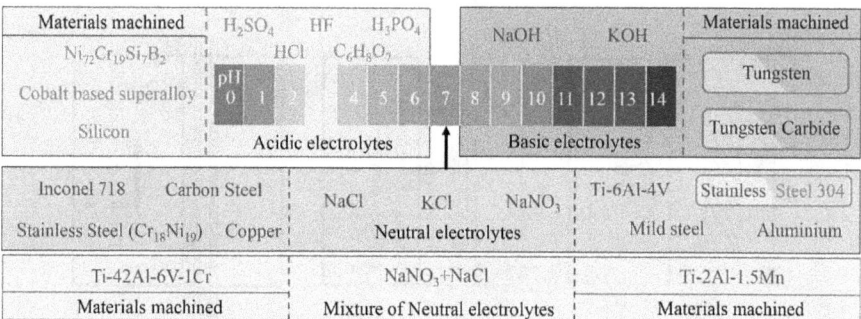

Materials machined	H_2SO_4 HF H_3PO_4	NaOH KOH	Materials machined
$Ni_{72}Cr_{19}Si_7B_2$	HCl $C_6H_8O_7$		
Cobalt based superalloy	pH		Tungsten
Silicon	0 1 2 4 5 6 7 8 9 10 11 12 13 14		
	Acidic electrolytes	Basic electrolytes	Tungsten Carbide
Inconel 718 Carbon Steel	NaCl KCl $NaNO_3$	Ti-6Al-4V	Stainless Steel 304
Stainless Steel ($Cr_{18}Ni_{19}$) Copper	Neutral electrolytes	Mild steel Aluminium	
Ti-42Al-6V-1Cr	$NaNO_3$+NaCl	Ti-2Al-1.5Mn	
Materials machined	Mixture of Neutral electrolytes	Materials machined	

FIGURE 11.8 Classification of the types of electrolytes used in Wire-ECMM. Reproduced with permission from [15].

clog the IEG and sometimes may even lead to the termination of machining [45]. However, KOH (0.5 M), a basic electrolyte, is used for machining pure tungsten [39]. Passivating tungsten oxide (WO_3) film breaks readily in the presence of $(OH)^-$ ions.

11.4.1.3   Acidic Electrolytes

A major advantage of using acidic electrolytes for machining is that reaction products remain dissolved in the solution. This is because hydroxide ions produced at the cathode are neutralized by the high concentration of hydrogen ions in the solution [45]. The machining of nickel-based metallic glass ($Ni_{72}Cr_{19}Si_7B_2$) at micro and nanoscale is a topical field of research because of the high hardness and strength offered by this material. Attempts have been made for the micro shaping of this glassy alloy using Wire-ECMM. A major bottleneck in its electrochemical treatment is the passivating nature of nickel and chromium. To overcome this problem, the use of acidic electrolytes such as sulfuric acid (H_2SO_4) (0.1 M) and hydrochloric acid (HCl) (0.1 M) is proposed [46]. The machined surface produced in this case is highly uniform and polished. In the presence of aqueous HCl as the electrolyte, pitting corrosion of metallic glass usually occurs. Passive film breaks down due to the adsorption of chlorine ions. HCl is also used for the machining of cobalt-based super alloy (0.01 M) [47] and stainless steel 304 (0.1 M). The major problem in machining using HCl is the blackening of the wire due to the precipitation of metallic iron on wire (caused by the migration of Fe ions toward the cathode). Therefore, when machining with HCl, high frequency vibrations of tool and/or workpiece is recommended for efficient flush of the reaction products. Hydrofluoric acid (HF) (1 M) is used for the machining of silicon. The passive film of silicon oxide easily gets dissolved using this strong acid [48]. Anhydrous citric acid ($C_6H_8O_7$) (0.3 M) is also used as an ecofriendly solution for the machining of stainless steel 304 [49]. Citrate ions chelate with iron ions to form citrate complex ions. These complex ions accelerate the dissolution, and a high material removal rate is achieved.

11.4.1.4   Mixture of Electrolytes

Electrolyte prepared by the mixture of two neutral salts ($NaNO_3$ + NaCl) in equal weight % is used for the machining of titanium alloys such as Ti-2Al-1.5Mn [50] (2.5% by weight of each salt) and Ti-42Al-6V-1Cr (5% by weight of each salt) [24]. However, the existing literature is silent about the reasoning behind selecting these electrolytes as the mixture. Another titanium alloy (Ti-6Al-4V) which is a multiphase alloy, is usually machined with $NaNO_3$ [51] and NaCl [52] solutions individually. Electrolyte solutions of sodium halides are found to give a better machining rate than that of sodium nitrate. However, not much discussion is available on the microstructure and roughness of the machined surface using these electrolytes. A mixture of phosphoric acid in ethanol and water (H_3PO_4: C_2H_5OH: H_2O = 2:6:2) is used for the machining of cobalt-based alloy [53]. The addition of ethanol reduces the viscosity of the solution and electrolysis products are easily flushed out of the machining region. The machined surface of high quality (S_a (area surface roughness) = 0.039 μm) is obtained. Authors believe that this area needs further research to support the claim of the experimental findings that the electrolyte mixture usually gives better results in the case of an alloy dissolution. One cannot make conclusions based on the isolated experimental results.

11.4.2 METHODS OF ELECTROLYTE SUPPLY

A key area of research in Wire-ECMM is stabilizing the machining process by enhancing electrolyte flushing efficiency. Generally, there are three methods of supplying electrolyte in the IEG. In the first and the most primitive method, electrolyte at high pressure and flow rate is flushed at an inclination to the axis of the wire (Figure 11.9(a)) [19, 32, 54]. This is analogous to the delivery of coolant in the machining zone in conventional turning and milling operations. However, the momentum of a high-pressure jet may cause the wire to swing radially. This may lead to frequent short-circuiting, tool breakage, and poor machining accuracy. Therefore, the wire must be kept under tension. The only advantage of this type of electrolyte delivery system is that the tool holder assembly is simple, and the wire can be changed rapidly in the event of a breakage. With this type of delivery system, the wire can be stationary, mono-directional traveling, or vibrating along its axis. Monitoring the tension in the wire is easy in this case. However, the influence of angle of inclination of electrolyte supply on process performance is yet to be explored in Wire-ECMM. Although, some preliminary investigations have been performed [55], but a detailed investigation is due on the influence of angle of inclination of electrolyte jet on the process performance in Wire-ECMM. In the second method, the electrolyte is flushed in the direction along the axis of the tool (Figure 11.9(b)) [24, 50]. By using this method, radial swing of the wire due to the momentum of high-speed electrolyte jet can be avoided and a thick workpiece can be machined. This type of electrolyte delivery system requires a specially designed tool-holding system which includes a nozzle with an electrolyte inlet passage. Generally, a stationary wire is used in this type of arrangement and vibrations may be given to the workpiece. In the third method of electrolyte supply, the tool and workpiece are submerged in the stagnant electrolyte bath (Figure 11.9(c)) [23, 40]. However, due to the stagnant state

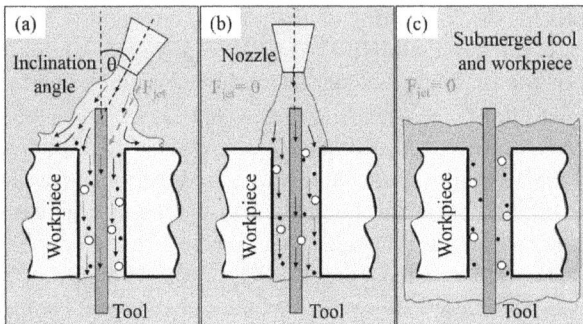

FIGURE 11.9 Schematic showing (a) electrolyte supply at an inclination angle (θ) to the wire axis, (b) flushing of electrolyte along the axis of the tool, and (c) submerging tool and workpiece inside the electrolyte. Reproduced with permission from [15].

of electrolyte, removal of electrolysis products and gas bubbles from the narrow IEG is not spontaneous. Usually, tool/workpiece vibration or a wire traveling perpendicular to the direction of feed is used to accelerate the flushing of electrolysis products from the gap.

11.5 PARAMETRIC ANALYSIS OF WIRE-ECMM

11.5.1 Influence of Process Parameters on Machining Efficiency

11.5.1.1 Effect of Electrolytic Factors

In Wire-ECMM, value of the maximum permissible tool/workpiece feed rate (m/s) (f_m) can be increased by the operator as the electrolyte flow rate increases up to a certain limit [56]. As the flow rate increases, the dispersion of gas bubbles and electrolysis products becomes more efficient. A uniform current density in the frontal and side IEGs is established and higher value of f_m can be used for machining. But as f_m is increased, frontal IEG decreases and the replenishment of fresh electrolytes in this gap becomes difficult. Consequently, f_m cannot be increased any further with an increase in electrolyte flow rate. This flow rate varies typically between 0.2 m/s [55] and 87 m/s [50] for axial flow. For inclined flow, it is usually up to 2 m/s [19]. High electrolyte pressure also assists in the dispersion of electrolysis products from the machining zone and machining can be performed at a higher f_m [33]. In the case of inclined flow, after a certain value of electrolyte pressure, the amplitude of vibration in the tool increases to a limit that frequent short circuits start occurring. Therefore, at high electrolyte pressure and flow rate, the tension in the wire must be increased for stable machining. Therefore, wires made of high strength material such as tungsten are generally used.

11.5.1.2 Effect of Power Factors

Increasing the value of applied potential leads to an increase in the current flowing in the circuit and consequently, MRR_l increases [46, 57] and higher f_m can be used for machining. Potential of 4 V [55] and above is generally used for micromachining operations depending upon anode material type, thickness, electrolyte concentration, and desired machining rate. In the operations where high MRR_l is required, the value of applied potential can go up to 50 V [33]. In pulsed voltage supply, pulse on-time plays a key role in determining f_m [33]. Reducing the pulse on-time decreases MRR_v and consequently, the frontal IEG also reduces. If the feed rate is not reduced, a short circuit will occur. For a fixed value of pulse on-time, increasing the pulse duration (sum of pulse on-time and off-time) reduces the machining time and lowers MRR_v. However, for shorter pulse durations, smaller hydrogen bubbles are generated and they readily escape from the IEG, thereby, increasing MRR_v. Ultra-short voltage pulse of on-time 20 ns and duration of 160 ns at a potential of 25 V is tested in Wire-ECMM [44].

11.5.1.3 Effect of Tool Rotation and Tool/Workpiece Vibration

Tool rotational speed (usually 1000 [40]–50,000 [28] rpm) also plays an important role in determining f_m when rotating tools are used. Rotation initiates the stirring action of electrolyte and electrolysis products are flushed out easily [28, 58]. Consequently, a uniform current density is established in the machining zone and MRR_l improves. Apart from the tool rotation, relative motion between the tool (wire) and workpiece along the direction of the axis of the wire also causes the electrolyte to flow in the IEG due to viscosity.

11.5.1.4 Effect of Workpiece Thickness

As the workpiece thickness increases, the narrow passage of frontal and side IEG becomes wider. Electrolyte pressure along the workpiece thickness decreases due to drag force and its replenishment becomes difficult in the deeper machining regions. Therefore, it becomes necessary to reduce the tool feed rate in order to avoid short-circuiting [33]. The maximum thickness of the workpiece machined to date using Wire-ECMM is 4 cm [59].

11.5.2 Effects of Process Parameters on Machining Accuracy

11.5.2.1 Electrolytic Parameters

As the charge carriers in the cell increase, the anodic dissolution rate is acceler-ated. If the tool/workpiece feed rate is not increased proportionately, then, side IEG increases and the machining accuracy deteriorates. Generally, electrolytes with low concentrations are used in micromachining operations. When axial flushing is used, side IEG decreases with an increase in electrolyte flow rate. This is due to two main reasons. First, high flow rate accelerates the flushing of electrolysis products from the IEG and second, wire radial vibrations due to the pressure exerted by the gas bubbles also subside. Nozzle–workpiece distance also plays a key role in determin-ing machining accuracy in the process. As this distance increases, the flaring of the electrolyte jet also increases, resulting in an increase in side IEG.

11.5.2.2 Power Parameters

MRR_v is directly proportional to the value of applied potential. Any increase in the value of applied potential results in an increase in the magnitude of current density over anode surface. As applied potential increases, MRR_v also increases and wider kerfs are machined [19, 34, 48, 55–57]. Side IEG can be minimized by working at a potential at which, MRR_l is only marginally more than the tool feed rate to avoid a short circuit. Side IEG also increases as the pulse on-time increases [48, 55]. In pulsed Wire-ECMM with a single wire, MRR_l is given by Equation 11.4 [61]:

$$MRR_l = \frac{1}{t_{duration}} \int_0^{t_{on}} \frac{I}{EF} dt \qquad (11.4)$$

where $t_{duration}$ is the pulse duration, t_{on} is the pulse on-time, E is the gram equiva-lent weight of anode material, and I is the machining current. For a large pulse duration, relatively larger volume of electrolysis products is generated and reduces the electrolyte conductivity in the machining region. Rapid and efficient flushing of these products from the narrow IEG is difficult and inadequate flushing results in poor machined surface quality. However, a higher machining rate can be sustained without short-circuiting with larger pulse duration. However, the reduction of side IEG due to increased feed rate is less than its increment due to increased pulse duration. Experimental observations suggest that a smaller side IEG can be obtained at small pulse on-time with low feed rate [33]. Pulse duration also plays a key role in determining the machining accuracy. The current (i) flowing in the cell during pulse on-time is given by Equation 11.5 [24]:

$$i = i_o \exp\left(\alpha \frac{F}{RT} \frac{V}{2} \frac{t}{\tau} \right) \tag{11.5}$$

where τ is charging time constant for EDL, i_o is the exchange current, V is the applied potential, α is the transfer coefficient, R is the gas constant, and T is the temperature. As the pulse duration decreases (or pulse frequency increases), the number of times EDL near the two electrodes charges and discharges per unit time also increases. For a fixed value of duty ratio ($t_{on} \times 1/t_{duration}$), the total pulse on-time per unit time does not change with the reduction in pulse duration but the time used for the charging of EDL ($\tau \times 1/t_{duration}$) increases [61]. The machining current during EDL charging time is less than the normal value and the time-averaged current is also reduced. Therefore, MRR_v decreases and a smaller side IEG is obtained. But consequently, a decrease in f_m [36, 61] also becomes inevitable. Duty ratio $\left(\dfrac{t_{on}}{t_{duration}} \right)$ also plays a key role in determining machining accuracy in the process. Side IEG decreases when the duty ratio is reduced. For short pulse on-time, if the duty ratio is too small ($t_{on} \ll t_{duration}$), EDL loads a small amount of charge and unloads completely during the pulse off-time. In consequence, there is no charge accumulated in the EDL. This results in no Faradaic current and no material removal at the workpiece [64]. Hence, for stable machining, duty ratio cannot be reduced below a certain value. The minimum duty ratio used in Wire-ECMM is 0.1 [61]. However, the value of duty ratio as 0.1 only expresses the extent of research (due to hardware constraints) and it is not the limiting value. Further research can be conducted to explore the behavior of material dissolution at duty ratios smaller than 0.1. This may necessitate the application of low pulse frequency such that t_{on} is sufficient for the charging of EDL and the flow of Faradaic current. Machining accuracy certainly improves with pulse frequency but the machining efficiency is compromised. With a high-frequency pulse voltage, higher peak voltage is necessary to reduce the EDL charging time. This also increases the energy requirement for machining [17, 18, 65]. In operations where machining accuracy is weighed superior to its efficiency, the use of ultra-high-frequency pulsed voltage is recommended over constant DC voltage.

11.5.2.3 Tool/Workpiece Feed Rate

Increasing the value of the tool/workpiece feed rate improves the machining accuracy in Wire-ECMM [17, 34]. This is mainly due to the reduction in tool–workpiece interaction time at any point on the anode. This reduces material dissolution from the workpiece and side IEG as well. Machining accuracy can be increased by reducing the applied potential and increasing the tool/workpiece feed rate. But it is important to note that both of these parameters are counteracting. Stable machining conditions may not persist at high feed rates if the applied potential is too low. Hence, an optimum balance between the two must be maintained such that neither machining efficiency nor accuracy is compromised significantly.

11.5.2.4   Tool/Workpiece Vibration

Tool/workpiece vibration aids in uniform distribution of electrolysis products in the machining gap and improves MRR_v. This adversely affects the machining accuracy as the side IEG also increases with an increase in vibration frequency and amplitude [60]. Experimental investigations suggest that although the side IEG increases, the kerf uniformity also improves (reduction in the standard deviation of average slit width). However, it may not be very appropriate to conclude that tool/workpiece vibrations are detrimental for machining accuracy in Wire-ECMM. This is because by increasing the amplitude and frequency of tool or workpiece vibrations, f_m can be increased. Machining at higher feed rates leads to the reduction in the side IEG and diminishes the contribution of vibration frequency and amplitude in increasing it.

11.5.3   Effect of Process Parameters on Machining Quality

11.5.3.1   Power Factors

Surface roughness is also influenced by the pulse duration used during machining. For short pulse duration, large pits are formed due to high MRR_v and accumulation of electrolysis products in the machining region. As the pulse duration increases, such pits are removed, and a polished surface is obtained. However, for very long pulse duration, the surface roughness deteriorates because of frequent short circuits. As pulse duration increases, surface roughness decreases initially because the frequency of short circuits decreases. But when pulse duration increases further, the accumulation of a larger volume of electrolysis products adversely affects the machining quality. Therefore, optimal values of pulse on-time and pulse duration must be experimentally determined to achieve high surface quality. Surface roughness of 0.039 μm (39 nm) is achieved on cobalt-based alloy for the pulse duration and pulse on-time of 2 μs and 125 ns, respectively, using a phosphoric acid-based electrolyte [53].

11.5.3.2   Tool/Workpiece Feed Rate

Surface roughness in Wire-ECMM is significantly influenced by the tool/workpiece feed rate. When the feed rate is low, the time available for polishing the surface is ample and low surface roughness is observed. As the feed rate increases, this polishing time is reduced. A low tool/workpiece feed rate is preferred to achieve better surface quality, but then machining efficiency is usually compromised [53]. Therefore, an optimum balance between the two must be maintained.

11.5.3.3   Tool/Workpiece Vibration

Parameters such as tool/workpiece traveling speed and amplitude of vibration play a key role in determining the quality of the machined surface in Wire-ECMM [47]. A high amplitude of tool vibration assists in flushing electrolysis products out of the machining gap and renewing the electrolyte more efficiently. This decreases the roughness of the machined surface. Compared to Wire-EDM, where a recast layer of solidified metal forms over the machined surface, the quality of surface generated by Wire-ECMM is superior [13]. A smooth surface with sub-micron level roughness can be achieved using Wire-ECMM [66].

11.6   MATHEMATICAL MODELING OF WIRE-ECMM

11.6.1   ANALYTICAL MODELING

In Wire-ECMM, channel width can be written as shown in Figure 11.10 and expressed in Equation 11.6:

$$w = 2\Delta_s + D_w \tag{11.6}$$

where Δ_s is the side IEG and D_w is the wire diameter. MRR_l can be written in terms of the process parameters as shown in Equation 11.7 [13]:

$$MRR_l = \frac{\eta E V \kappa}{F \rho y} \tag{11.7}$$

where η is the system current efficiency, E is the gram equivalent weight of the anode material, V is the applied potential, κ is the electrical conductivity of the electrolyte, F is the Faraday's constant, ρ is the density of anode material, and y is the IEG. For the case when the tool is given a continuous feed motion at a constant rate of f, the rate of change of y with time (t) at any angle (θ) from the direction of tool feed can be written as shown in Equation 11.8:

$$\frac{dy}{dt}\left(\text{at an angle } \theta\right) = MRR_l - f\cos\left(\theta\right) \tag{11.8}$$

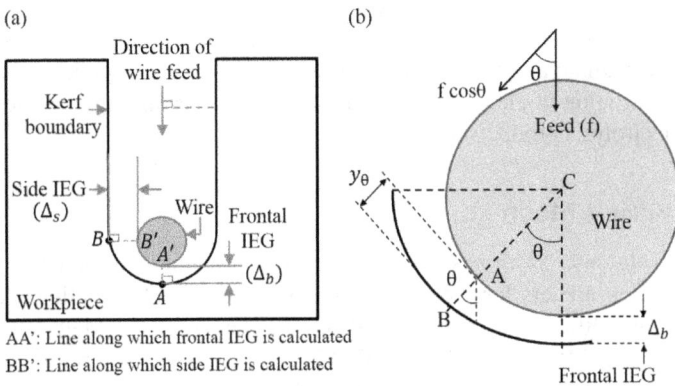

FIGURE 11.10   Schematic showing (a) a typical kerf profile machined using Wire-ECMM, (b) variation of IEG along the periphery of the wire. Reproduced with permission from [15].

For cases when $\theta = 0°$, the solution to Equation 11.8 under the boundary conditions: $y = y_0$ at time $t = 0$ s gives the expression for frontal IEG (the value of y for $\theta = 0°$ as shown in Equation 11.9:

$$y + \frac{\eta EV\kappa}{F\rho f} \ln\left|\left(\frac{\eta EV\kappa}{F\rho} - fy\right)\right| = y_0 + \frac{\eta EV\kappa}{F\rho f} \ln\left|\left(\frac{\eta EV\kappa}{F\rho} - fy_0\right)\right| - ft \quad (11.9)$$

here y is the initial gap maintained between the two electrodes before the machining starts. A generalized expression for calculating the IEG at any angle (θ) along the tool periphery is shown in Equation 11.10. This expression is valid for $0 < \theta < 90°$.

$$y + \frac{\eta EV\kappa}{F\rho f \cos(\theta)} \ln\left|\left(\frac{\eta EV\kappa}{F\rho} - f\cos(\theta)y\right)\right|$$
$$= y_0 + \frac{\eta EV\kappa}{F\rho f \cos(\theta)} \ln\left|\left(\frac{\eta EV\kappa}{F\rho} - f\cos(\theta)y_0\right)\right| - f\cos(\theta)t \quad (11.10)$$

Equation 11.9 can be numerically solved for y by substituting the values of different process parameters. The variation of y with t for different values of V, κ, f, and y_0 is shown in Figure 11.11. The value of η is assumed to be 1 (100% current efficiency for the system under consideration), F is considered as $96,500 \times 10^3$ C/kmol, E is considered as 27.56 kg/kmol (for iron), and ρ is taken as 7930 kg/m³.

It can be seen from this figure that y becomes constant after a certain time period. This state is called the state of equilibrium and is attained by the system after different time durations for different values of process parameters. Under the equilibrium conditions, MRR_l becomes equal to f and theoretically, the value of w becomes constant. As the system deviates from the equilibrium and y changes, MRR_l readjusts itself such that it again becomes equal to f and equilibrium is reattained. This feature is known as a "self-regulating feature" of ECM. It can be observed from Figure 11.10 that the equilibrium is attained by the system at lower values of y when the values of V, κ, and f are small. It is important to note that in actual machining, w is higher near the entry of the channel and lower at the bottom. This is primarily due to the stray corrosion and variation of electrolyte conductivity in the machining gap due to the presence of electrolysis products [68].

11.6.2 NUMERICAL MODELING

A substantial volume of work has been reported on the numerical modeling of the ECM process. Researchers have proposed the use of different modeling methods such as the Finite Difference Method (FDM), Boundary Element Method (BEM), and Finite Element Method (FEM). In FDM, IEG is discretized by equispaced square mesh, containing a set of grid points, which has difficulty in dealing with moving boundary conditions [69]. In case of a complex/non-linearly shaped IEG as in Wire-ECMM, the boundaries of tool and workpiece cannot be matched accurately

(a) Variation of frontal interelectrode gap with time for different values of applied potential

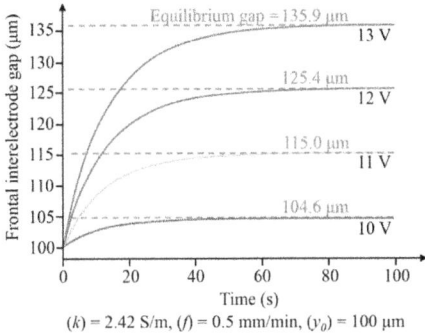

$(k) = 2.42$ S/m, $(f) = 0.5$ mm/min, $(v_0) = 100$ μm

(b) Variation of frontal interelectrode gap with time for different values of electrolyte electrical conductivity

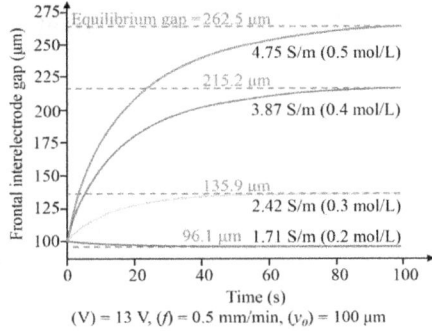

$(V) = 13$ V, $(f) = 0.5$ mm/min, $(v_0) = 100$ μm

(c) Variation of frontal interelectrode gap with time for different values of tool feed rate

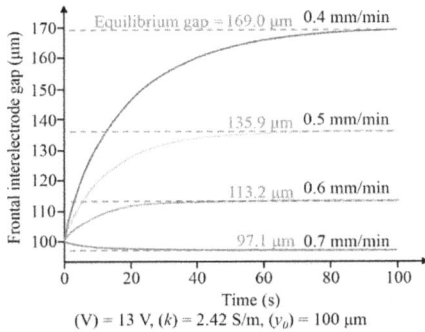

$(V) = 13$ V, $(k) = 2.42$ S/m, $(v_0) = 100$ μm

(d) Variation of frontal interelectrode gap with time for different values of initial interelectrode gap

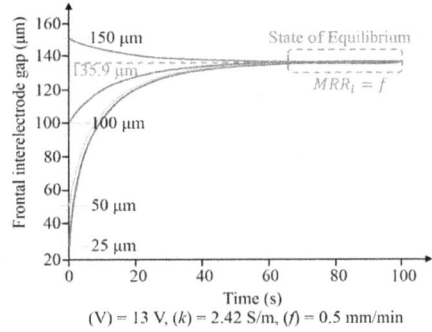

$(V) = 13$ V, $(k) = 2.42$ S/m, $(f) = 0.5$ mm/min

FIGURE 11.11 Graphs showing the variation of frontal IEG with (a) applied potential, (b) electrolyte electrical conductivity, (c) tool feed rate, and (d) initial IEG. Reproduced with permission from [67].

using square mesh which further incorporates errors. BEM with linear and quadratic isoparametric elements can be used for modeling but this method cannot adequately account for the non-linearity and anisotropy of the IEG which occurs due to the variation in the machining parameters. FEM overcomes these problems as mesh elements of different shapes and sizes can be used simultaneously. This makes it very convenient to incorporate different boundary conditions and analyze non-homogeneous conditions [70]. Numerically modeling the phenomenon of electrochemical dissolution using FEM requires correct determination of spatial electric potential distribution in the machining region. This is obtained by solving the elliptical Laplace equation in 2-D as shown in Equation 11.11 [71–76].

$$\frac{\partial^2 \left[\phi(x,y)\right]}{\partial x^2} + \frac{\partial^2 \left[\phi(x,y)\right]}{\partial y^2} = 0 \tag{11.11}$$

where $\phi(x, y)$ is potential distribution in the simulation domain. This equation can be solved for different boundary conditions according to the geometry of the simulation domain. A schematic diagram of a simulation domain in Wire-ECMM along with different boundary conditions is shown in Figure 11.12. n_x and n_y represent units outwards normal to the simulation domain in x and y directions, respectively. The boundary conditions are as follows:

$$\text{Boundary 1: } \phi = \phi_{applied} \text{ (Equipotential surface)} \tag{11.12}$$

$$\text{Boundary 2, 3, 4:} \frac{\partial \phi}{\partial n} = 0 \tag{11.13}$$

$$\text{Boundary 5: } \phi = 0 \text{ (Equipotential surface)} \tag{11.14}$$

The condition $\frac{\partial \phi}{\partial n} = 0$ on boundaries 2, 3, and 4 indicates that no electric field flux escapes these boundaries. Boundaries 1 and 5 are equipotential surfaces. The determination of shape and dimensions of the machined surface in the process using FEM requires a solution to a non-steady-state problem which further requires a large amount of computation time.

The potential distribution in the simulation domain under the application of a constant DC potential of 9 V is shown in Figure 11.13(a). The predicted anode profiles at different time instances are shown in Figure 11.13(b).

It is important to note here that some of the key aspects of anodic dissolution such as potential drop due to charging and discharging of EDL [71], reduction in material dissolution rate due to the formation of passive oxide film, and variation in the instantaneous current density distribution in machining gap due to the formation of hydrogen gas bubbles and electrolysis products are yet to be included in numerical models [77, 78, 79]. In practice, tool designers in ECM rely more on an empirical approach which involves the collection of a large number of data points. Using this approach, different aforementioned phenomena, which are otherwise difficult to incorporate in a numerical model due to their limited understanding, can also be included [75]. A nomographic approach can also be developed specifically for Wire-ECMM.

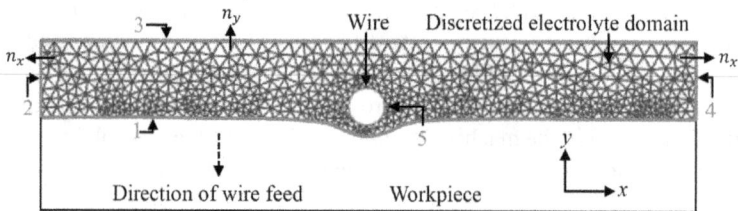

FIGURE 11.12 Schematic of a simulation model used for anode shape prediction in Wire-ECMM using FEM.

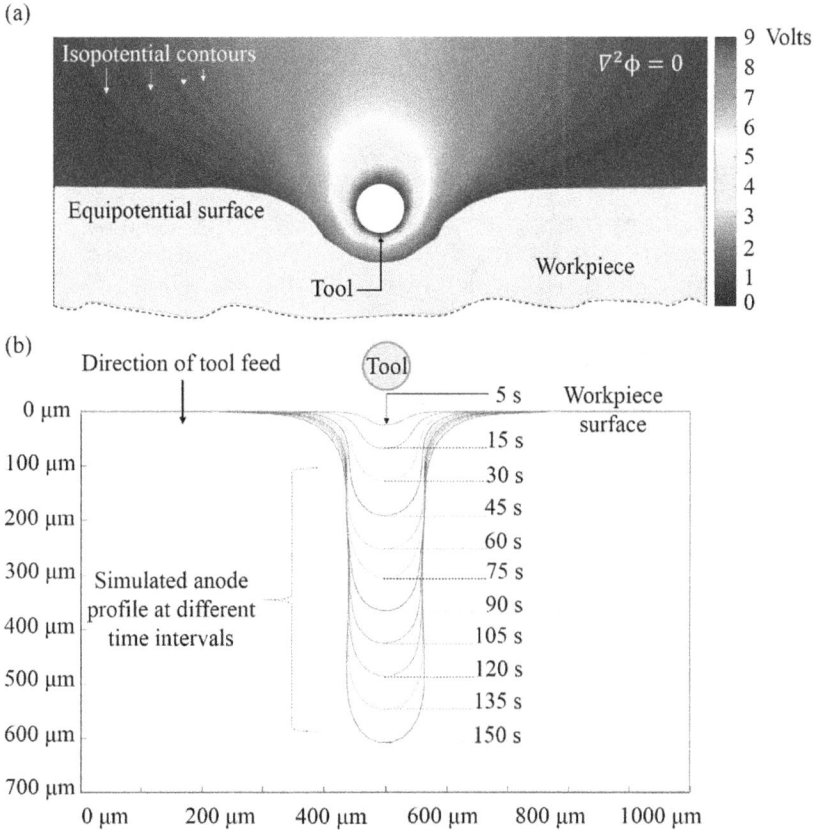

FIGURE 11.13 (a) Potential distribution in the simulation domain and (b) predicted anode profile at different time intervals. Reproduced with permission from [15].

11.7 WIRE ELECTROCHEMICAL TURNING OPERATION

In recent years, Wire-ECMM is employed for the machining of cylindrical work-pieces. The operation is called Wire Electrochemical Turning (Wire-ECTrg), and it is analogous to the conventional turning operation in terms of relative motion between the tool and workpiece [3]. An electrolyte is continuously supplied between the two electrodes, and the workpiece rotates about its axis as shown in Figure 11.14.

Preliminary investigations suggest that machining parameters such as electrolyte concentration, wire radial feed rate, applied potential, and workpiece RPM play a key role in determining the final anode profile which is characterized by groove width, taper angle, and corner radius. The introduction of Wire-ECTrg operation has opened up various new areas of application for Wire-ECMM. With wire as a tool, operations such as step turning (Figure 11.15(a)) [81], taper turning (Figure 11.15(b)) [81], threading (Figure 11.15(c)) [82], grooving (Figure 11.15(d)) [83], indexed cutting (Figure 11.15(e)) [84], and microturning (Figure 11.15(f)) [85] can now be

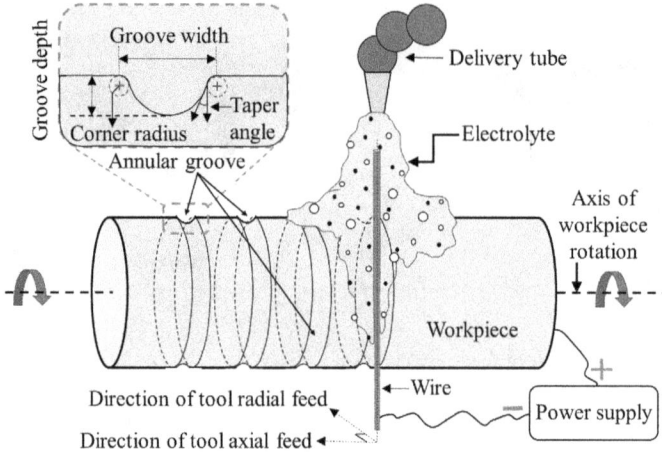

FIGURE 11.14 Schematic showing the principle of Wire-ECTrg operation. Reproduced with permission from [15].

performed at macro/micro level. Wire-ECTrg can prove to be especially useful in the machining/finishing of very thin-walled tubes which may not be able to withstand the forces generated during conventional turning. High aspect ratio microtools (for EDM/ECDM) with tight tolerances can be produced using this process.

11.8 SCOPE OF WIRE-ECMM

In the past few decades, the research community has worked extensively to show the capabilities of Wire-ECMM for different applications. There is a need to show-case the competence of this process by providing detailed research articles and a material-specific database to industry for the selection of parameters.

Features with a high aspect ratio (an aspect ratio of 31 [56]) can be machined using this process and may find application in a wide range of industries. For example, microsplines and curved flow channels can be used in microfluidic devices and actuators used in Micro Electromechanical Systems (MEMS) [86]. In biomedical industries, there are many applications, such as a microcantilever can be used weighing of microspecies, periodic microstructures can be used in an x-ray absorption contrast system of imaging [87], microtextures on human body implants can be used to induce desired tribological properties [44], grooving on hypodermic needles can be used to reduce insertion force [80], and microthreads can be machined on dental implants. In the communication industry, terahertz hollow core metal used as a rectangular waveguide can be machined using this process [88]. In the energy sector, the process can be used to slice thin wafers from silicon ingots used for making solar panels [26]. Turbine disk tenons [23] and high aspect ratio cylindrical microtools [89] to drill deep holes in turbine blades can also be machined using this process. For industries manufacturing microassemblies, this process can be used to make micropropeller blade structures [29], square and circular spiral springs [55, 63], microcams [47], and microgears [62]. Figure 11.16 shows some of the interesting features machined using the Wire-ECMM process.

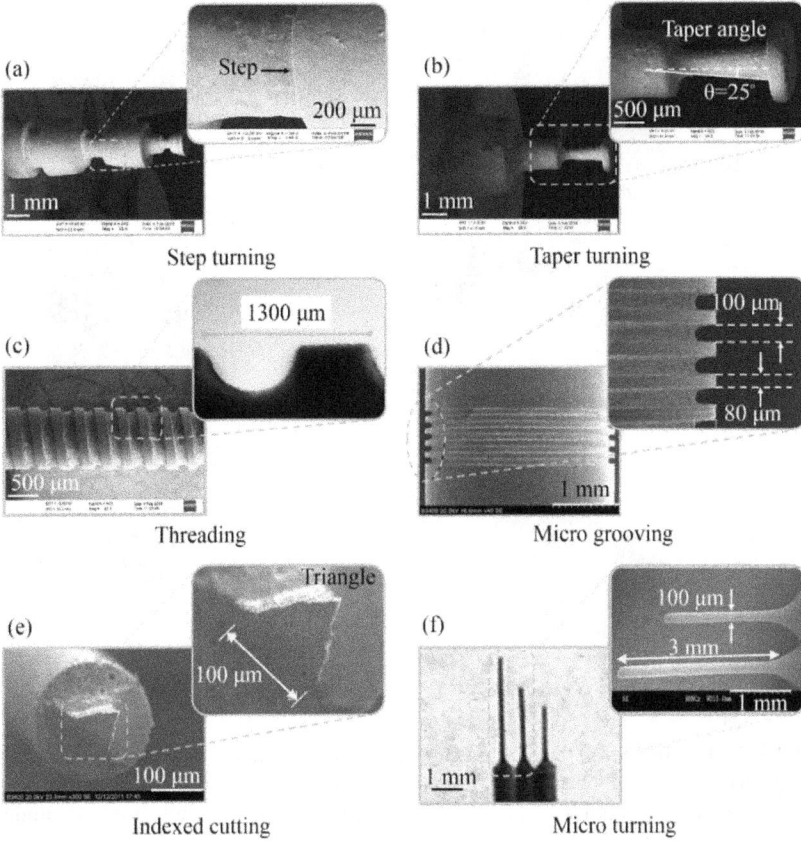

FIGURE 11.15 Images of (a) multi-stepped turning, (b) taper turning, (c) threading, (d) microgrooving, (e) indexed cutting, and (f) microturning. Reproduced with permission from [15].

11.9 RESEARCH OPPORTUNITIES

Wire-ECMM has been a topical field of research for the past two decades. This is primarily because of the advantages it offers, especially in performing micromachining operations. Although process developments in Wire-ECMM have attained a decent level of maturity, there are still many gaps where researchers can focus their study. Some of them are highlighted as follows:

11.9.1 MACHINING OF HOMOGENEOUS AND HETEROGENEOUS ALLOYS USING ELECTROLYTE MIXTURE

Homogeneous alloys may exhibit either selective dissolution or simultaneous dissolution. Most of the homogeneous alloys machined using ECM exhibit simultaneous dissolution. This means that the alloy constituents go into the solution in their stoichiometric proportion [90]. Heterogeneous alloys generally exhibit selective dissolution. This means that the less noble constituents dissolve preferentially, leaving

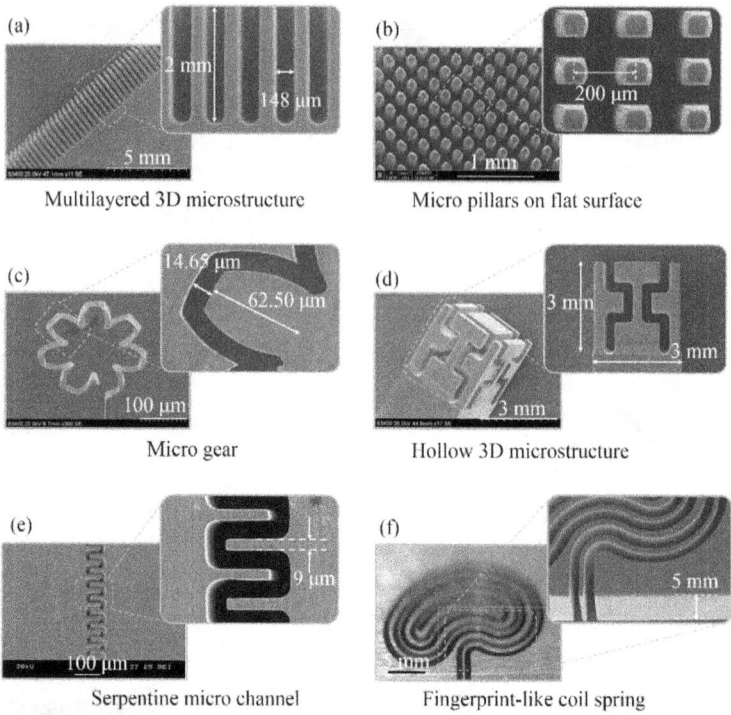

FIGURE 11.16 Image of (a) multi-layered 3-D microelectrodes, (b) square micropillar-shaped surface textures, (c) microspur gear, (d) 3-D hollow microstructure, (e) serpentine microchannel for microfluidic applications, and (f) fingerprint-like coil spring. Reproduced with permission from [15].

behind a porous metal phase enriched in more noble components. Anodic dissolution of such alloys may lead to the selective dissolution of one of its constituents, resulting in a very rough surface [45]. A popular example of such an alloy is Nitinol, also known as shape memory alloy. Different constituents of the electrolyte mixture can be determined by analyzing the polarization behavior of an individual alloy's components in different electrolytes and their proportions can be empirically determined. For example, simultaneous dissolution of Sn (tin)-Al (aluminum) multi-phase alloy requires a mixture of a strong alkaline solution and an acidic potassium sulfate solution. This is because in a strong alkaline solution, Al is selectively dissolved, and Sn remains passive, whereas in acidic solution, Al is protected by the formation of oxide film and Sn is dissolved [91].

11.9.2 INFLUENCE OF ELECTROLYTE JET INCLINATION ANGLE ON PROCESS PERFORMANCE

A study is due to investigate the influence of the electrolyte jet inclination angle on the process performance in Wire-ECMM. A recent publication on the influence of the electrolyte jet inclination angle [92] in the Jet-ECM process suggests that the resulting anode profile also changes with the inclination angle. This is due to the variation of current density distribution pattern. However, in Wire-ECMM, the vibration of a wire due to the momentum of the jet will also play a role in determining the homogeneity of the machined kerf. It is a field worth exploring.

11.9.3 NUMERICAL MODELS FOR ANODE SHAPE PREDICTION IN WIRE-ECMM

A simplified model of primary current distribution is generally used for the numerical modeling of Wire-ECMM. This model does not incorporate the potential drop in the machining region due to different resistances. Also, the electrolyte is usually modeled as a two-phase fluid, whereas it consists of three phases in the machining zone. Consequently, the results from the existing models deviate significantly from the experimental observations and have great scope for improvement. In view of this, a comprehensive analysis of three-phase electrolyte conductivity during electrochemical macromachining/micromachining has been proposed [66]. It has shown much better results compared to two-phase model results. The expression for calculating the effective electrical conductivity of electrolyte (κ_{eff}) by considering all three phases to be present is shown in Equation 11.15:

$$\kappa_{eff} = \kappa_0 \left(1 + \alpha \Delta T\right)\left(1 - \alpha_v\right)^a \left[1 - \beta\left(1 - \frac{\kappa_d}{\kappa_0}\right)\alpha_s\right]^{\gamma} \qquad (11.15)$$

where κ_0 is the initial electrolyte conductivity, α is the temperature coefficient of resistance of electrolyte, ΔT is the temperature difference, α_s is the void fraction of sludge, α_v is the void fraction of bubbles, and κ_d is the electrical conductivity of the dispersed medium. β and γ are constants and their values are 0.994 and 1.5 for the given combination, respectively. Experimental investigations carried out in this study also suggest that electrolyte conductivity decreases significantly (>20%) when the IEG is less than 200 μm. This is because for a small IEG, the effect of sludge and gas bubbles dominates the effect of heat generation which is responsible for the increment in electrolyte conductivity. In Wire-ECMM, the machining gap usually remains less than 200 μm. Any significant variation in electrolyte conductivity may lead to inhomogeneity in the kerf width. Therefore, an effective removal of sludge particles and gas bubbles from the machining zone is very important.

11.9.4 EXPERIMENTATION WITH A PARTIALLY INSULATED WIRE

Although it is not very convenient to place a thin, uniform, and stable layer of insulating material on the surface of a thin wire, it can, however, prove to be an effective solution for minimizing stray corrosion and improving machining accuracy. Simulation results with partially insulated tools [81] suggest significant improvement in machining accuracy. Experiments can be conducted to investigate the effect of tool insulation on machining accuracy, surface quality, and in the regions near to the machined zone.

11.9.5   Prospective Areas of Application

More application-based research can be carried out for the fabrication of critical components, such as surgical blades, micropin fin heat-exchangers, tooth implant screw threads, small crankshafts, etc. with low energy consumption. Such components should have a high surface finish and no surface defects. As there is no burr formation in the process, Wire-ECMM can be used to manufacture sharp cutting edges of surgical tools.

11.10 SUMMARY

Wire-ECMM, a variant of the ECM process, is proving to be a promising technique for micromachining operations. This is because the thermal and mechanical properties of the parent material remain unaltered after machining. In this chapter, various aspects of this process are discussed from a fundamental viewpoint. Different factors which influence the process performance are identified and discussed explicitly. The selection of electrolytes in this process depends upon the type of material being machined. The requirements for the selection of tools are also highlighted. Analytical models to predict the machining gap and material removal rate in the process are presented. The process is also modeled numerically using FEM to predict the shape and dimensions of the resulting anode profile. This chapter also presents an incisive discussion of the scope of the process and the possible areas where future research activities can be focused.

ACRONYMS

AJM	Abrasive Jet Machining
AWJM	Abrasive Water Jet Machining
BEM	Boundary Element Method
CNTs	Carbon Nanotubes
DC	Direct Current
EBM	Electron Beam Machining
ECM	Electrochemical Machining
ECMM	Electrochemical Micromachining
ECG	Electrochemical Grinding
ECTrg	Electrochemical Turning
EDL	Electrical Double Layer
FDM	Finite Difference Method
FEM	Finite Element Method
HAZ	Heat Affected Zone
IEG	Interelectrode Gap
Jet-ECM	Jet Electrochemical Machining
LBM	Laser Beam Machining
MEMS	Micro Electromechanical Systems
MRR$_v$	Volumetric Material Removal Rate
MRR$_l$	Linear Material Removal Rate

SEM	Scanning Electron Microscopy
SS 304	Stainless Steel – Grade 304
STED	Shaped Tube Electrochemical Drilling
USM	Ultrasonic Machining
Wire-ECMM	Wire Electrochemical Micromachining
Wire-EDM	Wire-Electric Discharge Machining

SYMBOLS

A	Atomic mass (g)
a	Transfer coefficient
C_{dl}	Electrical double layer capacitance (F)
D_w	Wire diameter (m)
E	Gram equivalent weight of anode material (g)
F	Faraday's constant (96,000 C/kmol)
F_p	Pulse frequency (Hz)
f_m	Maximum permissible tool/workpiece feed rate (m/s)
I	Circuit current (A)
i_o	Exchange current (A)
M	Mass of material dissolved (g)
n	Number of valence electrons
Q	Charge (C)
R	Universal gas constant
R_a	Profile roughness (m)
R_{ct}	Charge transfer resistance (Ω)
R_{el}	Resistance of the electrolyte band (Ω)
R_w	Warburg impedance (Ω)
S_a	Area surface roughness (m)
T	Temperature (°C)
t	Time interval (s)
$t_{duration}$	Sum of pulse on-time and off-time (s)
t_{off}	Pulse off-time (s)
t_{on}	Pulse on-time (s)
V	Magnitude of applied DC potential (V)
y_o	Interelectrode gap at time t = 0 (m)
Z	Electrochemical valency

GREEK LETTERS

Δ_s	Side interelectrode gap (m)
κ	Electrical conductivity of electrolyte (S/m)
κ_o	Initial electrical conductivity of electrolyte (S/m)
κ_{eff}	Effective electrical conductivity of electrolyte (S/m)
κ_d	Electrical conductivity of the dispersed medium (S/m)
Δ_b	Frontal interelectrode gap (m)

ρ Density of anode material (kg/m³)
ϕ Potential distribution function
η Current efficiency
τ Charging time constant of EDL capacitor (s)
a Transfer coefficient
α Temperature coefficient of resistance of electrolyte
α_s Void fraction of sludge
α_v Void fraction of bubble
β, γ Constants

BIBLIOGRAPHY

1. J.W. Schultze, A. Heidelberg, C. Rosenkranz, T. Schäpers, G. Staikov. Principles of electrochemical nanotechnology and their application for materials and systems. *Electrochimica Acta* 51 (2005): 775–786. doi:10.1016/j.electacta.2005.04.073.
2. Y. Ye, H. Lianhuan, H. Di, S. Jian-jia, T. Zhong-qun, T. Zhao-wu, Z. Dongping. Electrochemical micromachining under mechanical motion mode. *Electrochimica Acta* 183 (2015): 3–7. doi:10.1016/j.electacta.2015.04.046.
3. W. Han, M. Kunieda. Wire electrochemical grinding of tungsten micro-rod with electrostatic induction feeding method. *Procedia CIRP* 68 (2018): 699–703. doi:10.1016/j.procir.2017.12.140.
4. D.S. Patel, V.K. Jain, J. Ramkumar. Electrochemical grinding. In: V.K. Jain (Ed.), *Nanofinishing Science and Technology*, 1st ed. CRC Press, Boca Raton, 2016, pp. 321–352.
5. P. Taylor, S. Pathak, N.K. Jain, I.A. Palani. On use of pulsed-electrochemical honing to improve micro-geometry of bevel gears on use of pulsed-electrochemical honing to improve micro-geometry of bevel gears. *Materials and Manufacturing Processes* 29 (2014): 1461–1469. doi:10.1080/10426914.2014.952032.
6. V.K. Jain, P.M. Dixit, P.M. Pandey. On the analysis of the electrochemical spark machining process. *International Journal of Machine Tools and Manufacture* 39 (1999): 165–186. doi:10.1016/S0890-6955(98)00010-8.
7. T. Singh, A. Dvivedi. Developments in electrochemical discharge machining: A review on electrochemical discharge machining, process variants and their hybrid methods. *International Journal of Machine Tools and Manufacture* 105 (2016): 1–13. doi:10.1016/j.ijmachtools.2016.03.004.
8. K.K. Saxena, J. Qian, D. Reynaerts. A review on process capabilites of electrochemical micromachining and its hybrid variants. *International Journal of Machine Tools & Manufacture* 127 (2018): 28–56. doi:10.1016/j.ijmachtools.2018.01.004.
9. D. Patel, V. Jain, J. Ramkumar. Micro texturing on metallic surfaces: State of the art. *Proceedings of the Institution of Mechanical Engineers, Part B: Journal of Engineering Manufacture* (2016). doi:10.1177/0954405416661583.
10. Y. Ge, Z. Zhu, Z. Ma, D. Wang. Tool design and experimental study on electrochemical turning of nickel-based cast superalloy. *Journal of the Electrochemical Society* 165 (2018): 162–170. doi:10.1149/2.0371805jes.
11. A. Martin, M. Hackert-Oschätzchen, A. Schubert. Design and realisation of a specific device for jet electrochemical machining of rotating workpieces. In: *International Symposium on Electrochemical Machining Technology*, 2016, p. 39.
12. V. Sharma, D.S. Patel, V. Agrawal, V.K. Jain, J. Ramkumar. Investigations into machining accuracy and quality in wire electrochemical micromachining under sinusoidal and triangular voltage pulse condition. *Journal of Manufacturing Processes* 62 (2021): 348–367. doi:10.1016/j.jmapro.2020.12.010.

13. V.K. Jain. *Advanced Machining Processes*, 2nd ed. New Delhi: Allied Publisher, 2002.

14. M.M. Lohrengel, K.P. Rataj, T. Münninghoff. Electrochemical machining—Mechanisms of anodic dissolution. *Electrochimica Acta* 201 (2016): 348–353.

15. V. Sharma, D.S. Patel, V.K. Jain, J. Ramkumar. Wire electrochemical micromachining: An overview. *International Journal of Machine Tools and Manufacture* 155 (2020): 103579. doi:10.1016/j.ijmachtools.2020.103579.

16. S.R. Taylor, E. Gileadi. Physical interpretation of the Warburg impedance. *Corrosion* 51 (1995): 664–671. doi:10.5006/1.3293628.

17. Z. Stojek. The electrical double layer and its structure. *Electroanalytical Methods: Guide to Experiments and Applications* (2010): 3–8. doi:10.1007/978-3-642-02915-8_1.

18. H. Wang, L. Pilon. Accurate simulations of electric double layer capacitance of ultramicroelectrodes. *Journal of Physical Chemistry C* 115 (2011): 16711–16719. doi:10.1021/jp204498e.

19. A. Tyagi, V. Sharma, V.K. Jain, J. Ramkumar. Investigations into side gap in wire electrochemical micromachining (wire-ECMM). *The International Journal of Advanced Manufacturing Technology* 94 (2017): 4469–4478. doi:10.1007/s00170-017-1150-z.

20. B. Bhattacharyya. *Electrochemical Micromachining for Nanofabrication, MEMS and Nanotechnology*, 1st ed. William Andrew, 2015. doi:10.1016/C2014-0-00027-5.

21. H. Rasmussen, J.A. McGeough. Theory of overpotentials in electrochemical micromachining. *Journal of Materials Processing Technology* 149 (2004): 504–505. doi:10.1016/j.jmatprotec.2004.04.003.

22. J. Lei, X. Wu, B. Wu, B. Xu, D. Guo, J. Zhong. Fabrication of 3D microelectrodes by combining wire electrochemical micromachining and micro-electric resistance slip welding. *Procedia CIRP* 42 (2016): 825–830. doi:10.1016/j.procir.2016.03.002.

23. X.L. Fang, X.H. Zou, M. Chen, D. Zhu. Study on wire electrochemical machining assisted with large-amplitude vibrations of ribbed wire electrodes. *CIRP Annals – Manufacturing Technology* 66 (2017): 205–208. doi:10.1016/j.cirp.2017.04.135.

24. H. He, N. Qu, Y. Zeng, X. Fang, Y. Yao. Machining accuracy in pulsed wire electrochemical machining of γ-TiAl alloy. *International Journal of Advanced Manufacturing Technology* 86 (2016): 2353–2359. doi:10.1007/s00170-016-8402-1.

25. L. Meng, Y. Zeng, D. Zhu. Helical carbon nanotube fiber tool cathode for wire. *Electrochemical Micromachining* 165 (2018): 665–673. doi:10.1149/2.0421813jes.

26. C. Lee, Y. Kanda, S. Ikeda, M. Matsumura. Electrochemical method for slicing Si blocks into wafers using platinum wire electrodes. *Solar Energy Materials and Solar Cells* 95 (2011): 716–720. doi:10.1016/j.solmat.2010.10.009.

27. T.A. El-Taweel, S.A. Gouda. Performance analysis of wire electrochemical turning process-RSM approach. *International Journal of Advanced Manufacturing Technology* 53 (2011): 181–190. doi:10.1007/s00170-010-2809-x.

28. F. Xiaolong, Z. Xianghe, Z. Pengfei, Z. Yongbin, Q. Ningsong. Enhancement of performance of wire electrochemical micromachining using a rotary helical electrode. *International Journal of Advanced Manufacturing Technology* 84 (2016): 929–939. doi:10.1007/s00170-015-7755-1.

29. L. Meng, Y. Zeng, X. Fang, D. Zhu. Wire electrochemical micromachining of metallic glass using a carbon nanotube fiber electrode. *Journal of Alloys and Compounds* 709 (2017): 760–771. doi:10.1016/j.jallcom.2017.03.198.

30. J. Sprovieri. httpswww.assemblymag.comarticles93180-can-carbon-nanotubes-replace-copper (2016).

31. S. Wang, D. Zhu, Y. Zeng, Y. Liu. Micro wire electrode electrochemical cutting with low frequency and small amplitude tool vibration. *The International Journal of Advanced Manufacturing Technology* 53 (2011): 535–544. doi:10.1007/s00170-010-2835-8.

32. T.A. El-Taweel, S.A. Gouda. Study on the wire electrochemical groove turning process. *Journal of Applied Electrochemistry* 41 (2011): 161–171. doi:10.1007/s10800-010-0220-9.

33. R. Maeda, K. Chikamori, H. Yamamoto. Feed rate of wire electrochemical machining using pulsed current. *Precision Engineering* 6 (1984): 193–199.

34. D. Zhu, K. Wang, N.S. Qu. Micro wire electrochemical cutting by using in situ fabricated wire electrode. *CIRP Annals – Manufacturing Technology* 56 (2007): 241–244. doi:10.1016/j.cirp.2007.05.057.

35. K. Xu, Y. Zeng, P. Li, X. Fang, D. Zhu. Effect of wire cathode surface hydrophilia when using a travelling wire in wire electrochemical micro machining. *Journal of Materials Processing Technology* 235 (2016): 68–74. doi:10.1016/j.jmatprotec.2016.04.008.

36. H.D. He, N.S. Qu, Y.B. Zeng, P.Z. Tong. Improvement of hydrogen bubbles detaching from the tool surface in micro wire electrochemical machining by applying surface microstructures. *Journal of the Electrochemical Society* 164 (2017): 248–259. doi:10.1149/2.1131709jes.

37. P.C. Wayner. The interfacial profile in the contact line region and the Young-Dupré equation. *Journal of Colloid and Interface Science* 88 (1982): 294–295. doi:10.1016/0021-9797(82)90175-8.

38. Z. Xianghe, F. Xiaolong, C. Mi, Z. Di. Investigation on mass transfer and dissolution localization of wire electrochemical machining using vibratory ribbed wire tools. *Precision Engineering* 51 (2018): 597–603. doi:10.1016/j.precisioneng.2017.10.015.

39. X. Qi, X. Fang, D. Zhu. Investigation of electrochemical micromachining of tungsten microtools. *International Journal of Refractory Metals & Hard Materials* 71 (2018): 307–314. doi:10.1016/j.ijrmhm.2017.11.045.

40. Z. Xianghe, F. Xiaolong, Z. Yongbin, Z. Di. A high efficiency approach for wire electrochemical micromachining using cutting edge tools. *The International Journal of Advanced Manufacturing Technology* 91 (2017): 3943–3952. doi:10.1007/s00170-017-0063-1.

41. T. Yang, Y. Zeng, Y. Hang. Workpiece reciprocating movement aided wire electrochemical machining using a tube electrode with an array of holes. *Journal of Materials Processing Technology* 271 (2019): 634–644. doi:10.1016/j.jmatprotec.2019.04.044.

42. C. Xu, X. Fang, Z. Han, D. Zhu. Wire electrochemical machining with pulsating radial electrolyte supply and preparation of its tube electrode with micro-holes. *Applied Sciences (Switzerland)* 10 (2020). doi:10.3390/app10010331.

43. H. Wu, S. Xiao, Q. Li, Y. Zhang. Research of high efficiency combined wire electrochemical machining technology and its application. *Advanced Materials Research* 316 (2011): 1015–1019. doi:10.4028/www.scientific.net/AMR.314-316.1015.

44. K. Jiang, X. Wu, J. Lei, Z. Wu, W. Wu, W. Li, D. Diao. Vibration-assisted wire electrochemical micromachining with a suspension of B 4 C particles in the electrolyte. *The International Journal of Advanced Manufacturing Technology* (2018): 3565–3574.

45. R.J. Leese, A. Ivanov. Electrochemical micromachining: An introduction. *Advances in Mechanical Engineering* 8 (2016). doi:10.1177/1687814015626860.

46. L. Meng, Y. Zeng, X. Fang, D. Zhu. Micro-shaping of metallic glass by wire electrochemical micro-machining with a reciprocating traveling workpiece. *Journal of Alloys and Compounds* 739 (2018): 235–248. doi:10.1016/j.jallcom.2017.12.157.

47. K. Xu, Y. Zeng, P. Li, D. Zhu. Study of surface roughness in wire electrochemical micro machining. *Journal of Materials Processing Technology* 222 (2015): 103–109. doi:10.1016/j.jmatprotec.2015.03.007.

48. C. Lee, Y. Kanda, T. Hirai, S. Ikeda, M. Matsumura. Electrochemical grooving of Si wafers using catalytic wire electrodes in HF solution. *Journal of the Electrochemical Society* 156 (2009): 134–137. doi:10.1149/1.3033735.

49. S.H. Ryu. Eco-friendly ECM in citric acid electrolyte with microwire and microfoil electrodes. *International Journal of Precision Engineering and Manufacturing* 16 (2015): 233–239. doi:10.1007/s12541-015-0031-3.

50. Q. Ningsong, F. Xiaolong, L. Wei, Z. Yongbin, Z. Di. Wire electrochemical machining with axial electrolyte flushing for titanium alloy. *Chinese Journal of Aeronautics* 26 (2013): 224–229. doi:10.1016/j.cja.2012.12.026.

51. H. Li, C. Gao, G. Wang, N. Qu, D. Zhu. A study of electrochemical machining of Ti-6Al-4V in $NaNO_3$ solution. *Nature Scientific Reports* 6 (2016): 1–11. doi:10.1038/srep35013.

52. Y. He, J. Zhao, H. Xiao, W. Lu, W. Gan. Electrochemical machining of titanium alloy based on NaCl electrolyte solution. *International Journal of Electrochemical Science* 13 (2018): 5736–5747. doi:10.20964/2018.06.31.

53. S. Li, Y. Zeng. Improving surface quality and machining efficiency of microgrooves by WECMM in H_3PO_4-C_2H_5OH solution. *Journal of the Electrochemical Society* 166 (2019): 584–593. doi:10.1149/2.0731916jes.

54. V. Sharma, I. Srivastava, V.K. Jain, J. Ramkumar. Modelling of wire electrochemical micromachining (wire-ECMM) process for anode shape prediction using finite element method. *Electrochimica Acta* 312 (2019): 329–341. doi:10.1016/j.electacta.2019.04.165.

55. Y. Bin Zeng, Q. Yu, S.H. Wang, D. Zhu. Enhancement of mass transport in micro wire electrochemical machining. *CIRP Annals – Manufacturing Technology* 61 (2012): 195–198. doi:10.1016/j.cirp.2012.03.082.

56. S. Wang, Y. Zeng, Y. Liu, D. Zhu. Micro wire electrochemical machining with an axial electrolyte flow. *International Journal of Advanced Manufacturing Technology* 63 (2012): 25–32. doi:10.1007/s00170-011-3858-5.

57. S. Debnath, J. Kundu, B. Bhattacharyya. Modeling and influence of voltage and duty ratio on wire feed in WECM : Possible alternative of WEDM. *Journal of the Electrochemical Society* 165 (2018). doi:10.1149/2.0601802jes.

58. F. Klocke, T. Herrig, M. Zeis, A. Klink. Modeling and simulation of the fluid flow in wire electrochemical machining with rotating tool (wire ECM). *Esaform* 050014 (2017): 2–7. doi:10.1063/1.5008059.

59. T. Herrig, K. Oßwald, I. Lochmahr, A. Klink, F. Klocke, F. Bergs. High speed wire EDM flushing approach for wire electrochemical machining. In: *International Symposium on Electrochemical Machining Technology*, Aachen, Germany, 2018, p. 8.

60. C. Gao, N. Qu. Wire electrochemical micromachining of high-aspect ratio microstructures on stainless steel 304 with 270-µm thickness. *The International Journal of Advanced Manufacturing Technology* 100 (2018): 263–272.

61. H. He, Y. Zeng, N. Qu. An investigation into wire electrochemical micro machining of pure tungsten. *Precision Engineering* 45 (2016): 285–291. doi:10.1016/j.precisioneng.2016.03.005.

62. H.S. Shin, B.H. Kim, C.N. Chu. Analysis of the side gap resulting from micro electrochemical machining with a tungsten wire and ultrashort voltage pulses. *Journal of Micromechanics and Microengineering* 18 (2008): 075009. doi:10.1088/0960-1317/18/7/075009.

63. Y. Zeng, Q. Yu, X. Fang, K. Xu, H. Li, N. Qu. Wire electrochemical machining with monodirectional traveling wire. *International Journal of Advanced Manufacturing Technology* 78 (2015): 1251–1257. doi:10.1007/s00170-014-6745-z.

64. E.L. Hotoiu, S. Van Damme, C. Albu, D. Deconinck, A. Demeter, J. Deconinck. Simulation of nano-second pulsed phenomena in electrochemical micromachining processes – Effects of the signal and double layer properties. *Electrochimica Acta* 93 (2013): 8–16. doi:10.1016/j.electacta.2013.01.093.

65. P. Rodriguez, D. Hidalgo, J.E. Labarga. Optimization of pulsed electrochemical micromachining in stainless steel. *Procedia CIRP* 68 (2018): 426–431. doi:10.1016/j.procir.2017.12.090.

66. X. Fang, Y. Wu, P. Tong, Y. Zeng. Multiple slit electrochemical micromachining using a single wire and a constant inter-electrode voltage. *The International Journal of Advanced Manufacturing Technology* (2019). doi:10.1007/s00170-019-04143-w.

67. V. Sharma, D. Singh, M. Gyanprakash, J. Ramkumar. On altering the wetting behaviour and corrosion resistance of a large metallic surface area by wire electrochemical texturing. *Surface & Coatings Technology* 422 (2021): 127533. doi:10.1016/j.surfcoat.2021.127533.

68. V.K. Jain, A.K. Chouksey. A comprehensive analysis of three-phase electrolyte conductivity during electrochemical macromachining/micromachining. *Proceedings of the Institution of Mechanical Engineers, Part B: Journal of Engineering Manufacture* 232 (2018): 2449–2461. doi:10.1177/0954405417690558.

69. S. Hinduja, M. Kunieda. Modelling of ECM and EDM processes. *CIRP Annals – Manufacturing Technology* 62 (2013): 775–797. doi:10.1016/j.cirp.2013.05.011.

70. V.K. Jain, K.P. Rajurkar. An integrated approach for tool design in ECM. *Precision Engineering* 13 (1991): 111–124. doi:10.1016/0141-6359(91)90502-A.

71. V.M. Volgin, V.V. Lyubimov, I.V. Gnidina. Simulation of ion transfer during electrochemical shaping by ultrashort pulses. In: *Proceedings of the 4th International Conference on Industrial Engineering*, Springer International Publishing, 2019. doi:10.1007/978-3-319-95630-5.

72. V.K. Jain, V.N. Nanda. Analysis of taper produced in side zone during ECD. *Precision Engineering* 8 (1986): 27–33. doi:10.1016/0141-6359(86)90007-3.

73. M.S. Reddy, V.K. Jain, G.K. Lal. Tool design for ECM: correction factor method. *Journal of Engineering for Industry* 110 (1988): 111–118. doi:10.1115/1.3187858.

74. V.K. Jain, P.G. Yogindra, S. Murugan. Prediction of anode profile in ECBD and ECD operations. *International Journal of Machine Tools and Manufacture* 27 (1987): 113–134. doi:10.1016/S0890-6955(87)80044-5.

75. V.K. Jain, P.C. Pandey. Tooling design for ECM. *Precision Engineering* 2 (1980): 195–206. doi:10.1016/0141-6359(80)90012-4.

76. V.K. Jain, D. Gehlot. Anode shape prediction in through-mask-ECMM using FEM. *Machining Science and Technology* 19 (2015): 286–312. doi:10.1080/10910344.2015.1018533.

77. K. Prashanth, D.S. Patel, V.K. Jain, J. Ramkumar. Numerical modelling of ECMM of micro- dimples considering the effect of 3-phase electrolyte. *Journal of Micromanufacturing* (2019): 1–15. doi:10.1177/2516598419852208.

78. Y. Ge, Z. Zhu, Z. Ma, D. Wang. Large allowance electrochemical turning of revolving parts using a universal cylindrical electrode. *Journal of Materials Processing Technology* 258 (2018): 89–96. doi:10.1016/j.jmatprotec.2018.03.013.

79. Z. Sun, H. Wang, Y. Ye, Z. Xu, G. Tang. Effects of electropulsing on the machinability and microstructure of GH4169 superalloy during turning process. *The International Journal of Advanced Manufacturing Technology* (2018): 2835–2842. doi:10.1007/s00170-017-1407-6.

80. D.S. Patel, A. Singh, V.K. Jain, J. Ramkumar, A. Shrivastava. Investigations into insertion force of electrochemically micro-textured hypodermic needles. *The International Journal of Advanced Manufacturing Technology* (2019): 1311–1326. doi:10.1007/s00170-017-1265-2.

81. A. Tyagi, V. Sharma, D.S. Patel, V.K. Jain, J. Ramkumar. Experimental and analytical investigations into wire electrochemical micro turning. *Journal of Micromanufacturing* (2019): 1–17. doi:10.1177/2516598419827130.

82. V. Sharma, D.S. Patel, V.K. Jain, J. Ramkumar, A. Tyagi. Wire electrochemical threading: A technique for fabricating macro/micro thread profiles. *Journal of the Electrochemical Society* 165 (2018): 397–405. doi:10.1149/2.1181809jes.

83. W. Xiangyang, F. Xiaolong, Z. Yongbin, Q. Ningsong. Fabrication of micro annular grooves on a cylindrical surface in aluminum alloys by wire electrochemical micromachining. *International Journal of Electrochemical Science* 11 (2016): 7216–7229. doi:10.20964/2016.08.62.

84. Z. Liu, Y. Zeng, W. Zhang. Fabrication of metal microtool applying wire electrochemical machining. *Advances in Mechanical Engineering* 6 (2014): 1–7. doi:10.1155/2014/382105.

85. Z. Fan, L. Hourng, C. Wang. Fabrication of tungsten microelectrodes using pulsed electrochemical machining. *Precision Engineering* 34 (2010): 489–496. doi:10.1016/j.precisioneng.2010.01.001.

86. J. Singh, S. Bhattacharya. Fabrication of micro-mixer on printed circuit board using electrochemical micromachining. *Journal of Micromanufacturing* (2019): 1–10. doi:10.1177/2516598419838660.

87. H. He, Y. Zeng, Y. Yao, N. Qu. Improving machining efficiency in wire electrochemical micromachining of array microstructures using axial vibration-assisted multi-wire electrodes. *Journal of Manufacturing Processes* 25 (2017): 452–460. doi:10.1016/j.jmapro.2017.01.004.

88. X. Bi, Y. Zeng, X. Dai, N. Qu. Integral fabrication of terahertz hollow-core metal rectangular waveguides with a combined process using wire electrochemical micromachining, electrochemical deposition, and selective chemical dissolution. *The International Journal of Advanced Manufacturing Technology* (2019): 1–10. doi:10.1007/s00170-019-04680-4.

89. V. Sharma, A. Tyagi, I. Srivastava, M. Thalkar, J. Ramkumar, V.K. Jain. Modeling and simulation of wire electrochemical turning (wire-EC-Trg) process. In: *Proceedings of 10th International Conference on Precision, Meso, Micro and Nano Engineering (COPEN 10)*, 2017, pp. 636–641.

90. R.F. Steigerwald, N.D. Greene. The anodic dissolution of binary alloys. *Journal of the Electrochemical Society* (1962): 1026–1034.

91. M. Datta. Anodic dissolution of metals at high rates. *IBM Journal of Research and Development* 37 (1993): 207–226. doi:10.1147/rd.372.0207.

92. J. Mitchell-Smith, A. Speidel, A.T. Clare. Advancing electrochemical jet methods through manipulation of the angle of address. *Journal of Materials Processing Technology* 255 (2018): 364–372. doi:10.1016/j.jmatprotec.2017.12.026.

12 Anode Shape Prediction and Cathode Shape Design for ECM Processes

Jinming Lu and Ewald A. Werner

CONTENTS

12.1 INTRODUCTION

12.1.1 THE DIRECT AND THE INVERSE PROBLEM OF ECM

As described in the previous chapters, electrochemical machining (ECM) is a non-conventional material removal process that can be employed to shape workpieces made of metallic materials with low machinability. In this process, a pre-shaped metallic tool is polarized as the cathode of an electrolytic cell and advanced towards

DOI: 10.1201/9780429160011-12

the workpiece at a defined feed rate. The metal of the anodically polarized work-piece is oxidized to ions that dissolve into the electrolyte. In ECM, the local material dissolution rate at the workpiece (anode) increases with increasing current density. When the process is in a steady state, the shape of the anode becomes approximately the negative image of that of the cathode (Figure 12.1).

An ECM-related topic that is important for both fundamental research and industry is to find an accurate relationship between the shape of the workpiece and that of the tool, since they are not exactly the negative image of each other. In this context, two problems exist:

1. The *direct* or *anode shape prediction problem* of ECM refers to anticipating the evolution of the anode shape in a defined standard ECM process with a given set of process parameters and a known shape of the cathode.
2. In industrial applications, however, the known information often is the desired shape of the workpiece. Whenever a new workpiece has to be manufactured by ECM, the necessary shape of the tool has to be determined in advance, which often requires costly and time consuming experimental prestudies. To reduce the number of the preliminary experiments, much effort is made to compute the shape of the tool required to produce the workpiece with the desired shape in a defined ECM process – a problem often referred to as the *inverse* or *cathode (shape) design problem* of ECM.

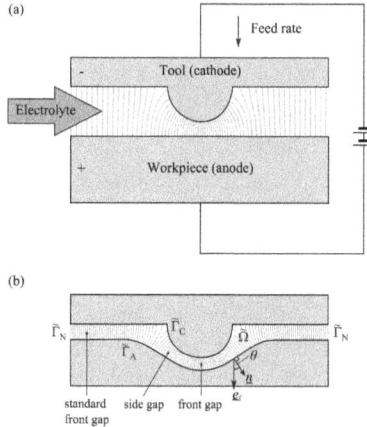

FIGURE 12.1 Basic principle of electrochemical machining (ECM): Initial (a) and steady-state (b) configuration consisting of a tool (cathode) of a defined shape and a plane anode workpiece to be machined. The thin lines in the gap between the electrodes illustrate the current density. For each point on the workpiece surface, the angle θ is measured between the outer surface unit normal vector $\underline{n}$ and the feed direction unit vector $\underline{e}_f$. In this chapter, "front gap" refers to an interelectrode gap region where the workpiece surface is perpendicular to the feed direction (i.e., $\theta = 0$). "Standard front gap" is a front gap region where the electrodes are plane. All other regions of the interelectrode gap are called "side gap". The other symbols are explained in Section 12.2.3.

12.1.2 MODELING OF THE ECM PROCESS

A central aspect to both the direct and the inverse problems is the appropriate modeling of the ECM process. The main challenge is the correct description of the interdependencies between the physical quantities involved (see Figure 12.2).

Despite the high number of interdependencies and the complex distribution of the electric potential within the interelectrode gap, several sources in literature (e.g., McGeough and Rasmussen, 1974; Deconinck, 1992; Zhou and Derby, 1995; Klocke and König, 2006) use the so-called *potential model* to describe ECM. This physical model is based on the following assumptions (cf. Zhou and Derby, 1995):

1. The rate of the electrolyte flow is assumed to be high enough to provide effective flushing, resulting in spatially and temporally constant electrolyte conductivity within the interelectrode gap.
2. The specific volumetric material removal rate, $\tilde{K}_V$, defined by*

$$\dot{\tilde{V}} = \tilde{K}_V \tilde{I},$$ (12.1)

is constant, where $\dot{\tilde{V}}$ is the volume of dissolved anode material per unit of time and $\tilde{I}$ is the total current.

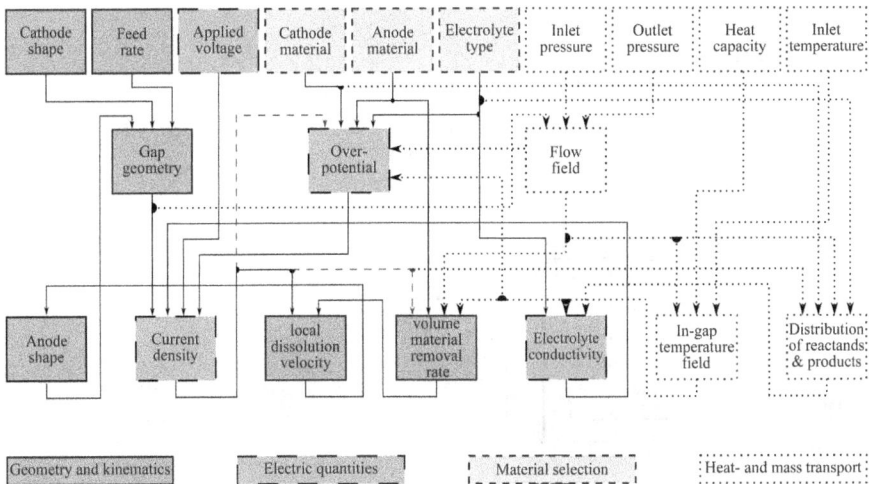

FIGURE 12.2 Interdependencies between predetermined machining parameters (upper row) and process variables (center and lower row) in ECM. The quantities are classified into four groups according to their nature. A connection pointing from a quantity A to a quantity B means that A has an influence on B. Interdependencies illustrated by dotted lines are related to heat and mass transport. Dashed lines illustrate the current-density-dependence of the overpotential and the effective volume material removal rate.

---

* In this chapter, a dimensioned quantity is marked with a tilde $\tilde{\ }$.

3. Overpotential effects and the thickness of the surface film are negligible; the electric potential drop within the electrolyte, $\tilde{U}$, is given by

$$\tilde{U} = \tilde{U}_{gap} - \Delta\tilde{U}, \tag{12.2}$$

where $\tilde{U}_{gap}$ is the electric voltage applied between cathode and anode and $\Delta\tilde{U}$ is the sum of electrode overpotentials, which is assumed to be constant within this model.

4. Metal dissolution at the anode occurs in the direction perpendicular to its surface and is independent of local material properties.

The interdepencies between the machining parameters and the process variables in the potential model of ECM are sketched in Figure 12.3. Within the potential model, the so-called *primary current distribution* is considered, which is a function of the applied voltage and the gap geometry only (Gileadi, 2011). Compared with the comprehensive interdependency map (cf. Figure 12.2), all physical quantities related to heat and mass transport are assumed constant, resulting in constant electrolyte conductivity. Furthermore, the current density dependence

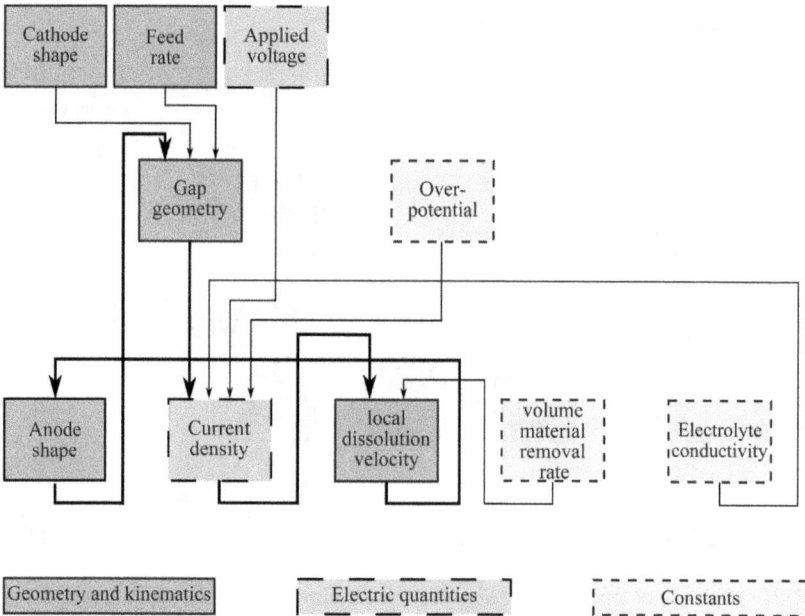

FIGURE 12.3 Interdependency map of the potential model. The map is obtained from Figure 12.2 through neglecting the influence of physical quantities related to heat and mass transport, and the current-density-dependence of overpotential and volume material removal rate. The thick connections illustrate the interplay between the anode shape and the current density.

of overpotential and volume material removal rate (dashed connections in Figure 12.2) are neglected. These simplifications are justified if heat generated by electric current flow and the products of electrochemical reactions are effectively removed from the interelectrode gap.

Several authors use more sophisticated multiphysics process models to describe the ECM process. Compared with the potential model, further aspects depicted in Figure 12.2 are taken into account. A review of these models is provided by Hinduja and Kunieda (2013).

One aspect is the variable electrolyte conductivity distribution $\tilde{\kappa}(\tilde{\underline{x}})$ that can be modeled by (Thorpe and Zerkle, 1969)

$$\tilde{\kappa}(\tilde{\underline{x}}) = \tilde{\kappa}_0(1 - \alpha_{void}(\tilde{\underline{x}}))^m\left[1 + \tilde{\gamma}(\tilde{T}(\tilde{\underline{x}}) - \tilde{T}_0)\right], \tag{12.3}$$

where $\alpha_{void}(\tilde{\underline{x}})$ is the void fraction (i.e., volume fraction of gaseous reaction products in the electrolyte), $\tilde{T}(\tilde{\underline{x}})$ is the electrolyte's temperature distribution, $\tilde{\gamma}$ is the conductivity constant of the electrolyte, m is a constant varying between 1.5 and 2 for the heterogeneous mixture of liquid and gas, and $\tilde{T}_0$ and $\tilde{\kappa}_0$ are temperature and conductivity of the electrolyte at the inlet, respectively. The use of Equation 12.3 requires the knowledge of the temperature and void fraction distribution within the electrolyte. Advanced models further take into account the influence of sludge due to the dissolution of the workpiece material (Jain and Chouksey, 2018). Hence, heat transfer and multiphase flow of the electrolyte within the interelectrode gap have to be modeled.

Another aspect is the electrode overpotential $\Delta\tilde{U}$ (cf. Equation 12.2), which depends on the current density and the flow field in general. Since the activation, concentration and resistance overpotentials are difficult to characterize separatedly (Hinduja and Kunieda, 2013) and metal dissolution often occurs in the transpassive range when using a passivating electrolyte (Datta and Landolt, 1982), an empirical modeling approach based on experimental results such as

$$\Delta\tilde{U} = \tilde{a} + \tilde{b}\tilde{j} \tag{12.4}$$

proposed by De Silva et al. (2000) can be reasonable, where $\tilde{a}$ and $\tilde{b}$ are parameters obtained from experiments, and $\tilde{j}$ is the current density.

Depending on the type of the electrolyte, the (variable) current efficiency

$$\eta(\tilde{j}) = \tilde{K}_{V,eff}(\tilde{j})/\tilde{K}_V \tag{12.5}$$

can play an important role in modeling. When a passivating electrolyte and pulsed current are used, the current efficiency is a function of the current density, pulse time, and electrolyte concentration. As shown by De Silva et al. (2000), the current efficiency can be modeled empirically by a hyperbolic tangent function of the current density.

12.2 ANODE SHAPE PREDICTION

12.2.1 CHARACTERISTICS OF THE INTERELECTRODE GAP IN SINGLE-AXIS ECM USING PLANE ELECTRODES

To understand the dynamics of an ECM process, this section considers a basic single-axis ECM experiment using electrodes with plane parallel surfaces of area $\tilde{A}$ (see Figure 12.4(b)) that are inclined by the angle θ (see Figure 12.4(a)). With constant feed rate $\tilde{v}_f$ and constant machining voltage $\tilde{U}_{gap}$, the potential model can be formulated mathematically in the following way:

1. The magnitude of the uniform current density within the interelectrode gap of width $\tilde{h}$ is given by

$$\tilde{j} := \left|\tilde{\underline{j}}\right| = \frac{\tilde{I}}{\tilde{A}} = \tilde{\kappa}\frac{\tilde{U}}{\tilde{h}}, \tag{12.6}$$

where $\tilde{\kappa}$ is the constant electrolyte conductivity and $\tilde{U}$ is the potential drop within the electrolyte as defined in Equation 12.2. In Equation 12.6, Ohm's law $\tilde{U} = \tilde{R}\tilde{I}$ is applied, where $\tilde{R} = \tilde{h}/(\tilde{\kappa}\tilde{A})$ is the Ohmic resistance of the interelectrode gap modeled as a rectangular block of uniform conductivity (see Figure 12.4(b)).

2. Material dissolution can be described by the *anode dissolution velocity* $\tilde{\underline{v}}_A$, whose magnitude $\tilde{v}_A$ is linked to that of the current density at the anode, $\tilde{j}_A$, by

$$\tilde{v}_A = \tilde{K}_V \tilde{j}_A, \tag{12.7}$$

which is obtained through dividing both sides of Equation 12.1 by $\tilde{A}$.

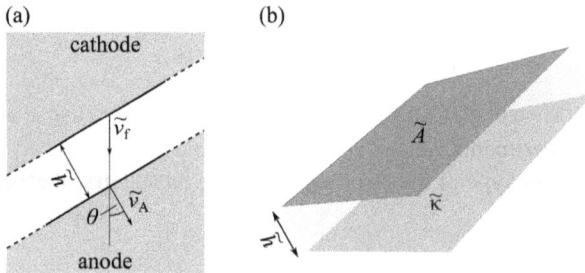

FIGURE 12.4 Single-axis ECM using plane electrodes. (a) Nomenclature of the kinematic quantities used in Equations 12.6–12.9. The angle θ is measured between the surface normal and the feed direction. (b) Schematic illustration of the interelectrode gap domain of width $\tilde{h}$ between plane electrode surfaces that are parallel to each other. The area of the surface of each electrode is $\tilde{A}$; the variable $\tilde{\kappa}$ denotes the conductivity of the electrolyte that is assumed spacially uniform and constant in time.

Since the shape of the workpiece does not change, the kinematics of the process is described by the change of the interelectrode gap width $\tilde{h}$ with machining time $\tilde{t}$,

$$\frac{d\tilde{h}}{d\tilde{t}} = \tilde{v}_A - \tilde{v}_{f\perp}, \tag{12.8}$$

where $\tilde{v}_{f\perp} = \tilde{v}_f \cos\theta$ is the component of the feed velocity normal to the electrode surface. Taking Equations 12.7 and 12.6 into account, Equation 12.8 transforms to the differential equation

$$\frac{d\tilde{h}}{d\tilde{t}} = \frac{\tilde{\kappa}\tilde{K}_V\tilde{U}}{\tilde{h}} - \tilde{v}_f \cos\theta, \tag{12.9}$$

which can be solved by separation of variables.

In the case of $\tilde{v}_{f\perp} = 0$, i.e., the tool does not move $\left(\tilde{v}_f = 0\right)$ or the electrode surface is parallel to the feed direction ($\theta = 90°$), Equation 12.9 transforms to

$$\int_{\tilde{h}_0}^{\tilde{h}(t)} h' dh' = \int_0^{\tilde{t}} \tilde{\kappa}\tilde{K}_V\tilde{U} dt',$$

resulting in parabolical growth of the gap width with time,

$$\tilde{h}(\tilde{t}) = \sqrt{\tilde{h}_0^2 + 2\tilde{\kappa}\tilde{K}_V\tilde{U}\tilde{t}}, \tag{12.10}$$

where $\tilde{h}_0$ is the initial width of the gap.

In the case of $\tilde{v}_{f\perp} \neq 0$, Equation 12.9 can be used to analyze the self-regulating feature of ECM.

A stationary, i.e., time-independent solution is obtained for

$$0 = \frac{d\tilde{h}}{d\tilde{t}} = \frac{\tilde{\kappa}\tilde{K}_V\tilde{U}}{\tilde{h}} - \tilde{v}_f \cos\theta \quad \Rightarrow \quad \tilde{h}_e = \frac{\tilde{\kappa}\tilde{K}_V\tilde{U}}{\tilde{v}_f \cos\theta}. \tag{12.11}$$

The quantity $\tilde{h}_e$ is the equilibrium gap width in steady-state ECM when the anodic dissolution velocity is exactly compensated by the component of the feed velocity normal to the electrode surface. In the case of $\theta = 0$, i.e. in the so-called standard front gap region (see Figure 12.1), the quantity defined by

$$\tilde{h}_e^{std} := \frac{\tilde{\kappa}\tilde{K}_V\tilde{U}}{\tilde{v}_f} \tag{12.12}$$

will be referred to as the *standard equilibrium front gap width*.

When the gap width is larger than the equilibrium gap width, i.e., $\tilde{h} > \tilde{h}_e$, the right hand side of Equation 12.9 is negative, resulting in a decrease in $\tilde{h}$ due to $d\tilde{h}/d\tilde{t} < 0$. In the case of $\tilde{h} < \tilde{h}_e$ the width of the gap $\tilde{h}$ increases since $d\tilde{h}/d\tilde{t} > 0$.

With $\tilde{h}_e$ defined in Equation 12.11, Equation 12.9 is transformed to

$$\frac{d\tilde{h}}{d\tilde{t}} = \tilde{v}_f \cos\theta \left(\frac{\tilde{h}_e}{\tilde{h}} - 1 \right) = \tilde{v}_f \cos\theta \frac{\tilde{h}_e - \tilde{h}}{\tilde{h}}, \text{ or }$$

$$\int_0^{\tilde{t}} \tilde{v}_f \cos\theta \, dt' = \int_{\tilde{h}_0}^{\tilde{h}(t)} \frac{h'}{\tilde{h}_e - h'} dh' = \int_{\tilde{h}_0}^{\tilde{h}(t)} \left(\frac{\tilde{h}_e}{\tilde{h}_e - h'} - 1 \right) dh'.$$

Solving for the dissolution time yields

$$\tilde{t} = \frac{1}{\tilde{v}_f \cos\theta} \left(\tilde{h}_0 - \tilde{h}(\tilde{t}) + \tilde{h}_e \ln \frac{\tilde{h}_0 - \tilde{h}_e}{\tilde{h}(\tilde{t}) - \tilde{h}_e} \right). \tag{12.13}$$

Equation 12.13 represents implicitly the evolution of the width of the interelectrode gap that approaches the equilibrium gap width $\tilde{h}_e$ asymptotically.

Problem

Starting with an initial gap width $\tilde{h}_0 = 2\tilde{h}_e$, compute the time $\tilde{t}_E$ (in units of time to machine exactly one equilibrium gap width) required to reach a state in that the deviation of the gap width from the equilibrium gap width is 5% (i.e., $\left(\tilde{h}(\tilde{t}_E) - \tilde{h}_e \right)/\tilde{h}_e = 0.05$)).

Solution

Equation 12.13 is transformed into a dimensionless form in that time is represented in units of time to machine one equilibrium gap width

$$\frac{\tilde{t}\tilde{v}_f \cos\theta}{\tilde{h}_e} = \frac{\tilde{h}_0}{\tilde{h}_e} - \frac{\tilde{h}(t)}{\tilde{h}_e} + \ln \frac{\tilde{h}_0/\tilde{h}_e - 1}{\tilde{h}(\tilde{t})/\tilde{h}_e - 1}.$$

With $\tilde{h}_0/\tilde{h}_e = 2$ and $\tilde{h}(\tilde{t}_E)/\tilde{h}_e = 1.05$

$$\frac{\tilde{t}_E\tilde{v}_f \cos\theta}{\tilde{h}_e} = 2 - 1.05 + \ln \frac{1}{0.05} \approx 3.95,$$

i.e., after a time required to machine about four equilibrium gap widths, the deviation of the gap width from the equilibrium gap width is less than 5%.

12.2.2 COS θ-METHOD

The so-called cos θ-method is an approximative method applied to both anode shape prediction and cathode shape design in ECM. The method was first pioneered by Tipton (1964). Apart from the simplifications assumed by the potential model (see Figure 12.3), in which overpotentials, fluid flow, heat and mass transport, and the resulting variation in electrolyte conductivity are neglected, the cos θ-method further assumes that current flux lines are straight and parallel to each other (Tipton, 1971). These assumptions are only valid within the plane segments of the electrodes. The basic idea of the cos θ-method can be derived from Equations 12.11 and 12.12, resulting in the following relation of the width of the interelectrode gap:

$$\tilde{h}_e(\theta) = \frac{\tilde{h}_e^{std}}{\cos\theta}. \tag{12.14}$$

In the literature, several opinions about the reference surface for the measurement of the angle θ exist (Jain and Pandey, 1980b). For the application of the cos θ-method to the anode shape prediction problem only (i.e., without considering cathode shape design), the cathode surface is usually the reference surface (Dietz et al., 1973; Hinduja and Kunieda, 2013).

Due to the simplified assumptions, the cos θ-method is inaccurate when applying to interelectrode gap regions with $\theta > 45°$ or where the cathode has sharp corners (McGeough, 1974; Jain and Pandey, 1980b).

12.2.3 NUMERICAL METHODS

While in ECM using plane electrodes (see Section 12.2.1) the only changing geometric quantity is the width of the interelectrode gap, the shape of anode changes during machining when using curved cathodes. In general, the evolution of the anode shape cannot be described by analytic expressions, making numerical methods indispensable for computations.

Due to the interdependent physical and (electro)chemical processes involved in ECM (see Figure 12.2), models for simulating the ECM process can become very complex. In the following, the computation of the anode shape evolution is described exemplarily for the potential model (see Secton 12.1.2), taking only the following two aspects into account (cf. Figure 12.3):

1. The computation of the primary current density within the interelectrode gap.
2. The computation of the changes in the shape of the anode.

Within the framework of the potential model, the current density $\tilde{\underline{j}}$ is given by Ohm's law in the local form,

$$\tilde{\underline{j}} = -\tilde{\kappa}\underline{\tilde{\nabla}}\tilde{\Phi}, \tag{12.15}$$

which is a generalization of Equation 12.6. In Equation 12.15, the variable $\tilde{\Phi}$ denotes the electric potential that is determined by solving the Laplace equation

$$\underline{\nabla}\cdot\left(\tilde{\kappa}\underline{\nabla}\tilde{\Phi}\right) = 0 \quad \text{in } \tilde{\Omega}, \tag{12.16}$$

$$\frac{\partial \tilde{\Phi}}{\partial \tilde{n}} = 0 \quad \text{on } \tilde{\Gamma}_N, \tag{12.17}$$

$$\tilde{\Phi} = 0 \quad \text{on } \tilde{\Gamma}_C, \tag{12.18}$$

$$\tilde{\Phi} = \tilde{U} \quad \text{on } \tilde{\Gamma}_A, \tag{12.19}$$

posed on the computational domain $\tilde{\Omega}$ that represents the interelectrode gap. The boundaries $\tilde{\Gamma}_N$, $\tilde{\Gamma}_C$, and $\tilde{\Gamma}_A$ are the non-metallic side boundaries, and the boundaries representing the cathode and the anode surface, respectively, as illustrated in Figure 12.1(b). The anode dissolution velocity (cf. Equation 12.7) at every point of the anode boundary $\tilde{\Gamma}_A$ has to be computed in the vector form

$$\tilde{\underline{v}}_A = \tilde{K}_V \left| \tilde{\underline{j}}_A \right| \underline{n}, \tag{12.20}$$

where $\underline{n}$ denotes the outer unit surface normal of $\tilde{\Omega}$ at the considered point on $\tilde{\Gamma}_A$ and $\tilde{\underline{j}}_A$ denotes the current density at the anode surface. In each time step, the boundary value problem of Equations 12.16–12.19 is solved numerically for the electric potential $\tilde{\Phi}$. The current density $\tilde{\underline{j}}$ is then calculated by Ohm's law (Equation 12.15). After computation of $\tilde{\underline{v}}_A$ according to Equation 12.20, the geometry of the computational domain is updated for the next time step, i.e., each point $\tilde{\underline{x}}_A$ on the anode boundary $\tilde{\Gamma}_A$ is moved according to

$$\tilde{\underline{x}}_A \to \tilde{\underline{x}}_A + \Delta \tilde{t} \tilde{\underline{v}}_A, \tag{12.21}$$

and each point $\tilde{\underline{x}}_C$ on the cathode boundary $\tilde{\Gamma}_C$ is moved according to $\tilde{\underline{x}}_C \to \tilde{\underline{x}}_C + \Delta \tilde{t} \tilde{\underline{v}}_f$, where $\Delta \tilde{t}$ denotes the width of the time step.

To solve the boundary value problem defined by Equations 12.16–12.19, several numerical methods are possible. Purcar et al. (2004) or Narayanan et al. (1986b) applied the boundary element method (BEM) that is shown to be more accurate than comparable computations using the finite difference or finite element methods Narayanan et al., 1986b. According to Hinduja and Kunieda (2013), BEM cannot be extended to solve more sophisticated ECM models with variable electrolyte conductivity.

This disadvantage is overcome by the finite element method (FEM) that was first applied to ECM by Jain and Pandey (1980b), including the computation of the temperature distribution within the interelectrode gap. Further studies include the computation of the flow field (Jain et al., 1987; Chang and Hourng, 2001).

Numerical solutions to the anode shape prediction problem applying multiphysics models are reviewed by Hinduja and Kunieda (2013) or Xu and Wang (2021). Recently, multiphysics simulations (Liu et al., 2019) and complex simulations of the precise ECM with oscillating tool cathodes (Klocke et al., 2018) have been presented.

12.3   CATHODE SHAPE DESIGN

12.3.1   CATHODE SHAPE DESIGN AS INVERSE PROBLEM

In the basic ECM experiment (Section 12.2.1), the ECM process approaches a steady state in which the width of the gap between two plane and parallel electrodes does not change. In ECM using curved electrodes, a steady state is reached when the interelectrode gap domain moves at the feed rate without changing its shape (Nilson and Tsuei, 1974). In this situation, the normal component of the velocity of the anode boundary $\tilde{\Gamma}_A$ in a coordinate system tied to the cathode boundary $\tilde{\Gamma}_C$ must vanish, i.e.,

$$\left(\underline{\tilde{v}}_A - \tilde{v}_f \underline{e}_f\right)^T \cdot \underline{n} = \tilde{\kappa}\tilde{K}_V \frac{\partial \tilde{\Phi}}{\partial \tilde{n}}\bigg|_{\tilde{\Gamma}_A} - \tilde{v}_f \cos\theta = 0, \tag{12.22}$$

where $\underline{e}_f$ is the unit vector in feed direction (see Figure 12.1(b)).

When the shape of the cathode is to be designed to produce an anode of a desired shape in steady-state ECM, the following mathematical problem has to be solved:

$$\underline{\nabla} \cdot \left(\tilde{\kappa} \underline{\nabla} \tilde{\Phi}\right) = 0 \quad \text{in } \tilde{\Omega}, \tag{12.23}$$

$$\frac{\partial \tilde{\Phi}}{\partial \tilde{n}} = 0 \quad \text{on } \tilde{\Gamma}_N, \tag{12.24}$$

$$\tilde{\Phi} = 0 \quad \text{on } \tilde{\Gamma}_C \text{ (unknown)}, \tag{12.25}$$

$$\tilde{\Phi} = \tilde{U} \quad \text{on } \tilde{\Gamma}_A, \tag{12.26}$$

$$\frac{\tilde{\kappa}\tilde{K}_V}{\tilde{v}_f} \frac{\partial \tilde{\Phi}}{\partial \tilde{n}} = \cos\theta \quad \text{on } \tilde{\Gamma}_A, \tag{12.27}$$

i.e., the shape of the cathode boundary $\tilde{\Gamma}_C$ has to be determined in a way that the additional boundary condition given by Equation 12.27 is satisfied. The search for $\tilde{\Phi}$ from Equations 12.23, 12.26, and 12.27 can be seen as a Cauchy problem of the Laplace equation and is known to be ill-posed (Hadamard, 1923).

In the context of cathode shape design, it is important to note that there are configurations in which a steady state does not exist, such as the side gap in electrochemical drilling without cathode insulation (Reddy et al., 1988) or the electrochemical trepanning of turbine blades (Gu et al., 2019). In these cases, the shape of the cathode can be designed by correction methods (Section 12.3.6).

12.3.2   NONDIMENSIONALIZATION

Nondimensionalization of equations, as e.g., carried out for the Navier–Stokes equations in fluid mechanics, helps to reduce the number of free parameters in

TABLE 12.1

Nondimensionalization of Variables in the Potential Model

Physical Quantity	Characteristic Scale	Dimensionless Variable(s)
Length	$\tilde{h}_e^{std} = \tilde{\kappa}\tilde{K}_V\tilde{U} / \tilde{v}_f$	$\underline{x} = \underline{\tilde{x}} / \tilde{h}_e^{std}, h = \tilde{h} / \tilde{h}_e^{std}$
Electric potential	$\tilde{U}$	$\Phi = \tilde{\Phi} / \tilde{U}$
Dissolution velocity	$\tilde{v}_f$	$\underline{v}_A = \underline{\tilde{v}}_A / \tilde{v}_f$
Time	$\tilde{h}_e^{std} / \tilde{v}_f$	$t = \tilde{t}\tilde{v}_f / \tilde{h}_e^{std}$
Current density	$\tilde{\kappa}\tilde{U} / \tilde{h}_e^{std} = \tilde{v}_f / \tilde{K}_V$	$\underline{j} = \underline{\tilde{j}}\tilde{K}_V / \tilde{v}_f$

mathematical models and to generalize solutions to problems of similar physical situations. In the case of ECM, the cathode shape design methods introduced in the following sections (12.3.4–12.3.9) will be based on a dimensionless formulation of the potential model in which physical quantities are normalized by characteristic scales, as summarized in Table 12.1. Using the dimensionless variables therein, the inverse problem defined by Equations 12.23–12.27 can be expressed as

$$\underline{\nabla} \cdot \left(\underline{\nabla}\Phi \right) = 0 \quad \text{in } \Omega, \tag{12.28}$$

$$\frac{\partial \Phi}{\partial n} = 0 \quad \text{on } \Gamma_N, \tag{12.29}$$

$$\Phi = 0 \quad \text{on } \Gamma_C \text{ (shape unknown)}, \tag{12.30}$$

$$\Phi = 1 \quad \text{on } \Gamma_A, \text{and} \tag{12.31}$$

$$\frac{\partial \Phi}{\partial n} = \cos\theta \quad \text{on } \Gamma_A, \tag{12.32}$$

where Ω is the computational domain and $\Gamma_{N,C,A}$ are the boundaries given in units of the standard equilibrium front gap width (cf. Equation 12.12)

$$\tilde{h}_e^{std} = \frac{\tilde{\kappa}\tilde{K}_V\tilde{U}}{\tilde{v}_f}. \tag{12.33}$$

12.3.3 COS θ-METHOD

The cos θ-method introduced in Section 12.2.2 for anode shape prediction can also be applied to design the shape of the cathode. Under the condition that current flux lines are straight and parallel to each other (Tipton, 1971), the potential gradient in Equation 12.27 can be approximated by

$$\frac{\partial \tilde{\Phi}}{\partial \tilde{n}} \approx \frac{\tilde{U}}{\tilde{h}_e(\theta)},$$

resulting in

$$\tilde{h}_e(\theta) = \frac{\tilde{\kappa}\tilde{K}_V\tilde{U}}{\tilde{v}_f\cos\theta} = \frac{\tilde{h}_e^{std}}{\cos\theta}, \tag{12.34}$$

where $\tilde{h}_e^{std} = \tilde{\kappa}\tilde{K}_V\tilde{U}/\tilde{v}_f$ is the standard equilibrium front gap width defined in Equation 12.12.

Figure 12.5 shows the relation between a given point on the anode with coordinates $(\tilde{x}_A, \tilde{y}_A)$ and the corresponding point on the cathode with coordinates $(\tilde{x}_C, \tilde{y}_C)$ computed by applying the cos θ-method. The angle θ is measured between the feed direction and the outer surface normal of the anode boundary. The width of the inter-electrode gap given by Equation 12.34 is measured in the normal direction of the anode surface. According to Figure 12.5 and applying Equation 12.34, the following relations can be derived

$$\tilde{x}_C = \tilde{x}_A - \tilde{h}_e(\theta)\sin\theta = \tilde{x}_A - \tilde{h}_e^{std}\tan\theta, \tag{12.35}$$

$$\tilde{y}_C = \tilde{y}_A + \tilde{h}_e(\theta)\cos\theta = \tilde{y}_A + \tilde{h}_e^{std}. \tag{12.36}$$

For an anode shape described by $\tilde{y} = \tilde{F}(\tilde{x})$, the cathode shape computed by applying the cos θ-method can be expressed as a parametric curve by substituting $\tan\theta = d\tilde{F}/d\tilde{x}$ into Equations 12.35 and 12.36 and renaming $\tilde{x}_A$ to a curve parameter $\tilde{\xi}$, resulting in

$$\tilde{x}_C(\tilde{\xi}) = \tilde{\xi} - \tilde{h}_e^{std}\frac{d\tilde{F}(\tilde{\xi})}{d\tilde{\xi}}, \tag{12.37}$$

$$\tilde{y}_C(\tilde{\xi}) = \tilde{F}(\tilde{\xi}) + \tilde{h}_e^{std}. \tag{12.38}$$

As described in Section 12.3.3, the cos θ-method is inaccurate for electrode regions with $\theta > 45°$. In the case of electrodes with small radii of curvature, the results obtained by cos θ-method is even inaccurate for $\theta = 0$ (Lu et al., 2014). However, the method can be used to determine the starting shape when using an iterative numerical technique to determine the cathode shape.

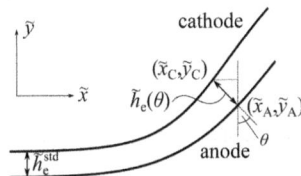

FIGURE 12.5 Principle of the cos θ-method for cathode shape design. The equilibrium gap width $\tilde{h}_e(\theta)$ is calculated applying Equation 12.34.

12.3.4 COMPLEX VARIABLE METHOD

For the cathode shape design problem in two dimensions, an explicit analytical solution can be found if the anode shape Γ_A can be described by an analytic function F. In these cases, the complex variable method (Krylov, 1968; Nilson and Tsuei, 1974) can be used to calculate the surface profile of the unknown cathode as a parametric curve in the x-y-plane.

The complex variable method makes use of the property of an analytic function $W(z) = \Phi(x, y) + i\Psi(x, y)$ of the complex variable $z = x + iy$ that its real and imaginary parts satisfy the Cauchy–Riemann differential equations and the two-dimensional Laplace equation. It further makes use that the inverse function $z(W) = x(\Phi, \Psi) + iy(\Phi, \Psi)$ is analytic in a neighborhood of a point where the derivative of W is not zero and an analytic function is completely determined in the complex plane, e.g., by its values on the real or imaginary axis (identity theorem).

For an anode shape that is described (after nondimensionalization) by $y = F(x)$ with an analytic function F, equipotential curves in the interelectrode gap are described in a complex form by (Nilson and Tsuei, 1974, Equation 12.26)

$$y(\xi) + ix(\xi) = F(\xi + i\Phi) + i\xi - \Phi, \tag{12.39}$$

where ξ is a curve parameter and Φ is the dimensionless electric potential with $\Phi = 0$ on the anode and $\Phi = -1$ on the cathode. The cathode shape described by a parametric curve $\left(x_C(\xi), y_C(\xi)\right)$ in the x-y-plane is obtained by substituting $\Phi = -1$ into Equation 12.39,

$$x_C(\xi) = \xi + \text{Im}\left[F(\xi - i)\right], \tag{12.40}$$

$$y_C(\xi) = 1 + \text{Re}\left[F(\xi - i)\right]. \tag{12.41}$$

Although workpiece shapes in practise are rarely represented by analytic functions, solutions computed using the complex variable method can serve as benchmarks for numerical methods.

Nonanalytic workpiece shapes can be represented by a Fourier series consisting of an infinite sum of sinusoid terms within an interval. The complex variable method can be applied to the Fourier series of the anode profile. However, it can be shown that for workpiece shapes that cannot be represented by analytic functions, exact solutions to the cathode shape design problem do not exist (Lu and Werner, 2021).

Problem 1

Compute the 2D cathode profile corresponding to the anode profile described by

$$y = F(x) = -A \exp\left(-\frac{x^2}{2\sigma^2}\right), \quad A > 0 \tag{12.42}$$

applying the complex variable method. Plot cathode and anode profiles for $A = 8$, $A = 15.51$, $A = 20$, and $\sigma = 4$, respectively. Interpret the results.

Solution to Problem 1
With

$$F(\xi - i) = -A \exp\left(-\frac{\xi^2 - 2i\xi - 1}{2\sigma^2}\right)$$

$$= -A \exp\left(\frac{1 - \xi^2}{2\sigma^2}\right)\left[\cos\left(\frac{\xi}{\sigma^2}\right) + i\sin\left(\frac{\xi}{\sigma^2}\right)\right],$$

the surface profile of the cathode can be represented as a parametric curve $\underline{x}_C(\xi) = (x_C(\xi), y_C(\xi))^T$ in the x-y plane with

$$x_C(\xi) = \xi + \text{Im}\left[F(\xi - i)\right] = \xi - A \exp\left(\frac{1 - \xi^2}{2\sigma^2}\right)\sin\left(\frac{\xi}{\sigma^2}\right), \quad (12.43)$$

$$y_C(\xi) = 1 + \text{Re}\left[F(\xi - i)\right] = 1 - A \exp\left(\frac{1 - \xi^2}{2\sigma^2}\right)\cos\left(\frac{\xi}{\sigma^2}\right). \quad (12.44)$$

Figure 12.6 illustrates surface profiles of anodes and the corresponding cathodes obtained for three combinations of the parameters A and σ. In panel (a), the cathode is regular and in line with ECM experience because the cathode is approximately the negative image of the anode. In panel (c), the cathode surface profile has a "loop" at $x = 0$, which is physically not meaningful as the cathode cannot penetrate itself. The cathode shape in panel (b) with a singularity at $x = 0$ seems to be the limit between physically meaningful (i.e., realizable) and meaningless cathode.

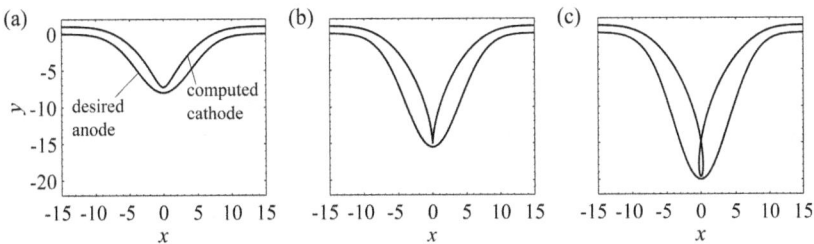

FIGURE 12.6 Cathode shapes (upper lines) required to produce Gaussian anode shapes (lower lines) defined by Equation 12.42 with $\sigma = 4$. The solutions in parametric form are given by Equations 12.43 and 12.44. The scale on the y-axis and the legend of (a) apply for all three panels. (a) shows the example with $A = 8 < A^* = 15.51$, resulting in a regular cathode shape. (b) shows the anode with $A = A^* = 15.51$, for which the computed cathode is singular and represents the limit of physically realizable cathode shapes. (c) shows a non-physical solution to a cathode shape obtained for a Gaussian anode with $A = 20$.

The qualitative behavior of the cathode shapes in Figure 12.6 can be understood by considering the derivative of $x_C(\xi)$ (Equation 12.43) with respect to ξ. For given width σ,

$$A^* = \sigma^2 \exp\left(-\frac{1}{2\sigma^2}\right) \qquad (12.45)$$

is a critical value of A. For $A < A^*$, $\dot{x}_C(\xi)$ is positive for all ξ, i.e., $x_C(\xi)$ increases monotonously. For $A > A^*$, $\dot{x}_C(\xi)$ is negative near $\xi = 0$, resulting in the physically meaningless loop of the cathode surface profile as in Figure 12.6(c).

To interpret the results shown in Figure 12.6 in terms of their significance for practice, the critical amplitude of the anode (cf. Equation 12.45) has to be expressed in the dimensioned form

$$\tilde{A}^* = \frac{\tilde{\sigma}^2}{\tilde{h}_e^{std}} \exp\left[-\frac{\left(\tilde{h}_e^{std}\right)^2}{2\tilde{\sigma}^2}\right]. \qquad (12.46)$$

According to Equation 12.46, the (dimensioned) critical amplitude $\left(\tilde{A}^*\right)$ increases with decreasing width of the standard equilibrium front gap $\left(\tilde{h}_e^{std}\right)$ for given horizontal dimension of the anode $\left(\tilde{\sigma}\right)$. Hence, for anode shapes with large ratios of vertical to horizontal dimensions (i.e., $\tilde{A}/\tilde{\sigma}$ in the examples) a physically realizable cathode shape only exists if $\tilde{h}_e^{std}$ is sufficiently small.

Problem 2

Compute the 2D cathode profile corresponding to the anode profile described by Equation 12.42 applying the cos θ-method. Compare the result with that obtained from the complex variable method with regard to the width of the interelectrode gap at $x = 0$. Plot both results for $A = \pm 8$ and $\sigma = 4$.

Solution to Problem 2

With $x = \tilde{x} / \tilde{h}_e^{std}$ and $h_e^{std} = \tilde{h}_e^{std} / \tilde{h}_e^{std} = 1$, Equations 12.37 and 12.38 are nondimensionalized to

$$x_C(\xi) = \xi - \frac{dF(\xi)}{d\xi}, \qquad (12.47)$$

$$y_C(\xi) = F(\xi) + 1. \qquad (12.48)$$

Hence, the cathode profile computed by applying the cos θ-method is expressed as

$$x_C(\xi) = \xi\left[1 - \frac{A}{\sigma^2}\exp\left(-\frac{\xi^2}{2\sigma^2}\right)\right],$$

$$y_C(\xi) = 1 - A \exp\left(-\frac{\xi^2}{2\sigma^2}\right).$$

At $x = 0$, the tangent to the anode surface is horizontal, i.e., the (dimensionless) width of the interelectrode gap computed applying the $\cos\theta$-method is $h_e^{\cos\theta} = 1$. As a result of the (exact) complex variable method, the width of the interelectrode gap at $x = 0$ is given by

$$h_e^{c.v.} = y_C - y_A = 1 - A\left[\exp\left(\frac{1}{2\sigma^2}\right) - 1\right].$$

Hence, the deviation of the results computed by applying the $\cos\theta$-method from the exact results becomes small if $|A|$ is small and σ is large. This is the case when the anode is nearly flat at $x = 0$, i.e., the curvature of the anode at $x = 0$ is small. The electrode shapes are illustrated in Figure 12.7.

12.3.5 Solving an "Initial Value Problem" Numerically

From a mathematical point of view, Equations 12.28–12.32 represent an "initial value problem" of the Laplace equation in that the value of Φ and its normal derivative $\partial\Phi / \partial n$ are specified on the anode boundary Γ_A. The position of the opposite cathode boundary Γ_C is unknown. In the one dimensional case, the ordinary equation $d^2\Phi / dx^2 = 0$ with Φ and $d\Phi / dx$ specified on one point can be solved easily. In two or higher dimensions, however, this so-called Cauchy problem of the Laplace equation is known to be ill-posed (Hadamard, 1923).

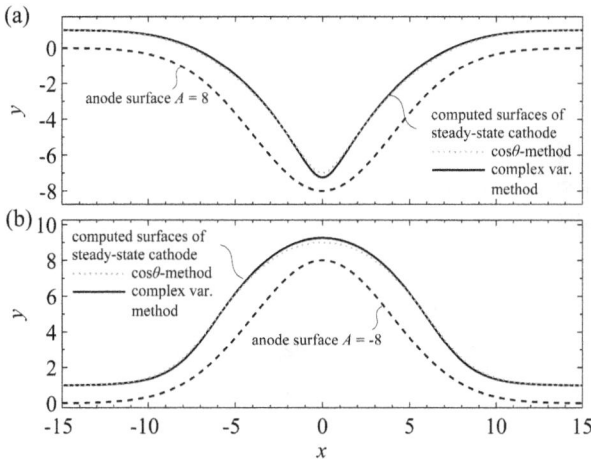

FIGURE 12.7 Cathode shapes computed for anode shapes defined by Equation 12.42 with $\sigma = 4$ and $A = 8$ (a) or $A = -8$ b) applying the $\cos\theta$-method (dotted curves) and the complex variable method (solid curves).

To solve the cathode shape design problem in ECM numerically, several studies propose to solve the "initial value problem" directly (Zhu et al., 2003; Sun et al., 2006; Demirtas et al., 2017). The equations are derived from the relation between the Laplace equation and variational principles, as described in the following.

The solution to Equation 12.28 with Neumann boundary condition defined by Equation 12.32 is equivalent to finding a function Φ that minimizes the functional (see Zienkiewicz et al. (2005), Sec. 3.7)

$$G(\Phi) = \frac{1}{2} \int_{\Omega} \left| \nabla \Phi \right|^2 dV - \int_{\Gamma_A} \Phi \cos \theta \, dS = G_{\Omega}(\Phi) + G_{\Gamma_A}(\Phi). \tag{12.49}$$

The problem is solved numerically by applying the finite element method. The gap region with an initial trial cathode shape is divided into a series of eight-node hexahedron elements. Inside each element Ω^e, a potential function is defined as

$$\varphi^e(\underline{x}) = \sum_{i=1}^{8} N_i^e(\underline{x}) \varphi_i^e, \tag{12.50}$$

where φ_i^e is the electric potential at the element node i and N_i is the corresponding shape function, which is a function of $\underline{x} = (x, y, z)$. Substitution of Equation 12.50 into Equation 12.49, the volume integral $G_{\Omega}(\Phi)$ in Equation 12.49 gives a contribution of

$$G_{\Omega}^e = \frac{1}{2} \sum_{i=1}^{8} \sum_{j=1}^{8} \varphi_i^e \varphi_j^e \int_{\Omega^e} \underline{\nabla N_i^e} \cdot \underline{\nabla N_j^e} \, dV = \frac{1}{2} [\varphi^e]^T \cdot [k^e] \cdot [\varphi^e], \tag{12.51}$$

where $\left[\varphi^e \right]$ is a column vector containing the nodal values φ_i^e and $[k^e]$ is the element stiffness matrix. For an element on the lowest layer, i.e., at the anode boundary (see Figure 12.8), the surface integral $G_{\Gamma_A}(\Phi)$ in Equation 12.49 gives a contribution of

$$G_{\Gamma_A}^e = -\sum_{i=1}^{8} \varphi_i^e \int_{\Gamma_A^e} N_i \, dS = -[b^e]^T \cdot [\varphi^e]. \tag{12.52}$$

The necessary condition to find a minimum of G^e is $\partial G^e / \partial \varphi_i^e = 0$ resulting in the element equation

$$[k^e] \cdot [\varphi^e] = [b^e]$$

for an element at the anode boundary. The element equations are assembled, resulting in

$$\begin{bmatrix} [k]_{11} & [k]_{12} & 0 & 0 & 0 \\ [k]_{21} & [k]_{22} & [k]_{23} & 0 & 0 \\ 0 & [k]_{32} & [k]_{33} & [k]_{34} & 0 \\ 0 & 0 & [k]_{43} & [k]_{44} & [k]_{45} \\ 0 & 0 & 0 & [k]_{54} & [k]_{55} \end{bmatrix} \cdot \begin{bmatrix} [\phi]_1 \\ [\phi]_2 \\ [\phi]_3 \\ [\phi]_4 \\ [\phi]_5 \end{bmatrix} = \begin{bmatrix} [b]_1 \\ 0 \\ 0 \\ 0 \\ 0 \end{bmatrix} \tag{12.53}$$

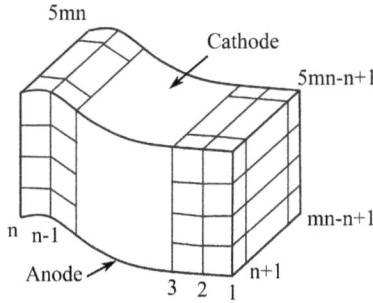

FIGURE 12.8 Meshing and node numbering used for solving the cathode shape design problem numerically as an "initial value problem". The gap region is divided into elements with $n \times m \times 5$ nodes (figure adapted from Sun et al. (2006)).

for the discretization with five layers of nodes (Figure 12.8). In Equation 12.53, $[k]_{ij}$ is an $L \times L$-matrix (with $L = mn$ being the number of nodes in each layer). $[\phi]_i$ is an L-dimensional column vector with nodal potential values of a layer to be computed. $[b]_1$ is an L-dimensional column vector originating from the surface integral term $G_{\Gamma_A}(\Phi)$ in Equation 12.49, thus reflecting the Neumann boundary condition at the anode boundary (Equation 12.32).

To meet the Dirichlet boundary condition at the anode boundary (Equation 12.31), the values of $[\phi]_1$ are set to 1 (i.e., $\tilde{\Phi} = \tilde{U}$). Equation 12.53 turns to a recursion formula

$$[\phi]_2 = -[k]_{12}^{-1} \cdot \left([k]_{11} \cdot [\phi]_1 - [b]_1 \right) \quad \text{for } i = 1, \text{ and} \tag{12.54}$$

$$[\phi]_{i+1} = -[k]_{i,i+1}^{-1} \cdot \left([k]_{i,i-1} \cdot [\phi]_{i-1} + [k]_{i,i} \cdot [\phi]_i \right) \quad \text{for } i > 1, \tag{12.55}$$

where $[k]_{i,j}^{-1}$ is the inverse matrix of $[k]_{i,j}$. After computing the potential distribution, the cathode surface is determined from the equipotential surface with $\Phi = 0$ (cf. Equation 12.30).

This method has been successfully applied to 3D problems (Sun et al., 2006; Demirtas et al., 2017). However, it requires special numerical schemes such as the layer-by-layer computation of the potential (Equations 12.54 and 12.55) and is mainly suitable for designing cathode shapes with gentle curvatures (Hinduja and Kunieda, 2013).

12.3.6 Correction Factor Method

The correction factor method proposed by Reddy et al. (1988) modifies, in different stages, cathode shapes assumed in the first design cycle. The modification continues until it produces a desired workpiece shape, within specified tolerances, while machining under given conditions. The iterative cathode shape design procedure starts with an assumed cathode shape and runs through the following cycle:

1. Applying the anode shape prediction procedure e.g., based on the FEM method (Jain and Pandey, 1980a), an anode profile is computed for the currently given cathode shape and specified machining conditions.
2. The anode profile computed in the cycle m is compared with the required anode profile to calculate the deviation $Er_m(i)$ between them at every node i.
3. If the deviation $Er_m(i)$ of the computing cycle m is larger than the specified tolerance Tz, the correction factor Cf_{m+1} for the next computing cycle $m + 1$ is calculated. For the second computing cycle, the correction factor for node i is

$$Cf_2(i) = Er_1(i).$$

For the third and subsequent cycles, the correction factor is

$$Cf_{m+1}(i) = -\frac{Y_m(i) - Y_{m-1}(i)}{Er_m(i) - Er_{m-1}(i)} Er_m(i) \quad \text{for } m \geq 2,$$

where $Y_m(i)$ denotes the width of the interelectrode gap in vertical direction at node i.
4. The cathode shape is modified by adding the correction factor to the width of the interelectrode gap at node i,

$$Y_{m+1}(i) = Y_m(i) + Cf_{m+1}(i).$$

The steps 1 to 4 are repeated until the difference between the predicted and the desired anode shapes is within the specified tolerances.

The correction factor method has been applied for designing one- and two-dimensional tool cathode shapes, taking temperature-dependent variations of electrolyte conductivity into account. Jain and Rajurkar (1991) included this method in an integrated tool design approach. The method can be applied to ECM configurations in which a steady state does not exist, such as in electrochemical drilling (Reddy et al., 1988), since the steady-state condition (Equation 12.32) is not required. The concept can be applied to three-dimensional problems and combined with more complex ECM process models and other numerical techniques for anode shape prediction (e.g., finite difference or boundary element method).

Similar methods based on empirical correction are proposed by Narayanan et al. (1986a) and Hardisty and Mileham (1999) for cathode shape design in steady-state ECM. In these approaches, the deviation of the actual current density from the desired value in steady state is used to correct the shape of the cathode. Gu et al. (2019) applied a correction method to design the cathode shape for electrochemical trepanning of turbine blades, which is a non steady-state ECM process.

12.3.7 DESIGN PARAMETER OPTIMIZATION METHOD

Starting with an initial cathode, Zhou and Derby (1995) use gradient-based optimization to correct the cathode shape iteratively. In this approach, the cathode is represented by a sum of basis functions with coefficients as design variables. In the 2D case, the cathode is represented by a sum of basis functions $\psi_i(x)$,

$$f(x) = \sum_{i=1}^{M} a_i \psi_i(x) \tag{12.56}$$

where the set of the corresponding coefficients a_i is the design parameter of the cathode. In a similar way, the required anode is represented by

$$r(x) = \sum_{i=1}^{M} r_i \psi_i(x). \tag{12.57}$$

With $f(x)$ and $r(x)$ as cathode and anode boundaries, the anode shape prediction problem (Section 12.2.3) is solved numerically until the interelectrode gap is in steady state. The computed anode shape is represented by

$$h(x) = \sum_{i=1}^{M} h_i \psi_i(x), \tag{12.58}$$

where the coefficients h_i are obtained by curve fitting.

To correct the cathode shape $f(x)$ that depends on the coefficient a_i, the objective function

$$O(a_1, ..., a_M) = \sum_{j=1}^{N} \left[h(x_j) - r(x_j) \right]^2 \tag{12.59}$$

is defined, summing up the squared differences between the computed anode $h(x)$ and the required anode $r(x)$ at each node x_j. To find the design variables $\underline{a} = (a_1, ..., a_M)$ that minimize the objective function $O(\underline{a})$, the gradient of $O(\underline{a})$ with respect to $\underline{a}$

$$\nabla O(\underline{a}) = \left[\frac{\partial O(\underline{a})}{\partial a_1}, ..., \frac{\partial O(\underline{a})}{\partial a_M} \right]^T$$

has to be computed. The partial derivatives are computed numerically by

$$\frac{\partial O(\underline{a})}{\partial a_i} \approx \frac{O(a_1, ..., a_i + \Delta_i, ..., a_M) - O(a_1, ..., a_i, ..., a_M)}{\Delta_i} \quad \text{for } 1 \leq i \leq M,$$

where Δ_i is a small disturbance imposed on the component i of vector $\underline{a}$.

A related optimization method is proposed by Das and Mitra (1992). Instead of Equation 12.59, they use the sum of squared differences between the current density at the computed anode and that at the required anode as the objective function. Both optimization methods are computationally intensive since for the gradient computation in each optimization step, the anode shape prediction problem has to be solved $M + 1$ times.

12.3.8 EMBEDDING METHOD

Similar to the method introduced in the previous section, the embedding method proposed by Hunt (1990) approximates the shape of the cathode $f(x)$ as well as that of the computed anode $h(x)$ and required anode $r(x)$ by sums of basis functions (cf. Equations 12.56–12.58)). Since the computed anode shape $h(x)$, characterized by h_i, is the result of ECM using the cathode shape $f(x)$ defined by a_i, the former can be seen as a function of the latter, i.e., $h_i = h_i(a_1,...,a_M)$. Instead of minimizing the difference between $h(x)$ and $r(x)$, as intended by Equation 12.59 of the previous section, the functions are set equal, resulting in

$$h_1(a_1,...,a_M) = r_1,$$

$$\vdots \quad \vdots \quad \vdots \tag{12.60}$$

$$h_M(a_1,...,a_M) = r_M,$$

which is a system of nonlinear equations.

The system of equations (Equation 12.60) is solved by applying the Newton method. With $\underline{a} = (a_1,...,a_M)$ and simular expressions for $\underline{h}$ and $\underline{r}$, the iteration equations can be written as

$$\underline{\underline{J}} \cdot \Delta \underline{a}^{(k)} = \underline{r} - \underline{h}^{(k)}$$

$$\underline{a}^{(k+1)} = \underline{a}^{(k)} + \Delta \underline{a}^{(k)},$$

where k is the iteration counter and $\underline{\underline{J}}$ is the Jacobian whose (i,j)-th element has to be calculated numerically by

$$\frac{\partial h_i}{\partial a_j} \approx \frac{h_i(a_1,...,a_j + \Delta_j,...,a_M) - h_i(a_1,...,a_j,...,a_M)}{\Delta_j},$$

where Δ_j is a small disturbance imposed on component j of vector $\underline{a}$. The method is also computationally intensive since in each Newton iteration step, the anode shape prediction problem has to be solved $M + 1$ times.

The embedding method has been applied by Chang and Hourng (2001) to solve the cathode shape design problem in two dimensions.

12.3.9 CONTINUOUS ADJOINT SHAPE OPTIMIZATION METHOD

The inverse problem (Equations 12.28–12.32) can be reformulated into a constrained shape optimization problem (Lu et al., 2014). This approach is used in similar mathematical problems in plasma engineering (see e.g., Fischer et al., 2012) or medical imaging (see e.g., Eppler, 2009).

In the shape optimization problem, an initial guess Γ_{C0} is assumed as the shape of the cathode. On the overdetermined boundary Γ_A, one of the boundary conditions, e.g. Equation 12.32, is transformed into an objective function J (Equation 12.61). The remainder of the equations form a boundary value problem for Φ posed on Ω, which represents an equality constraint. The shape of the cathode is optimized iteratively.

When the Neumann boundary condition (Equation 12.32) is transformed, the optimization problem reads as follows: Find the shape of Γ_C that minimizes the objective function

$$J_N(\Omega, \Phi) = \frac{1}{2} \int_{\Gamma_A} \left(\frac{\partial \Phi}{\partial n} - \cos\theta \right)^2 dS \tag{12.61}$$

subject to the so-called system equations

$$\underline{\nabla} \cdot \left(\underline{\nabla}\Phi \right) = 0 \quad \text{in } \Omega, \tag{12.62}$$

$$\frac{\partial \Phi}{\partial n} = 0 \quad \text{on } \Gamma_N, \tag{12.63}$$

$$\Phi = 0 \quad \text{on } \Gamma_C, \tag{12.64}$$

$$\Phi = 1 \quad \text{on } \Gamma_A. \tag{12.65}$$

As shown by Lu et al. (2014), this problem can by solved by introducing the Lagrangian

$$\begin{aligned}
\mathscr{L}_N\left(\Omega, \Phi, \underline{\lambda}\right) &= \frac{1}{2} \int_{\Gamma_A} \left(\frac{\partial \Phi}{\partial n} - \cos\theta \right)^2 dS + \int_{\Omega} \lambda_\Omega \underline{\nabla} \cdot \left(\underline{\nabla}\Phi \right) dV + \int_{\Gamma_C} \lambda_C \Phi \, dS \\
&+ \int_{\Gamma_N} \lambda_N \frac{\partial \Phi}{\partial n} dS + \int_{\Gamma_A} \lambda_A \left(\Phi - 1 \right) dS,
\end{aligned} \tag{12.66}$$

where $\underline{\lambda}$ is an abbreviation of the quadruple $\left(\lambda_\Omega, \lambda_C, \lambda_N, \lambda_A \right)^T$ of Lagrange multipliers consisting of scalar valued functions defined in Ω and on the boundaries, respectively.

The so-called adjoint equations are derived from

$$\partial_\Phi \mathscr{L}_N\left(\Omega, \Phi, \underline{\lambda}\right)[\psi] = 0,$$

where

$$\partial_\Phi \mathscr{L}_N\left(\Omega, \Phi, \underline{\lambda}\right)[\psi] = \lim_{\varepsilon \to 0} \frac{\mathscr{L}_N\left(\Omega, \Phi + \varepsilon\psi, \underline{\lambda}\right) - \mathscr{L}_N\left(\Omega, \Phi, \underline{\lambda}\right)}{\varepsilon} \tag{12.67}$$

is the Gâteaux derivative of $\mathscr{L}_N$ with respect to Φ in an arbitrary direction ψ, resulting in the so called adjoint equations

$$\nabla \cdot \left(\nabla \lambda_\Omega \right) = 0 \quad \text{in } \Omega, \tag{12.68}$$

$$\frac{\partial \lambda_\Omega}{\partial n} = 0 \quad \text{on } \Gamma_N, \tag{12.69}$$

$$\lambda_\Omega = 0 \quad \text{on } \Gamma_C, \tag{12.70}$$

$$\lambda_\Omega = -\left(\frac{\partial \Phi}{\partial n} - \cos \theta \right) \quad \text{on } \Gamma_A, \tag{12.71}$$

which is a boundary value problem for λ_Ω.

The gradient of the objective function with respect to variations of the shape of the domain Ω, is obtained by computing the shape derivative of $\mathscr{L}_N$ in direction of the so-called shape variation direction field $\underline{v}$,

$$\partial_\Omega \mathscr{L}_N \left(\Omega, \Phi, \underline{\lambda} \right) [\underline{v}] = \lim_{\varepsilon \to 0} \frac{1}{\varepsilon} \left[\mathscr{L}_N \left(\Omega_\varepsilon, \Phi, \underline{\lambda} \right) - \mathscr{L}_N \left(\Omega, \Phi, \underline{\lambda} \right) \right],$$

with domain variation

$$\Omega_\varepsilon = \left\{ \underline{x} + \varepsilon \underline{v}(\underline{x}); \underline{x} \in \Omega \right\},$$

resulting in

$$\partial_\Omega \mathscr{L}_N \left(\Omega, \Phi, \underline{\lambda} \right) [\underline{v}] = \int_{\Gamma_C} \left(\underline{v}^T \cdot \underline{n} \right) \frac{\partial \lambda_\Omega}{\partial n} \frac{\partial \Phi}{\partial n} dS = \int_{\Gamma_C} v_n G dS. \tag{12.72}$$

The term

$$G = \frac{\partial \lambda_\Omega}{\partial n} \frac{\partial \Phi}{\partial n} \bigg|_{\Gamma_C} \tag{12.73}$$

is called the shape gradient.

The solution to the cathode shape design problem starts with an initial guess Γ_{C0} of the cathode shape, e.g., by applying the $\cos \theta$-method. The shape of the cathode is then optimized iteratively by applying the following procedure:

1. Solve the system equations (Equations 12.62–12.65) for the dimensionless electric potential Φ.
2. Compute the normal derivative of Φ at Γ_A.
3. Solve the adjoint equations (Equations 12.68–12.71) for the Lagrange mulitplier λ_Ω, where the normal derivative of Φ at Γ_A computed in the previous step has to be specified as boundary condition (Equation 12.71).
4. Compute the normal derivatives of Φ and λ_Ω at Γ_C and the shape gradient according to Equation 12.73.

5. For optimization using the steepest descent method, the optimal shape variation direction field is given by the negative gradient, i.e.,

$$\underline{v}^{opt} = -G\underline{n}. \tag{12.74}$$

6. Update the shape of the cathode by moving each point $\underline{x}$ on Γ_C according to

$$\underline{x} \rightarrow \underline{x} + \tau \underline{v}^{opt}, \tag{12.75}$$

where $\tau > 0$ denotes the step length.

The advantage of the procedure is the following: In the design parameter optimization approach described in Section 12.3.7, the anode shape prediction problem involving the Laplace equation has to be solved $M + 1$ times per iteration, where M is the number of design parameters used to parametrize the shape of the cathode. This can be computationally intensive for large M. For the continuous adjoint shape optimization method, the design variable is the shape of the cathode, which contains as many degrees of freedom as the number of mesh nodes at the cathode boundary. However, the Laplace equation has only to be solved twice per iteration in the form of the system equation (Equation 12.62) and the adjoint equation (Equation 12.68), independent of the number of design variables.

To evaluate the convergence of the optimization process, two "indicators" are computed at the end of each iteration step. The maximum-norm of the shape gradient (cf. Equation 12.73) is used to assess whether additional iteration steps would further "improve" the cathode shape substantially. However, a zero gradient is not sufficient to assess whether the obtained shape for Γ_C is a solution to the original inverse problem (Equations 12.28–12.32), i.e., whether both conditions on the overdetermined anode boundary (Equations 12.31 and 12.32) are fulfilled. For the latter purpose, the quantity

$$\delta_{max,N} = \max_{\Gamma_A} \left| \cos\theta \left(\frac{\partial \Phi}{\partial n} \right)^{-1} - 1 \right| \tag{12.76}$$

derived from the integrand of the objective function J_N (cf. Equation 12.61) is defined as an additional convergence indicator for the shape optimization problem (Equations 12.61–12.65). By definition, $\delta_{max,N}$ is non-negative and only becomes zero if the additional boundary condition (Equation 12.32) on Γ_A of the original inverse problem is fulfilled.

Figure 12.9 shows the computed shape of a cathode necessary to produce the anode shape described by a Gaussian (Equation 12.42 with $\sigma = 4$ and $A = 8$). For the iterative computation, the offset curve of the anode with normal distance of 1 (dotted black line in Figure 12.9(a)) is assumed to be the surface profile of the initial cathode. The computational domain is discretized by an $N_x \times N_y = 200 \times 20$-mesh; the

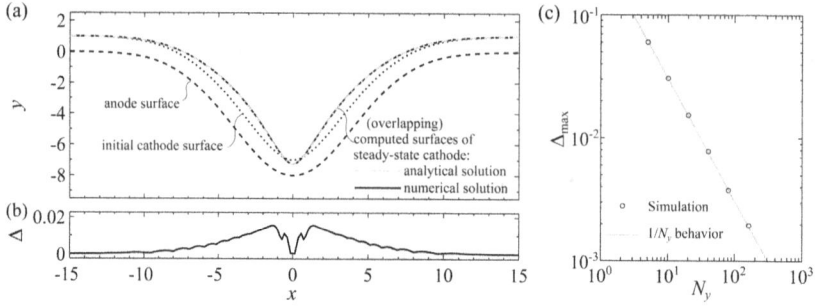

FIGURE 12.9 Two-dimensional cathode shape design example (Lu et al., 2014). (a) The black dashed line shows the anode shape defined by Equation 12.42 with $A = 8$ and $\sigma = 4$; the cathode shape obtained from iterative optimization (solid line) overlaps with the analytical solution (dash/dotted line) obtained from Equations 12.43 and 12.44. (b) shows the deviation Δ of the numerical result from the analytical solution in surface normal direction when using an $N_x \times N_y = 200 \times 20$-mesh. The dependence of the maximum of Δ on N_y shows a $1/N_y$-behavior (c).

computation is carried out for 200 iteration steps with constant step length $\tau = 0.4$ (see Equation 12.75).

The surface profile of the cathode obtained after 200 iterations is shown by the solid line in Figure 12.9(a), which overlaps with the analytical solution given by Equations 12.43 and 12.44 (dash dotted line) on the scale of the figure. For clearer visualization, the absolute deviation Δ between these solutions, measured in surface normal direction, is computed and shown in Figure 12.9(b). The maximum of Δ, denoted by Δ_{max}, is less than 0.02, i.e., 2% of the width of the standard equilibrium front gap. An increase in the number of cells N_y results in an decrease in Δ_{max}, as shown in Figure 12.9(c).

In cases where the desired anode shape has sharp corners, perimeter regularization with appropriate choice of the regularization parameter α is shown to enable reasonable results (Lu and Werner, 2021). In these cases, the objective function defined by Equation 12.61 is extended by an additional integral term that penalizes the total surface area of the unknown cathode, $|\Gamma_C|$, i.e.,

$$J_{N,reg}(\Omega, \Phi) = \frac{1}{2} \int_{\Gamma_A} \left(\frac{\partial \Phi}{\partial n} - \cos\theta \right)^2 dS + \alpha \int_{\Gamma_C} 1 dS, \qquad (12.77)$$

resulting in an optimal shape variation direction field is then given by

$$\underline{v}^{opt} = -\left(\frac{\partial \lambda_\Omega}{\partial n} \frac{\partial \Phi}{\partial n} \bigg|_{\Gamma_C} + \alpha H \right) \underline{n}, \qquad (12.78)$$

where H is the sum of principal curvatures.

The continuous adjoint shape optimization method can be realized in practice without sophisticated programming skills for implementing specific numerical schemes or optimization routines as required in the methods suggested by Reddy et al. (1988), Hunt (1990), Zhou and Derby (1995) or Sun et al. (2006), since the key steps only involve solving the Laplace equation or computing normal derivatives of computed scalar fields at boundaries. These are standard routines in many commercial or open source simulation tools, making the method accessible for a larger group of application engineers not directly related to research.

For practice, multiphysics simulation tools that enable equation-based modeling and mesh deformations should be used. As examples of starting points for implementations, the COMSOL tutorial "Pulse Reverse Plating" (COMSOL, 2017) or the OpenFOAM tutorial "overLaplacianDyMFoam" (OpenCFD, 2020) can be modified. Existing multiphysics models for solving the direct problem of ECM using commercial tools, such as that proposed by Liu et al. (2019) can be easily extended to solve the inverse problem by adding an additional Laplace equation. The solution to the cathode shape design problem presented by Lu et al. (2014) is implemented using OpenFOAM.

The compatibility of the continuous adjoint shape optimization method with commercial and open source multiphysics tools facilitates the use of more complex ECM process models. First extensions can include the following features:

1. The electrolyte conductivity can vary in space, assuming increasing temperature as well as increasing amount of gas and slag towards the electrolyte outlet of the interelectrode gap (Jain and Chouksey, 2018). Mathematically, the dimensionless electrolyte conductivity is described by a scalar field $\kappa(\underline{x}) = \tilde{\kappa}(\underline{x}) / \tilde{\kappa}_0$, where $\tilde{\kappa}_0$ is the conductivity of the electrolyte at the inlet of the interelectrode gap and $\tilde{\kappa}(\underline{x})$ can be modeled e.g. by Equation 12.3.
2. The current efficiency η for anodic dissolution can depend on the local current density (see Equation 12.5).
3. The variable overpotential can be taken into account in the form of linearized current-density–potential-relations (cf. Equation 12.4)

$$\Phi = 0 - q_C j \quad \text{on } \Gamma_C, \tag{12.79}$$

$$\Phi = 1 - q_A j \quad \text{on } \Gamma_A, \tag{12.80}$$

as experimentally observed by Lohrengel (2005) or De Silva et al. (2000), where q_C and q_A denote the inverse of the slopes of the cathodic and anodic polarization curves.

The formulation of the shape optimization problem would then be modified to minimizing the objective function (cf. Equation 12.61)

$$J_{N1}(\Omega, \Phi) = \frac{1}{2} \int_{\Gamma_A} \left(\kappa\eta \frac{\partial \Phi}{\partial n} - \cos\theta \right)^2 dS \tag{12.81}$$

subject to (cf. Equations 12.62–12.65)

$$\underline{\nabla} \cdot \left(\kappa \underline{\nabla} \Phi \right) = 0 \quad \text{in } \Omega, \tag{12.82}$$

$$\frac{\partial \Phi}{\partial n} = 0 \quad \text{on } \Gamma_N, \tag{12.83}$$

$$\Phi + q_C \kappa \frac{\partial \Phi}{\partial n} = 0 \quad \text{on } \Gamma_C, \tag{12.84}$$

$$\Phi + q_A \kappa \frac{\partial \Phi}{\partial n} = 1 \quad \text{on } \Gamma_A. \tag{12.85}$$

The adjoint equations,

$$\nabla \cdot \left(\kappa \nabla \lambda_\Omega \right) = 0 \quad \text{in } \Omega, \tag{12.86}$$

$$\frac{\partial \lambda_\Omega}{\partial n} = 0 \quad \text{on } \Gamma_N, \tag{12.87}$$

$$\lambda_\Omega + q_C \kappa \frac{\partial \lambda_\Omega}{\partial n} = 0 \quad \text{on } \Gamma_C, \tag{12.88}$$

$$\lambda_\Omega + q_A \kappa \frac{\partial \lambda_\Omega}{\partial n} = -\left(\kappa \eta \frac{\partial \Phi}{\partial n} - \cos\theta \right) \quad \text{on } \Gamma_A \tag{12.89}$$

and the shape derivative,

$$\partial_\Omega \mathcal{L}_N \left(\Omega, \Phi, \underline{\lambda} \right)[\underline{v}] = \int_{\Gamma_C} \kappa(\underline{v} \cdot \underline{n}) \frac{\partial \lambda_\Omega}{\partial n} \frac{\partial \Phi}{\partial n} dS$$
$$+ \int_{\Gamma_C} (\underline{v} \cdot \underline{n}) q_C \kappa \frac{\partial \lambda_\Omega}{\partial n} \left[\kappa \frac{\partial^2 \Phi}{\partial n^2} - \underline{\nabla} \cdot \left(\kappa \underline{\nabla} \Phi \right) \right] dS, \tag{12.90}$$

can be derived in a similar procedure. For cathode shape design problems using sophisticated ECM models, the dimensioned formulations can be more advantageous in practice.

FREQUENTLY USED SYMBOLS AND ABBREVIATIONS

ABBREVIATIONS

2D, 3D   two-, three-dimensional
ECM   Electrochemical machining

SYMBOLS, OVER AND UNDERSETS

$\tilde{\cdot}$ (tilde)	Dimensioned physical quantity
$\underline{\cdot}$ (underline)	(Column) spatial vector quantity or operator
$(\underline{\cdot})^T$	Transposed vector
∇	Nabla operator
$\overline{\partial}(\cdot)/\partial n$	Normal derivative on a boundary
$\partial_\Omega J(\Omega_0, \cdot)[\underline{v}]$	Partial shape derivative of J at Ω_0 in the direction of the deformation velocity field $\underline{v}$ of the deformation velocity field $\underline{v}$

UPPER-CASE ROMAN

$\tilde{A}$	Area of the plane electrode surface
A	Amplitude parameter characterizing the vertical extension of the anode profile
A^*	Critical value of A relevant for the physical realizability of solutions
$Cf_m(i)$	Correction factor in the m-th cycle for node i
$Er_m(i)$	Deviation between computed and required anode at node i in cycle m
$\tilde{F}(x), (F(x))$	(Dimensionless) function describing the anode profile
$G, G_\Omega, G_{\Gamma_A}$	Functional of the variational problem
$G, (G_{\max})$	(Maximum of the) shape gradient
H	Sum of principal curvatures
$\tilde{I}$	Total electrical current
$J_N, (J_{N,reg})$	(Regularized) objective function
$\tilde{K}_V, (\tilde{K}_{V,eff})$	(Effective) specific volumetric material removal rate
$\mathscr{L}_N$	Lagrangian of the shape optimization problem
$\tilde{R}$	Ohmic resistance of the interelectrode gap
$\tilde{U}$	Electric potential drop within the electrolyte
$\tilde{U}_{gap}$	Electric voltage applied between cathode and anode
$Y_m(i)$	Gap width in vertical direction at node i in cycle m

LOWER-CASE ROMAN

$\tilde{a}, \tilde{b}, q_C, q_A$	Empirical parameters describing the electrode overpotential
$[b], [b^e]$	(Element) right hand side vector in a linear system of equations
$\tilde{h}, (h)$	(Dimensionless) width of the interelectrode gap
$\tilde{h}^{std}, (h^{std})$	(Dimensionless) standard front gap width
$\tilde{h}_e, (\tilde{h}_e^{std})$	(Standard) equilibrium front gap width
$\tilde{j}, (j)$	(Dimensionless) current density at the anode surface
$\overline{j}_A$	Dimensionless current density at the anode surface
$[k], ([k^e])$	(Element) stiffness matrix
n	Index
$\underline{n}$	Outer unit normal of anode boundary

$\tilde{t}$	Time
$\underline{v}$	Deformation velocity field (continuous)
v_n	Component of deformation velocity field in boundary normal direction
$\underline{v}_A,(\underline{v}_A)$	(Dimensionless) anode dissolution velocity
$\tilde{\underline{v}}_f,(\underline{v}_f)$	(Dimensionless) cathode feed velocity
$\tilde{\underline{x}},(\underline{x})$	(Dimensionless) spatial coordinates
$x_{A,C}, y_{A,C}, z_{A,C}$	Dimensionless coordinates of points on cathode and anode boundaries

UPPER-CASE GREEK

Γ	Boundary (in general)
$\tilde{\Gamma}_A,(\Gamma_A)$	(Dimensionless) anode boundary
$\tilde{\Gamma}_C,(\Gamma_C)$	(Dimensionless) cathode boundary
$\tilde{\Gamma}_N,(\Gamma_N)$	(Dimensionless) non-metallic side boundary
Δ	Normal deviation of the computed electrode profile from the exact profile
Δ_{rms}	Mean normal deviation of computed electrode profile
$\Delta\tilde{U}$	Sum of electrode overpotentials
$\Delta\tilde{t}$	Time step
Δx	Mesh size along the gap direction
$\tilde{\Phi},(\Phi)$	(Dimensionless) electric potential
$\tilde{\Omega},(\Omega)$	(Dimensionless) computational domain
$\partial\Omega$	Total boundary of the dimensionless computational domain
Ω_ε	Deformed dimensionless computational domain

LOWER-CASE GREEK

$\alpha(\alpha_{opt})$	(Optimum) regularization parameter
α_{void}	volume fraction of gaseous reaction products in the electrolyte
$\tilde{\gamma}$	Conductivity constant of the electrolyte
$\delta_{\max}$	Maximum of the integrand of the shape optimization objective function
η	Current efficiency
θ	Angle between feed direction and anode surface normal
$\tilde{\kappa},(\kappa)$	(Dimensionless) electrolyte conductivity
$\tilde{\kappa}_0$	Electrolyte conductivity at the inlet
$\lambda_{\Omega,A,C,N}$	Lagrange multipliers for calculation domain, anode, cathode, and non-metallic side boundary
ξ	Curve parameter
$\tilde{\sigma},(\sigma)$	(Dimensionless) standard deviation of the Gaussian
τ	Iteration step length
$[\phi],([\varphi^e])$	(Element) solution vector

REFERENCES

Chang, C. and Hourng, L. (2001). Two-dimensional two-phase numerical model for tool design in electrochemical machining. *Journal of Applied Electrochemistry* 31(2):145–154.

COMSOL. (2017). *Comsol Multiphysics Application Gallery: Pulse Reverse Plating (Application ID: 61601)*. COMSOL Multiphysics 5.6 Edition.

Das, S. and Mitra, A. K. (1992). Use of boundary element method for the determination of tool shape in electrochemical machining. *International Journal for Numerical Methods in Engineering* 35(5):1045–1054.

Datta, M. and Landolt, D. (1982). High rate transpassive dissolution of nickel with pulsating current. *Electrochimica Acta* 27(3):385–390.

De Silva, A., Altena, H. and McGeough, J. (2000). Precision ECM by process characteristic modelling. *CIRP Annals – Manufacturing Technology* 49(1):151–155.

Deconinck, J. (1992). *Current Distributions and Electrode Shape Changes in Electrochemical Systems*. Springer, Berlin.

Demirtas, H., Yilmaz, O. and Kanber, B. (2017). A simplified mathematical model development for the design of free-form cathode surface in electrochemical machining. *Machining Science and Technology* 21(1):157–173.

Dietz, H., Guenther, K. G. and Otto, K. (1973). Reproduction accuracy with electrochemical machining: Determination of the side gap. *Annals of the CIRP* 22:61–62.

Eppler, K. (2009). A shape calculus analysis for tracking type formulations in electrical impedance tomography. *Journal of Inverse and Ill-Posed Problems* 17(8):733–751.

Fischer, Y., Marteau, B. and Privat, Y. (2012). Some inverse problems around the tokamak Tore Supra. *Communications on Pure and Applied Analysis* 11(6):2327–2349.

Gileadi, E. (2011). *Physical Electrochemistry*. Wiley, Weinheim.

Gu, Z., Zhu, W., Zheng, X., and Bai, X. (2019). Cathode tool design and experimental study on electrochemical trepanning of blades. *The International Journal of Advanced Manufacturing Technology* 100(1–4):857–863.

Hadamard, J. (1923). *Lectures on Cauchy's Problem in Linear Partial Differential Equations*. Mrs. Hepsa Ely Silliman Memorial Lectures. Yale University Press, New Haven.

Hardisty, H. and Mileham, A. R. (1999). Finite element computer investigation of the electrochemical machining process for a parabolically shaped moving tool erodingan arbitrarily shaped workpiece. *Proceedings of the Institution of Mechanical Engineers Part B: Journal of Engineering Manufacture* 213(8):787–798.

Hinduja, S. and Kunieda, M. (2013). Modelling of ECM and EDM processes. *CIRP Annals – Manufacturing Technology* 62(2):775–797.

Hunt, R. (1990). An embedding method for the numerical solution of the cathode design problem in electrochemical machining. *International Journal for Numerical Methods in Engineering* 29(6):1177–1192.

Jain, V. K. and Chouksey, A. K. (2018). A comprehensive analysis of three-phase electrolyte conductivity during electrochemical macromachining/micromachining. *Proceedings of the Institution of Mechanical Engineers, Part B: Journal of Engineering Manufacture* 232(14):2449–2461.

Jain, V. K. and Pandey, P. C. (1980a). Finite element approach to the two dimensional analysis of electrochemical machining. *Precision Engineering* 2:23–28.

Jain, V. K. and Pandey, P. C. (1980b). Tooling design for ECM. *Precision Engineering* 2(4):195–206.

Jain, V. K. and Rajurkar, K. P. (1991). An integrated approach for tool design in ECM. *Precision Engineering* 13(2):111–124.

Jain, V. K., Yogindra, P. and Murugan, S. (1987). Prediction of anode profile in ECBD and ECD operations. *International Journal of Machine Tools and Manufacture* 27(1):113–134.

Klocke, F., Heidemanns, L., Zeis, M., and Klink, A. (2018). A novel modeling approach for the simulation of precise electrochemical machining (PECM) with pulsed current and oscillating cathode. *Procedia CIRP* 68:499–504.

Klocke, F. and König, W. (2006). *Fertigungsverfahren 3: Abtragen, Generieren, Laser material bear beitung.* Springer, Berlin.

Krylov, A. L. (1968). The Cauchy problem for the Laplace equation in the theory of electrochemical metal machining. *Soviet Physics-Doklady* 13:15–17.

Liu, W., Ao, S., and Luo, Z. (2019). Multi-physics simulation of the surface polishing effect during electrochemical machining. *International Journal of Electrochemical Science* 14:7773–7789.

Lohrengel, M. M. (2005). Pulsed electrochemical machining of iron in $NaNO_3$: Fundamentals and new aspects. *Materials and Manufacturing Processes* 20(1):1–8.

Lu, J., Riedl, G., Kiniger, B., and Werner, E. A. (2014). Three-dimensional tool design for steady-state electrochemical machining by continuous adjoint-based shape optimization. *Chemical Engineering Science* 106:198–210.

Lu, J., and Werner, E. A. (2021). Cathode shape design for steady-state electrochemical machining. Submitted to Machining Science and Technology.

McGeough, J. A. (1974). *Principles of Electrochemical Machining.* Chapman and Hall, London.

McGeough, J. A., and Rasmussen, H. (1974). On the derivation of the quasi-steady model in electrochemical machining. *IMA Journal of Applied Mathematics* 13(1):13–21.

Narayanan, O. H., Hinduja, S., and Nobel, C. F. (1986a). Design of tools for electrochemical machining by the boundary element method. *Proceedings of the Institution of Mechanical Engineers Part C: Journal of Mechanical Engineering Science* 200(3):195–205.

Narayanan, O. H., Hinduja, S., and Noble, C. F. (1986b).The prediction of work-piece shape during electrochemical machining by the boundary element method. *International Journal of Machine Tool Design and Research* 26(3):323–338.

Nilson, R., and Tsuei, Y. (1974). Inverted Cauchy problem for the Laplace equation in engineering design. *Journal of Engineering Mathematics* 8(4):329–337.

Open CFD. (2020). Openfoam v2012 tutorial: overLaplacianDyMFoam. https://www.openfoam.com. Retrieved 05 May 2021.

Purcar, M., Bortels, L., den Bossche, B. V., and Deconinck, J. (2004). 3D electrochemical machining computer simulations. *Journal of Materials Processing Technology* 149(1–3): 472–478.

Reddy, M. S., Jain, V. K., and Lal, G. K. (1988). Tool design for ECM: Correction factor method. *Journal of Engineering for Industry* 110(2):111–118.

Sun, C., Zhu, D., Li, Z., and Wang, L. (2006). Application of FEM to tool designfor electrochemical machining freeform surface. *Finite Elements in Analysis and Design* 43(2):168–172.

Thorpe, J., and Zerkle, R. (1969). Analytic determination of the equilibrium electrode gap in electrochemical machining. *International Journal of Machine Tool Design and Research* 9(2):131–144.

Tipton, H. (1964). The dynamics of electrochemical machining. In: *Proceedings of the 5th International Conference on Machine Tool Design and Research*, September 1964, Birmingham, UK, pp. 509–522.

Tipton, H. (1971). Calculation of tool shape for ECM. In: Faust, C. L., editor, *Fundamentals of Electrochemical Machining*, pp. 87–102. Electrochemical Society Inc., Princeton, NJ.

Xu, Z., and Wang, Y. (2021). Electrochemical machining of complex components of aeroengines: Developments, trends, and technological advances. *Chinese Journal of Aeronautics* 34(2):28–53.

Zhou, Y., and Derby, J. J. (1995). The cathode design problem in electrochemical machining. *Chemical Engineering Science* 50(17):2679–2689.

Zhu, D., Wang, K., and Yang, J. (2003). Design of electrode profile in electrochemical manufacturing process. *CIRP Annals Manufacturing Technology* 52(1):169–172.

Zienkiewicz, O., Taylor, R., and Zhu, J. (2005). *Finite Element Method – Its Basis and Fundamentals*. Elsevier, Oxford.

13 Electrochemical Discharge Machining

Dileep Kumar Mishra,
Julfekar Arab, and Pradeep Dixit

CONTENTS

DOI: 10.1201/9780429160011-13

13.1   INTRODUCTION

Micromachining electrically non-conductive substrates such as glass, quartz, composites, etc., has been a challenging task. Methods that can be used to create microfeatures in non-conductive substrates are diamond grinding, ultrasonic machining (USM), abrasive jet machining (AJM), laser beam machining (LBM), ion beam machining (IBM), chemical machining/etching (CM), etc. [1]. Wet etching, as well as plasma-based dry etching, has been used to create microfeatures in glass; however, these methods have a lower etching rate. Plasma etching-based process, i.e., deep reactive ion etching (DRIE), can be used to create deep features but has a slower etch rate in high aspect ratio holes and requires a metal mask [2, 3]. Silica-based substrates, i.e., quartz, borosilicate glass, and sodalime glass, have excellent optical transparency, biocompatibility, and chemically resistant properties [4]. Due to these properties, the micromachining of silica-based substrates has wide applications in the field of biomedical devices, microfluidics, lab-on-a-chip, and microelectromechanical systems (MEMS) devices [5–7]. However, due to the brittle nature of the silica-based substrates, micromachining using conventional contact-based techniques is difficult.

Micromachining of glass substrates using non-conventional methods such as USM, AJM, IBM, or LBM has either a high setup cost or has a slow material removal rate; while CM requires the masking of surfaces not to be machined and has high infrastructure cost. A comparison between the different methods that are used to fabricate microfeatures in non-conductive substrates is shown in Table 13.1 [8]. It can be observed from Table 13.1 that electrochemical discharge machining (ECDM) is a relatively faster and economical method to fabricate microfeatures in electrically non-conductive substrates. Unlike chemical etching, no mask is required to create the desired microstructure in the glass workpiece during the ECDM process. ECDM is a hybrid of the electrochemical machining (ECM) and electric discharge machining (EDM) processes [9, 10]. Both the ECM and EDM processes can be used to

---

TABLE 13.1

Comparison of Different Machining Processes Used to Fabricate Microstructures in Electrically Non-Conductive Substrates [8]

Machining Processes	Characteristic Features
Laser beam machining (LBM) and ion beam machining (IBM)	High setup cost, microcracks, and thermal damage to the workpiece due to high energy intensity.
Chemical machining (CM)/ etching	Difficult to fabricate high aspect ratio, slow material removal rate (MRR), good surface finish.
Abrasive jet machining (AJM)	Low machining rate, high tool (nozzle) wear, tool (nozzle) blockage, and poor surface quality.
Ultrasonic machining (USM)	Low machining rate, high tool wear and poor surface quality.
Electrochemical discharge machining (ECDM)	Economical, high machining rate, low tool wear, and relatively good surface finish.

machine electrically conductive substrates only, whereas ECDM has been primarily used to machine electrically non-conductive substrates. However, the ECDM process has also been used in the past for electrically conductive materials to enhance the performance of the ECM process. The first attempt was made by McGeough et. al. of Edinburgh, UK [11]. Later, some work was done at IIT Kanpur by different research groups. However, in this chapter, the ECDM process is discussed emphasizing its applications only in electrically non-conductive materials. The ECDM process is also known by other names, such as electrochemical spark machining (ECSM), spark-assisted chemical engraving (SACE), etc. During the ECDM process, electrochemical discharges are generated between the tool electrode (cathode) and the electrolyte when a potential difference is applied between a tool electrode and a counter electrode (anode) [12]. Therefore, material removal occurs by melting and vaporization of the workpiece due to the electrochemical discharges and high-temperature chemical etching by the electrolyte [13].

13.1.1 SUMMARY OF PAST RESEARCH WORK

Drilling a microhole in a glass workpiece using the ECDM process was demonstrated for the first time by Kurafuji [14]. Afterwards, researchers explored the capabilities of the ECDM process to create microfeatures in various electrically non-conductive substrates such as composites [15–17], ceramics [18–27], etc., and electrically conductive substrates [28–30]. Jain et al. used traveling wire electrochemical spark machining (TW-ECSM) to cut the composites [17]. The insight of the ECDM process into the generation of the electrochemical discharge was studied by Basak and Ghosh [12, 31] and Jain et al. [32]. Yang et al. demonstrated that apart from melting and vaporization, chemical etching of a glass workpiece also contributes significantly to material removal during the ECDM process [33].

A research group consisting of Wüthrich and Fascio performed extensive work in exploring the fundamentals of gas film formation and electrochemical discharges [34, 35]. Later on, several researchers also explored the hybridization of the ECDM process with other machining techniques such as laser energy [36, 37], ultrasonic vibrations [38–42], magnetic field [43–45], etc. to enhance the material removal rate, improve the dimensional accuracy and the surface finish of the machined substrate. Wüthrich et al. reported a reduction in the machining time by employing ultrasonic vibrations to the tool electrode [41]. An increase in the machining depth was reported by Han et al. due to the application of ultrasonic vibrations to the electrolyte [42].

Fabrication of miniaturized microholes and microchannels with dimensions less than 100 µm was demonstrated by Cao et al. [46]. Patro et al. created high aspect ratio microtool electrodes using the controlled ECM process and later fabricated microholes using the ECDM process [47]. Over the years, several researchers investigated the effect of various ECDM process parameters, i.e., machining voltage, duty cycle, frequency, type of power supply, the polarity of power supply, electrolyte concentration, immersion depth, tool feed rate, tool–workpiece gap, tool texture/roughness, etc., that affect the material removal rate and the dimensional characteristics of the machined microfeatures [48–56].

13.1.2 APPLICATIONS OF THE ECDM PROCESS

Over recent decades, the ECDM process has been extensively used to fabricate microfeatures in electrically non-conductive substrates such as glass, composites, ceramics, and electrically conductive substrates such as steel for various applications. Initially, researchers used a single-tip tool and fabricated microholes and microchannels in silica-based substrates using the ECDM process. Fascio et al. demonstrated the fabrication of microstructures such as engraving and microchannels for reactors in the Pyrex glass, as shown in Figure 13.1 [35].

Cao et al. fabricated several 3D structures such as microchannels, micropillars, microwells in Pyrex glass with dimensions of less than 100 μm, as shown in Figure 13.2 [46].

Mishra et al. demonstrated the capabilities of the ECDM process in fabricating different microfeatures such as microchannels and letter engraving (Figure 13.3(a–d)) using a single-tip tool electrode [53].

Jui et al. fabricated a high aspect ratio (~11) microhole in a 1.2 mm-thick borosilicate glass substrate in a 1 M NaOH electrolyte using a 40 V DC power supply and 1 μm/s vertical tool feed rate as shown in Figure 13.4 [54]. The authors used a low concentration of electrolytes to reduce the tool wear and overcut. The top and bottom diameters were measured to be 180 μm and 40 μm, respectively.

Esashi et al. demonstrated the fabrication of electrical feed through structures for silicon pressure sensors [57]. Li et al. fabricated an array of microholes using a spiral electrode of tungsten carbide with 100 μm diameter and reported a significant reduction in the critical voltage by 22.8% and an improvement in the dimensional accuracy by 16.9% compared to normal ECDM [58].

Instead of a single-tip tool electrode, array-tool electrodes can be used to create multiple microfeatures with relatively higher productivity, location accuracy, and lower tool wear. The customized array-tool electrodes can be made using the wire-electric discharge machining (W-EDM) process with a varying number of tips and sizes [59]. For the first time, Arab et al. demonstrated the fabrication of multiple through holes in fused silica wafers using area array-tool electrodes of varying configurations, i.e., 3×3, 5×5, 2×5, and the optimized ECDM process parameters that have applications in radio frequency microelectromechanical systems (RF-MEMS) devices as shown in Figure 13.5 [60].

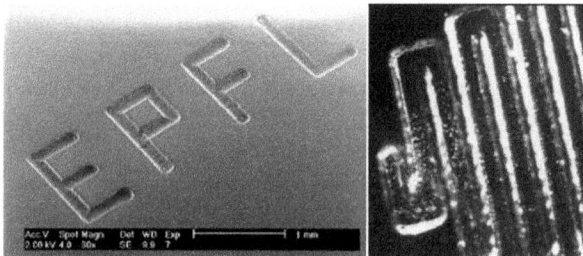

FIGURE 13.1 Fabrication of microchannels in Pyrex glass using the ECDM process. Reproduced with permission from [35].

FIGURE 13.2 Fabrication of 3D microstructures in Pyrex glass with dimensions less than 100 μm. Reproduced with permission from [46].

Later, Kannojia et al. filled the through-holes with electrically conductive copper material using the bottom-up electrodeposition method to obtain through-glass vias (TGVs), as shown in Figure 13.6 [61]. This was the first-time demonstration of the ECDM process in fabricating TGVs on a full-scale fused silica wafer. The electrodeposition is generally the preferred technique to fill the through holes for various 3D applications [62, 63]. Traditionally, these TGVs are required to develop 3D interconnects required for various RF-MEMS devices [64, 65]. Compared to conventional silicon substrate used in through-silicon vias (TSVs), glass substrates have superior electrical insulation, and therefore are becoming more popular for upcoming RF-MEMS applications.

Kannojia et al. have demonstrated the fabrication of TGV-based 3D toroidal and spiral inductors in a 2 inch-fused silica wafer, as shown in Figure 13.7 [66, 67]. Later, front and back redistribution lines were made on the top of the substrate using the through-resist electroplating technique. These 3D inductors have applications for RF-MEMS devices such as a low-noise amplifier, voltage-controlled oscillators, and RF filters [66].

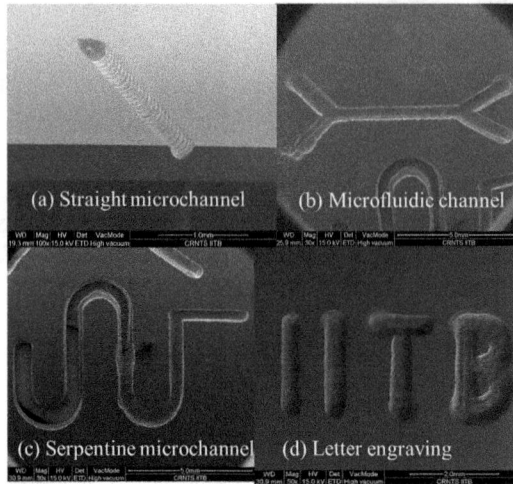

FIGURE 13.3 Fabrication of microchannels and letter engraving in the glass workpiece using the ECDM process. Reproduced with permission from [53].

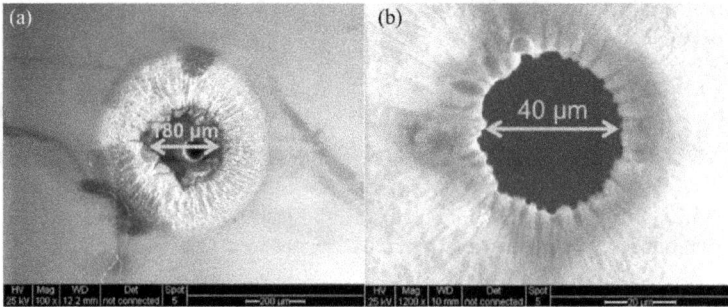

FIGURE 13.4 Fabrication of a high aspect ratio microhole by ECDM. Reproduced with permission from [54].

Moreover, the microchannels created by the ECDM process in the glass substrates can be used to create embedded redistribution lines (RDL) by filling the microchannels with electrically conductive materials, as shown in Figure 13.8 [68]. Unlike conventional RDLs, the embedded RDLs are formed inside the glass-based substrates resulting in flat surface topography and compact 3D devices.

Sharma et al. demonstrated the fabrication of deeper microholes in the alumina ceramic substrates using a combined feed mechanism and single-tip tool electrodes [69]. Later, the authors fabricated an array of through holes in 425 μm-thick alumina substrate using the 3 × 3 area array-tool electrode and the combined pulse-feed approach of the ECDM process, as shown in Figure 13.9 [18].

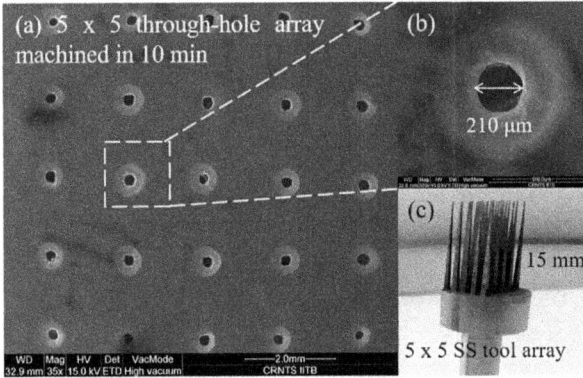

FIGURE 13.5 Fabrication of multiple through holes in the glass substrate using area array-tool electrodes. Reproduced with permission from [60].

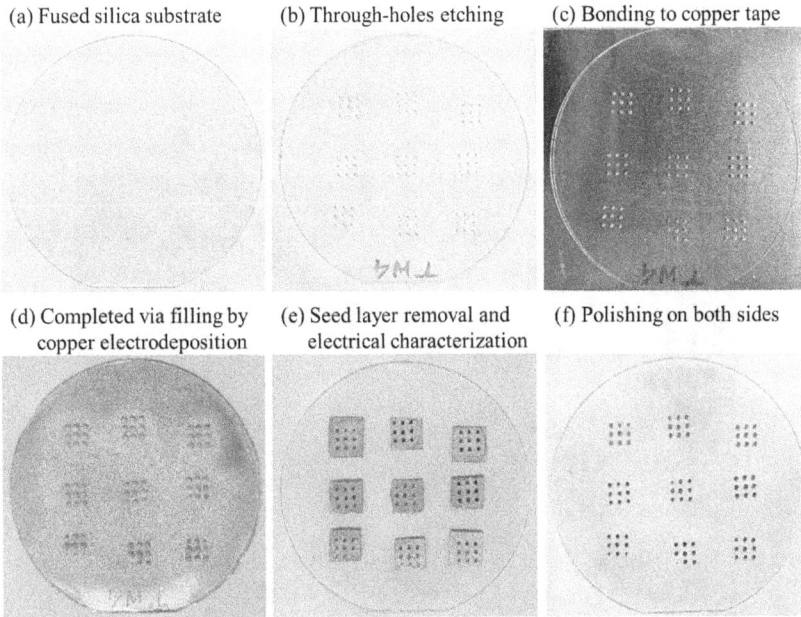

FIGURE 13.6 Fabrication of TGVs in fused silica wafer using the ECDM process followed by the copper electrodeposition technique. Reproduced with permission from [61].

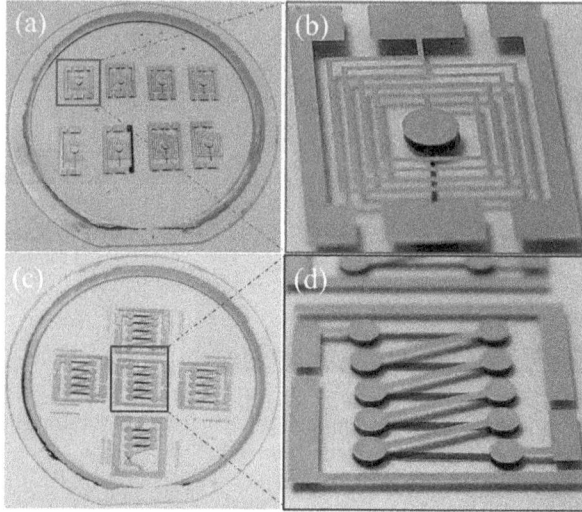

FIGURE 13.7 Fabrication of TGVs in fused silica wafer using the ECDM process followed by the copper electrodeposition technique. Reproduced with permission from [66].

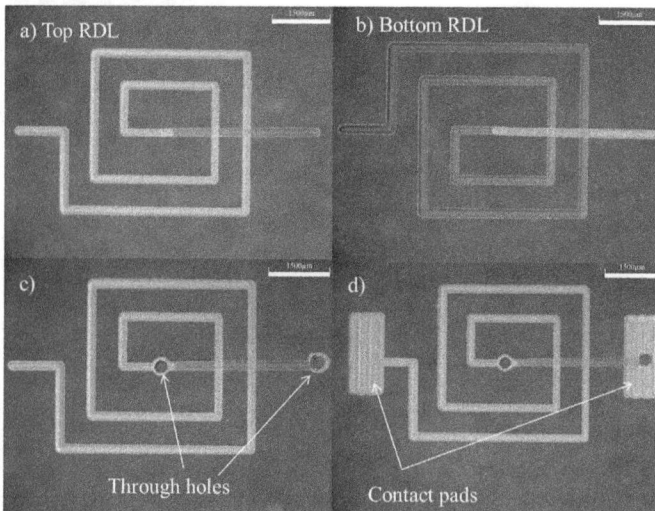

FIGURE 13.8 Stages for the fabrication of embedded RDLs and square inductor using the ECDM process: (a) Top microchannel, (b) bottom microchannel, (c) through holes, and (d) contact pads.

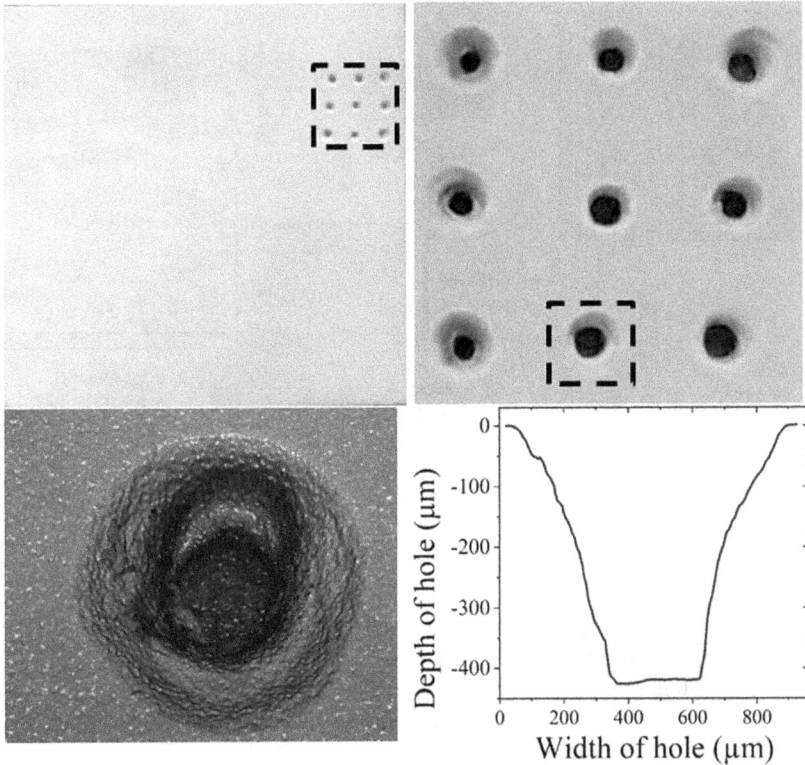

FIGURE 13.9 Fabrication of through holes in alumina substrate using the ECDM process and area array-tool electrodes. Reproduced with permission from [18].

13.2 DESCRIPTION OF THE ECDM PROCESS

The ECDM process is stochastic in nature and involves complex physics. Therefore, a complete description of the process physics would be beneficial to comprehend the process easily. In this section, different components of the ECDM setup are discussed. Later, the different stages of the gas film formation are described with the help of high-speed camera images and empirical relations.

13.2.1 COMPONENTS OF THE ECDM EXPERIMENTAL FACILITY

The development of the ECDM facility is relatively easy and economical. For the accurate microfabrication of microholes and microchannels, precise control over the motion of the workpiece and the tool feed rate is required, for which a computer numerical control (CNC) microcontroller is used. Figure 13.10 shows a schematic of the ECDM setup that consists of the following components:

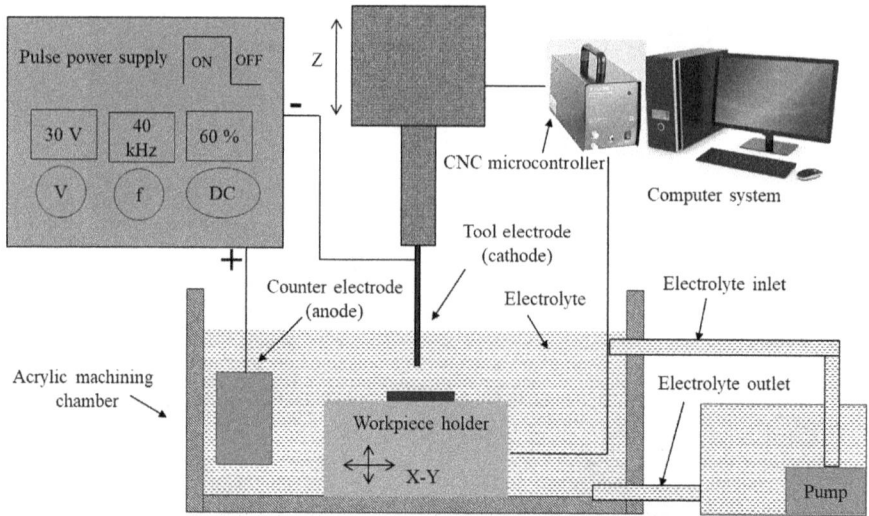

FIGURE 13.10 Schematic of different components of the ECDM setup.

(a) A pulse power supply to apply the required machining voltage and control the discharge frequency and time.
(b) X, Y, Z stages controlled by CNC microcontroller.
(c) The electrolyte circulation system consisting of a submersible pump and a valve to control the flow rate of the electrolyte.
(d) A processing chamber with the provision for holding the workpiece.

The tool electrode (generally cathode) is mounted on the Z stage while the processing chamber is placed over the X–Y stage and firmly secured. Depending on the requirement, the tool holder can be modified to incorporate spindle rotation and ultrasonic vibrations. A counter/auxiliary electrode (generally anode) with a comparatively higher surface area (~100 times) than the cathode is kept in the machining chamber immersed inside the electrolyte. The workpiece is positioned accurately in the workpiece holder placed inside the processing chamber. Compared to the stagnant electrolyte, continuous flow of electrolyte helps in keeping the electrolyte composition uniform, and also to maintain its uniform level using the flow control valve. Periodic filtering of the electrolyte can be performed to remove the accumulated sludge and debris.

13.2.2 Different Stages for the Formation of Hydrogen Gas Film

Different stages involved in the generation of hydrogen gas bubbles during the ECDM process are depicted in Figure 13.11. Application of a potential difference across the electrodes, i.e., the tool electrodes and the counter electrode, leads to the electrolysis of the aqueous electrolyte medium. Due to the electrolysis, gas bubbles are generated, i.e., hydrogen gas at the cathode and oxygen gas at the anode due to

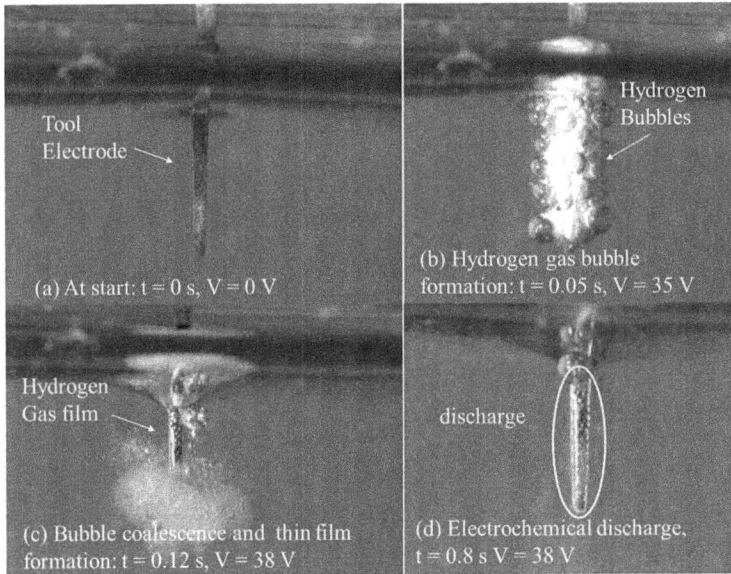

FIGURE 13.11 High-speed camera images showing different stages in the ECDM process: (a) Initial stage, (b) formation of hydrogen gas bubbles at the cathode, (c) hydrogen gas film formation, and (d) generation of electrochemical discharge.

the electrochemical reaction at the respective electrodes. With a further increase in the applied voltage, the rate of bubble generation increases. Due to higher bubble density near the tool electrode surface, the hydrogen gas bubbles coalesce together, forming a thin layer of hydrogen film. This thin layer of hydrogen acts as an insulator and hinders the passage of current between the electrodes. When the electric field strength due to the applied voltage across the hydrogen gas film becomes higher than the breakdown strength, an electrochemical discharge occurs between the tool electrode and the electrolyte. Due to the electrochemical discharge, high-velocity electrons strike the glass substrate (kept in the close vicinity of the cathode generally <20 μm) while moving toward the anode/counter electrode. Subsequently, material removal from silica-based materials occurs through the combined action of both melting/vaporization and chemical etching by the electrolyte medium. The chemical etching phenomenon depends on the interaction between the electrolyte and workpiece material and the temperature of the electrolyte.

Figure 13.12 depicts the typical voltage vs current (V-I) graph of the ECDM process obtained for a 10 wt.% KOH electrolyte [53]. Five distinct zones can be identified from the V-I curve [70].

The zone "ab'" is termed an over-potential region. In this zone, the applied voltage is less than 2.5 V; and the electrolysis process has not started. The zone "bc" is called an Ohmic zone in which bubbles are formed due to the electrolysis process of the electrolyte, and a linear trend was observed between the applied voltage and the measured circuit current. Afterwards, the slope of the curve changes, and a

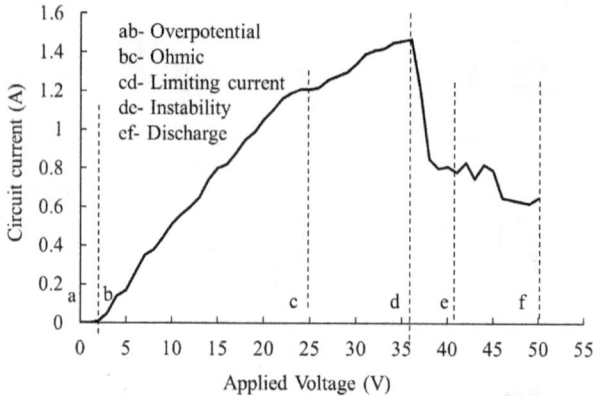

FIGURE 13.12   Voltage vs current graph obtained using 10% KOH electrolyte.

marginal increase in the current value is measured in the zone "cd". Bubbles start to coalesce together to form the thin hydrogen film around the tool electrode surface. This is followed by an instability zone "de" in which the circuit current drops suddenly with a small increase in the applied voltage. The tool electrode is completely covered with the hydrogen gas film. Therefore, the resistance offered to the flow of current increases, and thus, the current value drops suddenly. Beyond the "e", the electrochemical discharge zone "ef" is observed in which the measured circuit current is the bare minimum and discrete succession electrochemical discharges occur between the tool electrode surface and the electrolyte.

13.2.3   DESCRIPTION OF DIFFERENT STAGES INVOLVED IN THE FORMATION OF GAS FILM

The formation of hydrogen gas film is critical during the ECDM process. The different steps involved in gas film formation over the electrode surface are bubble generation, a saturation of the electrolyte with bubbles near the tool surface, heterogeneous nucleation at tool electrode cavities, bubble growth, bubble departure, and film formation. A detailed description of each of these steps is explained in the following sections.

13.2.3.1   Gas Bubble Generation

During the ECDM process, electrolysis takes place when a potential difference is applied across the electrodes. Subsequently, hydrogen and oxygen gas are generated in a dissolved form at the cathode and anode electrodes, respectively. The governing electrochemical reactions for the generation of the gas bubbles in different types of electrolyte medium are as follows:

At cathode or tool electrode:

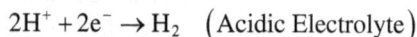

$$2H^+ + 2e^- \rightarrow H_2 \quad \left(Acidic\ Electrolyte \right)$$

$$2H_2O + 2e^- \rightarrow 2OH^- + H_2 \quad \left(\text{Basic Electrolyte}\right)$$

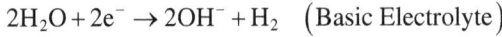

At anode or auxiliary electrode:

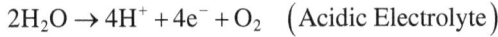

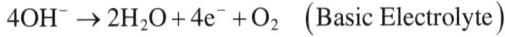

$$2H_2O \rightarrow 4H^+ + 4e^- + O_2 \quad \left(\text{Acidic Electrolyte}\right)$$

$$4OH^- \rightarrow 2H_2O + 4e^- + O_2 \quad \left(\text{Basic Electrolyte}\right)$$

In the neutral medium, the electrochemical reaction will be the same as that of the electrolysis of pure water. The gas bubbles generated by the above reaction are present in dissolved form near the tool electrode initially [70], and then evolve on the electrode surface by the heterogeneous nucleation process.

13.2.3.2 Bubble Nucleation, Growth, and Detachment

Once the electrolyte near the tool electrode surface is supersaturated with the hydrogen gas, heterogeneous nucleation of the gas bubbles starts from the nucleation sites on the electrode surfaces. For homogeneous nucleation, the critical radius $\left(r_c\right)$ of a bubble to grow over the electrode surface and the rate of nucleation (J), i.e., the number of bubbles formed per second per cubic centimeter of the liquid is given by Equations 13.1 and 13.2 respectively [71]

$$r_c = \frac{2\sigma}{P_b - P_l} \tag{13.1}$$

$$J = Z \exp\left(\frac{-16\pi\sigma^3}{3k_B T \left(P_b - P_l\right)^2}\right) \tag{13.2}$$

where σ is the surface tension between the liquid-gas interface, P_b and P_l are the pressure inside the bubble and total liquid pressure respectively, Z is the pre-exponential frequency factor, k_B is the Boltzmann constant, and T is the temperature.

Heterogeneous nucleation at the gas-evolving electrode requires much less dissolved gas supersaturation than required in the bulk liquid. The presence of an electrode surface reduces the energy barrier for the nucleation process due to variation in the contact angle θ. The contact angle determines the wettability of liquid over the solid surface and is defined as the angle between the plane of the solid surface and the tangent to the liquid surface at the contact line. Thus, in the case of heterogeneous nucleation, the surface tension (σ) is multiplied by a factor $f(\theta)$, which is given below by Equation 13.3.

$$f(\theta) = \frac{\left(1 + \cos\theta\right)^2 \left(2 - \cos\theta\right)}{4} \tag{13.3}$$

A nucleated bubble then starts growing over the electrode surface. Initially, bubble growth occurs by the internal pressure of the bubble. Later, the growth of the bubbles on the tool electrode surface occurs in two ways. The first method is by the diffusion or mass transfer of the dissolved gas to the gas–liquid interface. The generalized equation for the bubble growth with time (t) by the diffusion process was obtained by Scriven and is given by Equation 13.4 [72]

$$R_b = 2C_s(Dt)^{0.5} \tag{13.4}$$

where R_b is the bubble radius, C_s is the coefficient for the degree of supersaturation, and D is the diffusion coefficient.

The second method is by coalescence with the neighboring bubbles. The density of bubble nucleation sites increases with an increase in the current density, and the bubble nucleating sites become very close to each other [31]. In this situation, bubble growth occurs by the coalescence process with the neighboring bubbles over the electrode surfaces.

Finally, the bubble detaches from the electrode surface due to an imbalance of the forces acting on the bubble. A schematic of different forces, i.e., buoyancy force (F_B), the bubble pressure force (F_p), the inertia forces of the growing bubble (F_{TI}, F_{NI}), and the surface tension forces (F_{SX}, F_{SY}) acting on a gas bubble under stagnant electrolyte condition is shown in Figure 13.13(a).

From the equilibrium force balance criteria, Zhang and Zeng [73] obtained the critical bubble radius beyond which the bubbles leave the electrode surface for a stagnant electrolyte as given by Equation 13.5

$$R_d = \sqrt{\frac{6\sigma \sin\theta (\alpha - \beta)(\sin\alpha + \sin\beta)}{g(1+\cos\theta)^2 (2-\cos\theta)(\rho_l - \rho_b)\left[\pi^2 - (\alpha - \beta)^2\right]}} \tag{13.5}$$

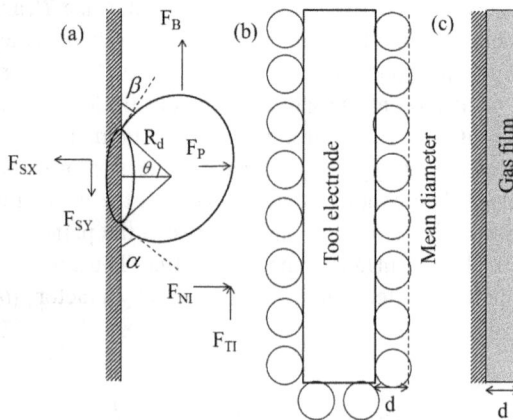

FIGURE 13.13 Formation of a gas film during the ECDM process: (a) Forces acting on the bubble, (b) evolution of the bubbles at the tool electrode surface, and (c) gas film formation by the coalescence of the bubbles.

where R_d is the bubble departure radius, σ is the surface tension between the liquid electrolyte and the gas bubble, θ is the contact angle between liquid-gas and liquid–solid interfaces, α and β are the advancing and receding contact angles caused due to upward tilting of the bubble ρ_l, and ρ_b is the density of the electrolyte and the gas bubble. Further, the surface roughness of the tool electrodes also changes the wettability of a liquid as described by the Wenzel model and expressed by Equation 13.6 [74]

$$\cos\theta = r\cos\theta_0 \tag{13.6}$$

where r is the roughness factor, i.e., area of the actual surface (cathode tool electrode) to the smooth surface with the same configurations, with θ_0 as the contact angle. Thus, as the roughness of the tool electrode surface increases, the contact angle θ also increases. The volume of a gas bubble (V_{bubble}) is equal to the volume of a truncated sphere with radius R_d and is given by Equation 13.7.

$$V_{bubble} = \frac{4\pi R_d^3}{3} f(\theta) \tag{13.7}$$

The thickness of the gas film can be determined by the mean departure diameter (d) of the bubbles, which coalesces with the neighboring bubbles to form the hydrogen film around the electrode surface (Figure 13.13(b)c)). It was also found that inflowing electrolytes, both drag and buoyancy forces, act in the same direction. This results in the reduction of R_d and hydrogen film thickness [75]. Further, Elhami and Razfar reported that the use of ultrasonic vibrations helps in reducing the gas film thickness, and it increases the machining speed in the discharge regime [40]. In another article, the authors reported a reduction in the entrance overcut of the drilled microhole when the amplitude of the ultrasonic vibrations was increased due to the formation of thin gas film and stable uniform discharges, as shown in Figure 13.14 [76]. However, when the amplitude was increased beyond 10 μm, non-uniform and unstable discharges were generated due to fluctuations in the gas films [76].

Rathore and Dvivedi developed a mathematical model for gas film thickness in the ultrasonic-assisted ECDM and reported a reduction in the bubble departure radius when ultrasonic vibration is used compared to normal ECDM [77]. Due to the application of the ultrasonic vibrations, the authors reported an improvement in material removal rate (MRR) by 11%, penetration depth by 27%, and reduction in overcutting of the machined hole by 23% compared to the normal ECDM process.

13.2.3.3 Electrochemical Discharge Generation Due to Hydrogen Film Breakdown

Once a complete hydrogen film is formed around the tool electrode, gas arc discharge occurs due to the application of an applied electric field reaching beyond the breakdown strength of the hydrogen gas film. Arc discharges in gases at low voltage (10–50 V), and high current (1–100 A) occurring at pressures ranging from 0.5 to 100 atm [70]. The emission of electrons from the cathode spot of the tool

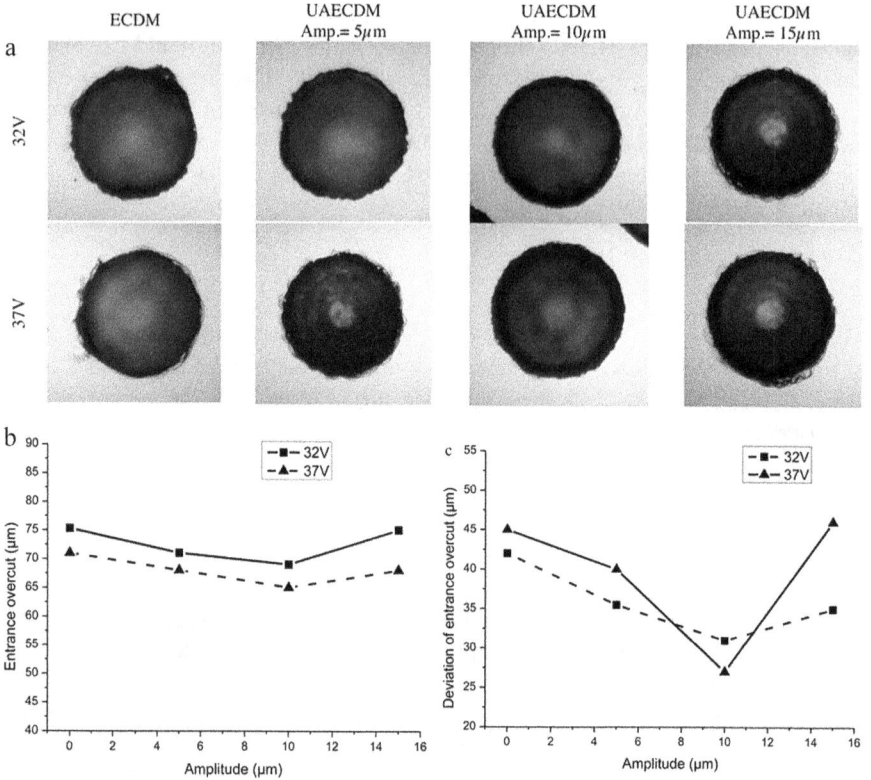

FIGURE 13.14 Effect of ultrasonic vibrations on the overcut of the microholes. Reproduced with permission from [76].

electrode occurs either by thermionic emission or field emission. The current density (J_e) due to thermionic emission and field emission of the electrons is given by the Richardson–Dushman and Fowler–Nordheim equations, respectively [70] as expressed by Equation 13.8

$$
J_e = \begin{cases}
A_R T^2 \exp\left(-\dfrac{e\varphi}{k_B T}\right) & \text{for themionic emission} \\[2em]
6.2 \times 10^{-6} \dfrac{\sqrt{E_F/e\varphi}}{E_F + e\varphi} E^2 \exp\left(-\dfrac{6.85 \times 10^9 (e\varphi)^{1.5}}{E}\right) & \text{for field emission}
\end{cases}
\tag{13.8}
$$

where A_R is a constant, k_B is the Boltzmann constant, $e\phi$ is the work function in electron volts, E_F is the Fermi level of the tool electrode, E is the electric field strength, and T is the temperature. The formation of hydrogen plasma correlation reactions can be grouped into five types, i.e., excitation, de-excitation, ionization, attachment, and elastic collision reactions [78]. After the formation of plasma, the avalanche motion of electrons occurs toward the counter electrode/anode through the electro-lyte medium. Since the glass workpiece is kept very close to the cathode (<20 μm), the avalanching electrons strike the workpiece surface and heat up the electrolyte medium locally.

13.2.4 MATERIAL REMOVAL DUE TO CHEMICAL ETCHING OF THE WORKPIECE BY THE ELECTROLYTE

As discussed earlier, chemical etching of the glass workpiece at elevated electrolyte temperature also contributes to material removal. The concentration and composi-tion of the electrolyte influence its electrical conductivity, which further affects the rate of bubble generation, film thickness, and electrochemical discharge frequency. The number of anions, cations, and charge present in the electrolyte influences the electrical conductivity (κ_e) of the electrolyte expressed by Equation 13.9 [70]

$$\kappa_e = e(n^+ z^+ u^+ + n^- z^- u^-) \tag{13.9}$$

where e is the numerical value charge of an electron, n^+ and n^- number density of the cations and anions having z^+ and z^- as charges with u^+ and u^- as ions mobility. Thus, increasing the electrolyte concentration increases the electrical conductivity of the electrolyte, and therefore increases the rate of bubble generation and electrochemical discharge frequency. An increase in the discharge frequency increases the energy released and the material removal rate.

The rate of chemical etching of the glass (R_{etch}) as a function of the alkaline elec-trolyte temperature (T) is expressed by Equation 13.10 [79]

$$R_{etch} = Ke^{-E_a/RT}e^{-G_0/RT} \tag{13.10}$$

where K is the proportionality constant, R is the universal gas constant, and G_0 is the standard Gibbs free energy for the chemical etching reaction. From Equation 13.10, it can be observed that with an increase in the electrolyte temperature, the rate of glass etching increases. Charles reported the different steps involved in the chemical etching of the glass as described below [80].

(a) An alkali (R) structure at the end of the silica network undergoes decompo-sition and is replaced by the hydrogen ion, thereby forming a hydroxyl ion and an alkali ion (R^+).

$$\equiv Si - o - R + H_2O \rightarrow \equiv Si - o - H + R^+ + OH^-$$

(b) The siloxane bond is broken by hydroxyl ion, forming a silanol group from one end, and the other end produces a structure that can further dissociate with the water molecule.

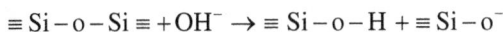

$$\equiv Si - o - Si \equiv + OH^- \rightarrow \equiv Si - o - H + \equiv Si - o^-$$

(c) The product formed in stage (b) reacts with the water molecule to produce hydroxyl ions, and the chain reaction continues. Thus, the presence of the hydroxyl ion enhances the chemical etching of the glass substrates.

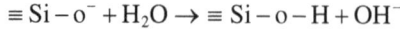

$$\equiv Si - o^- + H_2O \rightarrow \equiv Si - o - H + OH^-$$

The overall etching reaction of the glass in the alkaline electrolyte NaOH is given by the reaction shown below [10, 81].

$$2NaOH + SiO_2 \rightarrow Na_2SiO_3 + H_2O$$

The variation in the solubility of the glass with varying temperature and pH value of the electrolyte is shown in Figure 13.15 [33]. It can be clearly observed from Figure 13.15 that the rate of glass etching increases with an increase in the electrolyte temperature and pH. Fascio et al. reported that the rate of chemical etching in alkaline electrolytes is doubled with an increase in the pH by one unit or with an increase in the electrolyte temperature by 10°C [81].

13.3 DIFFERENT SPARKING THEORIES AND NUMERICAL MODELING APPROACHES OF THE ECDM PROCESS

The ECDM process consists of complex physical processes such as electrochemical reactions and electrochemical discharge generation. Researchers have proposed

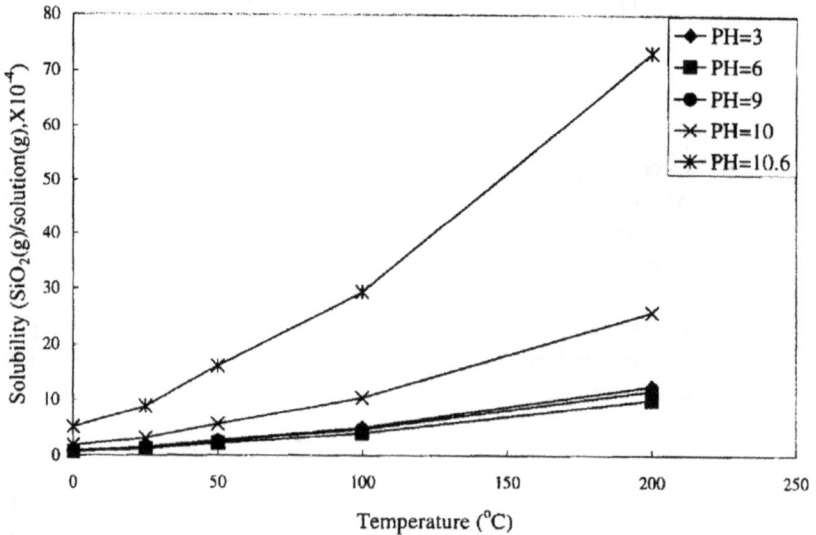

FIGURE 13.15 Variation in the solubility of borosilicate glass with temperature and pH value of the electrolyte. Reproduced with permission from [33].

different theories for electrochemical discharge formation and material removal, and have developed a theoretical and numerical model to predict the material removal rate and characteristics of the fabricated microfeatures. In this section, the different sparking theories and numerical model approaches are discussed.

13.3.1   THEORIES OF SPARKING AND MATERIAL REMOVAL MECHANISM

Researchers have explored the mechanisms for the occurrence of electrochemical discharge and material removal during the ECDM process. Crichton and McGeough used high-speed camera images and reported that both spark and arc discharges could be generated in electrolytes [11]. Basak and Ghosh proposed a "switching-off" mechanism for the discharge generation in which initially the tool electrode and electrolyte interface resistance increases due to the accumulation of a large number of bubbles, and the constriction effect occurs for the flow of the current [31]. Later, the cathode surface becomes completely covered by the fully grown bubbles, and the current value suddenly drops to zero, i.e., switches off like an electric circuit. In this short time, the induced switching, electromotive force (e.m.f.), is sufficient to generate a discharge between the tool electrode and the electrolyte. After this, the connection between the tool electrode and the electrolyte is re-established due to dislodging of the bubbles, and the cycle repeats itself. When the applied voltage is higher than the critical voltage, stable and continuous discharging takes place [31, 33].

In another work, the authors proposed the material removal mechanism during the ECDM process. Material removal occurs by the melting of the workpiece due to the fraction of spark energy transferred into it and subsequent removal of the melted portion by mechanical and electrical shock generated due to phase change and electrochemical discharge, respectively [12]. Afterwards, Jain et al. proposed a new theory for electrochemical discharges and called it "valve theory". In this theory, each bubble is proposed as a valve that produces the electrochemical discharge after its breakdown under a high electric field [32]. Yang et al. reported that apart from melting, chemical etching of the workpiece by the heated electrolyte also contributes to the material removal during the ECDM process [33].

13.3.2   THEORETICAL AND NUMERICAL MODELING OF THE ECDM PROCESS

Basak and Ghosh developed a theoretical model to predict the critical current and critical voltage for different combinations of electrode and electrolyte using cylindrical tool electrodes [31]. The authors also developed a mathematical model to predict the material removal rate and reported a good match between the predicted and observed experimental results of earlier researchers [12]. Using the "valve theory", Jain et al. developed a finite element method (FEM) model to predict the material removal rate during the ECDM process [32]. Later, researchers numerically predicted the distribution of the temperature inside the substrates using a simplified thermal model (without modeling the chemical etching phenomenon) and Gaussian heat flux. Goud and Sharma modeled MRR for electrochemical discharge drilling of sodalime glass and alumina using the 3D finite element method and 3D unsteady state heat diffusion equation [82].

Panda and Yadava developed a 3D finite element transient thermal model and estimated temperature distribution and MRR during TW-ECSM using Gaussian heat flux [83]. Recently, Elhami and Razfar simulated a semi-realistic ECDM model, i.e., plasma generation, using the tool geometry, anode, electrolyte, and hydrogen gas film [84]. Mishra et al. performed a numerical simulation of the ECDM process for the fabrication of microchannels considering 3D heat diffusion Equation 13.11 [53]. The governing boundary conditions used in the model are shown in Figure 13.16, and the Gaussian heat flux is represented by Equation 13.12

$$\frac{\partial^2 T}{\partial x^2} + \frac{\partial^2 T}{\partial y^2} + \frac{\partial^2 T}{\partial z^2} = \frac{1}{\alpha}\frac{\partial T}{\partial t} \tag{13.11}$$

$$Q_{Gaussian} = \begin{cases} \dfrac{4.45 C_{PE} VI}{\pi R_s^2}\exp\left\{-4.5\left(\dfrac{(x-x_0-ft)^2+(y-y_0)^2}{R_s^2}\right)\right\}, & \text{for}\left\{(x-x_0)^2+(y-y_0)^2\right\}^{0.5} < R_s \\ -h\Delta T, & \text{otherwise} \end{cases}$$

$$\tag{13.12}$$

where α is thermal diffusivity expressed as $\kappa/\rho c_p$ and κ, ρ, c_p are thermal conductivity, density, and specific heat of the workpiece, respectively. $Q_{Gaussian}$ is Gaussian heat flux applied in the discharge region, V is the applied machining voltage, I is the circuit current, C_{PE} is the energy partition, i.e., the fraction of the total energy transferred into the workpiece, R_s is the spark radius, and f is the tool feed rate. The

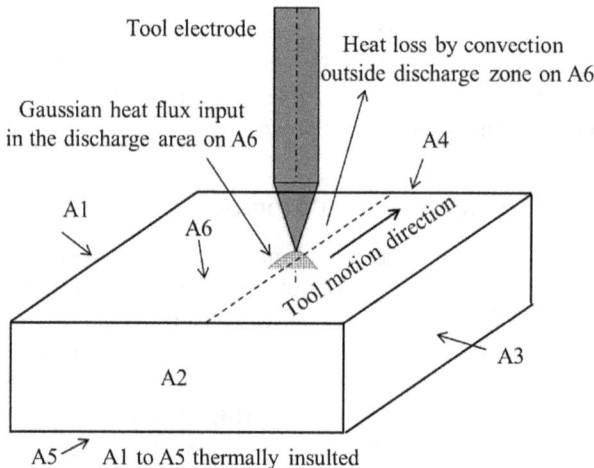

FIGURE 13.16 Different boundary conditions considered during numerical simulation of the ECDM process.

empirical formula of the circuit current I for different electrolytes as a function of electrolyte concentration C was developed by Goud and Sharma as expressed by Equation 13.13 [82].

$$I = \begin{cases} 0.4424 + 5.62 \times 10^{-2}C - 7 \times 10^{-4}C^2 + 3 \times 10^{-6}C^3 & \text{for} \quad \text{KOH} \\ 0.71429 + 9.13 \times 10^{-2}C - 2.7 \times 10^{-4}C^2 + 3.23 \times 10^{-6}C^3 & \text{for} \quad \text{NaOH} \end{cases} \quad (13.13)$$

Jain et al. developed an empirical relationship for the circuit current as a function of the applied machining voltage (V), electrolyte electrical conductivity (κ_e), tool electrode diameter (d), and tool immersion depth (d_{im}) as expressed by Equation 13.14 [32].

$$I = 0.1009V^{0.4815}\kappa_e^{0.342}d^{0.342}d_{im}^{0.2881} \quad (13.14)$$

The authors also estimated the electrolyte conductivity κ_e as a function of electrolyte concentration C expressed by Equation 13.15 [32].

$$\kappa_e = \begin{cases} -0.335781 + 3.86262C - 0.0680216C^2 & \text{for} \quad \text{KOH} \\ 0.885958 + 4.62384C - 0.193274C^2 & \text{for} \quad \text{NaOH} \\ -0.0154902 + 1.46172C - 0.0243105C^2 & \text{for} \quad \text{NaCl} \end{cases} \quad (13.15)$$

$$q_g = \left(\frac{4.45 \times P_e \times V \times I}{\pi \times (r_0 + z)^2} \right) \times \exp\left(\frac{-4.5 \times r^2}{r_0^2} \right) \quad (13.16)$$

Arab et al. modified the Gaussian heat flux equation to incorporate the reduction in the heat energy transferred into the workpiece due to an increase in the tool–workpiece gap as expressed by Equation 13.16 where q_g is the heat flux at z gap [52]. Due to an increase in the tool–workpiece gap, the maximum surface temperature and the depth of crater formed reduce due to a reduction in the heat energy transferred into the workpiece. In the ECDM process, material removal occurs by the combined effect of melting and vaporization as well as the chemical etching that is more enhanced in NaOH electrolyte than KOH electrolyte due to the relatively aggressive electrochemical discharge in the NaOH as captured by high-speed camera images shown in Figure 13.17 [52].

Figure 13.18 depicts the temperature contour and the cater depth obtained using the KOH and NaOH electrolyte at zero and 20 μm tool–workpiece gaps. The crater depth obtained after a single pulse in the NaOH electrolyte was higher compared to the KOH electrolyte due to higher discharge energy in the NaOH electrolyte. The film thickness formed in NaOH electrolyte is relatively higher compared to the KOH electrolyte, and thus the electrochemical discharges generated are more aggressive in the case of the NaOH electrolyte resulting in higher crater depth and MRR.

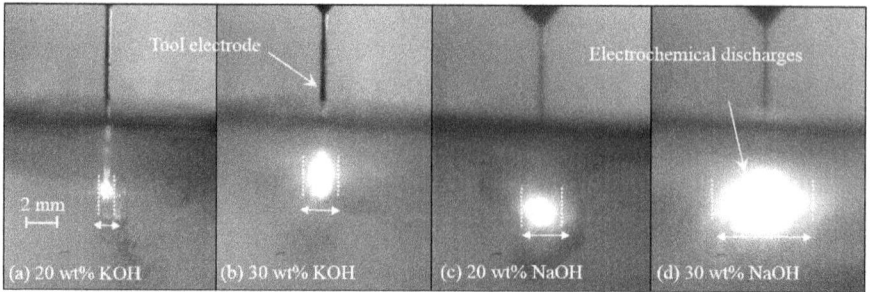

FIGURE 13.17 High-speed camera images for electrochemical discharges in KOH and NaOH electrolytes. Reproduced with permission from [52].

13.4 EFFECT OF THE ECDM PROCESS PARAMETERS

There are several process parameters that affect the ECDM process. The widely studied parameters are machining voltage, pulse frequency, duty cycle, electrolyte type, concentration, etc. Apart from these parameters, tool texture/roughness, tool–workpiece gap, and tool feed rate are also important parameters that affect the ECDM process and are discussed in the following section.

13.4.1 EFFECT OF THE TOOL ELECTRODE SURFACE ROUGHNESS

Tool electrode-related parameters affect dimensions as well as the surface quality of the microholes and microchannels in the ECDM process. Cylindrical and conical-shaped tool electrodes are predominantly utilized. Apart from these two shapes, the usage of flat sidewall and spherical tip yielded higher dimensional accuracy and machining efficiency [85, 86]. Nonetheless, the surface profile or textures are also equally important, which determine the machining performance [87, 88]. The tool electrode surface profile, which consists of peaks and valleys, affects the generation of gas bubbles and subsequently the gas film thickness. Singh and Dvivedi reported that textured tools generated thin and stable gas films resulting in uniform material removal, accuracy, and microchannel depth, as shown in Figure 13.19 [89]. The authors demonstrated that textured tools increase the MRR by 19% and microchannel depth by 65% compared to smooth surface tools [89].

Arab et al. demonstrated the influence of the surface roughness of the tool electrode on the hole overcut and the heat-affected zone (HAZ) width [48]. Tool electrodes of varying surface roughness were fabricated using wire-EDM and electrochemical finishing. A large amount of hydrogen gas bubbles of larger size and higher contact angles were produced on the tool electrode with higher surface roughness owing to the higher number of deeper valleys on the surface. These larger-sized gas bubbles merged together to form a gas film with a higher thickness which eventually led to microhole overcut with larger overcut as well as higher HAZs. Conversely, tool electrodes with a lower surface roughness resulted in a thin size gas film leading to

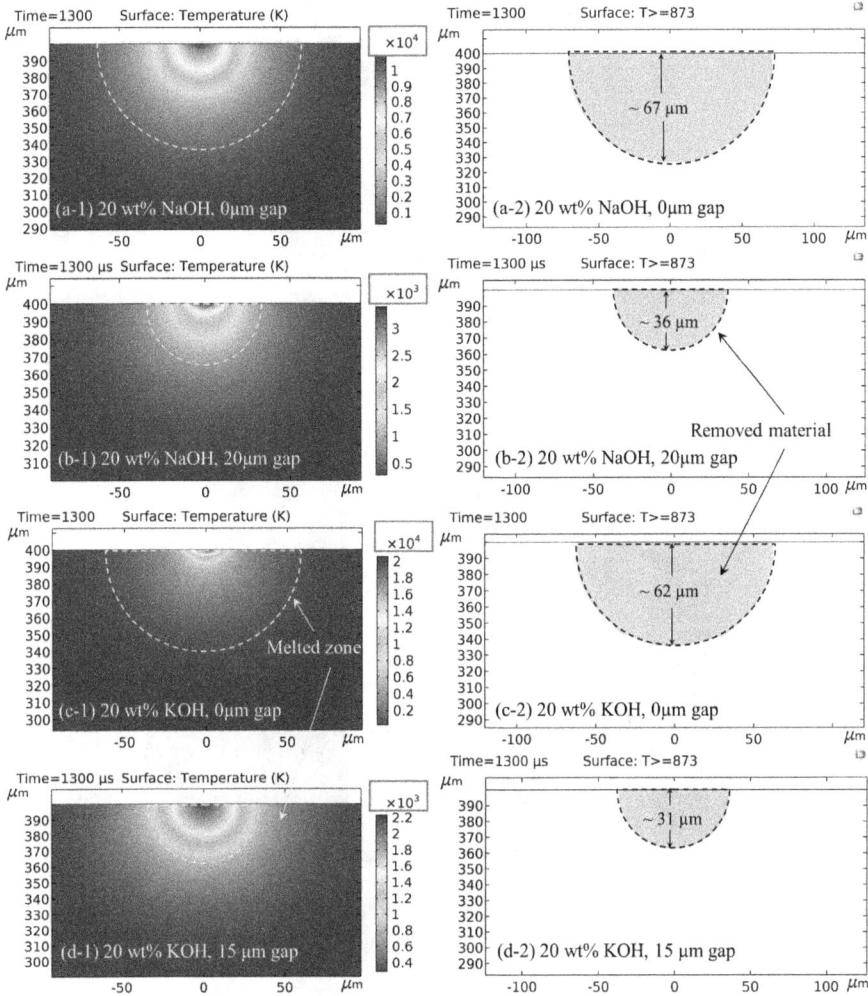

FIGURE 13.18 Temperature contours and crater profiles obtained using FEM simulation for KOH and NaOH electrolyte at zero and 20 μm gap. Reproduced with permission from [52].

electrochemical discharges of lower energy and lower mean current. These tool electrodes resulted in microholes with lower overcut and HAZ widths. The variation in the microhole size and HAZ for tool electrodes with higher and lower surface roughness are shown in Figure 13.20. Figure 13.20(a) and (d) show the rough and smooth tool electrodes, respectively. The average tool electrode length was 5 mm for both the tool electrodes. The surface textures of rough and smooth tools are depicted in Figure 13.20(b) and (e), respectively. As can be seen, the rough tool surface profile was observed to have hollow cavities and microcracks. The smooth tool surface was free of microcracks and microcavities. The microhole obtained with the rough and smooth

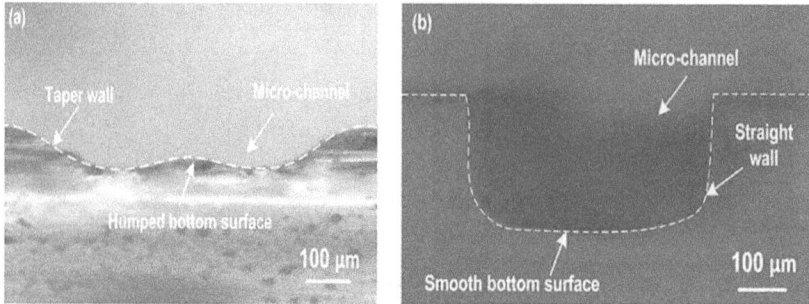

FIGURE 13.19 Images showing cross-section view of the microchannels fabricated using (a) smooth and (b) textured tool. Reproduced with permission from [89].

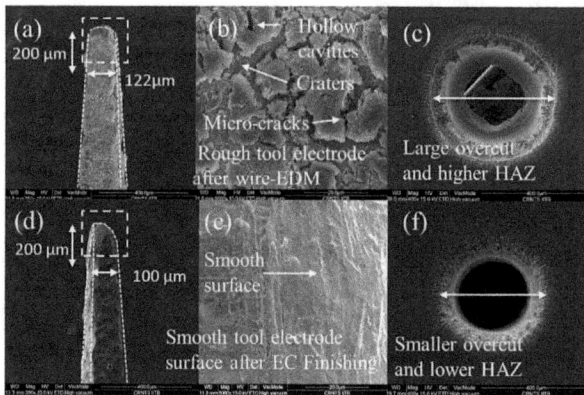

FIGURE 13.20 Tool electrodes (rough: a, smooth: d), Surface profiles of tools (rough: b, smooth: e), and microhole profiles obtained with rough (c) and smooth (f). Reproduced with permission from [48].

tools is shown in Figure 13.20(c) and (f), respectively. The microhole with a larger overcut and higher HAZ was made with the rough tool, whereas in the case of the smooth tool, the microhole was observed to have a smaller overcut and lower HAZ.

Yang et al. also reported the effect of the tool electrode surface roughness on ECDM machining characteristics using three different materials of the tool electrode and three different tool surface roughness values [8]. The authors observed that the tungsten carbide tool electrode with the least surface roughness produced a microhole of a smaller size compared to the other tool electrodes made from stainless steel and tungsten.

13.4.2 EFFECT OF THE TOOL-WORKPIECE GAP

The tool–workpiece gap is a critical parameter that directly affects the energy transferred into the workpiece during the ECDM process. The tool–workpiece gap should be <25 μm for effective heat transfer to the workpiece resulting in material removal [5]. Earlier reported works have primarily used a gravity feed mechanism to create microholes in the substrates. In the gravity feed mechanism, there is always a physical contact between the tool tip and the workpiece surface, thereby resulting in either the breakage of the substrate or drastic tool wear. Didar et al. reported that a tool–workpiece gap >15 μm resulted in shallow microchannel depth when the ECDM experiments were performed at a machining voltage of 30 V and the tool feed rate of 30 μm/s in a 30 wt.% NaOH electrolyte [7]. Ziki and Wüthrich used a force sensor to maintain the tool–workpiece gap during glass micro hole fabrication in a constant velocity-feed mechanism [90]. The authors first measured the force data obtained due to tool–workpiece contact during the downward feed of the tool electrode and subsequently retracted the tool intermittently to maintain the appropriate tool–workpiece gap throughout the process.

Mishra et al. investigated the effect of the tool–workpiece gap during the fabrication of the microchannel at varying tool feed rates, electrolyte type, and concentration [51]. The authors reported an increase in the microchannel width with an increase in the tool–workpiece gap up to a gap of 10 μm and 20 μm for KOH and NaOH electrolytes respectively, as shown in Figure 13.21.

Beyond the above-stated gap values, the microchannel width showed a decreasing trend. The microchannel's depth also showed a continuous decreasing trend with an increase in the tool–workpiece gap. With an increase in the tool–workpiece gap, the avalanching electrons collide with the workpiece on a relatively larger surface area. Although, at a higher gap, the net energy transferred into the workpiece reduces, which results in the reduction of the depth of the microchannel. The width of the microchannel increases up to a certain gap until the surface temperature is sufficient

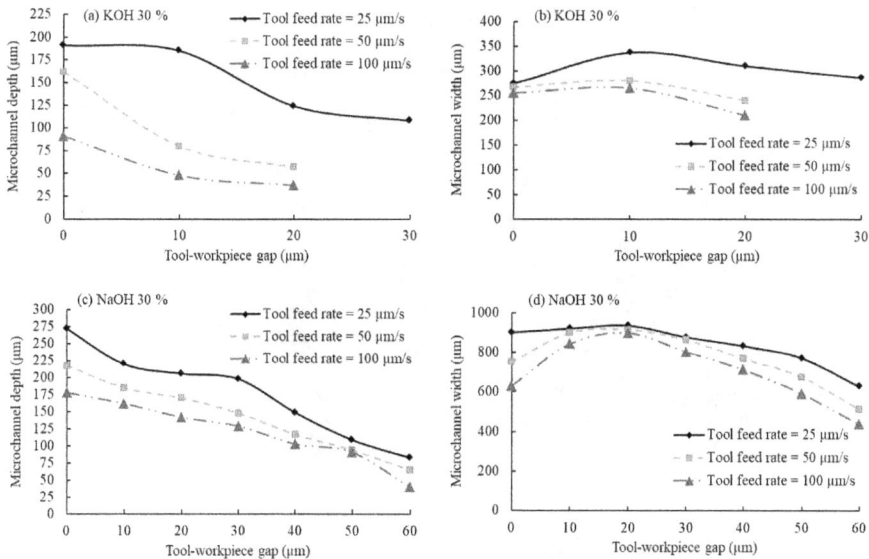

FIGURE 13.21 Effect of the tool–workpiece gap on the microchannel depth and width.

to remove the material. After the certain gap, the microchannel width also starts decreasing due to a reduction in the transferred heat energy (Figure 13.21). Moreover, after a gap of 30 μm and 60 μm for the KOH and NaOH electrolytes, respectively, no machining was observed because negligible electrons collide with the workpiece surface beyond these gaps while moving toward the counter electrode.

Arab et al. reported that the tool wear and hole depth reduced as the tool–workpiece gap was increased. The maximum gap up to which material removal occurs depends on the machining voltage, and electrolyte type, and concentration [52]. The optical images of microholes for varying tool–workpiece gaps with NaOH and KOH electrolytes are shown in Figure 13.22 [52].

In another work, Arab et al. reported a reduction in the overcut of the machined holes as the gap was increased [91]. From the above literature, it is clear that the tool–workpiece gap is an important parameter in the ECDM process, and the gap should be 10–20 μm to have higher material removal as well reduced tool wear.

13.4.2 EFFECT OF THE TOOL FEED RATE

Tool feed rate (TFR) is the magnitude of tool movement per unit of time. TFR is an important parameter in the ECDM process with a velocity-feed type system, where the tool electrode moves in a vertical and horizontal plane for the microhole drilling and milling process, respectively. Arab et al. demonstrated the effect of the TFR on the microholes drilled by the velocity-feed type ECDM [50]. Stable and uniform

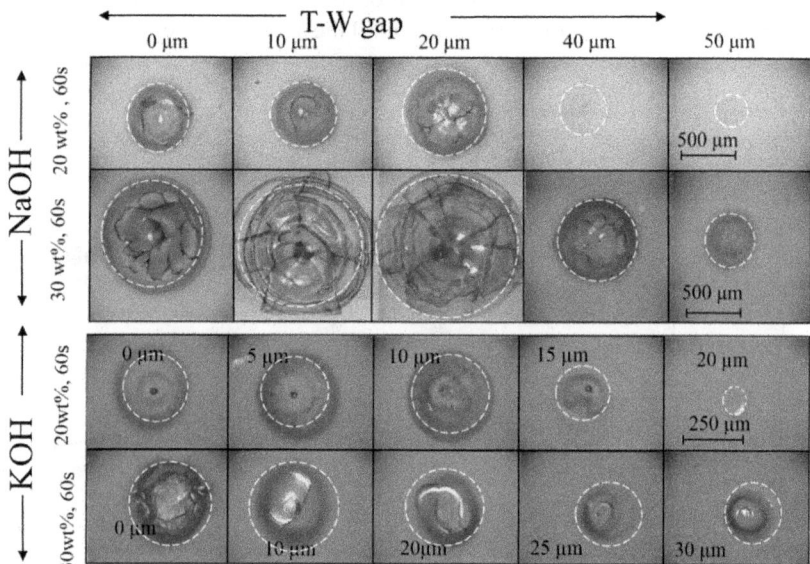

FIGURE 13.22 Microhole profiles for varying tool–workpiece gaps with NaOH and KOH electrolytes. Reproduced with permission from [52].

EC discharges were observed for the lower TFR (<5 µm/s) compared to higher TFR (≥5 µm/s) where non-uniform as well as unstable EC discharges with intermittent gas film formation were observed. The HAZs of the fabricated microholes were found to be relatively lower at the median value of TFR (2–4 µm/s). A median range of TFR was suggested for achieving optimum microhole size. At higher TFR (≥5 µm/s) values, the hole eccentricity was found to be severe. The circularity of the top and bottom holes was observed to decrease with increasing TFR values. Microhole profiles indicating the variation in overcut, circularity, and eccentricity are shown in Figure 13.23.

In the case of microchannel formation, researchers have shown the effect of the TFR on the dimensional and surface quality characteristics of the microchannels in the glass. Mishra et al. investigated the influence of TFR to fabricate precise microchannels [53]. It was noticed that the process gives higher material removal with optimum surface quality at faster TFR values of 100 µm/s for the NaOH electrolyte and 50 µm/s for the KOH electrolyte, respectively. Saranya et al. have utilized a side-insulated tool electrode and observed that higher microchannel depth was accomplished for the tool feed of 8.3 µm/s [92]. Singh et al. have created microchannels with a surface textured tool electrode to achieve higher material removal with lower microchannel width at a TFR value of 12 mm/min [89].

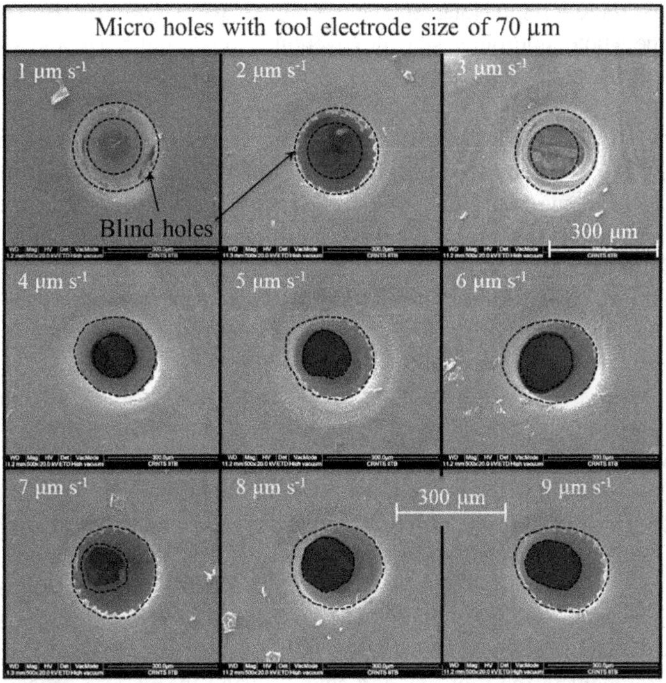

FIGURE 13.23 Microhole profiles for varying TFR values (1–9 µm/s). Reproduced with permission from [50].

13.5 FABRICATION OF MICROFEATURES USING THE ECDM PROCESS

The ECDM process can be used to create microholes and microchannels in electrically non-conductive substrates. Compared to deeper microholes, fabricating deeper microchannels using the ECDM process is relatively easier due to the sufficient availability of the electrolyte. Further, using the multi-pass approach, tool wear and the chance of tool–workpiece coming into contact are reduced, and the subsequent tool bending or glass breakage is also minimized. In this section, there is a detailed description of the creation of microfeatures using both the single and multi-tip tool electrodes.

13.5.1 MICROHOLE FORMATION USING SINGLE AND AREA ARRAY-TOOL ELECTRODE

Microholes in glass are widely required for various MEMS applications. In MEMS packaging, through-glass vias are required for the packaging of RF-MEMS devices [93]. Moreover, microholes are also needed in many microfluidic and lab-on-a-chip devices. The ECDM process is capable of fabricating precise microholes with appropriate dimensional accuracy [70]. Many researchers have explored the formation of a single microhole with various single-tip tool electrodes with different shapes and sizes. On the whole, microhole size and overcut depend on the size and shape of the tool electrode as well as input processing parameters such as machining voltage, tool feed, etc.

A few researchers have also used improvisations in the ECDM process to enhance the MRR without compromising the quality and precision of microholes. Researchers have suggested that usage of special shaped tool electrodes such as spherical and flat sidewall tools resulted in a higher material etch rate. Yang et al. [85] and Saranya et al. [94] have illustrated that usage of the spherical tool electrode reduced the time of machining of microholes which eventually improves the material etch rate, Figure 13.24. Electrolyte type and concentration are important factors that influence the opening size of the microholes in the glass.

Yang et al. have used different types of electrolytes to fabricate microholes in the borosilicate glass [33]. Neutral electrolytes like NaCl and KCl resulted in negligible material removal, which increased the machining time. Alkaline electrolytes (NaOH and KOH) were found to have significant material removal.

Recently Dixit et al. have utilized the customized area array or multi-tip tool electrode, which improvised the productivity of the ECDM compared to drilling the holes one by one with a single-tip electrode [60]. High aspect ratio and customized array-tool tips in stainless steel of 15 mm length, and a tool tip size of 150 µm was machined using the wire-EDM. In order to minimize hydrogen gas bubble entrapment in the tool tips and shank during the ECDM process, the tool tip dimensions were optimized. Experimental results suggested that the tool tips with a larger tip length were required to minimize the bubble entrapment, which led to stable machining with less hindrance. Experiments have shown that lower

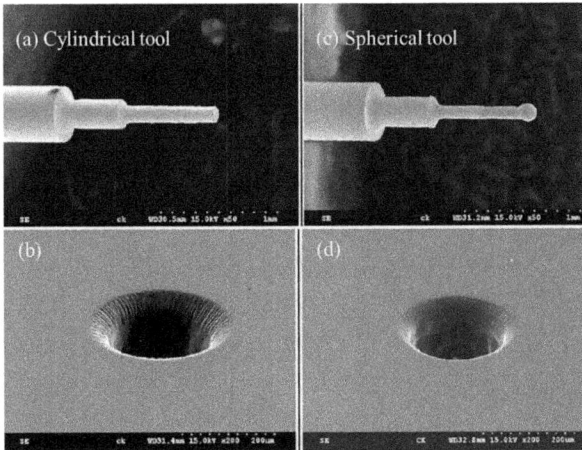

FIGURE 13.24 Fabricated microholes using spherical and flat sidewall tools. Reproduced with permission from [85].

viscosity and higher ion mobility are the two key aspects that resulted in superior microhole quality with lower overcut for KOH electrolyte compared to NaOH electrolyte. The effects of different combinations of pulse frequency and duty ratio on the top hole diameter, overcut, and the HAZ ware analyzed. A 3×3 array of through holes with vertical depth up to 520 μm were drilled in less than 10 minutes, which resulted in a higher etch rate than plasma etching and laser drilling. The machining process capability was demonstrated by the formation of a 2×5, 5×5 array of through holes in less than 10 minutes with the KOH electrolyte, as shown in Figure 13.25.

13.5.2 MICROCHANNEL FORMATION USING THE ECDM PROCESS

Microchannel formation in silica-based substrates has applications in biomedical and microfluidic devices. Fabrication of deeper microchannels can be achieved by two different methods of the tool motion in the X–Y plane. In the first method, initially, a drilling operation is performed up to the required depth of the microchannel, and then the tool electrode is moved in the X–Y plane for the desired channel length and channel configuration. In this case, there is always a glass workpiece ahead of the tool motion direction, and the maximum feed rate that can be used is always <3 μm/s; otherwise, mechanical tool–workpiece contact damages the glass workpiece or bends the tool tip while moving in the X–Y plane.

In the second method, a multi-pass technique is used in which there is always some gap (10–15 μm) between the tool electrode tip and the glass workpiece surface, as shown in Figure 13.26. Therefore, much higher tool feed rates (>50 μm) can be used to fabricate deeper and longer microchannels [95]. Moreover, in the second method, the tool wear is relatively lower compared to the first method due to the

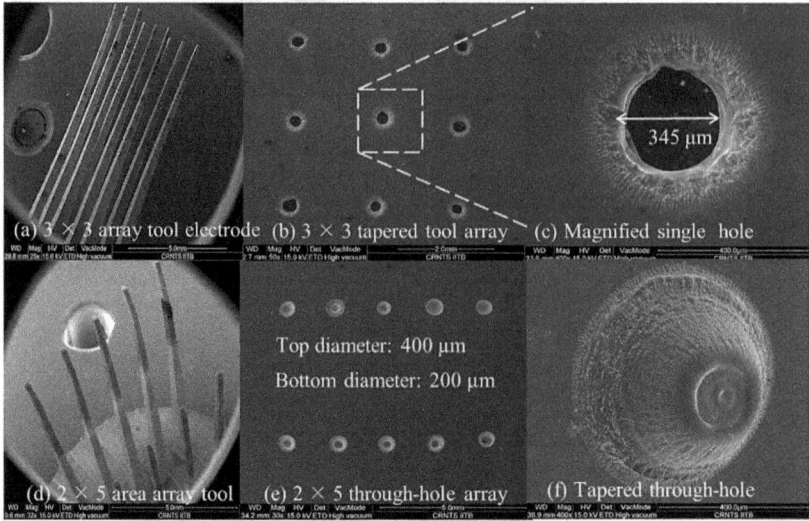

FIGURE 13.25 3 × 3 and 2 × 5 multi-tip array tools and fabricated arrayed microholes. Reproduced with permission from [60].

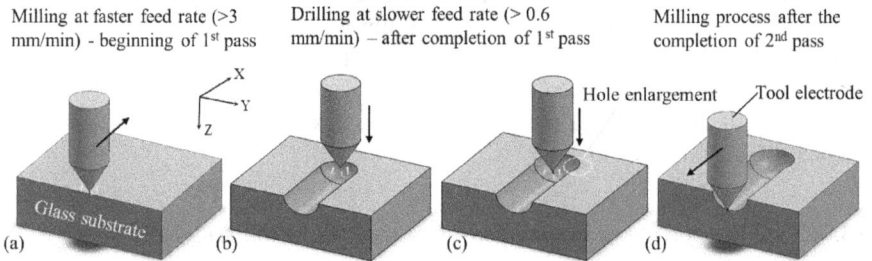

FIGURE 13.26 Schematic of multi-pass micromilling approach. Reproduced with permission from [95].

continuous and sufficient availability of the electrolyte around the tool electrode surface, resulting in a relatively higher heat dissipation rate.

As described earlier, the multi-pass technique has a lower machining time and tool wear rate to create deeper microchannels. Mishra et al. [95] fabricated through microchannels in a 400 μm-thick glass substrate using the multi-pass technique as shown in Figure 13.27. The authors reported a reduction in the number of passes required to create a through microchannel as the machining voltage was increased due to higher discharge energy and MRR at higher machining voltage.

Later, the authors also studied the effect of electrolyte concentration and the number of passes needed to create deeper microchannels in 1.1 mm-thick glass substrates.

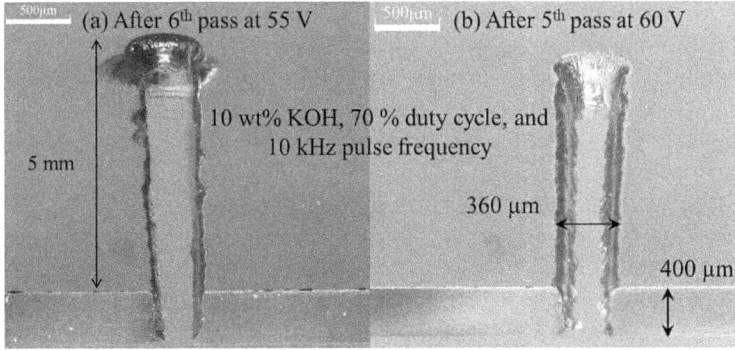

FIGURE 13.27 Fabrication of through microchannels (a) After 6th pass at 55 V and (b) after 5th pass at 60 V. Reproduced with permission from [95].

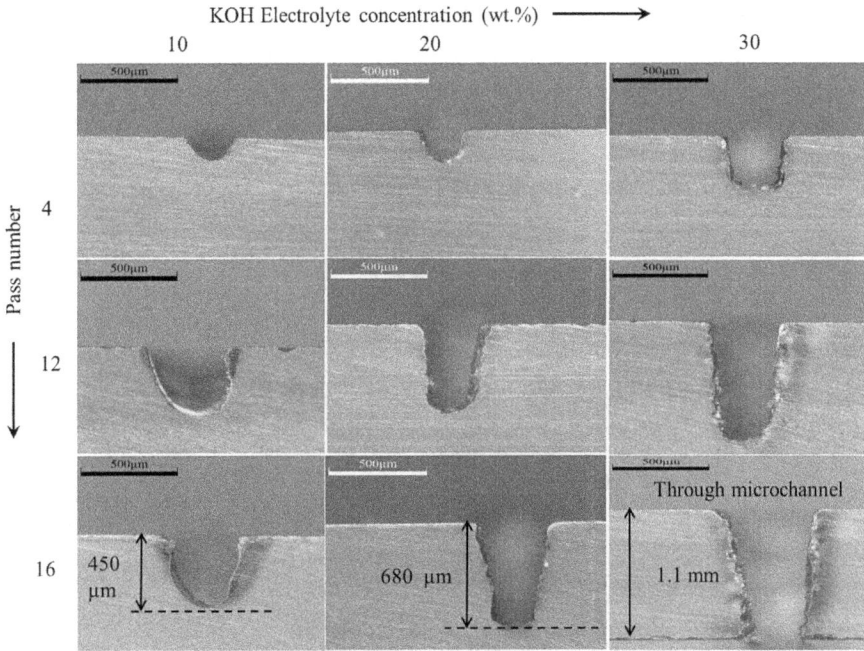

FIGURE 13.28 Optical image showing variation in the microchannel depth with varying passes and electrolyte concentrations. Reproduced with permission from [95].

Figure 13.28 shows the cross-section images of microchannels fabricated at varying KOH electrolyte concentrations and pass numbers.

The microchannel depth showed an increasing trend with an increase in the electrolyte concentration. As the electrolyte concentration increases, both the rate of chemical etching and the rate of electrochemical reaction increase, leading to a rise in the MRR and microchannel depth. Through microchannels with depths of >1100 μm were fabricated with 30% KOH electrolyte concentration in the 16th pass compared to a 680 μm channel depth obtained with 20% KOH electrolyte. Therefore, to create deeper microchannels, higher electrolyte concentration is recommended along with the multi-pass technique to reduce the tool wear and machining time.

Similar to microhole formation, fabrication of microchannels in the glass substrates can be achieved using either a single-tip or multi-tip line array-tool electrode. Fabrication of microchannels using multi-tip line array-tool electrodes increases the overall productivity of the ECDM process. Customized line array-tool electrodes can be made using wire-EDM with varying tip numbers, tip size, and pitch. Mishra et al. fabricated 1×10, 1×5, 1×2 with different tip sizes using the wire-EDM process as shown in Figure 13.29(a–d) [96].

Later, the authors used line array-tool electrodes in the ECDM process and demonstrated the fabrication of deeper/through array microchannels of different shapes, i.e., straight, spiral, and zigzag, using the customized line array-tool electrodes as shown in Figures 13.30 and 13.31. Therefore, depending on the requirement, a single-tip or line-array-tip tool electrode can be used to fabricate microchannels in the non-conductive substrates using the ECDM process. Further, for deeper microchannels, a multi-pass technique that has a lower tool wear rate and lower machining time is more suitable.

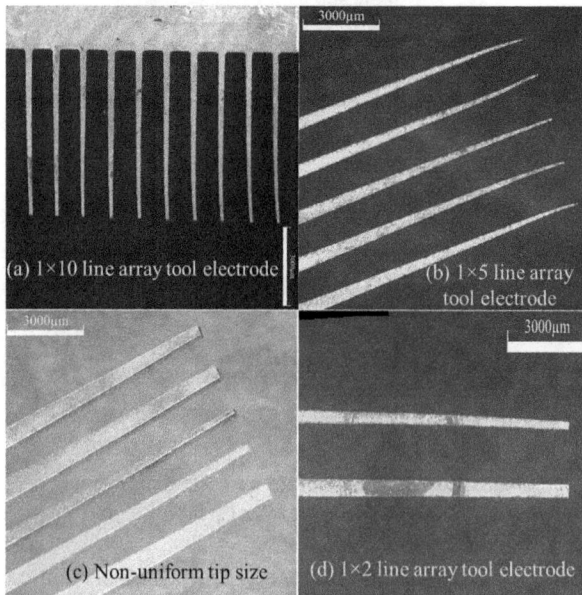

FIGURE 13.29 Fabrication of customized line-array tool electrodes using the wire-EDM process. Reproduced with permission from [96].

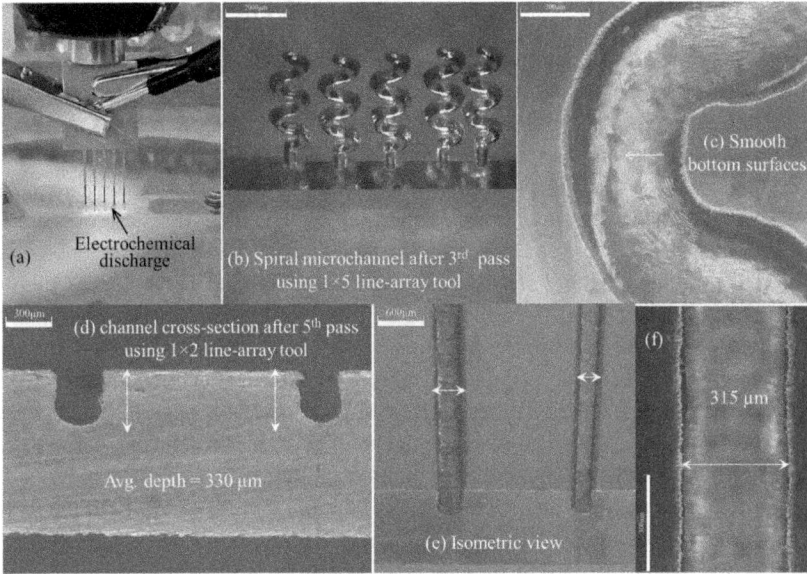

FIGURE 13.30 Fabrication of spiral and straight microchannels using the 1 × 5 and 1 × 2 line array-tool electrodes and multi-pass technique.

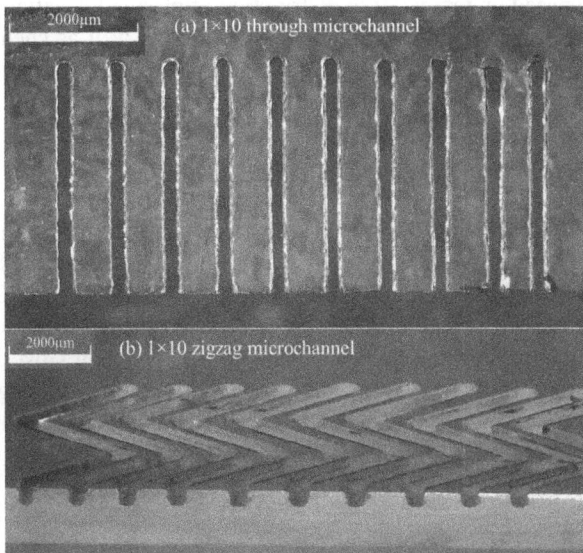

FIGURE 13.31 Fabrication of (a) through and (b) zigzag microchannels using 1 × 10 line-array tool electrodes. Reproduced with permission from [96].

13.6   CONCLUSIONS AND FUTURE SCOPE OF WORK

Over the last couple of decades, ECDM has emerged as a cost-effective and faster process to fabricate microfeatures in electrically non-conductive substrates, especially silica-based workpieces. The process can be used to fabricate microholes and micro-channels in glass substrates that have application in MEMS packaging and micro-fluidic devices. The ECDM process involves several complex physics phenomena such as bubble formation, gas film formation, electrochemical discharge generation, and chemical etching. A clear understanding of the physics of these phenomena is required to comprehend the ECDM process clearly and develop a complete mathematical/numerical model. Apart from the effects of the power supply parameters and the electrolyte type and concentration, a detailed understanding of the effect of the tool–workpiece gap, tool feed rate, and the tool texture/roughness is needed for effective utilization of the process. Increasing productivity and reducing the overall cost during the ECDM process can be achieved by using the array-tool electrodes wherever required.

Nevertheless, sludge removal during the fabrication of deep microfeatures is still very challenging due to an insufficient supply of electrolyte, especially in the formation of a hole. In situ and precise measurement of the tool–workpiece gap is yet to be explored. Tool wear during the ECDM process hampers the dimensional accuracy and increases the tool–workpiece gap. Therefore, a detailed study on tool wear is still an open area during the ECDM process. Further, micromachining of "difficult-to-cut materials" with high melting points such as silicon carbide is not well investigated.

The numerical/mathematical model developed so far does not include the contribution of the chemical etching phenomenon. Most of the numerical models developed so far are based on temperature distribution or plasma formation. Therefore, a model that can incorporate melting and vaporization, as well as chemical etching, is still an open research area. Further, fabrication of high aspect ratio microholes, i.e., >5, using the array-tool electrode has been a challenge and needs to be explored.

ACKNOWLEDGMENT

Pradeep Dixit would like to acknowledge the financial support from the Ministry of Human Resources and Development (MHRD) and the Department of Scientific and Industrial Research (DSIR) through (DSIR/PACE/TDDIMPRINT/7510). The majority of this work has been carried out as a part of the Impacting Research Innovation and Technology (Imprint) Project, initiated by MHRD, Indian Government, under the research grant 10007457. Financial support is also acknowledged from the Department of Science and Technology ((DST), Government of India, under the research grant through (DST/TDT/AMT/2017/181(G)). Julfekar Arab is supported by IIT Bombay – Institute Postdoctoral Fellowship.

ACRONYMS

AJM	Abrasive jet machining
CM	Chemical machining

CNC	Computer numerical control
DRIE	Deep reactive ion etching
ECDM	Electrochemical discharge machining
ECM	Electrochemical machining
ECSM	Electrochemical spark machining
EDM	Electric discharge machining
HAZ	Heat-affected zone
IBM	Ion beam machining
KOH	Potassium hydroxide
LBM	Laser beam machining
MRR	Material removal rate
NaOH	Sodium hydroxide
RF-MEMS	Radio-frequency microelectromechanical systems
SACE	Spark-assisted chemical engraving
TFR	Tool feed rate
TGVs	Through-glass vias
TSVs	Through-silicon vias
TW-ECSM	Traveling wire-ECSM
USM	Ultrasonic machining
W-EDM	Wire-EDM

SYMBOLS

α, β	Advancing and receding contact angles
σ	Surface tension between the liquid-gas interface
κ_e	Electrical conductivity
κ	Boltzmann constant
θ	Contact angle
ρ_e	Density of the electrolyte
ρ_g	Density of the gas bubble
$e\phi$	Work function
d	Mean departure diameter
n^+	Number density of the cations
n^-	Number density of the anions
r_c	Critical radius
u^+	Mobility of cation
u^-	Mobility of anion
z^+	Charge of a cation
z^-	Charge of an anion
C_s	Coefficient for the degree of supersaturation
D	Diffusion coefficient
E	Electric field strength
E_F	Fermi level of the tool electrode
F_p	Bubble pressure force
F_{TL}, F_{NL}	Inertia forces of the growing bubble

F_{SX}, F_{SY}	Surface tension forces
G_o	Standard Gibbs free energy
J	Rate of nucleation
J_e	Current density
P_b	Pressure inside the bubble
P_l	Total liquid pressure
R	Universal gas constant
R_b	Bubble radius
R_d	Bubble departure radius
R_{etch}	Rate of chemical etching of the glass
T	Temperature
V_{bubble}	Volume of a gas bubble
Z	Pre-exponential frequency

REFERENCES

1. Gautam, N.; Jain, V. K. Experimental Investigations into ECSD Process Using Various Tool Kinematics. *Int. J. Mach. Tools Manuf.*, 1998, *38* (1–2), 15–27.
2. Dixit, P.; Miao, J. High Aspect Ratio Vertical Through-Vias for 3D MEMS Packaging Applications by Optimized Three-Step Deep RIE. *J. Electrochem. Soc.*, 2007, *155* (2), H85.
3. Kolari, K.; Saarela, V.; Franssila, S. Deep Plasma Etching of Glass for Fluidic Devices With Different Mask Materials. 064010. https://doi.org/10.1088/0960-1317/18/6/064010.
4. Hou, X.; Zhang, Y. S.; Trujillo-de Santiago, G.; Alvarez, M. M.; Ribas, J.; Jonas, S. J.; Weiss, P. S.; Andrews, A. M.; Aizenberg, J.; Khademhosseini, A. Interplay Between Materials and Microfluidics. *Nat. Rev. Mater.*, 2017, *2* (5), 1–15.
5. Wüthrich, R.; Fascio, V. Machining of Non-Conducting Materials Using Electrochemical Discharge Phenomenon—An Overview. *Int. J. Mach. Tools Manuf.*, 2005, *45* (9), 1095–1108.
6. Ziki, J. D. A.; Didar, T. F.; Wüthrich, R. Micro-Texturing Channel Surfaces on Glass With Spark Assisted Chemical Engraving. *Int. J. Mach. Tools Manuf.*, 2012, *57*, 66–72.
7. Didar, T. F.; Dolatabadi, A.; Wüthrich, R. Characterization and Modeling of 2D-Glass Micro-Machining by Spark-Assisted Chemical Engraving (SACE) With Constant Velocity. *J. Micromech. Microeng.*, 2008, *18* (6), 65016.
8. Yang, C.-K.; Cheng, C.-P.; Mai, C.-C.; Wang, A. C.; Hung, J.-C.; Yan, B.-H. Effect of Surface Roughness of Tool Electrode Materials in ECDM Performance. *Int. J. Mach. Tools Manuf.*, 2010, *50* (12), 1088–1096.
9. Singh, T.; Dvivedi, A. Developments in Electrochemical Discharge Machining: A Review on Electrochemical Discharge Machining, Process Variants and Their Hybrid Methods. *Int. J. Mach. Tools Manuf.*, 2016, *105*, 1–13.
10. Jain, V. K.; Adhikary, S. On the Mechanism of Material Removal in Electrochemical Spark Machining of Quartz under Different Polarity Conditions. *J. Mater. Process. Technol.*, 2008, *200* (1–3), 460–470.
11. Crichton, I. M.; McGeough, J. A. Studies of the Discharge Mechanisms in Electrochemical Arc Machining. *J. Appl. Electrochem.*, 1985, *15* (1), 113–119.
12. Basak, I.; Ghosh, A. Mechanism of Material Removal in Electrochemical Discharge Machining: A Theoretical Model and Experimental Verification. *J. Mater. Process. Technol.*, 1997, *71* (3), 350–359.

13. Goud, M.; Sharma, A. K.; Jawalkar, C. A Review on Material Removal Mechanism in Electrochemical Discharge Machining (ECDM) and Possibilities to Enhance the Material Removal Rate. *Precis. Eng.*, 2016, *45*, 1–17.

14. Kurafuji, H. Electrical Discharge Drilling of Glass-I. *Ann. CIRP*, 1968, *16*, 415.

15. Tandon, S.; Jain, V. K.; Kumar, P.; Rajurkar, K. P. Investigations into Machining of Composites. *Precis. Eng.*, 1990, *12* (4), 227–238.

16. Jain, V. K.; Tandon, S.; Kumar, P. Experimental Investigations into Electrochemical Spark Machining of Composites. *J. Eng. Ind.*, 1990, *112* (2), 194–197.

17. Jain, V. K.; Sreenivasa Rao, P.; Choudhary, S. K.; Rajurkar, K. P. Experimental Investigations into Traveling Wire Electrochemical Spark Machining (TW-ECSM) of Composites. *J. Eng. Ind.*, 1991, *113* (1), 75–84.

18. Sharma, P.; Arab, J.; Dixit, P. Through-Holes Micromachining of Alumina Using a Combined Pulse-Feed Approach in ECDM. *Mater. Manuf. Process.*, 2021,36, 1501–1512.

19. Jain, V. K.; Choudhury, S. K.; Ramesh, K. M. On the Machining of Alumina and Glass. *Int. J. Mach. Tools Manuf.*, 2002, *42* (11), 1269–1276.

20. Bhattacharyya, B.; Doloi, B. N.; Sorkhel, S. K. Experimental Investigations into Electrochemical Discharge Machining (ECDM) of Non-Conductive Ceramic Materials. *J. Mater. Process. Technol.*, 1999, *95* (1–3), 145–154.

21. Jain, V. K.; Chak, S. K. Electrochemical Spark Trepanning of Alumina and Quartz. *Mach. Sci. Technol.*, 2000, *4* (2), 277–290.

22. Arab, J.; Dixit, P. Formation of macro-sized through-holes in glass using notch-shaped tubular electrodes in electrochemical discharge machining. *J. Manuf.. Process.* 2022, *78*, 92–106.

23. Singh, Y. P.; Jain, V. K.; Kumar, P.; Agrawal, D. C. Machining Piezoelectric (PZT) Ceramics Using an Electrochemical Spark Machining (ECSM) Process. *J. Mater. Process. Technol.*, 1996, *58* (1), 24–31.

24. Tokura, H.; Kondoh, I.; Yoshikswa, M. Ceramic Material Processing by Electrical Discharge in Electrolyte. *J. Mater. Sci.*, 1989, *24* (3), 991–998.

25. He, S.; Tong, H.; Liu, G. Spark Assisted Chemical Engraving (SACE) Mechanism on ZrO_2 Ceramics by Analyzing Processed Products. *Ceram. Int.*, 2018, *44* (7), 7967–7971.

26. Ji, B.; Tong, H.; Li, J.; Li, Y.; Xu, M. Scanning Process of Spark Assisted Chemical Engraving (SACE) on ZrO_2 Ceramics by Constraining Discharges to Tool Electrode End. *Ceram. Int.*, 2020, *46* (2), 1433–1441.

27. Bajpai, V.; Mishra, D.K.; Dixit, P. Fabrication of Through-glass Vias (TGV) based 3D microstructures in Glass Substrate by a lithography-free process for MEMS applications. *Appl. Surf. Sci.* 2022, *582*, 152494.

28. Coteaţă, M.; Schulze, H.-P.; Slătineanu, L. Drilling of Difficult-to-Cut Steel by Electrochemical Discharge Machining. *Mater. Manuf. Process.*, 2011, *26* (12), 1466–1472.

29. Huang, S. F.; Liu, Y.; Li, J.; Hu, H. X.; Sun, L. Y. Electrochemical Discharge Machining Micro-Hole in Stainless Steel With Tool Electrode High-Speed Rotating. *Mater. Manuf. Process.*, 2014, *29* (5), 634–637.

30. Krötz, H.; Roth, R.; Wegener, K. Experimental Investigation and Simulation of Heat Flux into Metallic Surfaces Due to Single Discharges in Micro-Electrochemical Arc Machining (Micro-ECAM). *Int. J. Adv. Manuf. Technol.*, 2013, *68* (5–8), 1267–1275.

31. Basak, I.; Ghosh, A. Mechanism of Spark Generation During Electrochemical Discharge Machining: A Theoretical Model and Experimental Verification. *J. Mater. Process. Technol.*, 1996, *62* (1–3), 46–53.

32. Jain, V. K.; Dixit, P. M.; Pandey, P. M. On the Analysis of the Electrochemical Spark Machining Process. *Int. J. Mach. Tools Manuf.*, 1999, *39* (1), 165–186.

33. Yang, C. T.; Ho, S. S.; Yan, B. H. Micro Hole Machining of Borosilicate Glass Through Electrochemical Discharge Machining (ECDM). In *Key Engineering Materials*; Trans Tech Publ, 2001; Vol. 196, pp. 149–166.

34. Wüthrich, R.; Hof, L. A. The Gas Film in Spark Assisted Chemical Engraving (SACE)—A Key Element for Micro-Machining Applications. *Int. J. Mach. Tools Manuf.*, 2006, *46* (7–8), 828–835.

35. Fascio, V.; Wüthrich, R.; Bleuler, H. Spark Assisted Chemical Engraving in the Light of Electrochemistry. *Electrochim. Acta*, 2004, *49* (22–23), 3997–4003.

36. Singh, M.; Singh, S.; Kumar, S. Investigating the Impact of LASER Assistance on the Accuracy of Micro-Holes Generated in Carbon Fiber Reinforced Polymer Composite by Electrochemical Discharge Machining. *J. Manuf. Process.*, 2020, *60*, 586–595.

37. Zhao, D.; Zhang, Z.; Zhu, H.; Cao, Z.; Xu, K. An Investigation into Laser-Assisted Electrochemical Discharge Machining of Transparent Insulating Hard-Brittle Material. *Micromachines*, 2021, *12* (1), 22.

38. Elhami, S.; Razfar, M. R. Effect of Ultrasonic Vibration on the Single Discharge of Electrochemical Discharge Machining. *Mater. Manuf. Process.*, 2018, *33* (4), 444–451.

39. Singh, T.; Dvivedi, A.; Shanu, A.; Dixit, P. Experimental Investigations of Energy Channelization Behavior in Ultrasonic Assisted Electrochemical Discharge Machining. *J. Mater. Process. Technol.*, 2021, 117084.

40. Elhami, S.; Razfar, M. R. Study of the Current Signal and Material Removal During Ultrasonic-Assisted Electrochemical Discharge Machining. *Int. J. Adv. Manuf. Technol.*, 2017, *92* (5), 1591–1599.

41. Wüthrich, R.; Despont, B.; Maillard, P.; Bleuler, H. Improving the Material Removal Rate in Spark-Assisted Chemical Engraving (SACE) Gravity-Feed Micro-Hole Drilling by Tool Vibration. *J. Micromech. Microeng.*, 2006, *16* (11), N28.

42. Han, M.-S.; Min, B.-K.; Lee, S. J. Geometric Improvement of Electrochemical Discharge Micro-Drilling Using an Ultrasonic-Vibrated Electrolyte. *J. Micromech. Microeng.*, 2009, *19* (6), 65004.

43. Hajian, M.; Razfar, M. R.; Movahed, S. An Experimental Study on the Effect of Magnetic Field Orientations and Electrolyte Concentrations on ECDM Milling Performance of Glass. *Precis. Eng.*, 2016, *45*, 322–331.

44. Cheng, C.-P.; Wu, K.-L.; Mai, C.-C.; Hsu, Y.-S.; Yan, B.-H. Magnetic Field-Assisted Electrochemical Discharge Machining. *J. Micromech. Microeng.*, 2010, *20* (7), 75019.

45. Xu, Y.; Chen, J.; Jiang, B.; Liu, Y.; Ni, J. Experimental Investigation of Magnetohydrodynamic Effect in Electrochemical Discharge Machining. *Int. J. Mech. Sci.*, 2018, *142*, 86–96.

46. Cao, X. D.; Kim, B. H.; Chu, C. N. Micro-Structuring of Glass With Features Less Than 100 Mm by Electrochemical Discharge Machining. *Precis. Eng.*, 2009, *33* (4), 459–465.

47. Patro, S. K.; Mishra, D. K.; Arab, J.; Dixit, P. Numerical and Experimental Analysis of High-Aspect-Ratio Micro-Tool Electrode Fabrication Using Controlled Electrochemical Machining. *J. Appl. Electrochem.*, 2020, *50* (2), 169–184.

48. Arab, J.; Kannojia, H. K.; Dixit, P. Effect of Tool Electrode Roughness on the Geometric Characteristics of Through-Holes Formed by ECDM. *Precis. Eng.*, 2019, *60*, 437–447.

49. Singh, T., Mishra, D.K.; Dixit, P. Effect of pulse frequency and duty cycle on electrochemical dissolution behavior of multi-tip array tool electrode for reusability in the ECDM process. *J. Appl. Electrochem.* 2022, 52, 667–682.

50. Arab, J.; Dixit, P. Influence of Tool Electrode Feed Rate in the Electrochemical Discharge Drilling of a Glass Substrate. *Mater. Manuf. Process.*, 2020, 35, 1749–1760.

51. Mishra, D. K.; Pawar, K.; Dixit, P. Effect of Tool Electrode-Workpiece Gap in the Microchannel Formation by Electrochemical Discharge Machining. *ECS J. Solid State Sci. Technol.*, 2020, *9* (3), 34011.

52. Arab, J.; Mishra, D. K.; Dixit, P. Measurement and Analysis of the Geometric Characteristics of Microholes and Tool Wear for Varying Tool-Workpiece Gaps in Electrochemical Discharge Drilling. *Measurement*, 2021, *168*, 108463.

53. Mishra, D. K.; Verma, A. K.; Arab, J.; Marla, D.; Dixit, P. Numerical and Experimental Investigations into Microchannel Formation in Glass Substrate Using Electrochemical Discharge Machining. *J. Micromech. Microeng.*, 2019, *29* (7), 75004.

54. Arab, J.; Pawar, K.; Dixit, P. Effect of tool-electrode material in through-hole formation using ECDM process. *Mater. Manuf. Process.* 2021, *36*, 1019–1027.

55. Mishra, D.K.; Dixit, P. Experimental investigation into tool wear behaviour of line-array tool electrode during the electrochemical discharge micromilling process. *J. Manuf. Process.* 2021, *72* (12), 93–104.

56. Singh, T.; Arab, J.; Dixit, P. A review on microhole formation in glass-based substrates by Electrochemical discharge drilling for MEMS applications. *Mach. Sci. Tech.* 2022, *26* (2), 276–337.

57. Esashi, M.; Matsumoto, Y.; Shoji, S. Absolute Pressure Sensors by Air-Tight Electrical Feedthrough Structure. *Sensors Actuators A Phys.*, 1990, *23* (1–3), 1048–1052.

58. Li, X.; Ren, Y.; Wei, Z.; Liu, Y. Development of Ultrasonic Vibration Assisted Micro Electrochemical Discharge Machining Tool. *Recent Patents Mech. Eng.*, 2019, *12* (4), 313–325.

59. Arab, J.; Adhale, P.; Mishra, D. K.; Dixit, P. Micro Array Hole Formation in Glass Using Electrochemical Discharge Machining. *Procedia Manuf.*, 2019, *34*, 349–354.

60. Arab, J.; Mishra, D. K.; Kannojia, H. K.; Adhale, P.; Dixit, P. Fabrication of Multiple Through-Holes in Non-Conductive Materials by Electrochemical Discharge Machining for RF MEMS Packaging. *J. Mater. Process. Technol.*, 2019, *271*, 542–553.

61. Kannojia, H. K.; Arab, J.; Pegu, B. J.; Dixit, P. Fabrication and Characterization of Through-Glass Vias by the ECDM Process. *J. Electrochem. Soc.*, 2019, *166* (13), D531–D538.

62. Xu, L.; Dixit, P.; Miao, J.; Pang, J. H. L.; Zhang, X.; Tu, K. N.; Preisser, R. Through-Wafer Electroplated Copper Interconnect With Ultrafine Grains and High Density of Nanotwins. *Appl. Phys. Lett.*, 2007, *90* (3), 33111.

63. Dixit, P.; Xu, L.; Miao, J.; Pang, J. H. L.; Preisser, R. Mechanical and Microstructural Characterization of High Aspect Ratio Through-Wafer Electroplated Copper Interconnects. *J. Micromech. Microeng.*, 2007, *17* (9), 1749.

64. Dixit, P.; Miao, J.; Preisser, R. Fabrication of High Aspect Ratio 35 Mm Pitch Through-Wafer Copper Interconnects by Electroplating for 3-D Wafer Stacking. *Electrochem. Solid State Lett.*, 2006, *9* (10), G305.

65. Dixit, P.; Miao, J. Aspect-Ratio-Dependent Copper Electrodeposition Technique for Very High Aspect-Ratio Through-Hole Plating. *J. Electrochem. Soc.*, 2006, *153* (6), G552–G559.

66. Kannojia, H. K.; Arab, J.; Sidhique, A.; Mishra, D. K.; Kumar, R.; Pednekar, J.; Dixit, P. Fabrication and Characterization of Through-Glass Vias (TGV) Based 3D Spiral and Toroidal Inductors by Cost-Effective ECDM Process. In *2020 IEEE 70th Electronic Components and Technology Conference (ECTC)*; 2020; pp. 1192–1198.

67. Kannojia, H. K.; Arab, J.; Kumar, R.; Pednekar, J.; Dixit, P. Formation of Through-Wafer 3-D Interconnects in Fused Silica Substrates by Electrochemical Discharge Machining. In *2019 IEEE 21st Electronics Packaging Technology Conference (EPTC)*; IEEE, 2019; pp. 253–257.

68. Verma, A. K.; Mishra, D. K.; Pawar, K.; Dixit, P. Investigations into Surface Topography of Glass Microfeatures Formed by Pulsed Electrochemical Discharge Milling for Microsystem Applications. *Microsyst. Technol.*, 2020, *26*, 2105–2116.

69. Sharma, P.; Mishra, D. K.; Dixit, P. Experimental Investigations into Alumina Ceramic Micromachining by Electrochemical Discharge Machining Process. *Procedia Manuf.*, 2020, *48*, 244–250.

70. Wuthrich, R.; Ziki, J. D. A. *Micromachining Using Electrochemical Discharge Phenomenon: Fundamentals and Application of Spark Assisted Chemical Engraving*; William Andrew, 2014.

71. Ward, C. A.; Balakrishnan, A.; Hooper, F. C. On the Thermodynamics of Nucleation in Weak Gas-Liquid Solutions. *J. Basic Eng.*, 1970, 695–704.

72. Scriven, L. E. On the Dynamics of Phase Growth. *Chem. Eng. Sci.*, 1959, *10* (1–2), 1–13.

73. Zhang, D.; Zeng, K. Evaluating the Behavior of Electrolytic Gas Bubbles and Their Effect on the Cell Voltage in Alkaline Water Electrolysis. *Ind. Eng. Chem. Res.*, 2012, *51* (42), 13825–13832.

74. Wenzel, R. N. Surface Roughness and Contact Angle. *J. Phys. Chem.*, 1949, *53* (9), 1466–1467.

75. Mehrabi, F.; Farahnakian, M.; Elhami, S.; Razfar, M. R. Application of Electrolyte Injection to the Electro-Chemical Discharge Machining (ECDM) on the Optical Glass. *J. Mater. Process. Technol.*, 2018, *255*, 665–672.

76. Elhami, S.; Razfar, M. R. Analytical and Experimental Study on the Integration of Ultrasonically Vibrated Tool into the Micro Electro-Chemical Discharge Drilling. *Precis. Eng.*, 2017, *47*, 424–433.

77. Rathore, R. S.; Dvivedi, A. Sonication of Tool Electrode for Utilizing High Discharge Energy During ECDM. *Mater. Manuf. Process.*, 2020, *35* (4), 415–429.

78. Liu, J.-Y.; Gao, Y.; Wang, G. Main Reaction Process Simulation of Hydrogen Gas Discharge in a Cold Cathode Electric Vacuum Device. *Pramana*, 2012, *79* (1), 113–124.

79. Paul, A. Chemical Durability of Glasses: A Thermodynamic Approach. *J. Mater. Sci.*, 1977, *12* (11), 2246–2268.

80. Charles, R. J. Static Fatigue of Glass. I. *J. Appl. Phys.*, 1958, *29* (11), 1549–1553.

81. Fascio, V.; Wuthrich, R.; Viquerat, D.; Langen, H. 3D Microstructuring of Glass Using Electrochemical Discharge Machining (ECDM). In *MHS'99. Proceedings of 1999 International Symposium on Micromechatronics and Human Science (Cat. No. 99TH8478)*; IEEE, 1999; pp 179–183.

82. Goud, M.; Sharma, A. K. A Three-Dimensional Finite Element Simulation Approach to Analyze Material Removal in Electrochemical Discharge Machining. *Proc. Inst. Mech. Eng. Part C*, 2017, *231* (13), 2417–2428.

83. Panda, M. C.; Yadava, V. Finite Element Prediction of Material Removal Rate Due to Traveling Wire Electrochemical Spark Machining. *Int. J. Adv. Manuf. Technol.*, 2009, *45* (5–6), 506.

84. Elhami, S.; Razfar, M. R. Numerical and Experimental Study of Discharge Mechanism in the Electrochemical Discharge Machining Process. *J. Manuf. Process.*, 2020, *50*, 192–203.

85. Yang, C.-K.; Wu, K.-L.; Hung, J.-C.; Lee, S.-M.; Lin, J.-C.; Yan, B.-H. Enhancement of ECDM Efficiency and Accuracy by Spherical Tool Electrode. *Int. J. Mach. Tools Manuf.*, 2011, *51* (6), 528–535.

86. Zheng, Z.-P.; Su, H.-C.; Huang, F.-Y.; Yan, B.-H. The Tool Geometrical Shape and Pulse-Off Time of Pulse Voltage Effects in a Pyrex Glass Electrochemical Discharge Microdrilling Process. *J. Micromech. Microeng.*, 2007, *17* (2), 265.

87. Han, M.; Chae, K. W.; Min, B. Fabrication of High-Aspect-Ratio Microgrooves Using an Electrochemical Discharge Micromilling Process. *J Micromech Microeng.*, 2017, *27* (5), 055004.

88. Han, M.-S.; Min, B.-K.; Lee, S. J. Micro-Electrochemical Discharge Cutting of Glass Using a Surface-Textured Tool. *CIRP J. Manuf. Sci. Technol.*, 2011, *4* (4), 362–369.

89. Singh, T.; Dvivedi, A. On Performance Evaluation of Textured Tools During Micro-Channeling With ECDM. *J. Manuf. Process.*, 2018, *32*, 699–713.

90. Abou Ziki, J. D.; Wüthrich, R. The Machining Gap During Constant Velocity-Feed Glass Micro-Drilling by Spark Assisted Chemical Engraving. *J. Manuf. Process.*, 2015, *19*, 87–94.

91. Arab, J.; Mishra, D. K.; Dixit, P. Role of Tool-Substrate Gap in the Micro-Holes Formation by Electrochemical Discharge Machining. *Procedia Manuf.*, 2020, *48*, 492–497.

92. Saranya, S.; Sankar, A. R. Fabrication of Precise Microchannels Using a Side-Insulated Tool in a Spark Assisted Chemical Engraving Process. *Mater. Manuf. Process.*, 2018, *33* (13), 1422–1428.

93. Sukumaran, V.; Bandyopadhyay, T.; Sundaram, V.; Tummala, R. Low-Cost Thin Glass Interposers as a Superior Alternative to Silicon and Organic Interposers for Packaging of 3-D ICs. *IEEE Trans. Components, Packag. Manuf. Technol.*, 2012, *2* (9), 1426–1433.

94. Saranya, S.; Sankar, A. R. Fabrication of Precise Micro-Holes on Quartz Substrates With Improved Aspect Ratio Using a Constant Velocity-Feed Drilling Technique of an ECDM Process. *J. Micromech. Microeng.*, 2018, *28* (12), 125009.

95. Mishra, D. K.; Arab, J.; Magar, Y.; Dixit, P. High Aspect Ratio Glass Micromachining by Multi-Pass Electrochemical Discharge Based Micromilling Technique. *ECS J. Solid State Sci. Technol.*, 2019, *8* (6), P322–P331.

96. Mishra, D. K.; Arab, J.; Pawar, K.; Dixit, P. Fabrication of Deep Microfeatures in Glass Substrate Using Electrochemical Discharge Machining for Biomedical and Microfluidic Applications. In *2019 IEEE 21st Electronics Packaging Technology Conference (EPTC)*; 2019; pp. 263–266.

14 Molecular Dynamics Simulation of Advanced Machining Processes

Xichun Luo, Xiaoguang Guo, Jian Gao,
Saurav Goel, and Saeed Zare Chavoshi

CONTENTS

DOI: 10.1201/9780429160011-14

14.1 INTRODUCTION: BACKGROUND AND DRIVING FORCES

Nowadays, advanced machining technologies are paving way for customized high precision 3D precision products made from a variety of materials, including difficult-to-machine materials, such as silicon, silicon carbide, glass, sapphire, hard steels, etc. This trend is largely driven by ever-increasing demands for ultra-precision products/components in energy, data processing, information technology, telecommunications, and astronomy applications, where high performance, multifunctional integration and/or miniaturization are very important. Examples can be seen in the manufacture of mirrors for inertial confinement for focusing lasers, substrates of IC chips, quantum optics, IR diffractive optics, microlens array for head-up displays, and x-ray mirrors, to name a few.

Figure 14.1 shows the evolution of achievable machining accuracy over the years, which keeps on improving with the development of ultra-precision machine tools, and new machining and measurement technologies. Ultra-precision diamond turning is at the pinnacle of the advanced machining process range in terms of accuracy and productivity, as it can create ultra-precision components with sub-micrometer form accuracy and nanometer surface finish by using a diamond tool in a single cut. One issue with diamond turning is that these processes leave tool feed marks on the machined surface which poses problems in using these components for extremely delicate jobs e.g., mirrors for x-rays and laser beam splitters produced by roll-to-roll embossing process. In such cases, ultra-precision polishing process needs to be adopted as a post-processing finishing operation to remove cutting tool feed marks and any subsurface damage left by previous ultra-precision manufacturing operations.

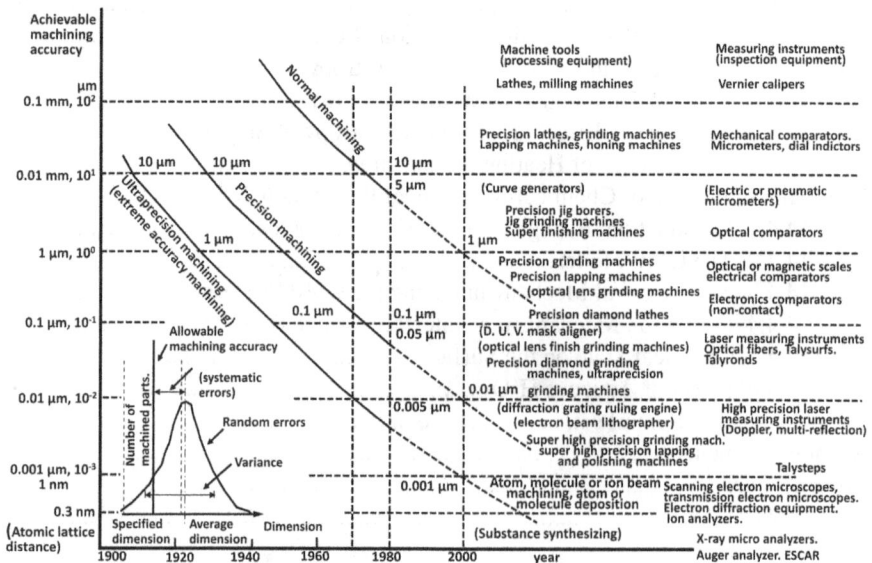

FIGURE 14.1 Evolution of ultra-precision manufacturing technologies. Reproduced with permission from [1], copyright Elsevier, 1983.

An understanding of the material removal mechanism and influencing factors affecting the machined surface quality is pivotal to gain greater determinism and attainable precision in these ultra-precision machining processes. As these processes may involve the removal of a few atomic layers from the workpiece surface, it is extremely difficult to perform *in-process* monitoring of the machining process. As such, molecular/atomic-level simulations are popular tools for studying ultra-precision machining processes. The Monte Carlo (MC) and molecular dynamics (MD) methods are two main computational techniques for molecular/atomic-level simulations. While the MC method is driven largely by the probability distributions in statistical physics, MD is a deterministic approach to obtain trajectories of atoms through solving classical equations of motion at each time step. Calculation of the heat capacity, compressibility, interfacial properties, and dynamic quantities can be done more efficiently and accurately using MD simulations. Therefore, MD simulation is an appropriate bottom-up simulation approach to investigate atomic scale events occurring during nanoscale manufacturing processes. A pioneer study of MD simulation of ultra-precision machining processes was performed by researchers in the US and Japan in the late 1980s in order to gain a fundamental understanding of material removal mechanism at sub-nano scales [2–4]. Since then, MD simulation has transcended the field of ultra-precision machining and has contributed significantly toward improved understanding of our knowledge in the inaccessible atomic scale regime.

This chapter introduces the fundamental theory of MD simulation and its applications in the study of ultra-precision diamond turning and chemical-mechanical polishing processes. Section 14.2 systematically introduces the working principle, development of force fields (potential function) and various case studies involving the use of MD simulation. Sections 14.3 and 14.4 introduce MD simulation studies on the nanoscale level of machining mechanics and mechanisms, including analysis of cutting forces, cutting temperature, stress distribution, high-pressure phase transformation, material removal, and surface generation mechanisms in ultra-precision diamond turning and chemical-mechanical polishing processes. Section 14.5 concludes with an assessment of the MD simulation study, its limitations, and scope of future work.

14.2 FUNDAMENTALS OF MOLECULAR DYNAMICS (MD) SIMULATION

14.2.1 CONCEPT OF MOLECULAR DYNAMICS (MD) SIMULATION

Molecular dynamics (MD) simulation is a combination of three distinct techniques, i.e., molecular modeling, computer simulation, and statistical mechanics. MD simulation starts with establishing the molecular model of the matter to be investigated according to its ground state atomic structures. Relevant force fields (potential functions) are then used to describe the interactions between the atoms. The interatomic forces between the atoms can be estimated directly from the potential function [5]. Newton's second law is usually used to describe the motion of the individual atoms [5]. The interfacial interaction is studied by a trajectory analysis of these atoms [6].

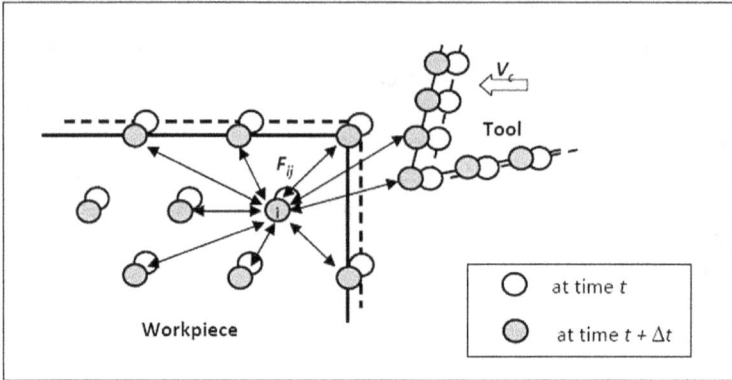

FIGURE 14.2 Principle of molecular dynamics simulation (Komanduri et al. [8]).

Taking the MD simulation model for ultra-precision diamond turning as an example (see Figure 14.2) the atoms within the workpiece and the cutting tool are at certain positions at time t. As the cutting tool moves by a certain distance in time interval Δt, an atom i, acted on by the force F_{ij}, changes its position in the time interval $(t + \Delta t)$. Consequently, the new position of atom i can be predicted using classical Newtonian mechanics [7]. The process can iteratively be solved at multiple time increments of Δt which is a molecular trajectory of the simulated cutting process.

The merit of MD simulation is that it is analyzed using Newton's equation thereby revealing discrete atomistic mechanics that are otherwise impossible to investigate using the conventional engineering tool, e.g., finite element analysis (FEA). One of the principal differences between FEA and MD simulation is that the nodes and the distances between the nodes in MD simulation are not selected arbitrarily, e.g., the position of atoms and interatomic distances used in atoms are governed by the fundamental of materials science as opposed to FEA where nodes are assigned randomly [9]. Also, the crystal structure in the MD simulation is dictated by the crystallographic structure of the material as opposed to FEA which uses random shape functions [9].

14.2.2 FORCE FIELD OR POTENTIAL ENERGY FUNCTIONS

A first step in performing an MD simulation is to choose a functional form for the intermolecular potential. Nearly always, we take the potential to be a pair-wise additive; that is, we assume the interaction energy among N atoms as the sum of the isolated two-body contributions [6], which can be expressed as:

$$U(r^N) = \sum_{i<j} \sum u(r_{i,j}) \tag{14.1}$$

where U – total potential energy among N atoms

r^N – a shorthand for the set of sphere position vectors, $r^N = \{r_1, r_2, r_3, \ldots, r_N\}$

u – a two-body potential

r_{ij} – the distance between the centers of atoms i and j.

The Lennard-Jones and Morse potential functions are among the earliest developed two-body potential functions. The Lennard-Jones potential is a useful model for modeling in interactions in liquids. For a pair of atoms i and j located at r_i and r_j, the potential energy can be expressed as [5]:

$$u(r_{ij}) = 4\varepsilon_{en}\left[\left(\frac{r_0}{r_{ij}}\right)^{12} - \left(\frac{r_0}{r_{ij}}\right)^{6}\right] \tag{14.2}$$

where ε_{en} – the energy at the minimum potential energy
  r_0 – the atomic distance at equilibrium
  r_c – the cut off distance beyond which the potential energy is approaching zero
  ε_{en} – governs the strength of the interaction.

Cut-off distance (r_c) is a very important criterion in establishing the occurrence of a brittle fracture in the material as this sets the separation criterion between the two atoms acted on by the stress.

Morse potential function is another popularly used model; it is usually used to model the interaction between metallic atoms as [5]:

$$u(r_{ij}) = D_e\left[\exp(-2a(r_{ij} - r_0)) - 2\exp(-a(r_{ij} - r_0))\right] \tag{14.3}$$

where D_e – cohesion energy
  A – width of the potential well expressed in distance units.

While the developed multi-body potentials provide more accuracy, they are sometimes computationally very expensive. Table 14.1 lists the evolutionary development of potential functions that are often used for MD simulation of ultra-precision machining processes.

14.2.3 INTERATOMIC FORCES

Since interatomic forces are necessarily conservative, the force that results from the potential can be expressed as:

$$F(r_{ij}) = -\frac{du(r_{ij})}{dr_{ij}} \tag{14.4}$$

The force applied on the i-th atom is the summation of forces acted by all its neighbor atoms, it is given by:

$$F_i = \sum_{j=1(j \neq i)}^{N} F(r_{ij}) \tag{14.5}$$

where N – the atoms number.

TABLE 14.1

Multi-Body Potential Functions with Respect to Time and Applications (Based on Goel [7])

S. No.	Year	Name of the Potential Function	Materials Suited
1	1984	EAM: embedded-atom method [10]	Cu
2	1985	Stillinger–Weber potential [11, 12]	Si
3	1987	SPC: simple point charge [13]	H_2O
4	1988	BOP: bond–order potential Tersoff-1 variant for silicon [14]	Si
	1988	Tersoff-2 for better elastic properties of silicon [15]	Si
	1989	Tersoff-3 for silicon, germanium, and carbon [16, 17]	Si, Ge, and C
	1990	Tersoff-4 for silicon and carbon [18]	Si and C
	1994	Tersoff-5 for amorphous silicon carbide [19] Refinements in Tersoff potential function [20, 21]	SiC
5	1989	MEAM: modified embedded-atom method [22]	Si and Ge
6	1990	REBO: reactive empirical bond order [23]	Carbon
7	2000	AIREBO: adaptive intermolecular reactive empirical bond order [24] (4 body potential function)	Hydrocarbons and carbon
8	2001	ReaxFF: reactive force field [25] (Capable of bond breaking and bond-formation during the simulation)	Universal
9	2005	ABOP: analytical bond order potential [26] (three-body potential function)	Si and C
10	2007	COMB: charge optimized many-body [27]	SiO_2, Cu, Ti
11	2008	EIM: Embedded-ion method [28]	Ionic e.g., NaCl
12	2010	GAP: Gaussian approximation potential [29]	Universal

14.2.4 BOUNDARY CONDITION AND ENSEMBLE

Period boundary condition (PBC) is the most popularly used boundary condition in MD [5]. To use PBC in a simulation of N atoms confined to a volume V, we imagine that volume V is only a small portion of the bulk material. The volume V is called the primary cell; it is representative of the bulk material to the extent that the bulk is assumed to be composed of the primary cell surrounded by exact replicas of itself. These replicas are called image cells. The image cells are of the same size and shape as the primary cell and each image cell contains N atoms, which are images of the atoms in the primary cell. Thus, the primary cell is imagined to be periodically replicated in all directions to form a macroscopic sample of the substance of interest. This periodicity extends to the positions and momenta of the images in image cells [6].

In MD simulation, the initial positions of these atoms are set by the unit cell (lattice parameter) of the material. This follows the process of energy minimization and then the velocities to the atoms are assigned using a Maxwell–Boltzmann distribution.

In an isolated system, three quantities exist, namely, the total energy (E), the total volume (V), and number of atoms (N). These quantities remain extensive, i.e., they remain proportional to the size of the system. An ensemble can simply be considered as a large number of replications of a cell which has limited number of atoms. Those repeating cells may not necessarily be the same until they have the same thermodynamic properties. It is therefore necessary to use the right ensemble. There are four common ensembles used at present, i.e., NVE, NVT, NPT, and NPH.

- Microcanonical ensemble (NVE): An isolation system with N atoms with a constant volume and the overall energy E of the system conserved.
- Canonical Ensemble (NVT): An isolation system with N atoms and the volume V and the temperature T of the system remains constant. This can be regarded as a system in a heat bath.
- Isobaric isothermal ensemble (NPT): An isolation system containing N atoms and the pressure P and the temperature T of the system are held constant. It can be regarded as a system in a heat and pressure bath.
- Isoenthalpic-isobaric ensemble (NPH): An isolation system with N atoms and constant enthalpy and pressure.

14.2.5 MOTION EQUATION AND ALGORITHM FOR MD SIMULATION

In MD simulation, the equations of motion of atoms are assumed to follow Newton's second law:

$$m_i \ddot{r}_i = F_i \quad (i = 1, 2, \ldots, N) \tag{14.6}$$

where m_i – mass of i-th-atom.

As there are N atoms in the MD simulation system, the equations of motion are a set of ordinary differential equations. They can be solved by finite-difference methods. In MD simulation, the calculation of intermolecular forces is by far the most time-consuming part. A trade-off between accuracy and stability of the algorithm to solve the motion equation has to be considered.

The simplest finite-difference method that has been widely used in MD is Verlet's algorithm because of its good stability, low storage space, and fast computational speed [30]. According to the algorithm, the vector of i-th atom at step $n + 1$ is [5]:

$$r_i^{n+1} = 2r_i^n - r_i^{n-1} + \frac{F_i^n h_s^2}{m_i} \tag{14.7}$$

where h_s – the integral time step.

From the position r_i^0 and r_i^1, the next position of i-th atom can be calculated by this equation.

The velocity can be calculated using the following equation:

$$V_i^n = \frac{r_i^{n+1} - r_i^{n-1}}{2h_s} \tag{14.8}$$

Equations 14.7 and 14.8 are composed of Verlet's algorithm. Other algorithms such as the Euler, Gear, Beeman, and Frog-Leap algorithms are also often used in MD simulations.

14.2.6 IMPLEMENTATION, VISUALIZATION, AND POST-PROCESSING

Currently, there are several open-source codes to perform the MD simulations. Among them the large-scale atomic/molecular massively parallel simulator (LAMMPS) [31] is one of the most popular MD simulation platforms for studying ultra-precision machining processes. While visual molecular dynamics (VMD) [32] and the open visualization tool (OVITO) [33] are used for the visualization and post-processing analysis of the atomic trajectories.

The cutting forces, temperature, and stresses in the machining zone can be calculated from the LAMMPS output data. There are several methods to characterize phase transformation in the machining process, such as calculation of the coordination number, common neighbor analysis (CNA), radial distribution function (RDF), angular distribution function [34], and the polyhedral template matching (PTM) method. CNA is mostly used to facilitate the characterization of phase transformation while the dislocation extraction algorithm (DXA) is a powerful tool to identify dislocation in the material removal process.

14.3 MD SIMULATION OF ULTRA-PRECISION DIAMOND TURNING OF HARD-BRITTLE MATERIALS

On account of their amazing engineering properties such as high refractive index, wide energy band gap, and low mass density, there is a demand for the manufacture hard-brittle materials such as silicon (Si), silicon carbide (SiC), and gallium arsenide (GaAs) for the optical, semiconductor, and opto-electronics industries. Unlike metals, brittle materials exhibit very low fracture toughness which limits their plastic deformation, thus making them difficult-to-machine materials.

Si is an archetypal semiconductor with physical and chemical properties that continue to draw massive research interest. SiC is a consummate candidate for several high temperature and durability applications for power electronics and heterogeneous catalyst supports. GaAs has a wide exploitation in photo-emitter and microwave devices, solar cells, wireless communication, as well as quantum computation.

14.3.1 MD SIMULATION OF ULTRA-PRECISION DIAMOND TURNING OF SI

In Goel et al.'s early-stage MD simulation study [35], the calculated stress state in the cutting zone during ultra-precision diamond turning of Si indicated Herzfeld–Mott transition (metallization) due to high-pressure phase transformation (HPPT) of silicon under the influence of deviatoric stress conditions. Consequently, the transformation of pristine silicon to β-silicon (Si-II) was suggested to be the likely reason for the observed ductility during its nanoscale cutting. A direct observation of this phase transition, however, has never been reported in any MD studies. Machining silicon in the ductile regime involves a number of simultaneously occurring processes, i.e., HPPT, wear of the cutting tool, movement of dislocations which are influenced by the crystal anisotropy of both the cutting tool and the workpiece.

14.3.1.1   Influence of Crystal Anisotropy

It has been found that the crystallographic directions corresponding to minimum values of the effective resolved shear stress are those with maximum hardness [36]. Also, the pressure required to drive phase transformation in Si is sensitive both to the crystal orientation and the cutting direction, i.e., <001> cutting direction produces better metallic response than <110> cutting direction on the (100) orientation of silicon whereas (111) orientation requires less transformation pressure than the (100) orientation. Jasinevicius et al. [37] conducted experimental trials to investigate the reason why the two different cutting directions on the same orientation gave different machining outcomes, especially <110> which produced a worse surface than the <100> direction on the same orientation (001). Based on the experimental results, they suggested that the difference in the HPPT in the <110> direction requires more energy to drive HPPT which makes the brittle mode dominant on this direction. Similarly, a diamond tool also shows a high degree of crystal anisotropy. Uddin et al. [38] have recommended the usage of dodecahedral orientation with a diamond tool while Cheng et al. [39] have suggested using the cubic orientation of the diamond tool in order to have better tool life.

In this chapter, three simulation cases are considered with different combinations of crystal orientations to investigate their influence on the thrust forces during nanometric cutting of silicon. A schematic diagram of the MD simulation model is shown in Figure 14.3. Both the nanocrystalline workpiece and the diamond cutting tool were modeled as deformable bodies in order to permit the tribological interactions between them. The model developed in this work is based on the boundary condition

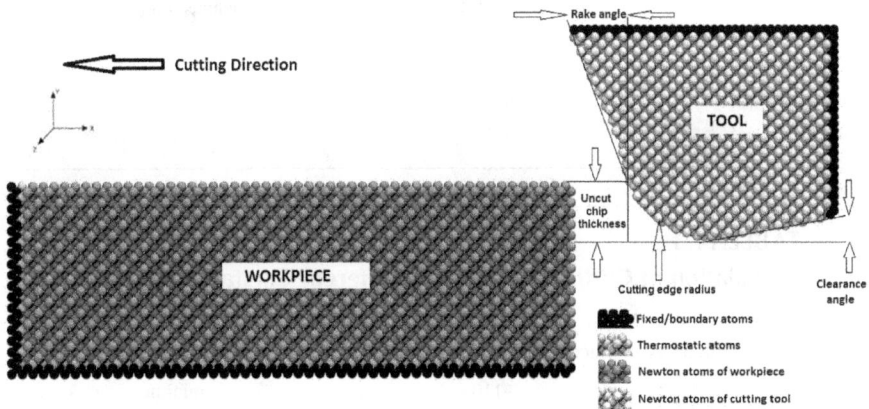

FIGURE 14.3   Schematic of MD simulation model. Reproduced with permission from [41], copyright Elsevier, 2012.

of fixing the bottom and outer sides of the substrate as it has been suggested to be an appropriate configuration to simulate a nanometric cutting process [40].

Also, the MD model incorporates a negative tool rake angle, as this is generally recommended for machining hard-brittle materials [42, 43]. The simulation was performed with the parameters shown in Table 14.2.

The evolution of thrust forces with a change in crystal orientation can be a good criterion to identify the appropriate crystal orientation for practical purposes. Table 14.3 and Figure 14.4 show the results obtained for the thrust forces while cutting silicon using different combinations of crystal orientations of tool and workpiece. It can be seen that the slope and amplitude of thrust forces were at its minimum while using the (111) orientation of the workpiece with cubic orientation of the diamond tool.

It is actually the magnitude of the negative rake angle which dictates whether or not dodecahedral or cubic orientation will perform better. One observation which

TABLE 14.2
Variables Used in the MD Simulation Model

Workpiece dimensions	42.0743 nm × 4.6353 nm × 3.5656 nm
Number of atoms in the workpiece and tool	36,657 and 6440, respectively
Cutting edge radius	1.313 nm
Uncut chip thickness/in-feed	1.313 nm
Crystal orientation	Three simulation cases were tested: i. Cubic orientation of tool with cutting direction <−110> while workpiece was machined on (111) orientation. ii. Cubic orientation of tool with cutting direction <100> while workpiece was machined on (010) orientation. iii. Dodecahedral orientation of tool with cutting direction <−110> while workpiece was machined on (111) orientation.
Tool rake and clearance angle	−25° and 10°
Equilibration temperature	300 K
Cutting velocity	100 m/s
Time step	0.5 fs

TABLE 14.3
Calculation of Cutting Forces with Different Crystal Orientation

S. No.	Orientation of Silicon Workpiece	Cutting Direction	Orientation of Diamond Tool	Magnitude of Thrust Forces Observed
1	(111)	<−110>	cubic	minimum
2	(010)	<100>	cubic	intermittent
3	(111)	<−110>	dodecahedral	maximum

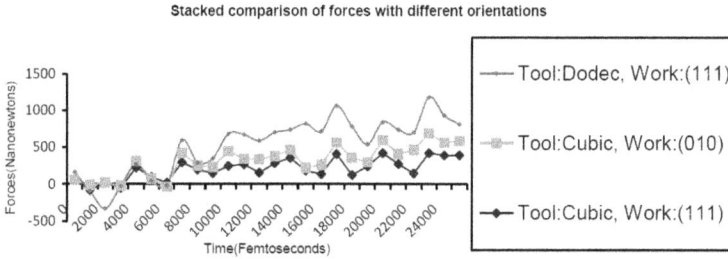

FIGURE 14.4 Variation in the thrust forces during nanometric cutting of silicon. Reproduced with permission from [41], copyright Elsevier, 2012.

is obviously clear from the MD simulation is that the machining of silicon on (111) plane along <–110> cutting direction consumes the least energy to cut which is in accordance with the wealth of publications.

14.3.1.2 Diamond Tool Wear Mechanism

A good understanding of the wear mechanism is an essential step in the identification of the measures needed to suppress wear to enhance tool life. Enormous work has been done in the past on the characterization of tool wear in single-point diamond turning (SPDT) through observations and measurements of worn tools following experimental machining trials. An important consideration neglected in the past is that at a constant spindle rotation speed, the surface cutting speed varies from the maximum on the outside to zero at the center. Thus, the obvious differences in wear behavior due to different cutting speeds were not hitherto accounted for.

Early characterization of tool wear was done as normal wear, chipping, setting problems (not related to diamond tools), line effects, chip dragging, and fracture [44].

A summary of tool wear studies performed through MD simulation made in the past and their conclusion is tabulated in Table 14.4.

As evident from Table 14.4, there is a considerable disagreement on the conclusion of these studies concerning tool wear during the SPDT of silicon. While Cheng et al. [39] have identified a thermochemical mechanism as governing wear, Maekawa et al. [47] have suggested interdiffusion and readhesion. A theory, in which formation of "dynamic hard particles" causes tool wear has also been proposed [48, 49], but it lacks experimental evidence. Thus, the MD simulations that have been applied so far have not elucidated a convincing mechanism of tool wear during the SPDT of silicon. It is also pertinent to note that the MD simulation performed by most of the researchers have used Morse potential function to describe tool–workpiece interactions [47, 48, 51] which is more appropriate for diatomic molecules. A study made by Komanduri *et al.* [8] used the Tersoff potential function but assumed the tool was a rigid body so wear process could not be studied. The simulation results shown in this chapter were developed using three-body potential function which was a better choice than the Morse potential. From the simulations, a RDF between the silicon workpiece and the diamond tool was extracted to investigate the chemical activity

TABLE 14.4

MD Simulation Studies on Tool Wear during SPDT

Potential Function for Tool–Workpiece interaction	Material Simulated	Author, Year, and Country	Tool Consideration	Conclusion of the Study Concerning Cause of Tool Wear
LJ	Silicon	J. Belak, 1990, USA [45, 46]	Deformable	SiC asperity was observed during SPDT of silicon
Morse	Copper	K. Maekawa, 1995, Japan [47]	Deformable	Interdiffusion and readhesion
Tersoff	Silicon	R. Komanduri, 1998, USA [8, 42]	Rigid	-
MEAM	Silicon	X. Luo, 2003, UK [39]	Deformable	Thermochemical mechanism
Morse	Silicon	M. B. Cai, 2007, Singapore [48, 49]	Deformable	Formation of dynamic hard particles
MEAM	Iron	R. Narulkar, 2008, USA [50]	Deformable	Graphitization of diamond
Morse	Silicon	Z. Wang, 2010, China [51]	Deformable	No mechanism has been described

between the diamond tool and silicon workpiece, and accordingly RDF for Si–C at regular intervals is shown in Figure 14.5.

In Figure 14.5, at time step 0, $g(r)$ is 0 which means there is no chemical bonding between Si and C before the physical contact between the tool and the workpiece. However, with a subsequent increase of time steps or tool advancement, it was observed that the first peak at an interatomic distance of 1.9 Å and second peak at around 3.08 Å continued to grow with the cutting duration. It is well known that tetrahedral silicon carbide (SiC) possesses the same bond length and interplanar spacing. Hence, this is a clear indication of the formation of silicon carbide during the SPDT of silicon.

The close contact between the workpiece and tool results in locally high temperature which, in the actual machining environment, is supplemented with the presence of ambient oxygen. The highly affinitive freshly generated dangling bonds of silicon will tend to combine with the atmospheric oxygen to form silicon dioxide as the free energy change in all cases remains negative [52]. However, the reaction mechanism thereafter may occur either through a single-phase solid-state reaction or through a multiphase reaction mechanism as shown in Table 14.5.

It depends on the free energy change (function of process temperature) which governs the precedence of a chemical reaction in accordance with the thermodynamics. In order to ascertain the reaction mechanism, a graph depicting the free energy change with respect to temperature was plotted and shown in Figure 14.6.

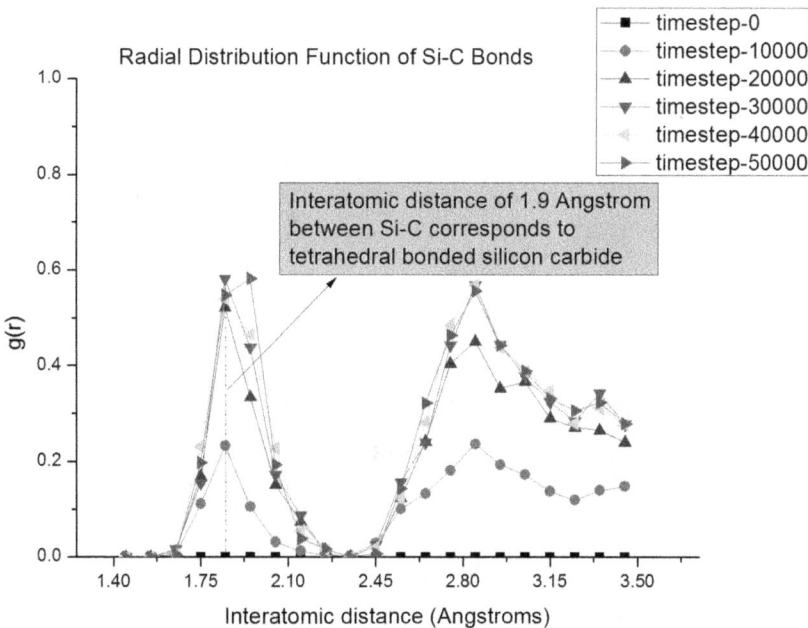

FIGURE 14.5 Radial distribution function (RDF) between Si–C during nanometric cutting of Si. Reproduced with permission from [41], copyright Elsevier, 2012.

TABLE 14.5

Reaction Mechanism for Formation of Silicon Carbide

Process	Chemical Reaction	Free Energy Change for the Reaction
Single-Phase Reaction		
Formation of silicon carbide	Si (s,l,g) + C → SiC [53]	$\Delta G_T^o = 499820 - 149T$ J/mol
Multiphase Reaction		
Formation of silicon dioxide	Si + O$_2$ → SiO$_2$ [53]	Free energy change in negative in all cases [52]
Formation of silicon oxide	SiO$_2$ + C → SiO + CO [53]	$\Delta G_T^o = 670402 - 327T$ J/mol
Formation of silicon carbide	SiO + 3CO → SiC + 2CO$_2$ [54, 55] or SiO (g) + C (s) → Si (g) + CO (g) [53] Si (g) + C (s) → SiC (S) [53]	

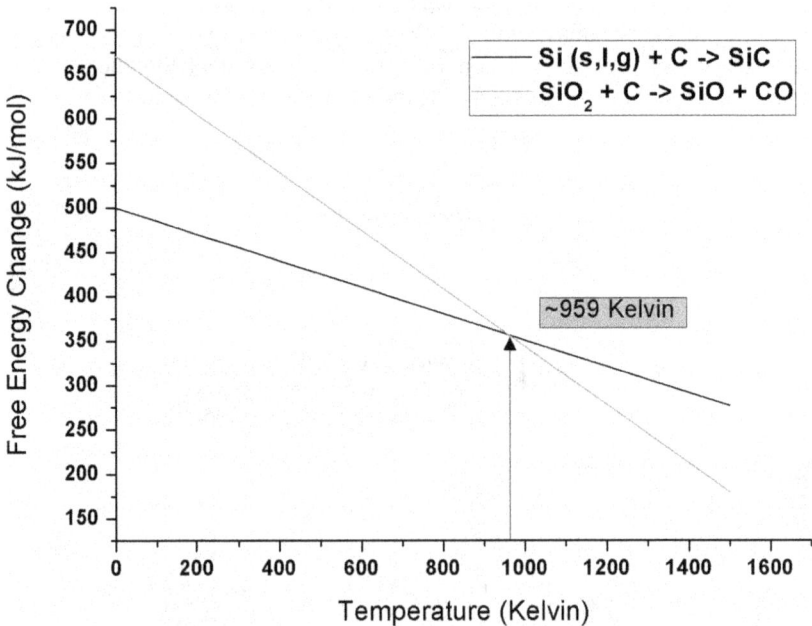

FIGURE 14.6 Gibb's free energy change for the formation of SiC.

It can be seen from Figure 14.6 that, in either case, the free energy change is positive and hence the reaction will not be spontaneous. It is further evident that the free energy change required for the activation of a solid-state single-phase reaction between silicon and carbon is thermodynamically more favorable up to a temperature of 959 K compared to a multiphase reaction via the formation of silicon oxide.

Conversely, beyond a temperature of 959 K, the silicon dioxide path is energetically more favorable toward the formation of silicon carbide, which implies that the presence of oxygen at a temperature above 959 K will accelerate the formation of silicon carbide. It is therefore imperative to know the cutting temperature in order to establish the route to the formation of silicon carbide. The temperature distribution on atoms obtained from the simulation of nanometric cutting of silicon is shown in Figure 14.7.

The maximum temperature on the tool tip was observed to be around 380 K while that on the workpiece was observed to be in the primary shear zone, and on the finished surface approaching almost 750 K at a high cutting speed of 100 m/s. Since the local temperature is well below 959 K even at such a high cutting speed, it is reasonable to conclude that a single-phase solid-state chemical reaction between dangling bonds of silicon with coordination number 1 or 2 and highly chemically active nascent surface/dangling bonds on the diamond tool will result in the formation of silicon carbide during their surface contact, which is stimulated further by the cutting stresses.

Similarly, Pastewka et al. [57] carried out an MD simulation of polishing of a diamond crystallite with another diamond crystallite at a sliding speed of 20 m/s. They concluded that it is the tribo-chemistry that plays a significant role in governing the diamond wear rate as shown in Figure 14.8. The various colors used in Figure 14.8 indicate whether the atom was initially bound to the top or bottom crystallite of the diamond while the black line shows the evolution of an amorphous interface of carbon atoms during the polishing process. A layer of "pilot" atoms that move around on the ordered phase repeatedly attracts the crystalline surface atoms. Since

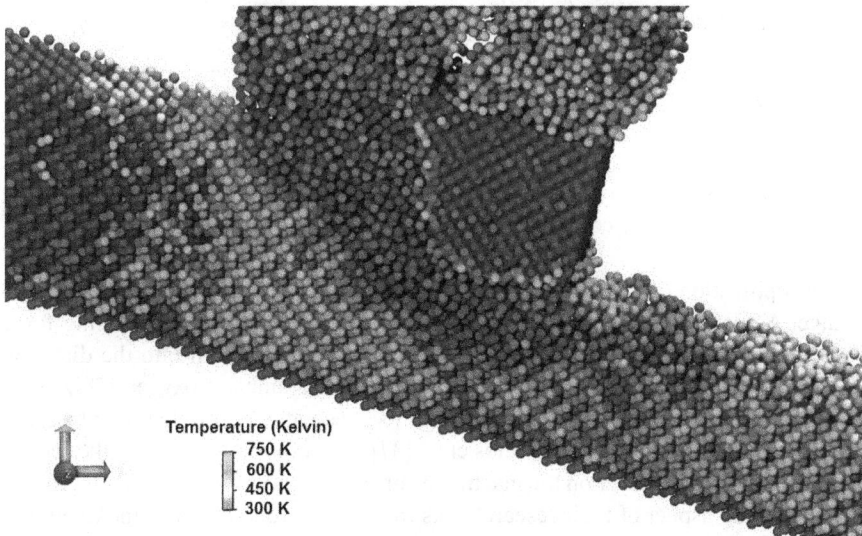

FIGURE 14.7 Temperature distribution on atoms during SPDT of silicon. Reproduced with permission from [41], copyright Elsevier, 2012.

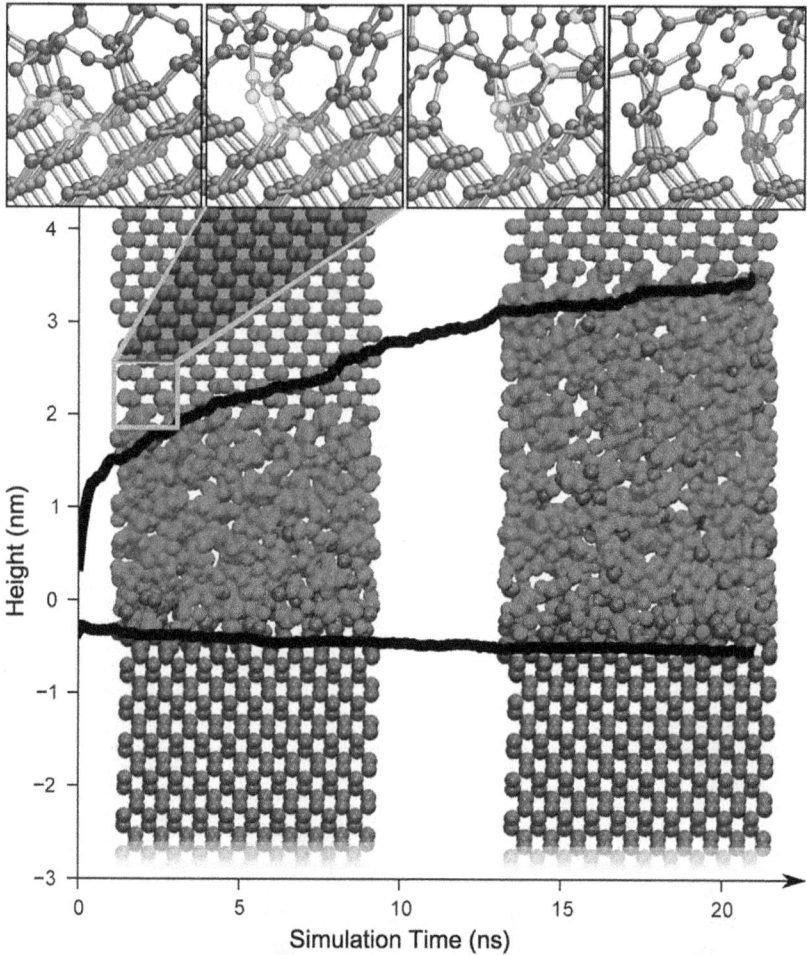

FIGURE 14.8   Sliding of a diamond over another diamond crystallite at 20 m/s. Reproduced with permission from [56], copyright The Materials Research Society, 2012.

the amorphization of the "pilot" atoms changes over time, the plucking forces also change. A surface atom is lifted into the amorphous phase when the pulling force becomes larger than the cohesive force holding the carbon atom into the diamond crystallite. This layer is subsequently removed by the ambient oxygen [58]. Quite similar to this phenomenon is the plucking of surface atoms from the diamond tool to form a thin film of SiC. Jasinevicius et al. [37] have reported observing the formation of an amorphous layer on the machined surface of silicon to an extent of 340 nm. An interesting aspect of their research was that the microhardness of the diamond-turned silicon was lower than that of the pristine silicon which was attributed to the presence of the amorphous layer. This presented an anomalous fact that compared to the machining of metals where the outermost layer of the machined surface always becomes harder due to work hardening; the machined silicon becomes softer. Mechanical machining thus ostensibly introduces a barrier layer, Beilby layer (tribomaterial) exhibiting a different refractive index from that of the substrate [59] as shown schematically in Figure 14.9.

Thus, SiC can either grow in the cutting chips or as a thin film on the surface of the diamond tool. In either of these cases, it will result in the formation of vacant sites on the diamond tool which were earlier identified as groove wear [61]. Also, the freshly formed SiC film will scrape off during continuous frictional and abrasion contact with the diamond tool during SPDT of silicon. It is also important to note here that, during nanoscale ductile cutting of brittle materials, the undeformed chip thickness varies from zero at the center of the tool tip to a maximum value at the top of the uncut shoulder. Thus, a "zero-cutting zone" exists within which no chips are produced. In this special zone, the tool acts more like a roller than a cutter and continuously slides on and burnishes the machined surface. A schematic diagram of this is shown in Figure 14.10 [38].

Figure 14.10 shows that the cutting edge of the tool continues to recede, and the flank wear region becomes predominant. This can be considered and understood from the perspective of the point of stagnation. As shown in Figure 14.11, there exists a point on the cutting-edge radius where the tangential velocity of the workpiece becomes zero [63]. It is of interest to note that below the point of stagnation the material gets compressed downward under the rake of the tool. Contrarily, above the point of stagnation, the shear of the material is more pronounced than compression.

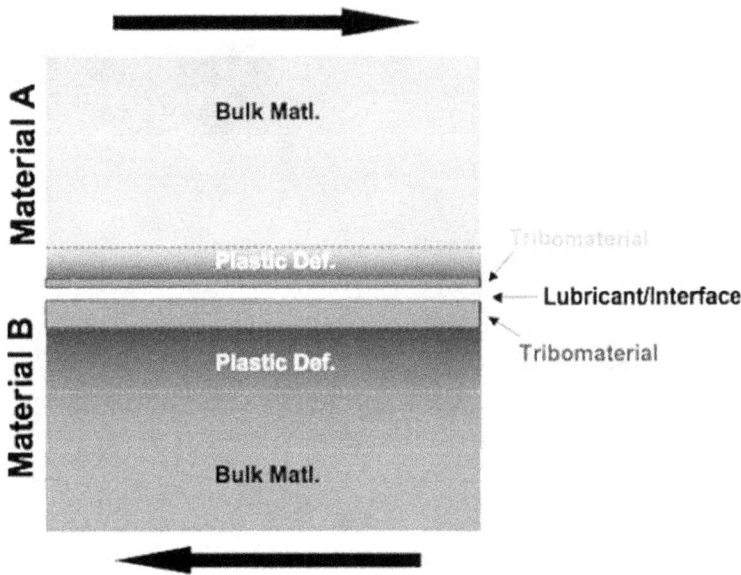

FIGURE 14.9 Schematic of a simple system consisting of a harder material "A" sliding on a softer material "B". Near to the sliding interface, a Beilby layer of tribomaterial develops. Reproduced with permission from [60], copyright Springer Nature, 2009.

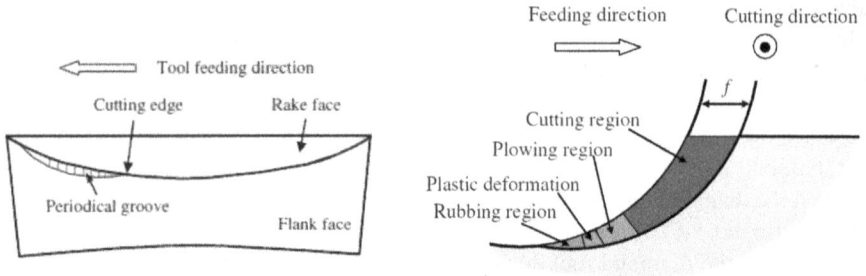

FIGURE 14.10 Schematic of the groove wear. Reproduced with permission from [62], copyright Springer Nature, 2011.

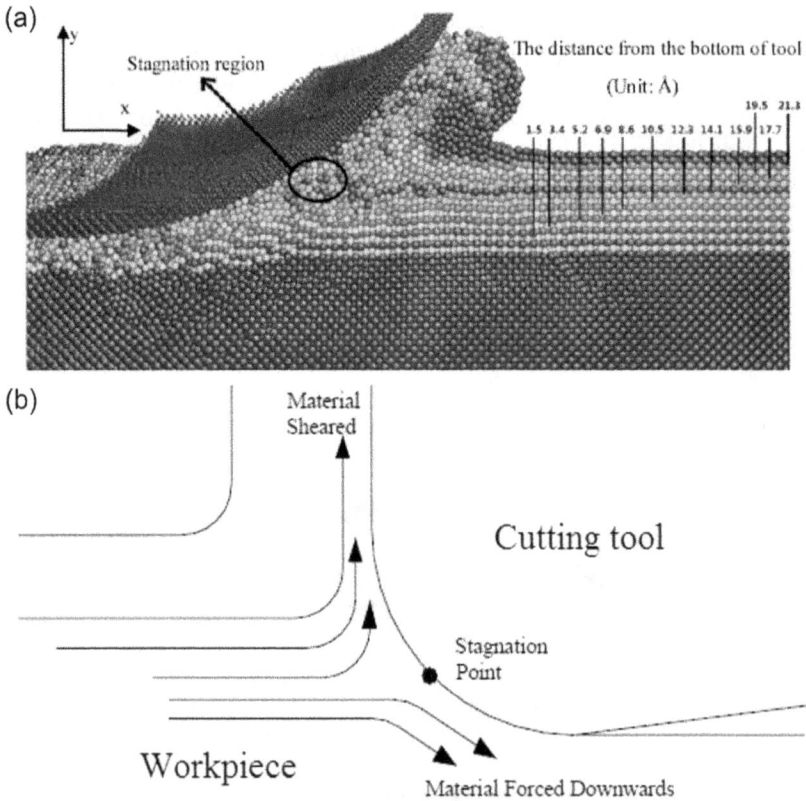

FIGURE 14.11 Stagnation point on the cutting tool during SPDT. (a) MD model. Reproduced with permission from [64], copyright Springer Nature, 2012. (b) Schematic model.

The flank wear region causes a reduction in the clearance angle, which gives rise to increased frictional resistance. This is the reason for the observation of relatively higher temperature at the tool flank than the tool rake face. This phenomenon is in contrast to the conventional machining operation where the tool rake face remains at a higher temperature than the tool flank face. This is because of the large amount of the energy released from the cutting chips and the consequent heat dissipated from the tool rake face. Contrarily, during SPDT, the effect of frictional heat between the tool flank face and the finished surface of the workpiece is more than that on the tool rake face. Due to the high temperature on the flank face, the chemical kinetics between silicon and carbon atoms become more favorable at the tool flank face than the rake face of the tool. Subsequently, abrasion due to continuous frictional contact with the flank face further enhances the wear rate at the tool flank face. This causes very low rake wear compared to the flank wear. The plucking of surface atoms from the diamond tool and subsequent abrasion between a thin layer of SiC and the cutting tool causes simultaneous graphitization $sp^3 – sp^2$ (Figure 14.12) making the diamond tool vulnerable to the wear process during cutting.

Figure 14.12 shows that the peak at the interatomic distance of 1.54 Å (the known bond length of diamond) decreases during the simulation with a corresponding increase at the bond length of 1.42 Å, confirming $sp^3 – sp^2$ disorder of the diamond. It is therefore reasonable to conclude that both processes, i.e., formation of SiC and $sp^3 – sp^2$ disorder of the diamond proceed in tandem with each other and represent the fundamental cycle of wear of diamond tools during SPDT of silicon.

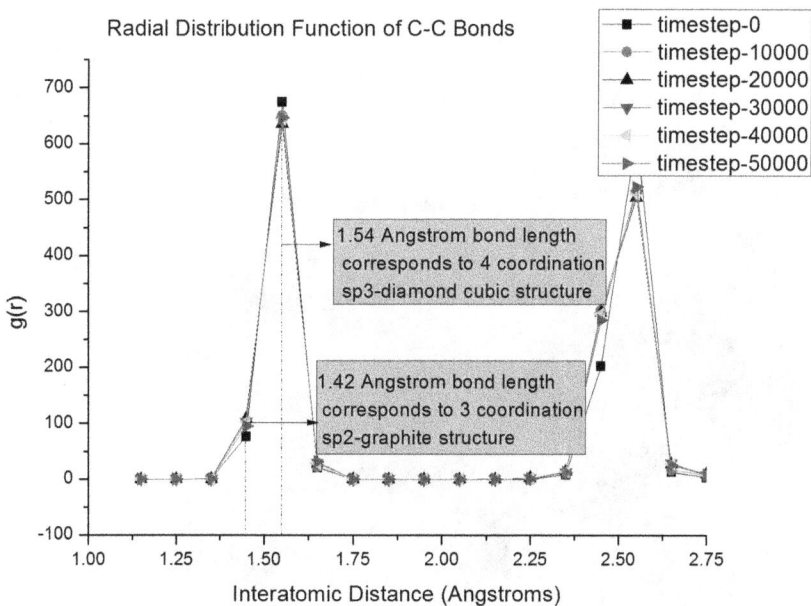

FIGURE 14.12 Radial distribution function of C-C bonds during SPDT of silicon.

It is worth noting that, although a high cutting speed was used in the MD simulation, the outcome of the process (i.e., the formation of SiC and sp^2 carbon) are also observed experimentally during machining of single crystal silicon [65] at a practical cutting speed.

14.3.2 Study of the Effect of Crystal Anisotropy of Si and the Influence of Heating the Substrate

A common understanding is that the yield strength and hardness of materials reduces with high temperatures. As such, the fracture toughness increases which in turn eases plastic deformation. In other words, at high temperatures, plasticity plays a greater role in the fracture and deformation processes of hard-brittle materials such as silicon and silicon carbide primarily as a result of thermally generated intrinsic defects and thermal softening processes. In such thermally enhanced processing, the workpiece can be heated by a continuous-wave fiber laser and therefore is thermally softened. A laser provides intense localized heating to the workpiece ahead of the cutting region. In this way, the strength of the workpiece reduces and therefore its machinability is improved. This section therefore presents MD simulation of ultra-precision diamond turning of Si at elevated temperatures.

A three-dimensional MD model used in this case study (which is basically a plane-stress representation of actual machining operation) is illustrated in Figure 14.13. In this model, the workpiece and the cutting tool are modeled as deformable bodies in

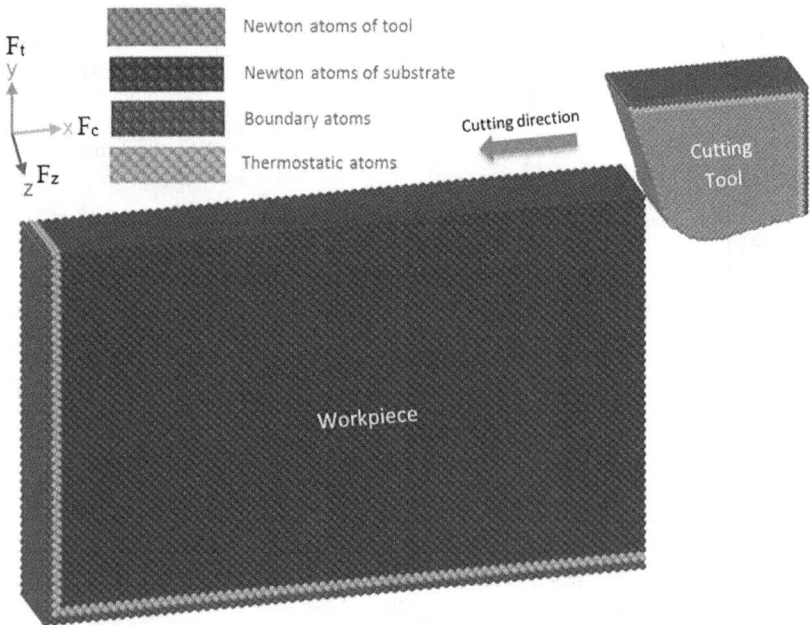

FIGURE 14.13 Schematic of the MD simulation model.

order to permit tribological interactions between them. However, if the workpiece is very soft (such as copper and brass), the diamond tool can be modeled as a rigid body since the cutting tool will not wear even after a cutting length of the order of 30 km [66]. Similar to the previous case, herein in the model (Figure 14.13) used a negative rake angle tool, as this is generally the recommendation for machining of hard and brittle materials such as silicon and silicon carbide. It should be noted that the force components along the x, y, and z directions referred to as tangential cutting force (F_c), thrust or normal force (F_t), and axial force (F_z), respectively. It may be noted that F_z is being introduced here only for awareness; however, since the simulation model is assumed to be a plane-stress condition, the average magnitude of F_z during cutting is expected to be zero.

In Figure 14.13, the region of atoms in the workpiece and tool are divided into three zones, namely, boundary atoms zone, thermostatic atoms zone, and Newtonian atoms zone. The boundary atoms are held rigidly to reduce the boundary effects and to ensure structure stability, and to maintain the symmetry of the lattice. The thermostatic zone is allowed to follow Berendsen thermostatic dynamics (LAMMPS NVT dynamics), so as to mimic dissipating the heat generated during cutting. The third category of atoms is called Newtonian atoms, which were allowed to follow the microcanonical (LAMMPS NVE dynamics) dynamics in accordance with the routine MD principles. The velocity Verlet algorithm [30] with the single time step of 1 femtosecond (fs) was employed for the time-marching method in the simulations. The MD model is based on the boundary condition of fixing the bottom and outer sides of the workpiece since it is recommended to be a suitable configuration to simulate a nanometric cutting process [40]. The PBC is imposed along the z direction of the simulation domain for the sake of reducing the effects of simulation scale and to mimic the plane-stress condition.

Table 14.6 shows the details of the MD model and the cutting parameters employed in the hot machining simulations. The high temperature nanometric cutting of single crystal silicon was performed at different temperatures using ABOP potential function.

The force components along the x, y, and z directions describe tangential, thrust, and axial forces, respectively. Surface form error is chiefly affected by thrust force since this component of the force has the tendency to separate the tool away from the workpiece. The tangential force brings about displacements in the direction of cut chip thickness and its variation therefore associates to chatter. Cutting forces change dynamically during the process, thus it is vital to monitor them so as to comprehend surface finish as well as tool wear phenomena. To do so, the total force exerted by the carbon atoms of the cutting tool on the workpiece is calculated. The average values of the forces are calculated only after the tool penetrated the workpiece by 15 to 25 nm, as shown in Figure 14.14. At the beginning stage of cutting, both the tangential and thrust forces were zero. As the diamond cutting tool starts moving toward the single crystal silicon workpiece, the negative forces were observed, as illustrated in Figure 14.14, signifying the effects of long-range attraction forces. Atoms attract each other when the instantaneous distance between two atoms is longer than the equilibrium distance between them. As the cutting tool advances more,

TABLE 14.6

Details of the MD Simulation Model and the Cutting Parameters for Hot Machining

Workpiece material	Single crystal silicon (Si)
Workpiece dimensions	Si: $38 \times 19 \times 5.4$ nm³
Tool material	Single crystal diamond
Cutting edge radius (tip radius)	3.5 nm
Uncut chip thickness (cutting depth in 2D)	3 nm
Cutting orientation and cutting direction	Case 1: (010) <100>
	Case 2: (110) <00−1>
	Case 3: (111) <−110>
Rake and clearance angle of the cutting tool	−25° and 10°
Workpiece temperature	300 K, 500 K, 750 K, 850 K, 1173 K, 1273 K, and 1500 K
Cutting speed	50 m/s
Time step	1 fs
Potential energy functions used for nanometric cutting	ABOP [26]

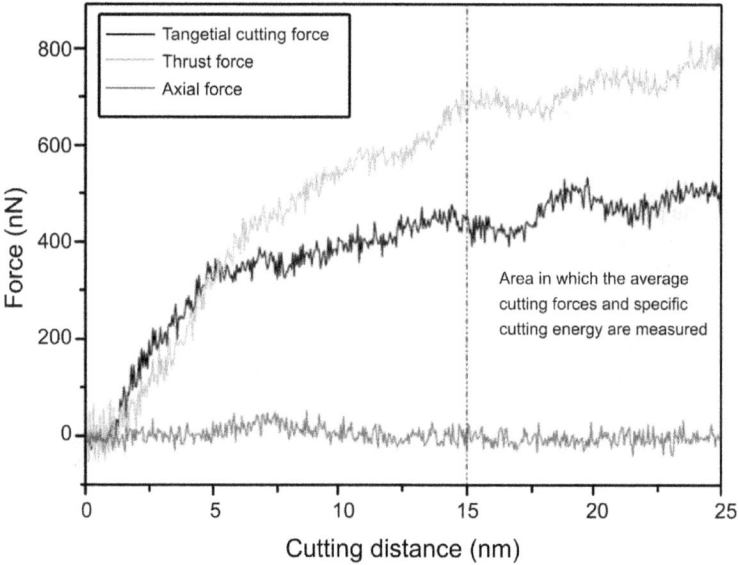

FIGURE 14.14 An MD output of the force plot showing the region where the average cutting forces and specific cutting energy were measured.

cutting forces on the tool atoms alter to repulsive and climb proportionally, in a sense of average, with the cutting length.

The magnitude of the average tangential cutting forces (F_c in nN), thrust forces (F_t in nN), resultant forces ($R = sqrt(Fc^2 + Ft^2)$ in nN), force ratio (F_c/F_t) and specific cutting energy (u in GPa) for all the simulated temperatures and crystal planes are calculated. The cutting resistance in general and is indicated by the term called "specific cutting energy". The specific cutting energy "u" expressed in N/m^2 or J/m^3, is defined as the work done by the cutting tool in removing the unit volume of material and is expressed as:

$$u = \frac{R \times v_c}{b \times t \times v_c} \qquad (14.9)$$

where R is the resultant force in nN equal to $sqrt(Fc^2 + Ft^2)$ in plane-stress condition, v_c is the cutting velocity (m/s), b is the width of cut (nm), and t is the uncut chip thickness (nm) or cutting depth (in plane-stress condition).

Figure 14.15 presents the variation of specific cutting energy at different cutting temperatures and crystal planes. It shows that low temperature machining leads to large specific cutting energy which is in accordance with the data of cutting forces. A common observation evident from Figure 14.15 is that anisotropy persists even at high temperatures. It may be seen that the (111) plane required the least specific cutting energy whereas the highest values appear on the (110) plane. As stated before, the slip in diamond cubic lattice is analogous to FCC crystals and occurs preferentially on the (111) slip planes, meaning thereby that the (111) plane should result in low specific cutting energy. Moreover, it has been experimentally demonstrated that the (111) silicon surface provides a finer quality of machined surface [35] and requires low specific cutting energy. Overall, it can be inferred that the (111)<–100> and

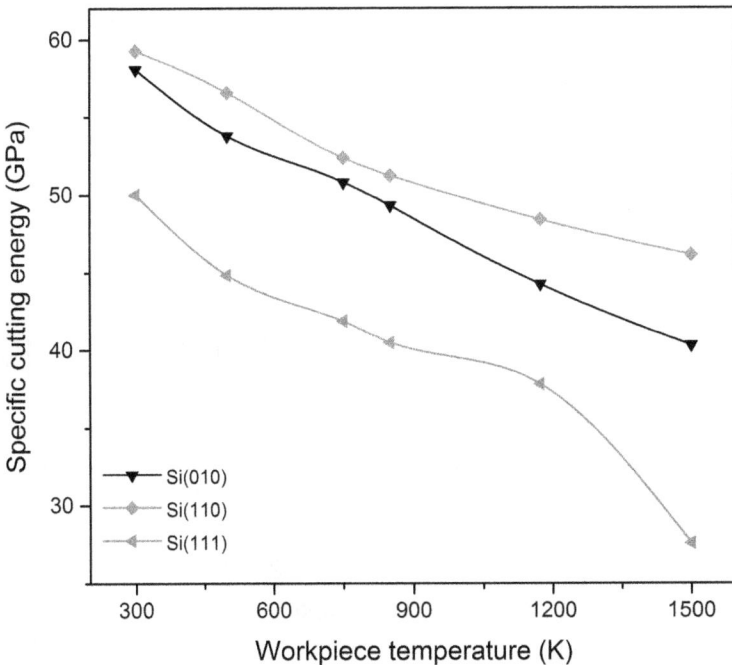

FIGURE 14.15 Specific cutting energy as a function of temperature and crystal plane.

(010)<100> crystal setups are the easy cutting combinations of plane and directions for cutting silicon which is in accord with the published experimental results [67].

The percentage reduction in the tangential force, thrust force, resultant force and specific cutting energy during elevated temperature cutting with respect to cutting at 300 K on different crystal planes obtained by ABOP is shown in Table 14.7. It was observed that the maximum reduction occurs on the (111) and (010) crystal planes, which is up to 25%. Furthermore, the anisotropy (cutting forces and specific cutting energy) was found to increase with the rise of temperature, i.e., it increases from ~16% at 300 K to ~20% at 1173 K and to ~40% at 1500 K, respectively.

Figure 14.16 presents variation in the average force ratio during nanometric cutting of silicon obtained from ABOP in the range of 300–1500 K for each crystal

TABLE 14.7

Percentage Reduction in Tangential, Thrust, Resultant Forces, and Specific Cutting Energy of Silicon at High Temperatures Relative to Room Temperature

Crystal Plane	% Reduction in Machining Energy at 850 K Compared to 300 K	% Reduction in Machining Energy at 1173 K Compared to 300 K
(010)	Up to 15%	Up to 24%
(110)	Up to 14%	Up to 19%
(111)	Up to 19%	Up to 25%

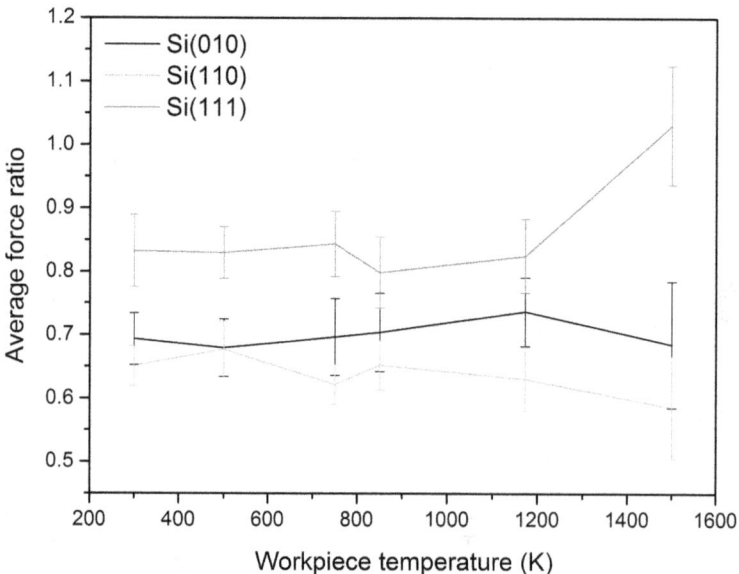

FIGURE 14.16 Variation in the average force ratio while cutting silicon on different crystallographic planes at various temperatures.

plane. The error bars in Figure 14.16 signify the magnitude of standard deviation and fluctuations in the average force ratio. The force ratio varies from 0.58 to 1.03 on different crystal planes and temperatures. It may be noted that the force ratio is maximum on the (111) plane while it was minimum on the (110) plane. A notable observation is that the force ratio remains almost unchanged up to 1173 K on all the three crystallographic planes and a sudden and abrupt change occurs beyond 1173 K.

14.4 MD SIMULATION OF CHEMICAL-MECHANICAL POLISHING (CMP)

The high calculation cost of quantum mechanics (QM) limits the time and scale, and classical MD lacks the description for chemical reactions. The reactive force field (ReaxFF) can be considered as a good remedy to study the chemical reactions and mechanical effects occurring during processes like CMP. This section discusses the CMP processes from the point of view of the micromechanical and chemical effects under the synergy of velocity, pressure, and polishing slurry as well as the influence of process parameters on the material removal and wear debris.

14.4.1 INTRODUCTION: BACKGROUND AND DRIVING FORCES

Due to the limitation of current computing power, it is particularly difficult to establish a model that complies with the actual polishing process. The polishing process using the MD simulation model is mainly divided into two kinds, one kind is the hemisphere wear particle model, using the hemisphere rather than the whole ball, so as to reduce the calculations. The other is a sandwich model. We know that in the actual polishing process, the contact and removal of abrasive particles and workpiece occur in a few convex peaks, and the contact time is very short. Therefore, the sandwich model assumes the simulation of the contact and removal of the convex peak part, and it has a good agreement with experiments. At present, scholars have studied the CMP process of Si [68, 69], Cu [70, 71], glass [72–74], KOH [75], and diamond [76–80] using the ReaxFF MD method.

14.4.2 THE CMP MODEL

The ReaxFF MD model of the CMP process is shown in Figure 14.17. From top to bottom, the model consists of rigid movable layer, diamond abrasive free layer, polishing slurry, diamond substrate layer, and fixed layer. The atoms in the fixed layer are similar to the diamond that is fixed on the experimental bench, and the atoms in free layer can move freely in simulation.

The whole simulation can be divided into three parts: (1) The rigid movable layer of the abrasive particles pressing down at a certain velocity until the rigid movable layer reaches the target pressure; (2) equilibrate pressure; (3) making the abrasive sliding along the x axis at a certain velocity.

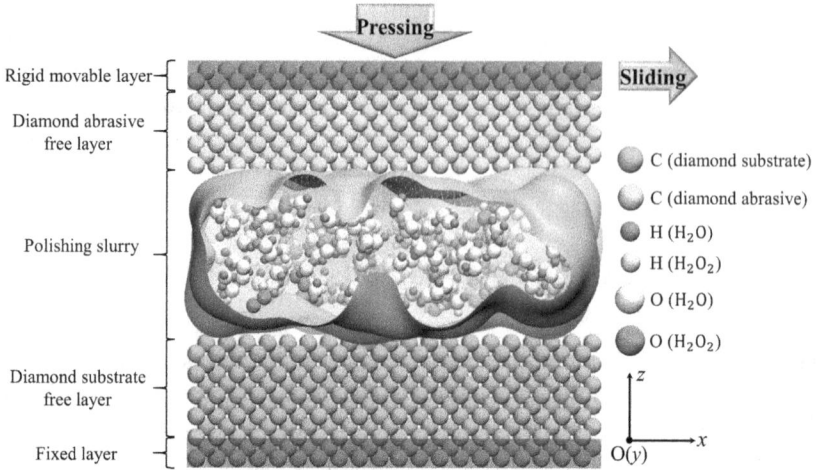

FIGURE 14.17 CMP model of diamond in aqueous H_2O_2. Reproduced with permission from [80], copyright Elsevier, 2020.

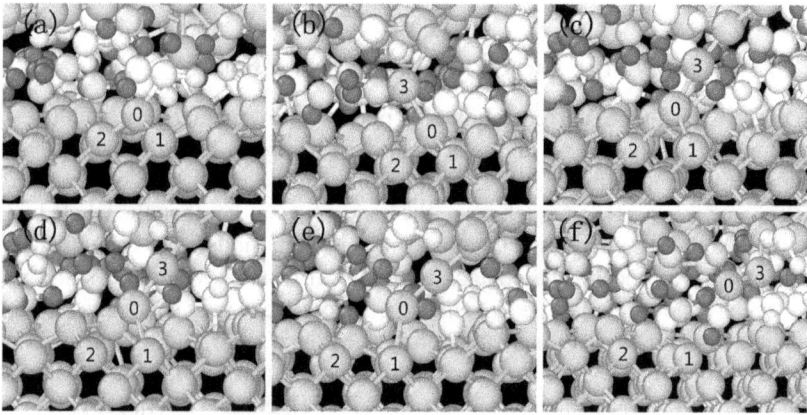

FIGURE 14.18 The removal of C atoms. (a) The adsorption of –OH. (b) The formation of C0–C3 bond. (c) The stretch of C0–C2 bond. (d) The breakage of C0–C2 bond. (e) The stretch of C0–C1 bond. (f) The breakage of C0–C1 bond. Reproduced with permission from [79], copyright Elsevier, 2019.

14.4.3 THE REMOVAL MECHANISM DURING THE CMP PROCESS

The removal details of C atoms during the CMP of diamond in aqueous H_2O_2 are shown in Figure 14.18. Firstly, a –OH from the decomposition of H_2O_2 adsorbed on the marked C_0 atom. The –OH is stretched under the action of abrasive, then the –OH gradually falls off from the C_0 atom under the shear action of the abrasive. Subsequently, the –OH drops off from the C_0 atom, and the C_3 atom from the diamond abrasive forms a bond with the C_0 atom. Finally, the C_0 atom is taken away by the abrasive.

Figure 14.19 represents the removal form of carbon atoms in CMP of diamond. Under low pressure, carbon atoms are mainly removed in the form of single C–C bridge bonds. With the increase in pressure, the single C–C bridge bonds translate into multiple bridge bonds, i.e., one carbon atom in the substrate forms interface bridge bonds with multiple carbon atoms in the abrasive.

Figure 14.20 shows the removal process of the atoms in CMP of fused glass. The Si^1 atom on the surface of fused glass is hydroxylated by the O^2H of H_2O, and the O^6 atom on the abrasive surface combines with the free H^{11} proton to form a Si–OH bond. A new chemical bond, Si^1–O^2–Si^{12}, is formed after the fracture of Si^1–O^2 bond. The chemical bonds between the silicon atoms and the atoms of the fused glass substrate are broken by mechanical forces transferred by the interfacial bridge bonds. Bonds are constantly formed and broken, and the synergy of several bridge bonds eventually removes the Si^1 atom.

Figure 14.21 depicts the removal forms of fused glass during CMP. With an increase of pressure, the removal method of fused glass changes from single atom removal to chain removal. When the pressure increases, cluster removal appears. When the pressure is 2 GPa, the fused glass surface was removed in the form of

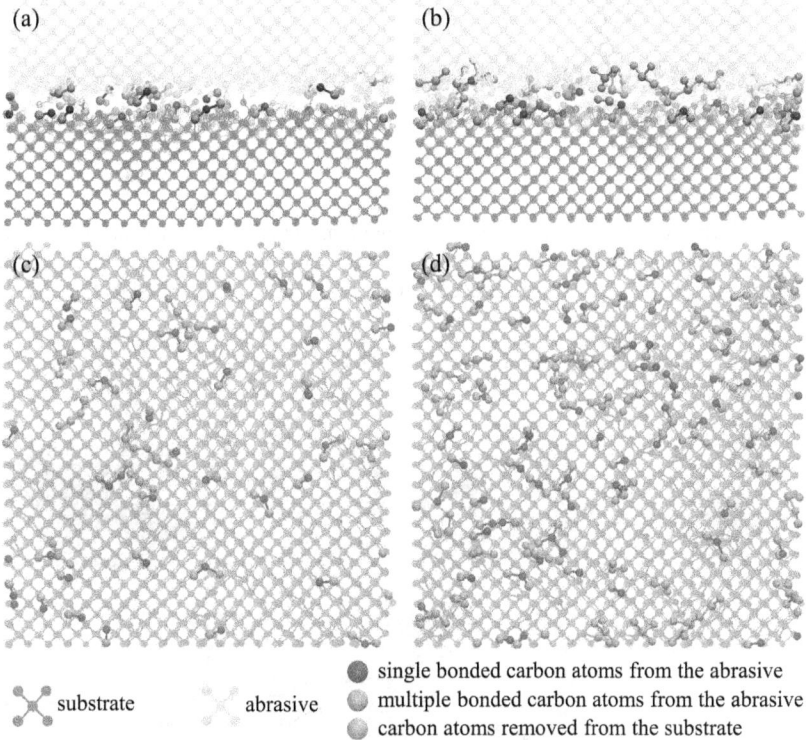

substrate   abrasive   ● single bonded carbon atoms from the abrasive
                                 ◐ multiple bonded carbon atoms from the abrasive
                                 ◌ carbon atoms removed from the substrate

FIGURE 14.19 Removal form of carbon atoms (a) single bridge bond removal under 12 GPa, (b) multiple bridge bonds removal under 14 GPa. Reproduced with permission from [80], copyright Elsevier, 2020.

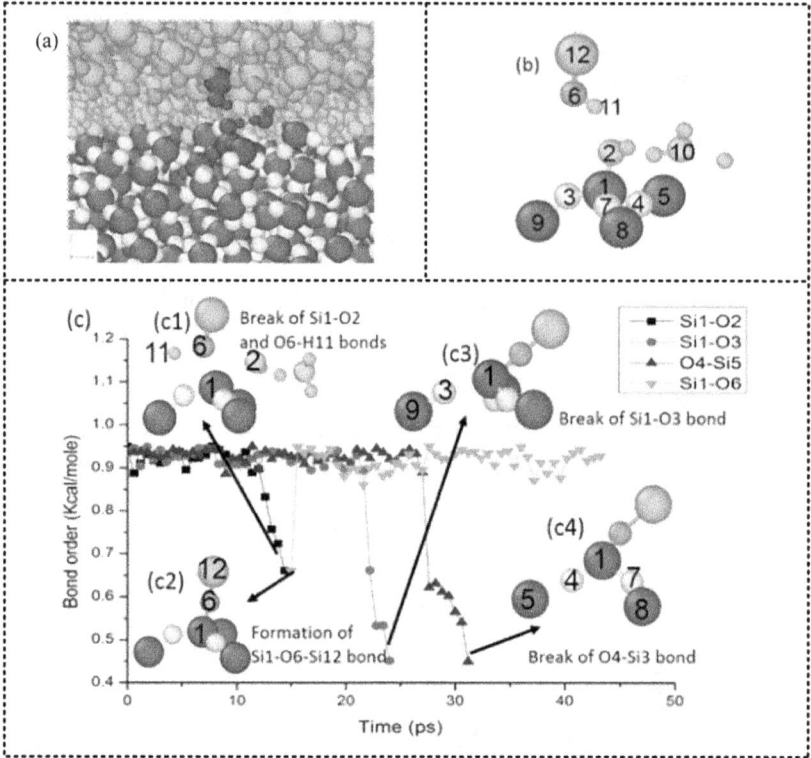

FIGURE 14.20 The formation of bridge bond and removal path of surface, (a) position of the tracked atoms during the sliding process, (b) initial state of the tracked atoms, (c) changes of bond order during the sliding process. Reproduced with permission from [72], copyright Elsevier, 2019.

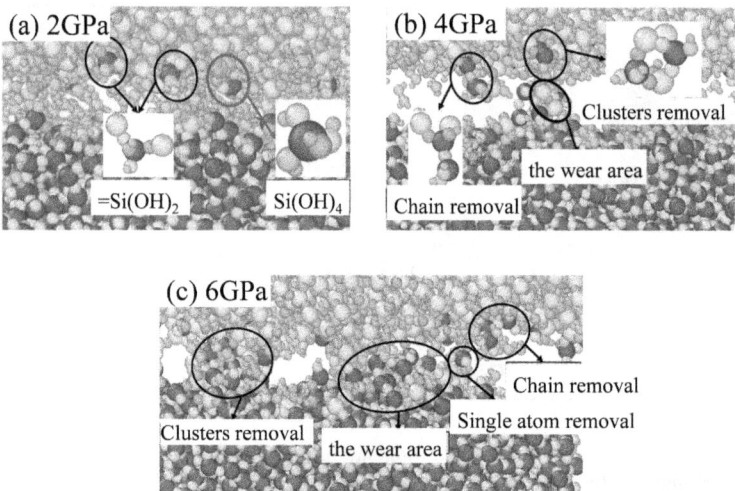

FIGURE 14.21 The removal forms of fused glass during CMP. (a) 2 GPa (b) 4 GPa(c) 6 GPa.

a single silicon atom. There are two forms, one is attached to the particle surface, which is bonded with oxygen atoms in the abrasive grain. Another is adsorbed on the abrasive grain in the form of $Si(OH)_2$ to be removed, and exists in solution in the form of $Si(OH)_4$. When the interfacial pressure is 4 GPa the removal mode of surface atoms changes, from single atom removal mode to chain removal mode. Most atoms are removed in linear or circular chain mode, and finally attach to the lower surface of the abrasion-based particles. A few atoms exist in the solution in the form of compounds. When the interfacial pressure is 6 GPa, the removal mode of surface atoms gradually changes from chain removal to cluster removal. The surface atoms of fused glass attach to the lower surface of the abrasive particles in the form of clusters, and a large number of surface atoms are removed. Meanwhile, chain removal and single atom removal also exist.

14.4.4 FORMATION MECHANISM OF INTERFACE BRIDGE BONDS

In CMP of diamond, there is growing evidence that the formation and breakage of interfacial bonds play a dual role in interface friction force. As seen from Figure 14.22, the numbers of interfacial C–O–C and C–C bonds increase with the

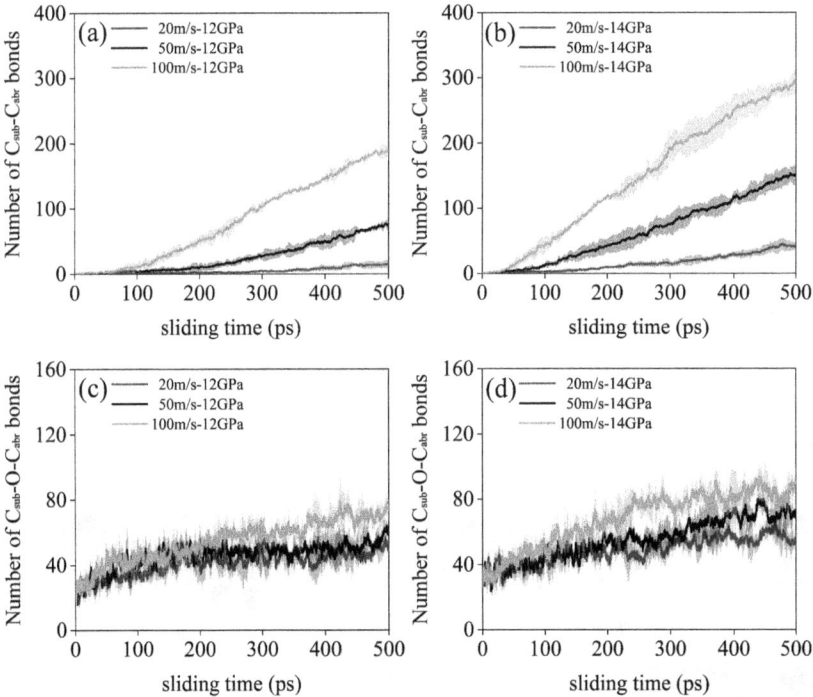

FIGURE 14.22 Numbers of the interfacial C_{sub}–C_{abr} bonds and C_{sub}–O–C_{abr} bonds over sliding time under different pressures and sliding velocities. (The C atoms from the substrate and abrasive are abbreviated as C_{sub} and C_{abr}, respectively.) The mean values (dark curve) and standard deviations (light shade) were obtained by the averaged data of three samples. Reproduced with permission from [80], copyright Elsevier, 2020.

increase of the sliding velocity, which is consistent with the initial trend of the friction force as shown in Figure 14.23. In the initial stage, the friction force depends on the interfacial bonding and varies proportionally with the numbers of C–O–C and C–C bonds. The continuous formation of the interfacial bonds strengthens the interface friction force.

Figure 14.24 shows the formation of an interface bridge bond in the CMP process of fused glass. In the initial state, the Si^1 atom bonded with O^2H^3 forms the surface hydroxyl structure. With the movement of abrasive, when the Si^1 atom gradually approaches the Si^4 atom on the fused glass surface, the dehydrogenation occurs on Si^1–O^2H^3 structure and the H^3 re-bonds with O^5 atom. Finally, the O^5H^3 is separated from the Si^4 atom under the action of abrasive, and interface Si^1–O^2–Si^4 bond forms through the dehydroxylation reaction.

14.4.5 CHEMICAL STATE ON THE SURFACE

Figure 14.25 depicts the chemical bonds formed on the diamond surface. In aqueous H_2O_2, the diamond surface reacts with polishing slurry, and the water molecules decompose into H and –OH, or H and two O. Ions adsorbed on the diamond surface,

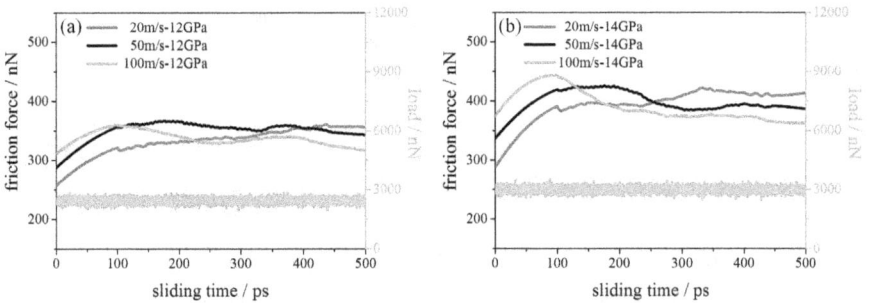

FIGURE 14.23 Evolutions of friction force over the sliding time under different pressures and velocities. (a) 12 GPa (b) 14 GPa. Reproduced with permission from [80], copyright Elsevier, 2020.

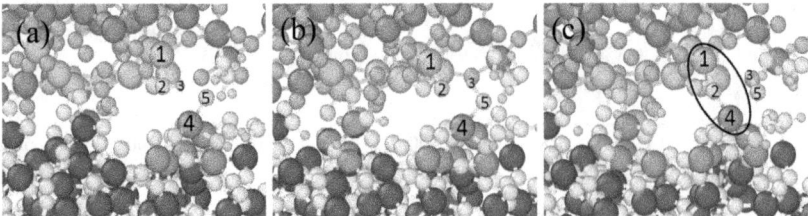

FIGURE 14.24 The formation process of interface bridge bond through dehydrogenation and dehydroxylation reactions (a) initial state, (b) dehydrogenation of Si^1–O^2H^3, (c) formation of Si^1–O^2–Si^4 bridge bond. Reproduced with permission from [72]. Copyright Elsevier, 2019.

(a) (b)

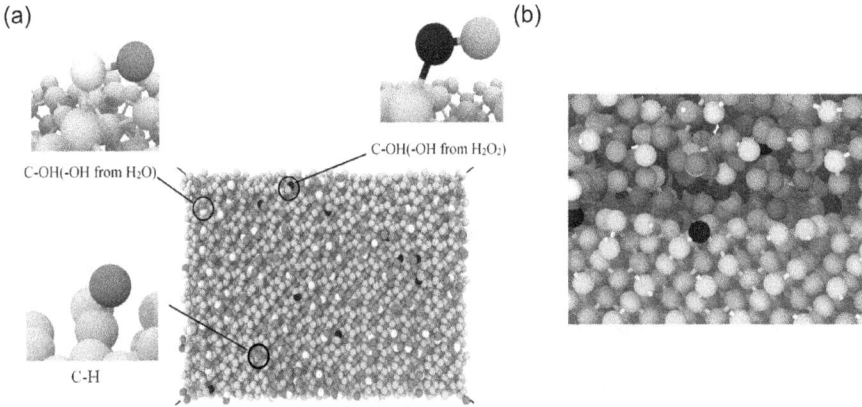

FIGURE 14.25 Chemical bonds formed on the diamond surface. (a) the formation of C–H, C–O–H, C–O; (b) the formation of C–O–C.

such as absorbing H to form C–H, adsorbing –OH to form C–OH, adsorbing O to form C–O, and O atoms "intrude" diamond to form C–O–C.

Figure 14.26 shows the surface of fused glass after reaction with H_2O. After reaction with H_2O, the chemical state of quartz glass surface has completely changed. H_2O has two adsorption forms on the quartz glass surface, namely molecular adsorption and dissociative adsorption. The main form of molecular adsorption is that H_2O adsorbs on the quartz glass surface. The process of dissociative adsorption is mainly that H_2O decomposes into H and OH, and then H and OH combine with the O atoms or under coordinated Si atoms on the quartz glass surface to form Si–OH group bonds, respectively.

14.4.6 EFFECT OF AQUEOUS H_2O_2

The degree of hydroxylation on the fused glass surface affects the removed forms of surface atoms. Figure 14.27 shows the distribution of atoms removed on the initial surface. When the ratio of Si–OH group is 0%, the surface atoms are mainly removed in the form of single Si atom along with the surrounding bonded O atoms, and the atoms removed are widely distributed. When the ratio of Si–OH group on the fused glass surface is 30%, the atoms removed are concentrated in one region. The removal form of surface atoms is chain removal at this situation. When the ratio is 50%, the atoms removed decrease obviously due to the decrease of bridge bonds.

14.5 CONCLUDING REMARKS

Unlike quantum mechanics, MD simulation discards the details of the electronic motion and only considers the movement of atomic nuclei which permits the study of relatively larger systems. While MD offers many other technical advantages, it is restricted by the size of the simulation and the time to perform that simulation.

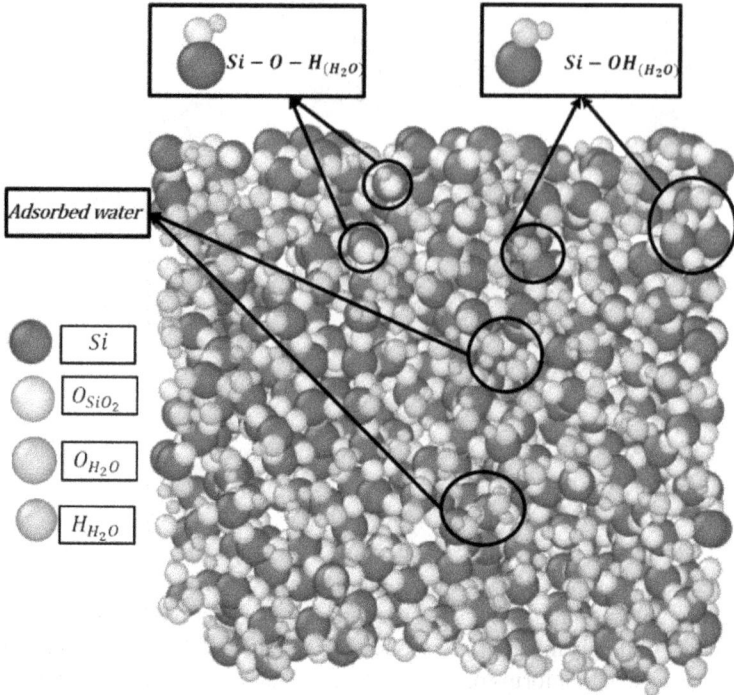

FIGURE 14.26 Surface of quartz glass after reaction with H_2O.

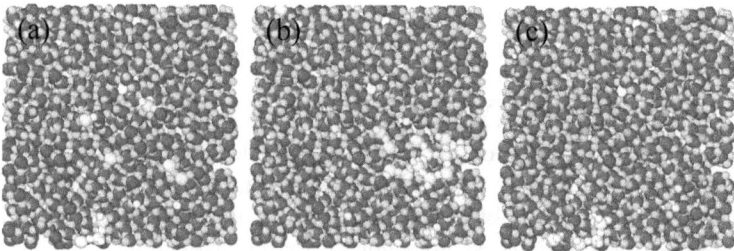

FIGURE 14.27 Distribution of atoms removed on the initial surface (red parts shown in the online published version of the paper represent the atoms removed) (a) 0% (b) 30% (c) 50%.

Particularly for simulators, analogous to the "Law of Constancy of Pain" is that while computing power has grown over time, the amount of wall-clock time available to an individual researcher on large computing platforms has not [81]. A glance at the comparison of the various key advantages and current limitations of MD in the context of nanometric cutting is presented in Table 14.8.

Based on the MD simulation presented in this chapter, the following conclusions can be drawn:

TABLE 14.8
Advantages and Limitations of the Molecular Dynamics Simulation

S. No.	Advantages	Limitations
1.	MD algorithm enables the taking into consideration of a more fundamental unit of matter i.e., atom and hence material properties are described naturally by their interaction potentials. Influence of crystal anisotropy, tribo-chemistry of the process, and basic mechanisms underlying a wear process can thus be suitably studied through MD. Also, MD permits investigation of the theoretical approachable limits.	MD cannot predict the attainable experimental measure of machined surface roughness which is a prime requirement governing the choice of a material in an industrial application. Even if a theoretical value is estimated, it will always remain an ideal limit which can only be attained under ideal set of machining conditions.
2.	MD permits online monitoring of the machining processes with good temporal and spatial resolution in a reversible manner. Any time step can simply be reversed through a computer program to analyze it at any given point of time.	Time to finish one simulation is a major challenge associated in performing a simulation with a realistic cutting speed and large size specimen.
3.	MD simulation avoids the use of expensive equipment and apparatus which are key requirements to perform nanometric cutting experiments. Besides, material, once consumed, will be required to reorder whereas MD can perform any number of trials with number of varying parameters.	Size of the workpiece and tool material cannot be varied to a larger (experimental) scale because of the current memory limitations associated with handling a large size data file.
4.	MD simulation offers repeatability of the process. This is to say that the type of work material, cutting tool material, and the environmental conditions can all be kept intact and maintained at a pre-decided value.	Ongoing work on the development of potential functions is still restricted to the use of a variety of coolants during a simulation which is often a prerequisite in a real experiment.
5.	MD simulation provides flexibility to perform the simulation at any place as a computer system is mobile whereas an ultra-precision machine tool (exhibiting high stiffness) demands static foundation and is thus static.	An advanced researcher can only perform an appropriate MD simulation as it requires accurate understanding of various disciplines whereas a machining trial can be performed by a relatively inferior knowledge.

- The machining of silicon on (111) plane along <–110> cutting direction consumes the least energy to cut which is in accordance with the wealth of publications.
- Tribo-chemistry (formation of silicon carbide) through a solid-state single-phase reaction up to a cutting temperature of 959 K between the diamond tool and silicon workpiece in tandem with graphitization (sp^3–sp^2) disordering of carbon atoms from the diamond tool has been explained as the mechanism by which a diamond cutting tool wears while cutting silicon.

- The resultant force exerted by the rake face of the tool on the chip is found to decrease by 24% when cutting the (111) surface at 1173 K compared to that at room temperature. Besides, smaller resultant force, friction coefficient at the tool/chip interface, and chip temperature are witnessed on the (111) crystal plane, as opposed to the other planes. MD simulation therefore approves that machinability of Si can be improved at elevated temperature.
- In the CMP process, H_2O has two adsorption forms on the quartz glass surface, namely molecular adsorption and dissociative adsorption. The main form of molecular adsorption is that H_2O adsorbs on the quartz glass surface. The process of dissociative adsorption is mainly that H_2O decomposes into H and OH, and then H and OH combine with the O atoms or under coordinated Si atoms on the quartz glass surface to form Si–OH group bonds, respectively.

The recommendations for future works are:

- **A user friendly and computing efficient multi-scale modeling approach** for the study of chip formation and removal process, including tool wear, from nanometer scale, micrometer scale to millimeter scale. The understanding of the material removal process will become more important when the machining capability of the ultra-precision machine tool is approaching its maximum.
- **Particle-fluid-solid simulation platform**. Use of coolant is a promising approach to suppress tool wear. However, due to the lack of potential functions, the effect of coolant during the ultra-precision machining process has not been full revealed.
- **MD simulation in air**. As ultra-precision machining processes take place in air, in the future work, MD simulations may be done for an in-depth study of its influence on the process.
- **Low cost nanometric sensors for in situ process monitoring.** This is much needed for the development of high-fidelity MD simulations to be experimentally validated with ease.

ACKNOWLEDGMENTS

The authors would like to thank the EPSRC (K018345/1, T024844/1 and V055208/1) and the National Natural Science Foundation of China (No. 52175382, 51975095 and 51991372) for their financial support of the MD simulations research.

Saurav Goel greatly acknowledges the financial support provided by the UKRI via Grants No. EP/L016567/1, EP/S013652/1, EP/S036180/1, EP/T001100/1 and EP/T024607/1, TFIN+Feasibility study award to LSBU (EP/V026402/1), the Royal Academy of Engineering via Grants No. IAPP18-19\295 and TSP1332, EURAMET EMPIR A185 (2018). Wherever applicable, the work made use of Isambard Bristol, UK supercomputing service accessed by a Resource Allocation Panel (RAP) grant as well as ARCHER resources (Project e648).

NOMENCLATURE

ABOP	analytical bond order potential
AIREBO	adaptive intermolecular reactive empirical bond order
BOP	bond–order potential
CMP	chemical-mechanical polishing
CNA	common neighbor analysis
COMB	charge optimized many-body
DXA	dislocation extraction algorithm
EAM	embedded-atom method
EIM	embedded-ion method
FEA	finite element analysis
GAP	Gaussian approximation potential
HPPT	high-pressure phase transformation
HX	hexagonal
LAMMPS	large scale atomic/molecular massively parallel simulator
MC	Monte Carlo
MD	molecular dynamics
MEAM	modified embedded-atom method
NPH	isoenthalpic-isobaric ensemble
NPT	isobaric isothermal ensemble
NVE	microcanonical ensemble
NVT	canonical ensemble
OVITO	open visualization tool
PBC	period boundary condition
QM	quantum mechanics
RDF	radial distribution functions
ReaxFF	reactive force field
REBO	reactive empirical bond order
SPC	simple point charge
SPDT	single-point diamond turning
VMD	visual molecular dynamics

REFERENCES

1. N. Taniguchi. "Current Status in, and Future Trends of, Ultraprecision Machining and Ultrafine Materials Processing." *CIRP Annals*, vol. 32, no. 2, pp. 573–582, 1983, doi: 10.1016/S0007-8506(07)60185-1.
2. J. Belak, W. Hoover, C. Hoover, A. De Groot, and I. Stowers. "Molecular Dynamics Modeling Applied to Indentation and Metal Cutting Problems." *Thrust Area Reps*, vol. 89, no. 1, pp. 4–8, 1990.
3. S. Shimada and N. Ikawa. "Molecular Dynamics Analysis as Compared With Experimental Results of Micromachining." *CIRP Annals*, vol. 41, no. 1, pp. 117–120, 1992, doi: 10.1016/S0007-8506(07)61165-2.
4. T. Inamura and N. Takezawa. "Cutting Experiments in a Computer Using Atomic Models of a Copper Crystal and a Diamond Tool." In *Progress in Precision Engineering*. Berlin: Heidelberg, 1991, pp. 231–242, doi: 10.1007/978-3-642-84494-2_25.

5. D. C. Rapaport. *The Art of Molecular Dynamics Simulation*. Cambridge: Cambridge University Press, 2004.

6. X. Luo. *High Precision Surfaces Generation: Modelling, Simulation and Machining Verification*. Chisinau, Moldova: LAMBERT Academic Publishing, 2011.

7. S. Goel. *An Atomistic Investigation on the Nanometric Cutting Mechanism of Hard, Brittle Materials*. Heriot-Watt University, 2013.

8. R. Komanduri, N. Chandrasekaran, and L. M. Raff. "Molecular Dynamics Simulation of the Nanometric Cutting of Silicon." *Philosophical Magazine B*, vol. 81, no. 12, pp. 1989–2019, 2001.

9. S. Z. Chavoshi. *An Investigation on the Mechanics of Nanometric Cutting for Hard-Brittle Materials at Elevated Temperatures*. University of Strathclyde, 2016.

10. M. S. Daw and M. I. Baskes. "Embedded-Atom Method: Derivation and Application to Impurities, Surfaces, and Other Defects in Metals." *Physical Review B*, vol. 29, no. 12, p. 6443, 1984.

11. F. H. Stillinger and T. A. Weber. "Computer Simulation of Local Order in Condensed Phases of Silicon." *Physical Review B*, vol. 31, no. 8, p. 5262, 1985.

12. F. H. Stillinger and T. A. Weber. "Erratum: Computer Simulation of Local Order in Condensed Phases of Silicon [*Phys. Rev. B* 31, 5262 (1985)]." *Physical Review B*, vol. 33, no. 2, pp. 1451–1451, 1986, doi: 10.1103/PhysRevB.33.1451.

13. H. J. C. Berendsen, J. R. Grigera, and T. P. Straatsma. "The Missing Term in Effective Pair Potentials." *Journal of Physical Chemistry*, vol. 91, no. 24, pp. 6269–6271, 1987.

14. J. Tersoff. "New Empirical Approach for the Structure and Energy of Covalent Systems." *Physical Review B*, vol. 37, no. 12, p. 6991, 1988.

15. J. Tersoff. "Empirical Interatomic Potential for Silicon With Improved Elastic Properties." *Physical Review B*, vol. 38, no. 14, p. 9902, 1988.

16. J. Tersoff. "Modeling Solid-State Chemistry: Interatomic Potentials for Multicomponent Systems." *Physical Review B*, vol. 39, no. 8, p. 5566, 1989.

17. J. Tersoff. "Erratum: Modeling Solid-State Chemistry: Interatomic Potentials for Multicomponent Systems." *Physical Review B*, vol. 41, no. 5, p. 3248, 1990.

18. J. Tersoff. "Carbon Defects and Defect Reactions in Silicon." *Physical Review Letters*, vol. 64, no. 15, p. 1757, 1990.

19. J. Tersoff. "Chemical Order in Amorphous Silicon Carbide." *Physical Review B*, vol. 49, no. 23, p. 16349, 1994.

20. P. M. Agrawal, L. M. Raff, and R. Komanduri. "Monte Carlo Simulations of Void-Nucleated Melting of Silicon Via Modification in the Tersoff Potential Parameters." *Physical Review B*, vol. 72, no. 12, p. 125206, 2005.

21. R. Devanathan, T. D. de la Rubia, and W. J. Weber. "Displacement Threshold Energies in β-SiC." *Journal of Nuclear Materials*, vol. 253, nos. 1–3, pp. 47–52, 1998.

22. M. I. Baskes, J. S. Nelson, and A. F. Wright. "Semiempirical Modified Embedded-Atom Potentials for Silicon and Germanium." *Physical Review B*, vol. 40, no. 9, p. 6085, 1989.

23. D. W. Brenner. "Empirical Potential for Hydrocarbons for Use in Simulating the Chemical Vapor Deposition of Diamond Films." *Physical Review B*, vol. 42, no. 15, p. 9458, 1990.

24. S. J. Stuart, A. B. Tutein, and J. A. Harrison. "A Reactive Potential for Hydrocarbons With Intermolecular Interactions." *The Journal of Chemical Physics*, vol. 112, no. 14, pp. 6472–6486, 2000.

25. A. C. Van Duin, S. Dasgupta, F. Lorant, and W. A. Goddard. "ReaxFF: A Reactive Force Field for Hydrocarbons." *The Journal of Physical Chemistry A*, vol. 105, no. 41, pp. 9396–9409, 2001.

26. P. Erhart and K. Albe. "Analytical Potential for Atomistic Simulations of Silicon, Carbon, and Silicon Carbide." *Physical Review B*, vol. 71, no. 3, p. 035211, 2005.

27. J. Yu, S. B. Sinnott, and S. R. Phillpot. "Charge Optimized Many-Body Potential for the Si/SiO$_2$ System." *Physical Review B*, vol. 75, no. 8, p. 085311, 2007.

28. X. W. Zhou and F. P. Doty. "Embedded-Ion Method: An Analytical Energy-Conserving Charge-Transfer Interatomic Potential and Its Application to the La-Br System." *Physical Review B*, vol. 78, no. 22, p. 224307, 2008.

29. A. P. Bartók, M. C. Payne, R. Kondor, and G. Csányi. "Gaussian Approximation Potentials: The Accuracy of Quantum Mechanics, Without the Electrons." *Physical Review Letters*, vol. 104, no. 13, p. 136403, 2010.

30. L. Verlet. "Computer "Experiments" on Classical Fluids. I. Thermodynamical Properties of Lennard-Jones Molecules." *Physical Review*, vol. 159, no. 1, p. 98, 1967.

31. S. Plimpton. "Fast Parallel Algorithms for Short-Range Molecular Dynamics." *Journal of Computational Physics*, vol. 117, no. 1, pp. 1–19, 1995, doi: 10.1006/jcph.1995.1039.

32. W. Humphrey, A. Dalke, and K. Schulten. "VMD: Visual Molecular Dynamics." *Journal of Molecular Graphics*, vol. 14, no. 1, pp. 33–38, 1996, doi: 10.1016/0263-7855 (96)00018-5.

33. A. Stukowski. "Visualization and Analysis of Atomistic Simulation Data With OVITO – The Open Visualization Tool." *Modelling and Simulation in Materials Science and Engineering*, vol. 18, no. 1, p. 015012, 2009.

34. G. J. Ackland and A. P. Jones. "Applications of Local Crystal Structure Measures in Experiment and Simulation." *Physical Review B*, vol. 73, no. 5, p. 054104, 2006.

35. S. Goel, X. Luo, A. Agrawal, and R. L. Reuben. "Diamond Machining of Silicon: A Review of Advances in Molecular Dynamics Simulation." *International Journal of Machine Tools and Manufacture*, vol. 88, pp. 131–164, 2015.

36. J. Yan, K. Syoji, and J. Tamaki. "Some Observations on the Wear of Diamond Tools in Ultra-Precision Cutting of Single-Crystal Silicon." *Wear*, vol. 255, nos. 7–12, pp. 1380–1387, 2003.

37. R. G. Jasinevicius, J. G. Duduch, L. Montanari, and P. S. Pizani. "Dependence of Brittle-to-Ductile Transition on Crystallographic Direction in Diamond Turning of Single-Crystal Silicon." *Proceedings of the Institution of Mechanical Engineers, Part B: Journal of Engineering Manufacture*, vol. 226, no. 3, pp. 445–458, 2012.

38. M. S. Uddin, K. H. W. Seah, X. P. Li, M. Rahman, and K. Liu. "Effect of Crystallographic Orientation on Wear of Diamond Tools for Nano-Scale Ductile Cutting of Silicon." *Wear*, vol. 257, nos. 7–8, pp. 751–759, 2004.

39. K. Cheng, X. Luo, R. Ward, and R. Holt. "Modeling and Simulation of the Tool Wear in Nanometric Cutting." *Wear*, vol. 255, nos. 7–12, pp. 1427–1432, 2003.

40. Z. G. Zhang, F. Z. Fang, X. T. Hu, and C. K. Sun. "Molecular Dynamics Study on Various Nanometric Cutting Boundary Conditions." *Journal of Vacuum Science & Technology B: Microelectronics and Nanometer Structures Processing, Measurement, and Phenomena*, vol. 27, no. 3, pp. 1355–1360, 2009.

41. S. Goel, X. Luo, R. L. Reuben, and H. Pen. "Influence of Temperature and Crystal Orientation on Tool Wear During Single Point Diamond Turning of Silicon." *Wear*, vol. 284, pp. 65–72, 2012.

42. R. Komanduri, N. Chandrasekaran, and L. M. Raff. "Effect of Tool Geometry in Nanometric Cutting: A Molecular Dynamics Simulation Approach." *Wear*, vol. 219, no. 1, pp. 84–97, 1998.

43. R. Komanduri, N. Chandrasekaran, and L. M. Raff. "Some Aspects of Machining With Negative-Rake Tools Simulating Grinding: A Molecular Dynamics Simulation Approach." *Philosophical Magazine B*, vol. 79, no. 7, pp. 955–968, 1999.

44. C. J. Wong. *"Fracture and Wear of Diamond Cutting Tools."* Journal of Engineering Materials and Technology, vol. 103, no. 4, pp. 341–345, 1981.

45. J. F. Belak, I. F. Stowers, and D. B. Boercker. "Simulation of Diamond Turning of Silicon Surfaces." In *Proceedings of 7th American Society Precision Engineering Annual Conference*, 1992, pp. 76–79.

46. J. F. Belak and I. F. Stowers. *A Molecular Dynamics Model of the Orthogonal Cutting Process.* Cambridge: Lawrence Livermore National Laboratory, 1990.

47. K. Maekawa and A. Itoh. "Friction and Tool Wear in Nano-Scale Machining—A Molecular Dynamics Approach." *Wear*, vol. 188, nos. 1–2, pp. 115–122, 1995.

48. M. B. Cai, X. P. Li, and M. Rahman. "Study of the Mechanism of Groove Wear of the Diamond Tool in Nanoscale Ductile Mode Cutting of Monocrystalline Silicon." 2007.

49. M. B. Cai, X. P. Li, and M. Rahman. "Characteristics of 'Dynamic Hard Particles' in Nanoscale Ductile Mode Cutting of Monocrystalline Silicon With Diamond Tools in Relation to Tool Groove Wear." *Wear*, vol. 263, nos. 7–12, pp. 1459–1466, 2007.

50. R. Narulkar, S. Bukkapatnam, L. M. Raff, and R. Komanduri. "Graphitization as a Precursor to Wear of Diamond in Machining Pure Iron: A Molecular Dynamics Investigation." *Computational Materials Science*, vol. 45, no. 2, pp. 358–366, 2009.

51. Z. Wang, Y. Liang, M. Chen, Z. Tong, and J. Chen. "Analysis About Diamond Tool Wear in Nano-Metric Cutting of Single Crystal Silicon Using Molecular Dynamics Method." In *5th International Symposium on Advanced Optical Manufacturing and Testing Technologies: Advanced Optical Manufacturing Technologies*, 2010, vol. 7655, p. 765500.

52. R. Beyers. "Thermodynamic Considerations in Refractory Metal-Silicon-Oxygen Systems." *Journal of Applied Physics*, vol. 56, no. 1, pp. 147–152, 1984.

53. Z. Cheng. *Reaction Kinetics and Structural Evolution for the Formation of Nanocrystalline Silicon Carbide Via Carbothermal Reduction.* Georgia: Georgia Institute of Technology, 2004.

54. F. Viscomi and L. Himmel. "Kinetic and Mechanistic Study on the Formation of Silicon Carbide From Silica Flour and Coke Breeze." *JOM*, vol. 30, no. 6, pp. 21–24, 1978.

55. A. W. Weimer, K. J. Nilsen, G. A. Cochran, and R. P. Roach. "Kinetics of Carbothermal Reduction Synthesis of Beta Silicon Carbide." *AIChE Journal*, vol. 39, no. 3, pp. 493–503, 1993.

56. L. Pastewka, M. Mrovec, M. Moseler, and P. Gumbsch. "Bond Order Potentials for Fracture, Wear, and Plasticity." *MRS Bulletin*, vol. 37, no. 5, pp. 493–503, 2012.

57. S. Arefin, X. P. Li, M. B. Cai, M. Rahman, K. Liu, and A. Tay. "The Effect of the Cutting Edge Radius on a Machined Surface in the Nanoscale Ductile Mode Cutting of Silicon Wafer." *Proceedings of the Institution of Mechanical Engineers, Part B: Journal of Engineering Manufacture*, vol. 221, no. 2, pp. 213–220, 2007.

58. Y. Gogotsi, G. Zhou, S.-S. Ku, and S. Cetinkunt. "Raman Microspectroscopy Analysis of Pressure-Induced Metallization in Scratching of Silicon." *Semiconductor Science and Technology*, vol. 16, no. 5, p. 345, 2001.

59. K. E. Puttick, L. C. Whitmore, P. Zhdan, A. E. Gee, and C. L. Chao. "Energy Scaling Transitions in Machining of Silicon by Diamond." *Tribology International*, vol. 28, no. 6, pp. 349–355, 1995.

60. D. A. Rigney and S. Karthikeyan. "The Evolution of Tribomaterial During Sliding: A Brief Introduction." *Tribology Letters*, vol. 39, no. 1, pp. 3–7, 2010.

61. X. P. Li, T. He, and M. Rahman. "Tool Wear Characteristics and Their Effects on Nanoscale Ductile Mode Cutting of Silicon Wafer." *Wear*, vol. 259, nos. 7–12, pp. 1207–1214, 2005.

62. Z. Zhang, J. Yan, and T. Kuriyagawa. "Study on Tool Wear Characteristics in Diamond Turning of Reaction-Bonded Silicon Carbide." *The International Journal of Advanced Manufacturing Technology*, vol. 57, nos. 1–4, pp. 117–125, 2011.

63. P. Albrecht. "New Developments in the Theory of the Metal-Cutting Process: Part I. The Ploughing Process in Metal Cutting." 1960.

64. M. Lai, X. D. Zhang, and F. Z. Fang. "Study on Critical Rake Angle in Nanometric Cutting." *Applied Physics A*, vol. 108, no. 4, pp. 809–818, 2012.

65. W. J. Zong, T. Sun, D. Li, K. Cheng, and Y. C. Liang. "XPS Analysis of the Groove Wearing Marks on Flank Face of Diamond Tool in Nanometric Cutting of Silicon Wafer." *International Journal of Machine Tools and Manufacture*, vol. 48, no. 15, pp. 1678–1687, 2008.

66. E. Brinksmeier and W. Preuss. "Micro-Machining." *Philosophical Transactions of the Royal Society A: Mathematical, Physical and Engineering Sciences*, vol. 370, no. 1973, pp. 3973–3992, 2012.

67. M. Wang, W. Wang, and Z. Lu. "Anisotropy of Machined Surfaces Involved in the Ultra-Precision Turning of Single-Crystal Silicon—A Simulation and Experimental Study." *The International Journal of Advanced Manufacturing Technology*, vol. 60, no. 5, pp. 473–485, 2012.

68. L. Chen, J. Wen, P. Zhang, B. Yu, C. Chen, T. Ma, X. Lu, S. H. Kim, and L. Qian. "Nanomanufacturing of Silicon Surface With a Single Atomic Layer Precision Via Mechanochemical Reactions." *Nature Communications*, vol. 9, no. 1, pp. 1–7, 2018.

69. J. Wen, T. Ma, W. Zhang, A. C. van Duin, and X. Lu. "Atomistic Mechanisms of Si Chemical Mechanical Polishing in Aqueous H_2O_2: ReaxFF Reactive Molecular Dynamics Simulations." *Computational Materials Science*, vol. 131, pp. 230–238, 2017.

70. X. Guo, X. Wang, Z. Jin, and R. Kang. "Atomistic Mechanisms of Cu CMP in Aqueous H_2O_2: Molecular Dynamics Simulations Using ReaxFF Reactive Force Field." *Computational Materials Science*, vol. 155, pp. 476–482, 2018.

71. X. Guo, S. Yuan, Y. Gou, X. Wang, J. Guo, Z. Jin, and R. Kang. "Study on Chemical Effects of H_2O_2 and Glycine in the Copper CMP Process Using ReaxFF MD." *Applied Surface Science*, vol. 508, p. 145262, 2020.

72. X. Guo, S. Yuan, J. Huang, C. Chen, R. Kang, Z. Jin, and D. Guo. "Effects of Pressure and Slurry on Removal Mechanism During the Chemical Mechanical Polishing of Quartz Glass Using ReaxFF MD." *Applied Surface Science*, vol. 505, p. 144610, 2020.

73. X. Guo, J. Huang, S. Yuan, C. Chen, Z. Jin, R. Kang, and D. Guo. "Effect of Surface Hydroxylation on Ultra-Precision Machining of Quartz Glass." *Applied Surface Science*, vol. 501, p. 144170, 2020.

74. D.-C. Yue, T.-B. Ma, Y.-Z. Hu, J. Yeon, A. C. T. van Duin, H. Wang, and J. Luo. "Tribochemistry of Phosphoric Acid Sheared Between Quartz Surfaces: A Reactive Molecular Dynamics Study." *The Journal of Physical Chemistry C*, vol. 117, no. 48, pp. 25604–25614, 2013.

75. P. Ranjan, R. Balasubramaniam, and V. K. Jain. "Mechanism of Material Removal During Nanofinishing of Aluminium in Aqueous KOH: A Reactive Molecular Dynamics Simulation Study." *Computational Materials Science*, vol. 156, pp. 35–46, 2019.

76. S. Yuan, X. Guo, J. Huang, Y. Gou, Z. Jin, R. Kang, and D. Guo. "Insight into the Mechanism of Low Friction and Wear During the Chemical Mechanical Polishing Process of Diamond: A Reactive Molecular Dynamics Simulation." *Tribology International*, vol. 148, p. 106308, 2020.

77. S. Yuan, X. Guo, J. Huang, M. Lu, Z. Jin, R. Kang, and D. Guo. "Sub-Nanoscale Polishing of Single Crystal Diamond (100) and the Chemical Behavior of Nanoparticles During the Polishing Process." *Diamond and Related Materials*, vol. 100, p. 107528, 2019.

78. X. Guo, S. Yuan, X. Wang, Z. Jin, and R. Kang. "Atomistic Mechanisms of Chemical Mechanical Polishing of Diamond (1 0 0) in Aqueous H_2O_2/Pure H_2O: Molecular Dynamics Simulations Using Reactive Force Field (ReaxFF)." *Computational Materials Science*, vol. 157, pp. 99–106, 2019.

79. S. Yuan, X. Guo, M. Lu, Z. Jin, R. Kang, and D. Guo. "Diamond Nanoscale Surface Processing and Tribochemical Wear Mechanism." *Diamond and Related Materials*, vol. 94, pp. 8–13, 2019.

80. S. Yuan, X. Guo, Q. Mao, J. Guo, A. C. T. van Duin, Z. Jin, R. Kang, and D. Guo. "Effects of Pressure and Velocity on the Interface Friction Behavior of Diamond Utilizing ReaxFF Simulations." *International Journal of Mechanical Sciences*, vol. 191, p. 106096, 2021.

81. S. J. Plimpton and A. P. Thompson. "Computational Aspects of Many-Body Potentials." *MRS Bulletin*, vol. 37, no. 5, pp. 513–521, 2012.

15 Sustainability Issues in Advanced Machining Processes

Palivela Bhargav Chandan, Aluri Manoj,
Kishor Kumar Gajrani, Shivansh Dhaka,
and Mamilla Ravi Sankar

CONTENTS

15.1 INTRODUCTION

Sustainability in machining is defined both within sustainability alone and as manufacturing sustainability. The word sustainability can be attributed to something that can be sustained (supported) and continued at that level indefinitely. On the other side of the coin, *sustainable manufacturing* can be attributed to the creation of products manufactured that employ processes that are energy conserving, non-polluting,

DOI: 10.1201/9780429160011-15

and preserve natural resources, besides being economically sound and also safer for employees, communities, and consumers. When applied to machining, sustainability refers to the making of products or components with the aid of subtractive processes dependent upon the removal of material by the action of a cutting tool usually using a machine tool to create the features and surfaces in a way that conserves and minimizes as well as being non-polluting [1]. Factors involved in a sustainable machining process are depicted in Figure 15.1.

Ever-increasing trends in the upcoming technology have also led to many concerns related to the environment, society, and economics in every field of engineering. In order to increase productivity and product quality as well as tool life, and to decrease production costs, the machining sectors have not paid enough attention to limiting environmental and health issues. A majority of the metal processing sector uses at least one type of machining process to remove material. A variety of manufacturing techniques is available; these are alternatives to machining techniques. Even then, machining is popular, and has a promising role in the shaping of products because of its capabilities in producing intricate geometries, excellent surface finish, and tremendous accuracy that other operations cannot readily produce. Machining is unusual as it is employed to manufacture both intermediate and finished items, either in small-scale or large-scale manufacturing. Nevertheless, machining creates a great deal of waste in terms of solid, liquid, and airborne particles that are hazardous to workplace safety, health, and environment. Machining also utilizes a considerable amount of electricity as a high-volume production process, which indeed constitutes an ancillary concern for the environment. In addition to increased global government restrictions, environmental and health risks connected with machining operations have also led to higher processing costs [2, 3].

Sustainability in any organization implies that something can be carried (sustained) indefinitely. Sustainability in manufacturing processes has drawn substantial attention in the last few years. The main drivers are steadily rising energy and

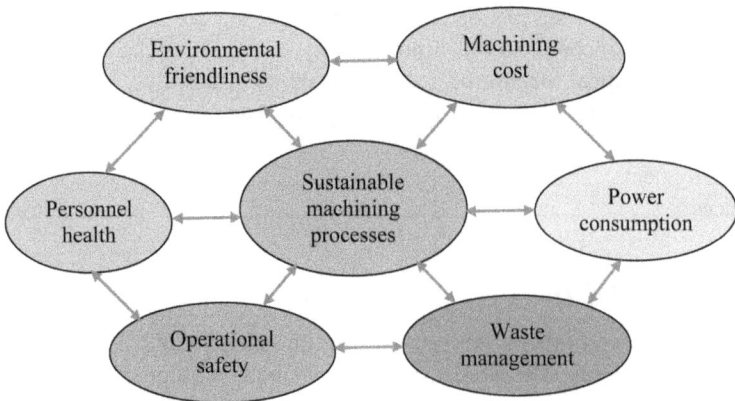

FIGURE 15.1 Factors involved in a sustainable machining process [1].

resource prices, government restrictions, and a desire to reduce the adverse health and environmental effects of the manufacturing processes [4, 5]. Recently, governments and world leaders have been advocating for *green and sustainable manufacturing* initiatives. To support and promote such initiatives, researchers, scientists, and engineers have the responsibility to make manufacturing processes green and sustainable to safeguard the present and future generations and the Earth in particular. Sustainable machining is defined as the creation of products by using methods that are environmentally friendly, economically sound, and safe for employees, consumers, and communities, as well as conserving less energy [6]. The first step toward sustainable manufacturing is to understand and characterize the energy consumption of any manufacturing system [7]. Sustainability in manufacturing can be achieved by focusing on three pillars, i.e., economy, society, and environment [8]. Conventionally, the 3-R concept of reduce, reuse, and recycle aims for a closed-loop cycle paradigm, but sustainable manufacturing focuses on the *6-R concept* which includes reduce, reuse, recover, redesign, remanufacture, and recycle [8]. Figure 15.2 illustrates a schematic of the 6-R concept of sustainability. Various researchers and industries are working toward achieving sustainability in manufacturing. Sustainability and energy efficiency-related issues in traditional methods of machining such as drilling, turning, and milling have already been studied and reported elsewhere [9, 10]. In this chapter, a systematic review of sustainability-related issues and energy consumption in advanced methods of machining such as electric discharge machining (EDM), ultrasonic machining (USM), and electrochemical machining (ECM), micromachining, etc., is discussed.

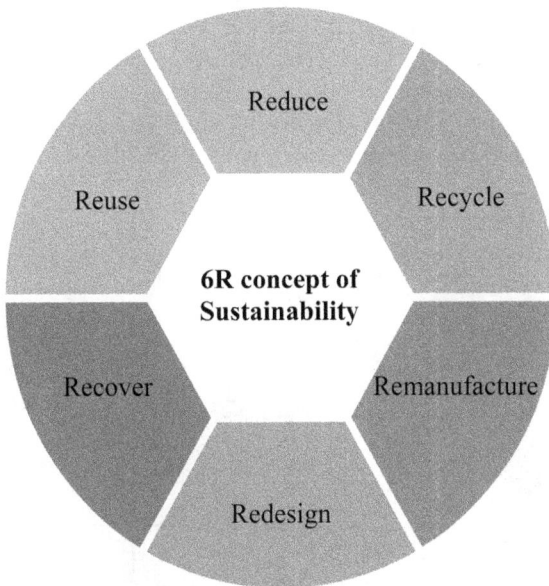

FIGURE 15.2 The schematic of the 6-R concept of sustainability.

In machining, the sustainability index is obtained from the study of relationships between the variables, between the ranges of sets of values of parameters of cutting and those values considered initially. So, the initial data must be recorded in all the cases under consideration. An improvement could lead to the sustainability index being attained regardless of the preliminary cut-off parameters that are taken into consideration.

15.2 MECHANICAL ADVANCED MACHINING PROCESSES AND SUSTAINABILITY ISSUES

15.2.1 ABRASIVE JET MACHINING AND SUSTAINABILITY ASPECTS

In abrasive jet machining (AJM), a high-velocity jet of abrasive(s) mixed gases/air is used to erode or chip off material from the workpiece under controlled conditions. Various types of abrasives such as silicon carbide, silica sand, olivine, aluminum oxide, garnet, etc. are commonly used in AJM. It is mostly used for cutting brittle and hard materials. Figure 15.3 shows a schematic of AJM, which has various advantages compared to conventional machining. The AJM process can cut complex contours on a wide range of materials. Also, the process is capable of cutting ultra-thin sheets with minimum bending in very little time.

However, there are some well-known environmental and health hazard issues associated with AJM. Any direct contact with the jet by the operator causes significant physical injury, such as unexpected removal of the operator's limb. The jet must be protected adequately, and the operator should maintain a safe distance from the jet [12].

The most hazardous issue is abrasive dust particles. A study revealed that a substantial threat could arise from the particles of the hazardous elements being cut like cadmium, zinc, arsenic, mercury, and lead thereby damaging the liver (as depicted in Figure 15.4) due to these toxins affecting the reproductive system, skeletal system, and kidneys as well as the respiratory system [13]. Another study revealed the sequence of hazards due to heavy metal sources in the vicinity of female operators

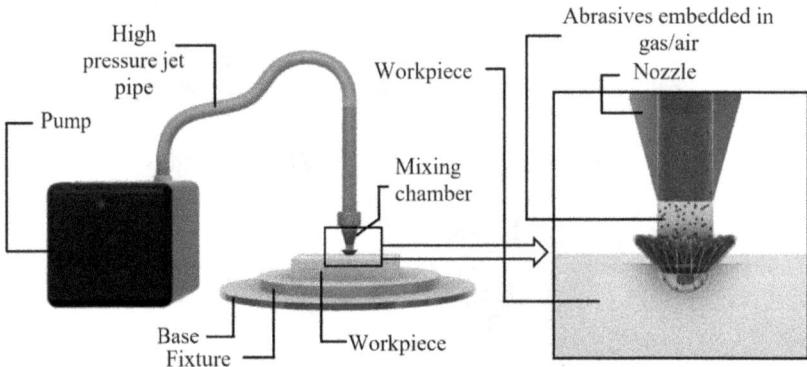

FIGURE 15.3   A schematic diagram of the abrasive jet machining process [11].

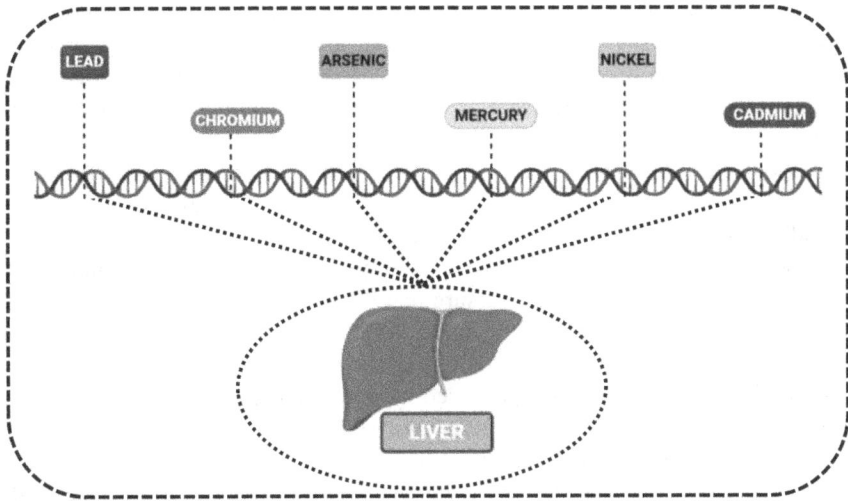

FIGURE 15.4   Toxicity by heavy metals (Pb, Cr, As, Hg, Ni, and Cd) in the liver tissue [13].

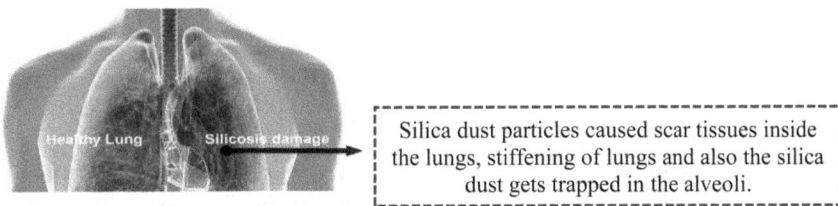

Silica dust particles caused scar tissues inside the lungs, stiffening of lungs and also the silica dust gets trapped in the alveoli.

FIGURE 15.5   Healthy vs silicosis-damaged lung [15].

indirectly affecting infants through breast milk. This also needs to be taken into consideration while assigning duties to female operators [14]. The worn-out particles of workpiece materials scattered due to high-velocity jets are also hazardous for the operators. Operators working in such environments for prolonged periods are at risk of inhalation of these fine particles, which can cause serious hazardous respiratory and lung diseases.

Whenever the medium containing abrasive particles or the material that is cut contains silica, it generates silica dust, which is a serious concern in AJM. Silica dust particles (silicosis) can cause scarring and stiffening of the lungs (as depicted in Figure 15.5), which can be carcinogenic if not attended to properly. Silicon exposure can raise the risk of lung cancer over a long period. Anyhow, adequate respiratory protection equipment is necessary to the protect workers against an atmosphere that contains hazardous chemicals when breathed in [16]. The particle size might be too large for the respiratory system or very few particles of respiratory size to be inhaled and collected. The skin can also absorb small metal particles or abrasive particles, leading to irritation or even infection. The abrasive used in the cutting process might get embedded into the workpiece, which results in poor quality output. Moreover, the dusty environment can reduce the production rate and produce low-quality products. To avoid any interaction with particles, it is necessary to keep the workstation clean. Cleaning methods should include dust wetting before cleaning. Workers are to be supplied with suitable respiratory protective equipment against substances present in the environment that are very harmful if inhaled. Symptoms include shortage of breath, pain in the chest, and coughing, thus leading to deterioration in the health of the individuals and also increased risks of the cancer [16].

Measures for controlling silica dust include availing the medium of jet machining that does not include silica, making sure that the appropriate or isolated strategies are accomplished besides machining, using the abrasive water jet methods; however, abrasive jet equipment should be kept in isolation.

As well as the health hazards produced by a dusty environment, the productivity of the workers will be lowered, higher mental or physical fatigue (several variables linked with fatigue [17] are as depicted in Figure 15.6), lower job satisfaction, and higher error rates are possible. The abrasives are embedded in the surface of the work inside the dusty atmosphere, thus leading to lower or poor quality of the product. Producing air that is dusty leads to consequences in the different machines in the workstation. Improvement in the condition of the workplace could be achieved via proper ventilation, thus implementing an environment air filtering setup as well as availing an appropriate dust collecting system [16].

15.2.2 ULTRASONIC MACHINING AND SUSTAINABILITY ASPECTS

In ultrasonic machining (USM), material is removed by a tool that is oscillating axially at high frequencies, which continuously pushes the abrasive mixed slurry toward the workpiece in a controlled manner. The material is eroded due to the collision of

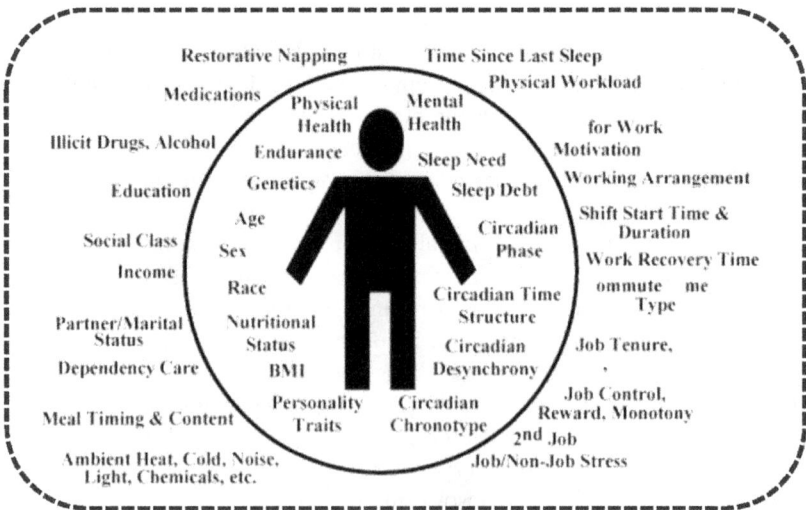

FIGURE 15.6   Range of variables that are linked with fatigue [17].

abrasive particles with the workpiece. The slurry may contain any type of abrasive particles like silicon carbide, aluminum oxide, boron carbide, etc. Usually, USM is used to machine brittle, hard materials like glass, ceramics, carbides, etc. Figure 15.7 depicts a schematic diagram of USM. USM has many sustainability issues, such as ultrasonic noise, electromagnetic field, abrasive slurry/fluids, etc., which are harmful to the environment and operators. Health issues arise with operators working for a prolonged duration in a location containing a high electromagnetic field (EMF). Nevertheless, the health risks can be prevented by staying a safe distance away from the source of EMF, as EMF intensity drops rapidly away from the source. A study revealed that reactive oxygen species (ROS) generated by the effects of exposure to EMF can damage various cellular structures in the neurons of the central nervous system as depicted in Figure 15.8 [18].

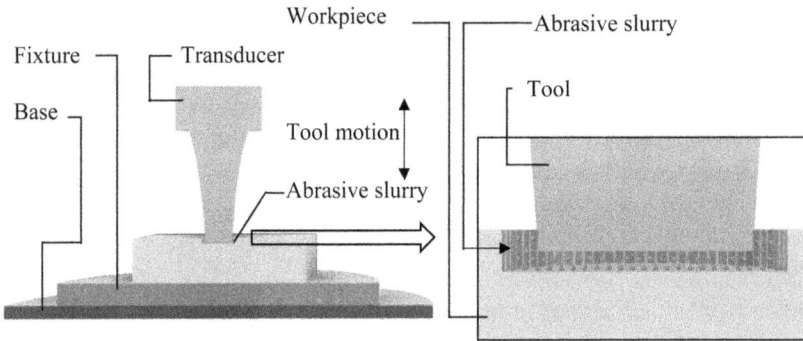

FIGURE 15.7 The schematic of ultrasonic machining [11].

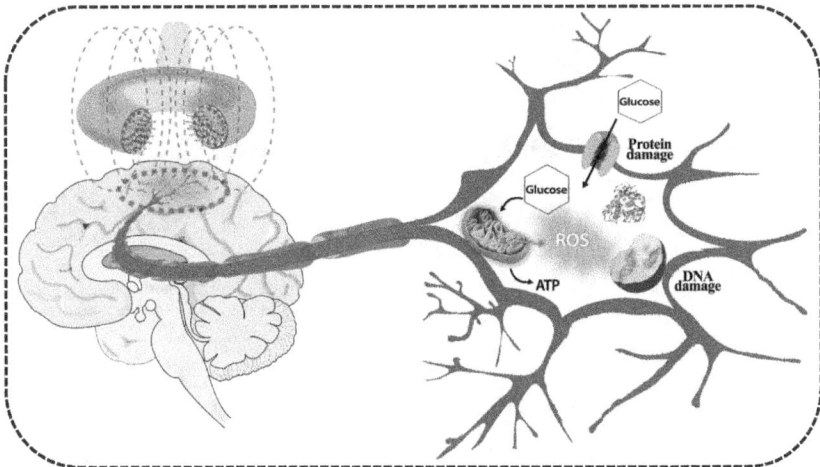

FIGURE 15.8 Reactive oxygen species (ROS) generated by the effects of exposure to electromagnetic field (EMF) can damage various cellular structures in neurons of the central nervous system. (Open access) [18].

In USM, noise production is in the audible range (96–105 dB). These noises/sounds are harmful to the operator's eardrum and hearing if exposed to them for a prolonged period. It can cause acoustic trauma, tinnitus, and noise-induced temporary or permanent threshold. Also, these noises are annoying and break the concentration and can cause hypertension, muscle tension, and stress [16]. In addition, abrasive slurry also causes comparable environmental and health consequences similar to those caused by cutting fluids. Slurry contact leads to skin disease or dermatitis. During the application of abrasive-based slurry, a mist is generated, causing small droplets of aerosols suspended in the environment for hours or possibly days. These microparticles' sizes are usually less than 10 μm, which can be inhaled and cause various respiratory diseases such as asthma, chronic bronchitis, or even laryngeal cancer in the long term. Engineering measures such as the installation of a mist collector or the use of anti-mist chemicals in the slurry can minimize mist exposure substantially [19]. Also, other hazards such as tinnitus (as depicted in Figure 15.9), tiredness, dizziness, and nausea are common due to exposure to ultrasonic frequencies.

Investigations are necessary on the economy of USM particularly from tool cost and production life point of view. To fabricate tools, cost-effective procedures are also needed. On the other hand, ultrasonic-aided machining has proven to be sustainable and green manufacturing technology owing to its lower power consumption, almost no or lower lubricant usage, a process that exhibits a pollution-free nature, very good quality of machining, and enhanced tool life.

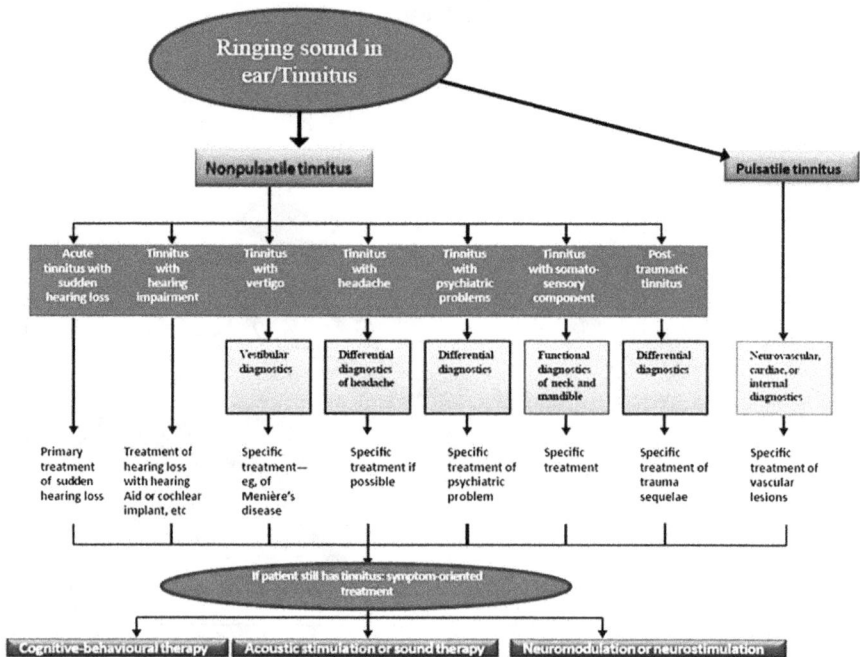

FIGURE 15.9 Tinnitus [20].

Ultrasonic-assisted machining system performance greatly depends upon the amplitude of the tool vibration that is achieved through an ultrasonic horn, the design of which is a challenge to researchers. Higher amplification of vibration aids in achieving good quality of the machined component, thereby minimizing wastage and creating an environment that is free from dust. Also, in turn it improves machining quality by reducing the forces in machining, formation of burrs, wastes, resin peeling off, walls tearing, and damage to the tool. It results in longer tool life in comparison to the traditional machining of the advanced materials. The design of the horn for the system of ultrasonic machining to achieve a higher amplitude of vibration at the end of the tool within the safer limits of working of the tool is very important for the improvement in the quality of machining of modern materials that are brittle. So, there is great scope for future research in the evolving field [21].

15.2.3   WATER JET MACHINING (WJM) AND SUSTAINABILITY ASPECTS

Water jet machining (WJM) which is a non-traditional machining process in which water at high velocity is availed to remove material from the required surface of the workpiece. WJM can be used to cut materials that are soft like rubber, plastic, or wood. It is mainly dependent upon the principle of erosion by water [22]. When a high-velocity jet of water strikes the surface from which the material is to be removed, material removal occurs. A hydraulic pump is availed for the circulation of water from an overhead tank containing water. The pump in turn directs water into the intensifier at low pressures, sometimes aided by a booster. The intensifier is used to increase water pressure to the higher pressure required. An accumulator temporarily stores the highly pressurized water, in turn supplying fluid when high-pressure energy is required to eliminate fluctuation in the pressure conditions in the process of machining. There is a valve for controlling the direction of flow and the pressure of the water. A flow regulator regulates the water flow. A nozzle aids in the conversion of pressure into useful kinetic energy (KE). After machining, the debris from the liquid is separated with the aid of the catcher, and the draining system cleans and sends it back into the parent reservoir for reuse purposes [11]. A schematic of WJM is depicted in Figure 15.10. One of the key issues in sustainability is that energy output and input are to be maintained as non-hazardous as possible.

It is wiser to prevent wastage than to clean up the wastage. However, in the case of WJM, this aspect is unavoidable and henceforth the concept of reusability should be employed [23].

Figure 15.11 depicts the contamination arising from a machining water jet that is disposed of. The purification of water and the sludge formed from material removal separation operations should be minimized in terms of energy consumption and usage of materials. The whole system should be to aim to maximize time efficiency [23].

Very often there are sustainability aspects relating to the corrosion of the material as the water is directly involved in machining. Also, the sludge mixed with post machining water causes water contamination and leads to water pollution (as depicted in Figures 15.11 and 15.12). After disposal, otherwise it leads to odor if

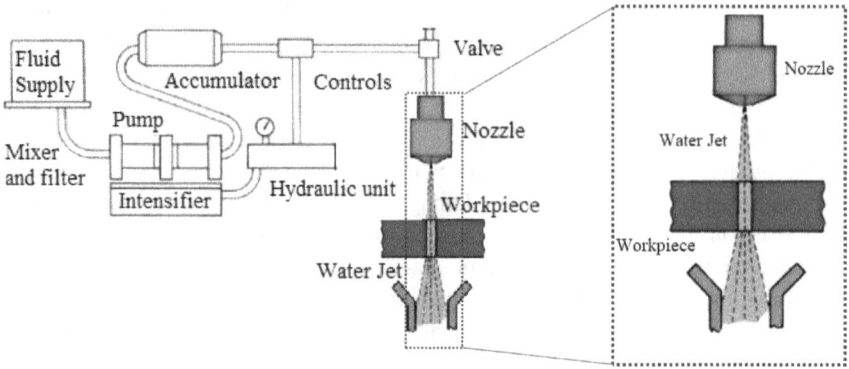

FIGURE 15.10 Schematic of water jet machining [11].

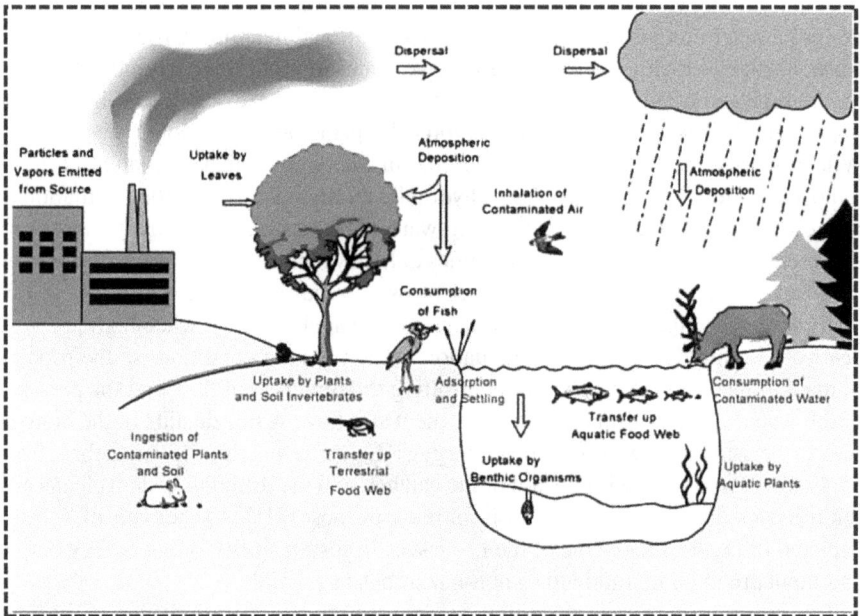

FIGURE 15.11 Contaminated water jet leading to pollution after disposal [24].

stored, and in cases where the operator inhales contaminated liquid for a long period, it may lead to antifungal issues. Rust inhibitors or additives can be added, however, economic aspects are to be observed [23]. Metal particles resulting from machining of the features may directly mix with the water jet and contaminate surface water after disposal, and it is toxic in the long run, as depicted in Figure 15.13.

15.2.4 Abrasive Water Jet Machining (AWJM) and Sustainability Aspects

Water jet machining with the aid of the abrasives is one of the most prominently and commonly used unconventional machining process. It works on the principle that

FIGURE 15.12 Discharge of effluents from machining water (a) contaminated machining water entering from drains into potable water due to cracks in pipelines (b) [25]. (Figure used for illustration purposes only.)

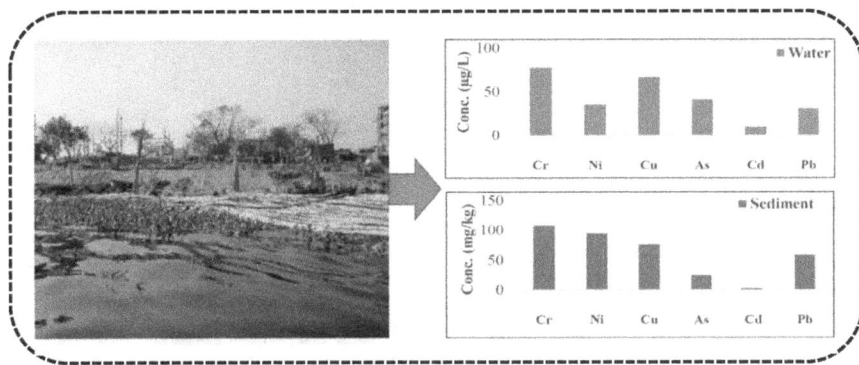

FIGURE 15.13 Metal pollution on the surface of the water. (Figure used for illustration purposes only [26].)

when a fine jet of highly pressurized water mixed with abrasive particles availed to accelerate the slurry strikes the surface of the targeted material, material removal takes place [27]. As shown in the schematic diagram (Figure 15.14), AWJM illustrates the assembly of the cutting head and the basic principle of machining.

The velocity of the abrasive particles mixed with the water jet is quite high and in special cases as high as 700 to 900 m/s jets are also used. Such a high jet velocity results in high KE leading to erosion of the material from the targeted workpiece. The pressure to operate the water jet is to be sufficient to produce the maximum possible velocity. When there is a further increase in the pressure of the water jet, the increased KE of the jet results in higher momentum which in turn is transferred to the abrasive particles so that the momentum changes and the impact of the abrasives leads to cutting of the target [27]. High-velocity jets are able to process a range of the materials such as composites, rocks, ceramics, and also metals [29]. The removal of the material takes place by the erosion wear on the top surface of the targeted material while following the wear deformation at the bottom portion of the targeted material being cut. At present, this process is far superior to many of the existing techniques of machining materials processing like composites, marbles, ceramics, etc. [27].

This process tends to pollute the environment. A dust collector system to collect the dust from the abrasives must be supplied to avoid health hazards and air

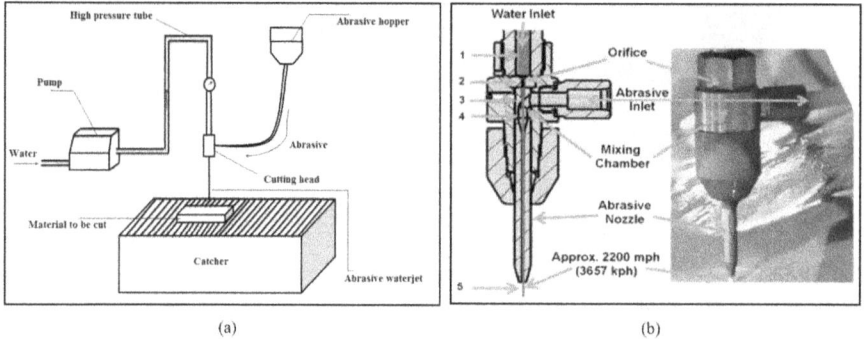

(a)   (b)

FIGURE 15.14 (a) Abrasive water jet machining process diagram. (b) Abrasive water jet machining head (water Inlet (1), orifice (2), abrasive inlet (3). Mixing Chamber (4). Abrasive Nozzle (5)) [28].

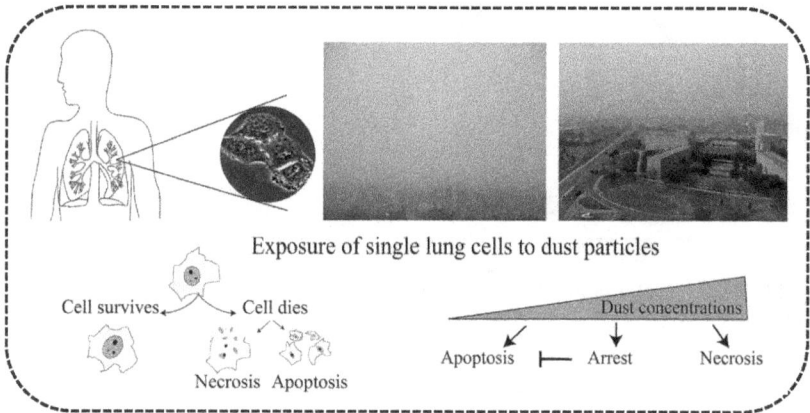

FIGURE 15.15 Single lung cells of the operator exposure to dust particles [30].

pollution, which can otherwise be inhaled by the operator and can cause various lung diseases as revealed in the study [30] (Figure 15.15).

Dustiness categories, and descriptions and related critical task exposure screening (CTES) solid fugacity categories are reported in Figure 15.16. The powder form of abrasive particles cannot be reused as its ability to cut decreases and it may get clogged in the nozzle orifice thereby leading to garbage disposal issues. However, some of the studies revealed that recycling of the grinding scrap can be used as alternative abrasive particles for the AWJM process. The studies conclude that the use of a grinding wheel rejects, via the process of mechanical crushing, and produces abrasive particles of good quality thus confirming that the abrasives of recycled alumina could be reused successfully in AWJM [32]. A critical task exposure screening tool developed for chemical assessment by a global chemical company classified and described whether the chemical exposure for a task is likely to be within the

Type of solid	Description	CTES solid fugacity
Extremely fine & light powder 	A powdered product containing very fine, free flowing, light particles. This category may also contain products with a mixture of very fine particles & large particles or granules. Handling the product in its dry form results in a dust cloud that remains airborne for a long time.	High
Fine dust 	A powdered product containing fine particles. This category may also contain products with a mixture of fine particles & large particles or granules. Handling the product in its dry form results in a dust cloud that is clearly visible for some time	
Coarse dust 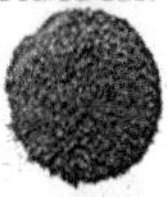	A powdered product containing coarse particles. Handling the product in its dry form results in a dust cloud that settles quickly due to gravity	Medium
Granules, flakes,or pellets 	Granules or flakes may fall apart & crumble, resulting in only a very limited amount of fine particles. Handling the product does not result in a visible dust cloud	Low
Firm granules, flakes, or pellets	Product does not result in dust emission without intentional breakage of products	

FIGURE 15.16 Dustiness categories and descriptions and related Critical Task Exposure *screening solid fugacity category* (Open access)[31].

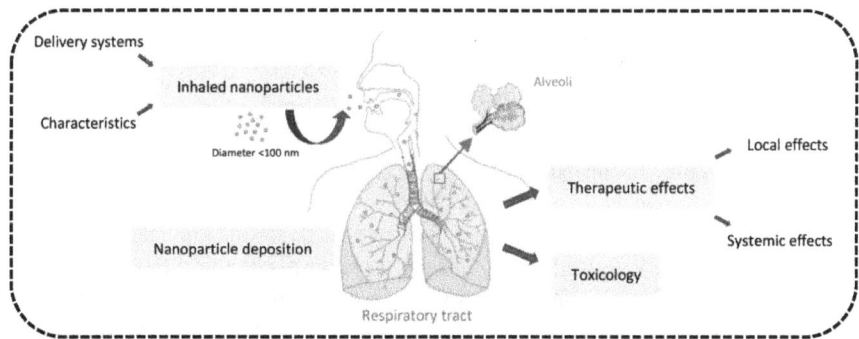

FIGURE 15.17 Inhalation of nanoparticles [33].

acceptable limits [31]. Nanoparticle inhalation also causes various local and systematic effects developed over the passage of time (Figure 15.17).

Besides consequences caused by nanoparticle inhalation (Figure 15.17), the inhalation of inorganic, particulate, and other organic pollutants over a prolonged period of time may develop oxidative stress and fetal lungs [34]. Reactive and cellular oxidative stresses are quite common to the operator in the case of inhaling abrasive particles suspended in air (Figure 15.18).

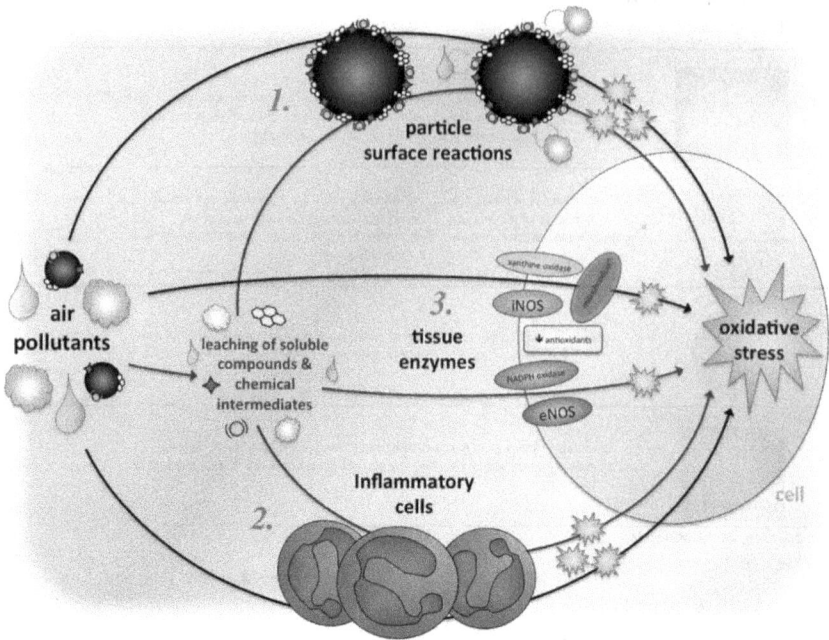

FIGURE 15.18 Mechanisms through which inhaled pollutants can induce cellular oxidative stress [35].

15.3 THERMOELECTRIC ADVANCED MACHINING PROCESSES AND SUSTAINABILITY ISSUES

15.3.1 Plasma Arc Machining (PAM)

In the process of machining using a plasma arc, a continuous arc is produced between the anode and cathode. Gas is sent around the cathode and moves to the anode. The temperature in the narrow region of the orifice that is around the cathode reaches as high as is required to produce a high temperature plasma arc. Several physical phenomena like radiation, convection, conduction, plasma transition, and mechanical deformation, etc., take place in PAM. The energy from the arc is absorbed by the workpiece material and it produces a high enough temperature so as to be above the melting point of the workpiece material. Under these conditions, the work material quickly melts and may even get vaporized [11].

High-velocity streams of the ionized gas flush off the machining debris from the top surface of the workpiece (Figure 15.19). For the organic materials machined with the aid of PAM, volatile product formation takes place and thereby material removal occurs. By the action of the high-velocity jets of plasma, it may even vaporize due to the heating action by the plasma jet. The plasma arc induced by the ionized gas is utilized in machining. Moreover, akin to electron beam machining, the dependency on the vacuum ambient limits the application of the plasma arc to large-scale

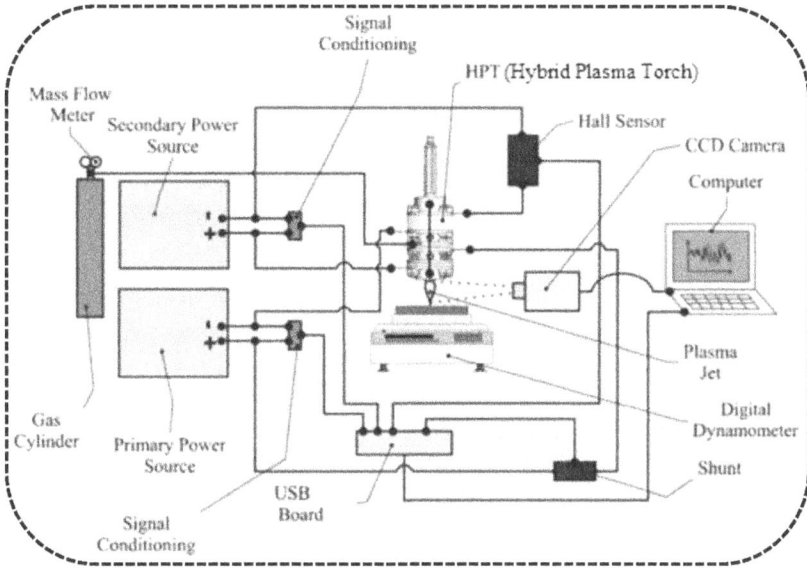

FIGURE 15.19   Schematic and setup for the plasma arc machining experiment [36].

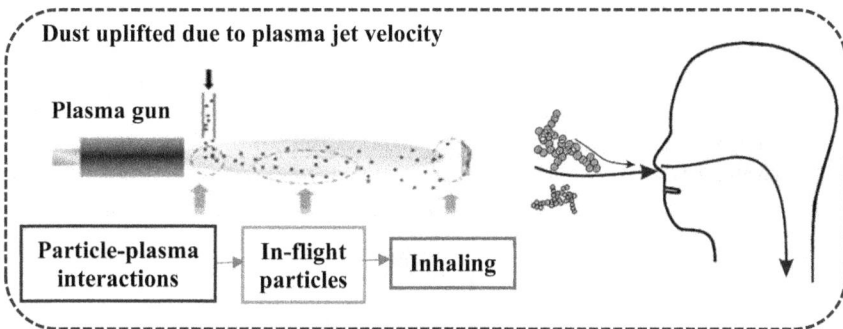

FIGURE 15.20   Dust inhaled due to plasma jet velocity.

machining, as the high cost aspects of machining come into picture [22]. The complete process of machining necessitates a large number of inert gases, which are harmful. The most harmful aspect of PAM is that the metallurgical changes take place on the surface of the workpiece.

The individual, worker or the operator handling the process must take precautionary measures. The process affects the eyes of the operator, so the operator must use proper goggles or helmet [37]. The hazard identification and risks involved in the workplaces involving plasma increases the possibility of threats to the staff in this environment. The dusty surroundings are one of the risks to the employees in the plasma workplace as depicted in Figure 15.20. The main contributor to dust is the velocity of plasma for cutting or machining operations in plasma environments in which the plasma ray comes into contact with the material thereby leading to inhalable fractions of solid aerosols [37].

As the plasma is very often involved in machining with the aid of the plasma jet, appropriate personal protective equipment should be worn by the operator wherever necessary. A real-time personal protection monitoring system, which is in use in the construction industry, should be used [38] and also protective shields and proper suits are mandatory as sparks are very often involved in PAM.

15.3.2 Laser Beam Machining (LBM)

LBM is the process where a highly intense laser beam is used to rapidly heat the surface of the targeted work and subsequently melts and or vaporizes the material of the target to create the necessary shape or geometry by removal of the appropriate amount of the material [39]. Laser is a massless tool, which does not undergo wear and tear, offering great flexibility. A schematic diagram of LBM is shown in Figure 15.21.

However, the major challenges are greater accumulation of heat in the workpiece resulting in the thermal damage such as microcracks, redeposition as recast structure formation, harmful gases, and particulate emissions compromising quality and productivity, as well as high energy consumption. In order to address these challenges, eco-LBM has been developed. On the other side of the coin, underwater machining with the aid of a laser beam (Figure 15.22) has been developed, which involves use of water in machining under the conditions of vaporization and melting of the workpiece. In addition to water, methanol, ethanol, and acetone are also used to create the wet surrounding conditions for improved LBM. There are three types of underwater LBM in the static mode or stagnant mode of water LBM, dynamic

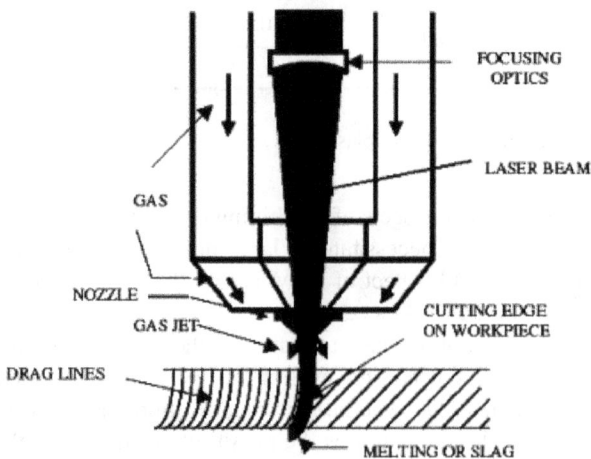

FIGURE 15.21 Laser beam machining schematic diagram [40].

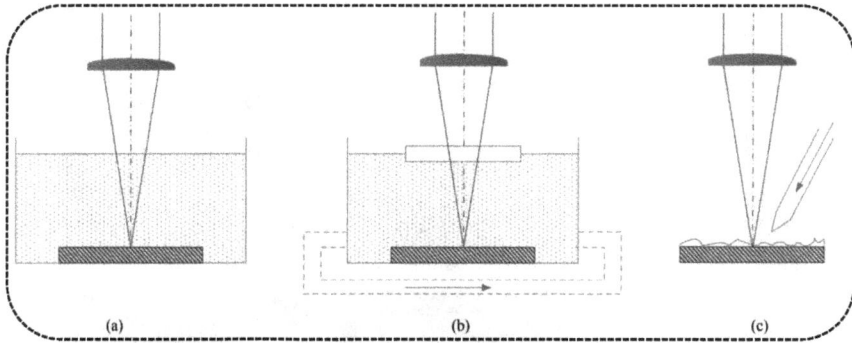

FIGURE 15.22 Illustration of underwater laser beam machining (a) static, (b) overflow and dynamic modes, (c) water jet aided underwater modes [41].

mode or flowing water LBM which is still water with the rate of flow as zero and water jet aided LBM [42].

Static modes of operating LBM aids in keeping the workpiece fully submerged in the water. The height of the level of the water above the workpiece surface varies in between 0.5 to 2 mm depending on the type of the laser and the requirements of the applications. Within the dynamic operation mode of LBM, distilled water at 10 to 30 ml/s flow rate is used. On the other hand, in water jet aided LBM, the laser beam including water is transmitted onto the workpiece surface for flushing out debris, that is melt, thereby preventing redeposition. Figure 15.22 shows a schematic representation of all three modes of underwater LBM techniques. The primary aim of underwater LBM [41, 42] is to minimize the molten material redeposition. It also includes cooling of the workpiece, hence a reduction in thermal damage to ensure sustainability and better quality of the machined surface to enhance the material removal rate (MRR). It also serves the purpose of reducing the gas emissions and the presence of particulates into the surrounding atmosphere. While LBM is going on, the temperature of the material is raised, partially vaporized, and chemically transformed.

Single or repeated exposure to certain wavelengths of the laser causes damage to the skin to various degrees and greater than that to other body parts. Injury (Figure 15.23) to the eyes is a result of a photochemical mechanism or thermal means that occurs when a laser beam interacts with the eyes [43].

If a beam enters the eyes, the power of the beam is concentrated with the aid of the lens within the eyes about 0.1 million times more at the retina. Hence, even a small amount of the laser beam leads to major damage to the eyes. The Principal American National Standards Institute (ANSI, Z136.1) gives a recommendation, necessary for safely using lasers, which are part of well-defined research and industrial practices.

The engineered features and controls are designed inside laser-aided working machines to minimize the risks associated with exposure to hazardous laser beams. The most commonly engineered controls are the housings for the protection and enclosures that protect equipment in the path of the beam. Interlocking is often

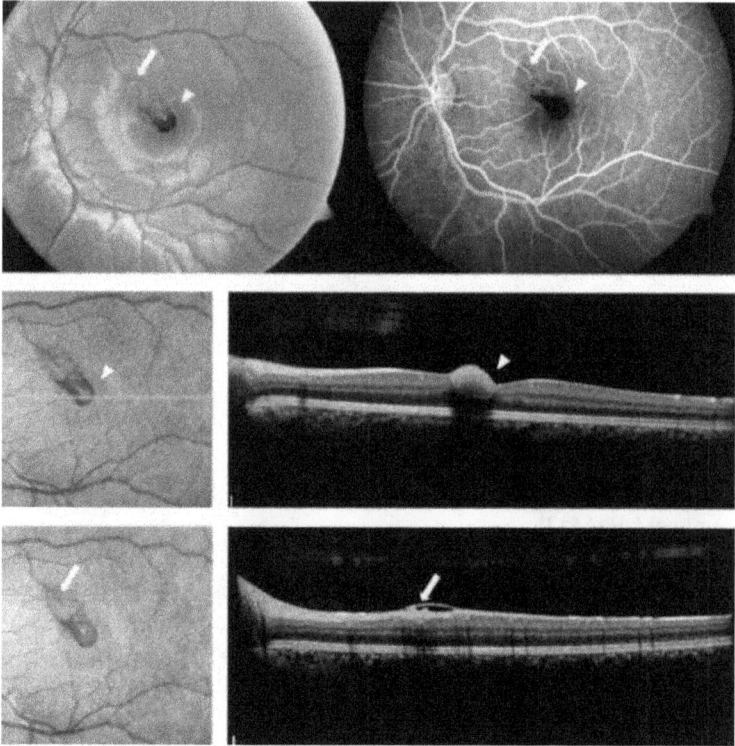

FIGURE 15.23   Laser exposures leading to retinal injury [43].

placed onto the housings for protection so that the removed beam is then shut off. The beam stopper is provided to collect accelerated particles and can be mounted on the end of a beam line. Sticking labels and the signs give the notice about the laser beams within the area of operation.

Protective equipment such as clothing, overall, or eyewear should be used if other controlled steps do not give the adequate protection. The protocols and regulations, which are designed to minimize the risks to laser beam exposure should be strictly followed. One of the most prominent controls in the administration is training [16]. Some of the potential hazards of laser beam exposure are depicted in Figure 15.24.

15.3.3 ELECTRON BEAM MACHINING

In electron beam machining (EBM), a high voltage of the order of 120 kV is used to accelerate electrons from 50 to 80% of the speed of the light toward the workpiece for material removal [44]. A schematic and features of the EBM are shown in Figure 15.25. In the EBM process, the cathode is first heated to a high temperature such that sufficient speed is acquired by the electrons to escape easily out into the space surrounding the cathode. A stream of a large number of electrons moves as a

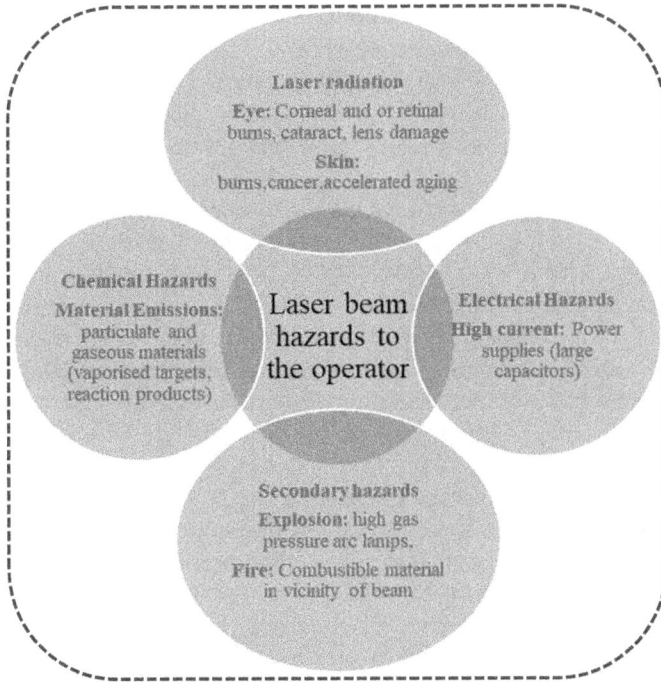

FIGURE 15.24   Laser hazards to the operator.

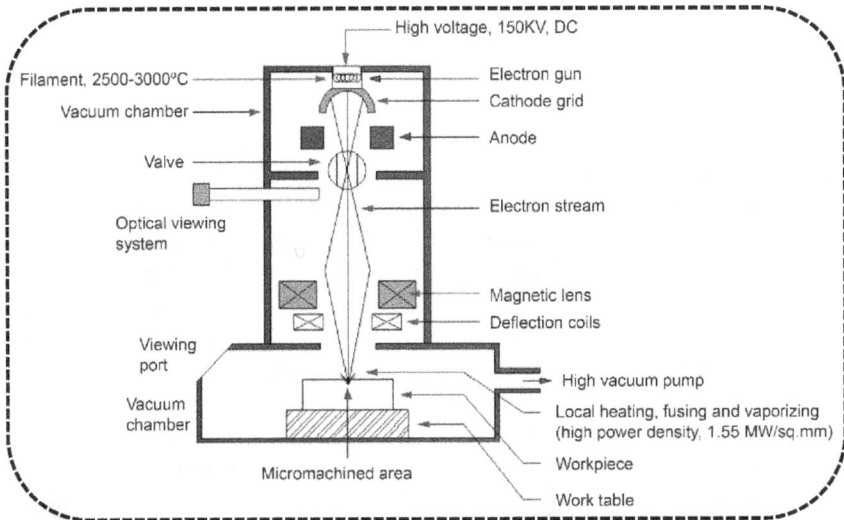

FIGURE 15.25   Schematic of electron beam machining setup and its features [45].

small diameter beam of electrons toward the anode (workpiece). As an outcome, the workpiece attains very high temperature by the bombardment of electrons in the surrounding areas such that it gets melted and/or vaporized in the bombardment area. When the high-velocity electron beam strikes the workpiece the KE of the electrons gets converted into heat. This in turn is responsible for the melting and vaporization of the workpiece material [11].

An electron beam gun is used for the production of an electron beam of the required shape and for focusing on the predetermined location. The gun used in the EBM is operated in the pulsed mode. A tungsten filament type of cathode, when superheated, produces a cloud of electrons. Sometimes, a cathode is used as a solid block heated indirectly by the emitted radiation from the filament. Due to the forces of the repulsion from the cathode, the electrons displace at very high accelerations toward the anode that attracts them. The velocity at which the electrons pass via the anode is approximately 66% of that of light. In the path of the electrons, a kind of bias electrode switch generates pulses. A magnetic lens is utilized to shape the beam of electrons into a converging beam. The beam is passed via an aperture that is variable to reduce the focused beam diameter by the removal of stray electrons. Magnetic lenses are used to locate the beam exactly, deflect the beam, and also make the beam round that is falling on to the workpiece. The density of the power at the surface of the work is very high such that it is capable of melting and vaporizing the material of the workpiece. Hence, the removal of the material in the EBM process is mainly due to vaporization. In the vacuum and also the machining chamber, the generation of the beam of electrons, its traveling in space, and the remaining mechanisms take place in the chamber maintained under vacuum. The vacuum chamber does not allow quick oxidation of the incandescent filament and hence there is no significant loss of energy of electrons as a result of colliding with the molecules of the air [11].

Although EBM is a cleaner machining process, the interactions between the workpiece and the beam of electrons generates x-rays which are hazardous and may lead to DNA damage, that is one of the primary concerns for the health and safety of the operators in case there is any excessive leakage. A number of measures could be taken into consideration to avoid any hazards to health like properly installing the setup within an enclosure, glass-leaded viewing ports, interlocking doors, following the methodology correctly, and only trained personnel allowed to operate it. Another potential hazard associated with EBM is the generation of dust due to vaporization of metallic workpiece material. Metal dust is deposited on the inside walls of the chamber. The operator is exposed to the dusty environment when the workpiece is positioned before machining, and also when collecting the workpiece after machining. High voltages that are associated with the electron gun, as well as the residual heat that is present within the workpiece and could also present potential health issues due to electric shock. The application of a strategy to control the dust and the use of personal protection equipment would lower the exposure time to the dusty environment [46].

15.3.4 ELECTRIC DISCHARGE MACHINING

Among all advanced machining processes, electric discharge machining (EDM) is one of the most widely used methods [47]. In this process, sparks are generated through discrete electrical discharges, which ionize the medium, which is usually a dielectric, to generate a plasma column between the electrodes. The sparks, that is plasma, result in extremely high temperature in the range of 8000–12,000 K, which locally melts and vaporizes the workpiece to remove the material [48].

Figure 15.26 illustrates the schematic and principal mechanisms (Figure 15.27) of EDM. Besides exceeding presumed constraints such as surface quality, geometrical and dimensional accuracies, material conductivity, and processing speed, the technique has overcome many technological constraints. Conversely, some of the concerns still remain, predominately in the environmental impact of the EDM process. It is due to the operational safety of the operator's health concerns related to the release of toxic fumes, vapors, and aerosols during the process, poor functional safety which may result in fire hazards, and also electromagnetic radiation. In the EDM process, ionization and deionization of the dielectric medium are controlled by regulating the voltage. Every cycle helps to remove a small amount of material from

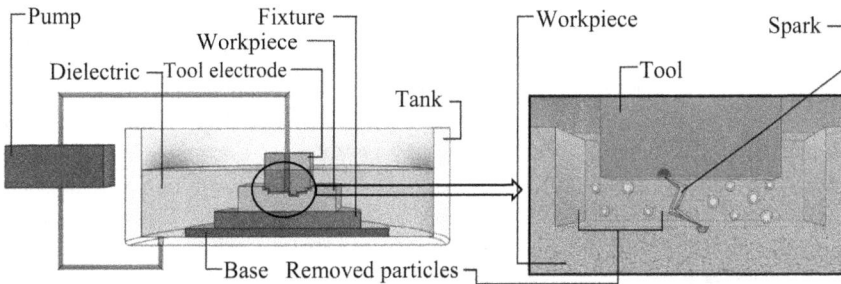

FIGURE 15.26 The schematic and principle of electric discharge machining.

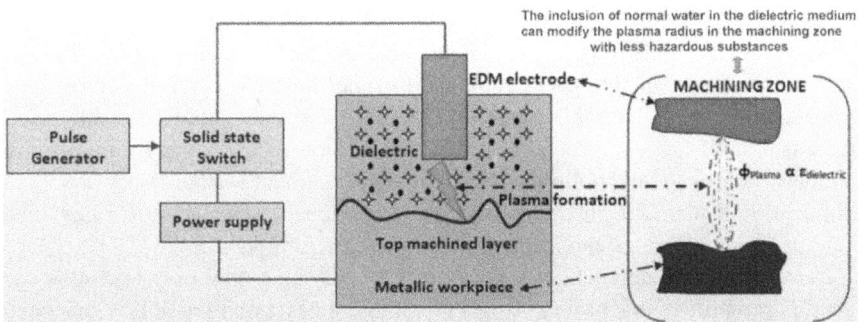

FIGURE 15.27 Schematic representation of electric discharge machining process mechanism [49].

the workpiece. However, the EDM process is very complex, stochastic, and highly dynamic as a number of variables of the process are involved. Decomposition of the dielectric fluid, high temperature, high frequency of discharge, and pressure waves are some of the factors which make it more complex. These result in unwanted outcomes such as radiation, noise, eroded particles from the workpiece, and emissions [50]. Generally, the EDM dielectric medium is liquid.

Mostly, hydrocarbon-dependent fluids like synthetic oils, kerosene, mineral oils, etc. are used as dielectric fluids. These dielectrics cause various issues broadly categorized in environmental and economic concerns as the emergence of poisonous gases, liquid, and solid waste. The dielectric fluids of synthetic and hydrocarbon-based oil emit benzene fumes and vapors, polycyclic aromatic hydrocarbons, and other harmful byproducts that include nonspecific aliphatic hydrocarbons, mineral aerosol vapor due to the degradation of oil and dissociation of their additives. Other contaminants such as nitrous oxide, carbon monoxide, ozone, and dangerous aerosols are also released while using commonly employed water-based solvents for wire electric discharge machining (wire-EDM) [51]. Dielectrics are responsible for the risk of fire explosion, hazardous emissions, sludge generation, toxic waste, high specific energy, uneven electrode wear, poor surface characteristics, and slow machining rate. It is also evident that in EDM, dielectric fluid accounts for around 43% of its environmental impact [52].

Dielectric fire is the most common hazard, especially in confined and inadequately ventilated spaces, when the temperature is near the ignition point. A cooling system must be installed in effort to stop dielectric fluid self-ignition. Widespread hydrocarbon oils necessitate forced ventilation systems to limit the quantity of hazardous gases and vapor that might cause a dielectric fluid to self-ignite. Research into potential fire causes owing to lower ignition temperature and intense electromagnetic radiation has shown that the dielectric fluids with an ignition temperature under 65°C are not recommended to be used in EDM for safety reasons. It is advisable to consider the dielectric fluids with the ignition point of over 100°C. The existence of a magnetic field at the electric impulses and electromagnetic radiation from creating plasma are indirect hazards to the operator [53]. Various sustainability aspects in the EDM process are shown in Figure 15.28.

An additional concern is chemical detergents used during cleaning of the bacteria and dielectric system that get accumulated inside the tank of the dielectric fluid, thus requiring regular cleaning and fluid replacement. The deionized resin, dielectric waste, and sludge produced after EDM operation require appropriate disposal to reduce water and land pollution [53]. Waste oil treatment is a necessary process to minimize its hazardous potential. Volatile and toxic emissions (Figure 15.29) from manufacturing activities can lead to health concerns for operators.

Aerosols include about 12.2% of the hydrocarbons and 69% of the metallic particles. During rough machining, a higher quantity of gases and aerosols is generated. The majority of the rest include carbon dust and unidentifiable tool and workpiece particles. They are between 25 and 29 nm, and they are typically spherical [55]. Therefore, the maximum emission values must be restricted within the permissible range [56]. Exposure to these emissions, finely suspended metallic particles, toxic

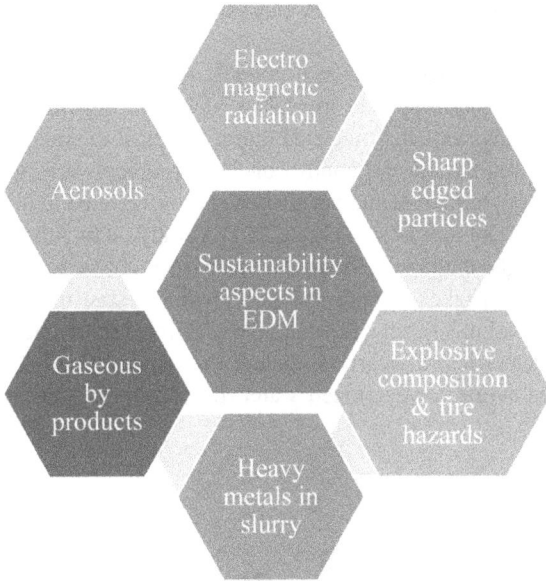

FIGURE 15.28 Various sustainability aspects in electric discharge machining.

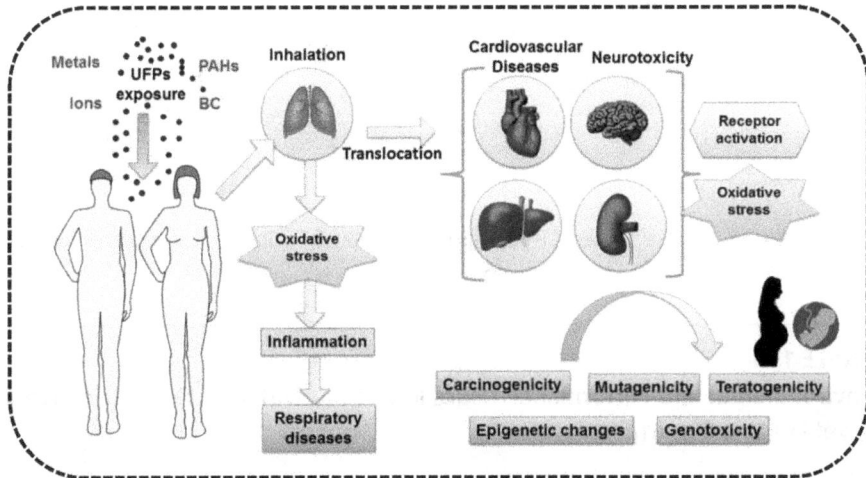

FIGURE 15.29 Toxicological effects from exposure to ultrafine particles (UFPs) (Open access) [54].

aerosols, and harmful gases for a prolonged period can cause serious health issues to the operators such as occupational dermatitis, respiration-based diseases, and these can be carcinogenic [57–59].

Flammable and combustible dielectric characteristics and possibly dangerous magnetic radiation are additional sources of danger. The design and application associated features of EDM machinery must be focused on while establishing the working environment.

EDM is also explored with dry/gaseous dielectrics which have numerous environmental and health concerns such as micro/nanodebris in breathing the environment sometimes interacting with the mucus (Figure 15.30), foul smell of dielectric etc. Therefore, proper regulation and control of these problems are needed. The summary of various environmental and performance issues related to different dielectric media used in EDM are tabulated in Table 15.1. A study on the ultrafine particle suspension in air revealed damage to the protection barrier mucus membrane [60]. EDM is a slow method and takes roughly around 50 times more machining time compared to conventional machining. Also, it takes 40 times more energy [62]. In EDM with oil-based dielectrics, the carbon content on the machined surface is roughly four times higher, while for deionized water, the carbon content is 50% less [63].

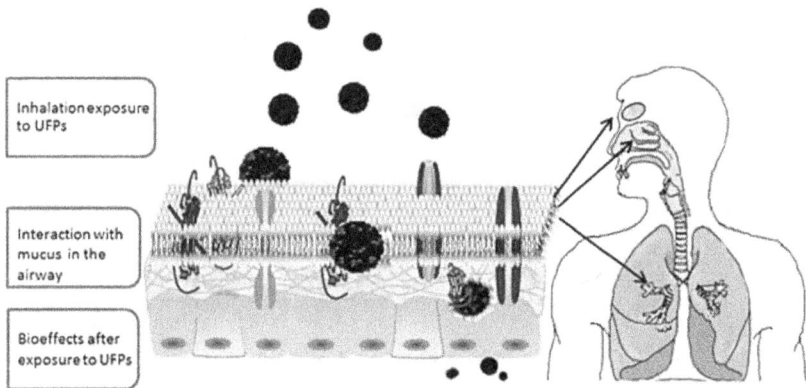

FIGURE 15.30 Mucosa covered by mucus in the airway constitutes the first protection barrier to outer UFPs [60].

TABLE 15.1

Environmental and Performance Issues Related to Different Dielectric Media Used in EDM [50, 61]

Dry/Gaseous Dielectrics	Water-Based Dielectrics	Oil-Based Dielectrics	Near-Dry Dielectrics
Poor geometric and dimensional accuracies, poor surface roughness, unstable spark generation, welding of debris, poor flushing.	Higher possibilities of microcrack formation, electrolysis of dielectric, decarburization, corrosion of workpiece materials.	Non-biodegradable and toxic waste generation, carburization, release of hazardous emissions, risk of fire explosion, harmful radiation, high specific energy.	High aerosol generation, more aerosol concentration.

Elimination of the dielectric fluid from sintered carbides in dry EDM resulted in increased erosion time and poorer surface finish due to the adhered erosion products with concomitant electrode wear reduction. Nevertheless, considering environmental constraints, even though duration has risen, dry EDM appears as an alternative to traditional EDM. Researchers focus on the EDM with a minimum dielectric mix in the air to decrease the negative consequences of dry EDM such as the poor efficiency of the process and inferior quality of the surface [64, 65].

In the case of near-dry EDM, some serious concerns such as rugged and blackish-machined surface, high concentration of aerosols in the environment, electrolytic contamination, and scattering of debris are observed [66].

Hybrid dielectric EDM was also explored as a possible solution [67]; however, all the three modes of dielectrics have some serious sustainability issues. A report on environmental assessment on the EDM gives an acquisition of data efforts to compute the total impact on the environment out of the three techniques such as electric discharge machining, ED die sinking (Figure 15.31) [68], microelectric discharge machining (micro-EDM) as well as wire-EDM. Reports reveal that during one hour of rough machining, the electrical energy of the EDM process and the hydrocarbon oil utilized as a dielectric are the primary contributors to the total environmental impact attributing to a percentage as high as 47.3% and 23.1%, respectively. Also, the energy consumption to lower the temperature of the machine amounts to 19.4% and exhaust system consumption of energy is as high as 3.9%. It amounts in total to 70.6% of the whole environmental impact.

Dielectric fluid accounts for one fourth of the impact due to production. It is also reported that hydrocarbon-dependent dielectrics contribute significantly to complete environmental hazards. Henceforth, alternative dielectrics such as water-dependent, gas-based and dry EDM techniques also have been researched [69]. Jatropha oil-based dielectric fluid in EDM revealed a higher MRR, lower roughness of the surface, and improved surface hardness [70]. Valaki et al. [71] revealed that palm oil as a dielectric enhanced the MRR but retained surface hardness and roughness as

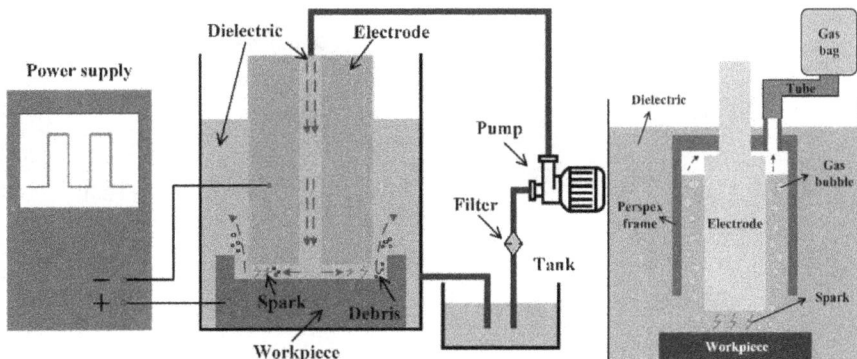

FIGURE 15.31 Sustainable electrical discharge machining using water in oil nanoemulsions that is schematic of sinking electric discharge machining (left) and a diagram of the gas collection device [68].

obtained in the case of kerosene as a dielectric. Pongamia Pinnata bio-based oil revealed 32% higher MRR in comparison to that of hydrocarbon oils [72]. Canola oil displayed a 21% higher rate of tool wear whereas it was 8% in the case of sunflower oil while machining titanium alloy [73]. Dry EDM was evolved in order to decrease pollution levels and harmful gas generation during the operation of EDM, (shown in Figure 15.32) and some of the benefits are decreased coolant cost, job satisfaction, and being less hazardous to the environment [75].

Dry EDM replaced gas dielectric fluid as an alternative to hydrocarbon oil in the conventional EDM process. High-pressure gas is supplied through the tubular electrode where the gas that is compressed cools the tool and prevents the eroded particles attaching on the surface of the tool. On the other hand, in the case of a near-dry EDM process, a mixture of gas and liquid as the dielectric replaces hydrocarbon oil inside the conventional EDM [76]. Reactive products are generated due to high pressure and temperature during discharge of the hydrocarbon-dependent dielectric, causing health issues to the operator [77]. Boubekri and Vasim [78] revealed that inhalation of the vaporized dielectric during machining causes adverse health effects including toxicity, respiratory disorders, and dermatitis as well as cancer. Goh and Ho [79] indicated that hydrocarbon-dependent dielectric fluid causes skin irritation, explosion, and fire risks in the working environment when the zone of the process is closed and not ventilated [80]. There are also the risks of explosion and electromagnetic radiation as the flash points are lower. These are the serious issues related to the safety of the operator in the work environment. From the above discussion, it can be concluded that vegetable oils exhibit higher values of MRR. Also, high viscosity of the dielectric fluid creates an obstacle for the heat transfer to the workpiece and accumulates more heat, thereby increasing the erosion in the electrode. Due to high viscosity and the content of the oxygen, there is an increase in the rate of tool wear. Higher hardness of the surface is observed with the vegetable oil-dependent

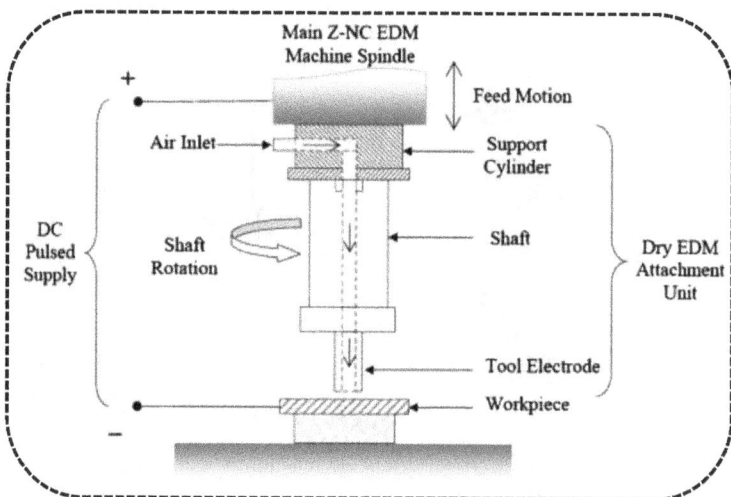

FIGURE 15.32 Dry-electric discharge machining schematic diagram [74].

dielectric fluids due to their higher thermal conductivity viscosity, and lower specific heat [81]. *Azadirachta indica* (neem) oil as a dielectric minimizes the chances of emissions of gas during machining and it aids in the transformation of an oxide-free hygienic breathing environment. With a 4.5–10 A pulse current, neem gives a 6.2–15.6% higher rate of removal and 12.25–15.45 % lower surface roughness than that of kerosene [82]. Selecting a suitable ratio of deionized water and tap water to protect the machine tools from corrosion is a challenge in the real-time green EDM process. Also, the reconditioning and/or recycling of biodiesels are also real-time obstacles in implementing the green EDM process.

Hybridization of EDM with traditional or other advanced machining processes brings enhanced productivity, that is higher MRR and better quality of the part, especially in the case of machining hard materials. Electrochemical discharge machining, electric discharge grinding, ultrasonic vibrations aided electric discharge machining (ultrasonic EDM), magnetic and abrasive field aided electric discharge machining, and electrochemical discharge grinding are few of the prominent hybrid versions of electric discharge machining. Hybridization facilitates machining as it minimizes the wear of the tool electrode besides effectively flushing out the debris from within the machining zone. As a result, it improves efficiency of the process by enhancing productivity and surface integrity which eliminates subsequent finishing of the machined part [83]. EDM has several hazardous potentials such as vapors, aerosols, smoke, heavier decomposition of metals. Skin is affected by the dielectrics of hydrocarbon, sharp metallic edged particles, plausible explosions and fire hazards, and electromagnetic radiation. For comparison purposes, die sinking creates more aerosols and fumes than wire-EDM. Composition of the materials including toxi-cants or health-impacting substances such as the nickel, is also responsible for the health hazards. The types of dielectrics, their viscosity, and composition influence the vapor and fume. Lowering the viscosity generates fewer vapors and fumes. To reduce the potential hazards during EDM, measures should be implemented such as using appropriate filters, incorporating the device to clean the dielectric, and keeping the temperature of the medium at least 15°C below the flash point of the dielectric.

Measures should be implemented for reducing radiation (electromagnetically emitted), using appropriate shielding of the machine, reducing the possibilities of fire hazard, using the dielectric fluids' level sensors, avoiding the use of dielectrics with flash points of around 65°C, disposing of wastes appropriately, raising the awareness of the operators for not using high voltages resulting in severe injuries or sometimes even death [16]. A study was also reported on the sustainable electric discharge mill-ing process as depicted in Figure 15.33 [84].

15.4 ELECTROCHEMICAL AND CHEMICAL MACHINING PROCESSES AND SUSTAINABILITY ISSUES

15.4.1 ELECTROCHEMICAL MACHINING

Electrochemical machining (ECM) which is one of the prominent advanced mate-rial removal techniques used to obtain maximum precision and accuracy on the machined component made of electrically conductive hard materials. It works on

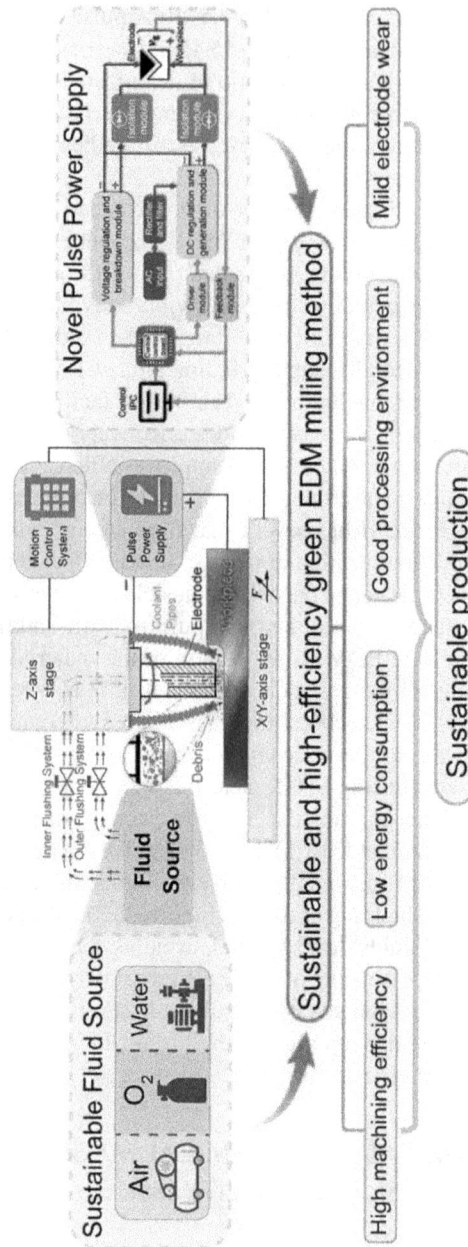

FIGURE 15.33 Sustainable and high-efficiency green electrical discharge milling method [84].

the principle of material removal by a localized anodic electrochemical dissolution process using an electrolyte [85–89].

The most common electrolyte used in ECM is a diluted salt solution of sodium chlorate, sodium chloride, or sodium nitrate. However, acidic electrolyte like nitric acid (HNO_3) is more common for micro-ECM [90]. Other electrolytes such as citric acid could be used as an environment friendly alternative [91]. Figure 15.34 illustrates a schematic of a ECM setup and the mechanism of material removal. Despite providing several advantages like stress-free surfaces, high productivity, excellent surface finish, etc., it tends to suffer from certain intrinsic restrictions such as the formation of a passive oxide layer on the working surface, problems with sludge flushing, the electrolyte's corrosive and toxic nature, and high process sensitivity of the electrolyte [93]. The ECM process generates many byproducts such as sludge, which may contain traces of heavy metals, metal ions, oils, nitrates, and harmful acids. Unfortunately, the electrolyte is contaminated by heavy metals accumulated from the components. In the electrolyte, metal hydroxides can be adsorbed to chemically altered byproducts such as hazardous ammonia, chromate, and nitrate [94]. However, use of fresh electrolytes in ECM produces fewer environmental issues.

Hydroxide slurry detoxification and disposal expenses are restricting variables that influence ECM's economic efficiency. During the electrolyzing process, explosive

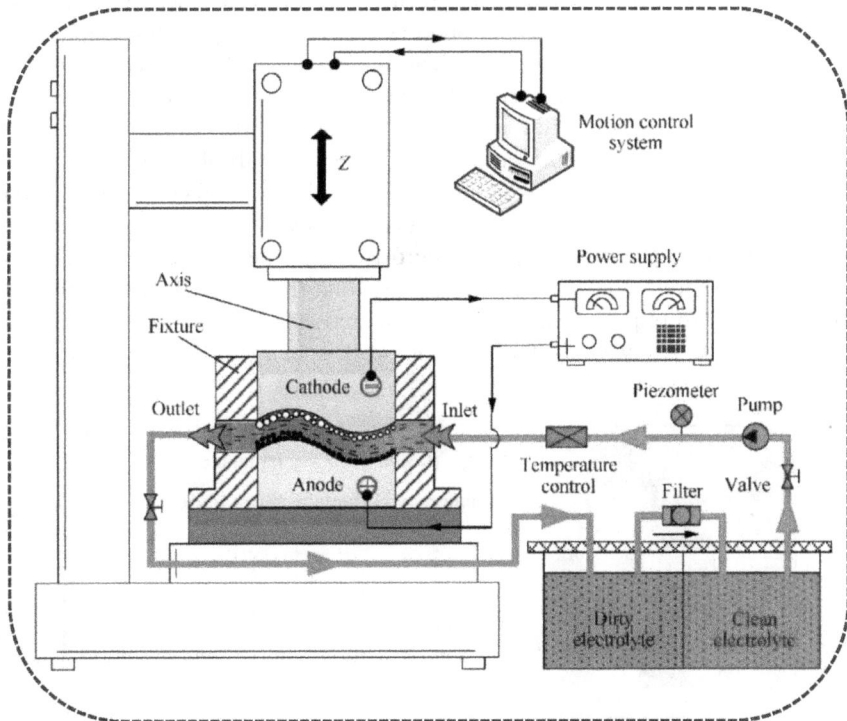

FIGURE 15.34 A schematic of electrochemical machining [92].

hydrogen gas is produced. To keep the hydrogen gas from exceeding its lower limit of flammability as well as keeping aerosols out of the operators' respiratory tract, a local exhaust system should be efficient enough. Inhalation, ingestion, skin and eye contact are all possible health concerns if exposed to trivalent chromium compounds. A study also revealed the adverse effects of soot particles (Figure 15.35) [95]. In humans, this produces contact dermatitis and affects the skin, liver, and kidneys [96]. Nitrites are known for causing numerous health consequences such as vitamin A shortage, reduced thyroid gland function, and blood hemoglobin reactions.

A local exhaust system for ventilation, enclosure of the process setup, and personal protection equipment are efficient methods in limiting the workers' exposure to nitrogen and trivalent chromium. Moreover, toxic vapors, electrolyte contamination and splashing are various occupational health hazards [97]. In this process, acidic electrolytes such as $HClO_4$, HCl, H_2SO_4, and HF are used, which are poisonous and hazardous for the machine operators. Such electrolytes also harm components of the machine tool and eventually destroy it. Electrolyte splashing can cause skin irritation, eye contamination, dermatitis, etc. Inhalation of smoke or gases produced by the use of chloride- and nitrate-based electrolytes over a long period can cause ingestion, kidney, liver, skin, and eye-related diseases. Table 15.2 shows the common sustainability issues relating to different advanced machining processes [97]. For scientists and practitioners, it seems to be vital to develop a safe and green ECM method. A pulsed power supply, use of greener or ecologically friendly electrolytes as well as process hybridization and optimization are the key approaches in making the ECM process ecologically sustainable. Micro-ECM leads to environmental issues, possibly from harmful chemical products, poorer fatigue resistance, and its lack of the capability to create sharp corners. In addition, if the corners are too sharp as depicted in Figure 15.36, the corners may pierce the hands of the operator.

Micro-ECM is well known for its burr-free machined surfaces with no physical and thermal effects. As a result, it leads to less wastage and would increase the sustainability. Also, theoretically micro-ECM has negligible or no tool wear as there is no direct contact between the tool and the targeted work specimen. Micro-ECM also requires electrolyte for the activation of the process and creation of current pathways in the tool and work specimen. Passive electrolytes

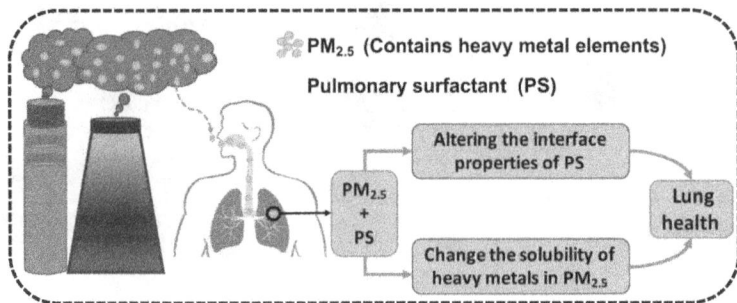

FIGURE 15.35 Effects of inhaling soot particles by the operator [93].

TABLE 15.2

Common Sustainability Issues in Some Advanced Machining Processes [96]

Machining Type	Media/Electrolyte	Sustainability Issues and Health Hazards
EDM (initial days)	Kerosene	High risk of fire, bad odor, skin irritation, harmful gases
EDM (initial days)	Mineral seal	Low life span, carcinogenic
EDM	Transformer oil	Frequent sludge removal, high rate of sludge accumulation, high oxidation rate
Wire-EDM	Deionized water	Can cause skin and eye irritation
EDM	Oil	High cost, harmful gases, skin diseases
ECM	NaCl, NaNO$_3$, acids	Liver, kidney, and skin diseases, breathing issues, electrolyte disposal issues, sludge generation and management issues, corrosion risk, high energy consumption
USM	Abrasive slurry (SiC, B$_4$C, Al$_2$O$_3$)	Slurry fluids can cause dermatitis, tinnitus, acoustic trauma, health issues due to ultrasound

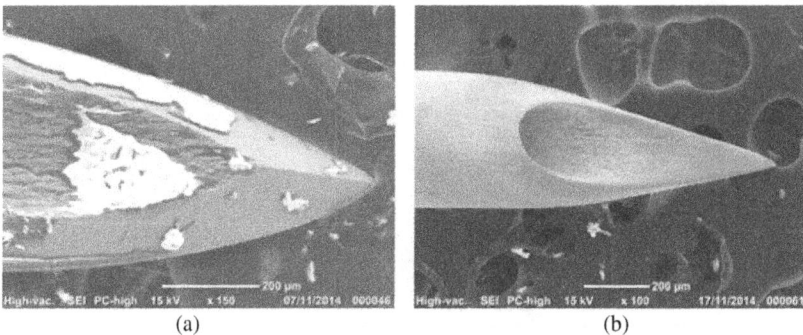

FIGURE 15.36 Possibility of sharp tip piercing the operator's hands, a) ground needle tip, b) micro electrochemical machined needle tip [98].

containing oxidizing ions are known for precision material removal whereas the electrolytes that are non-passive contain aggressive anions and they form soluble products which can be completely swept away from the gap between the electrodes. Machining performance is affected by sludge which may create sparks. Figure 15.37 represents a micro-ECM sustainability assessment flow chart [98]. Yang et al. [99] used normal mineral water as an electrolyte in the ECM process, which is a superior substitution of conventional acidic or basic electrolyte. It is completely eco-friendly and non-corrosive in nature resulting in greener ECM with higher efficiency as well as precision at a considerably lower cost. For sustainable ECM, sodium carbonate is an often-recommended electrolyte that is environmentally friendly [100].

Hybridizing the ECM process with other traditional machining methods is one of the ways to make this process sustainable overall. Considerable attempts were

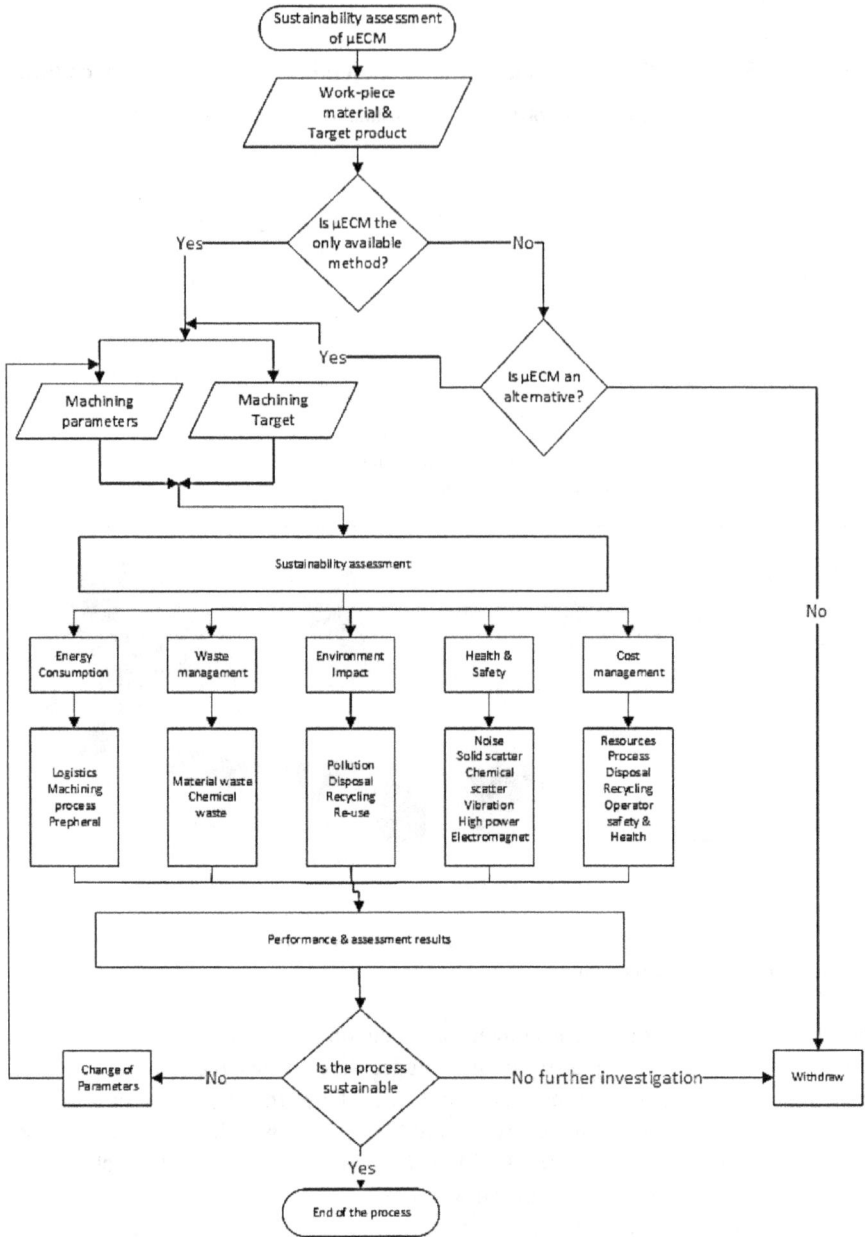

FIGURE 15.37　Micro electrochemical machining sustainability assessment flow chart [98].

made in the past to increase local electrolyte dissolution, flushing efficiency, MRR, and the quality of the machined part by combining it with honing, grinding, EDM and the use of a laser beam, ultrasonic vibration, and abrasives. Nanofluid ECM with suspended particles of nanocopper in sodium chloride electrolyte was used for ECM of highly hardened high chromium die steels resulting in enhanced MRR, and surface finishing enhancement in comparison to plain electrolyte-aided ECM. During the process of electrolyzing, explosive hydrogen gas is generated as depicted in Equations 15.1 and 15.2.

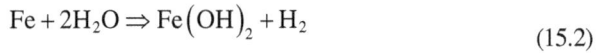

$$H_2O + 2e^- \Rightarrow H_2 + 2OH \tag{15.1}$$

$$Fe + 2H_2O \Rightarrow Fe(OH)_2 + H_2 \tag{15.2}$$

Local exhaust is provided to prevent hydrogen gas concentration to the limit of flammability, besides removing the mists (volatile organic compounds inhalation) from the breathing zones of workers. Other methods of control include enclosing the process setup, local exhaust for ventilation, generalized ventilation diluting, and protection equipment [101] (Figure 15.38).

15.4.2 CHEMICAL MACHINING AND SUSTAINABILITY ISSUES

Chemical machining (CM) uses acid as an etchant, which has adverse consequences on the surrounding environment, shortcomings in storage and handling, as well as the consequences of damaging various materials. A CM setup and its material removal mechanisms are shown in Figures 15.39 and 15.40, respectively. The etchant's acidity is normally measured by pH number. Solutions having pH values less than 7 are described as acidic. Photochemical machining (PCM) is carried out with the aid of a solution of ferric chloride in an aqueous state at temperatures over 50°C. Ferric chloride is relatively cheaper, acidic, and readily available besides being

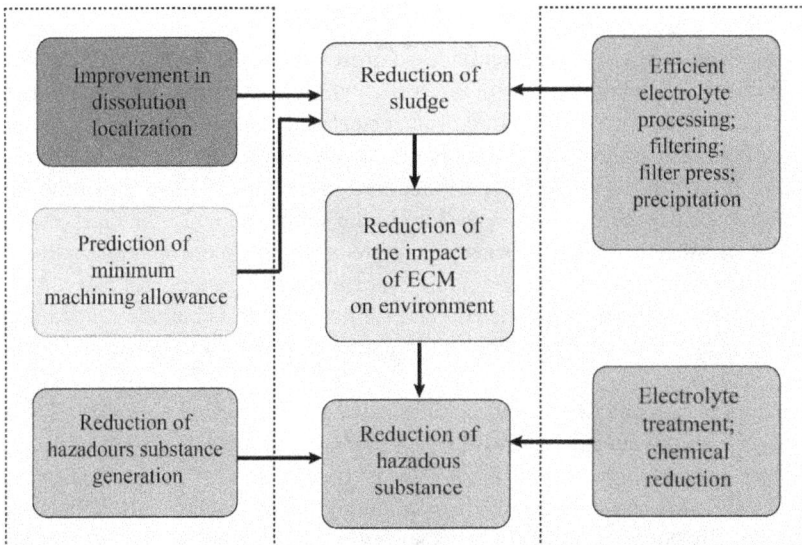

FIGURE 15.38 Electrochemical machining techniques to reduce environmental consequences [102].

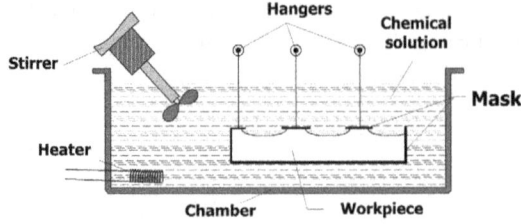

FIGURE 15.39 Overview of chemical machining process [103].

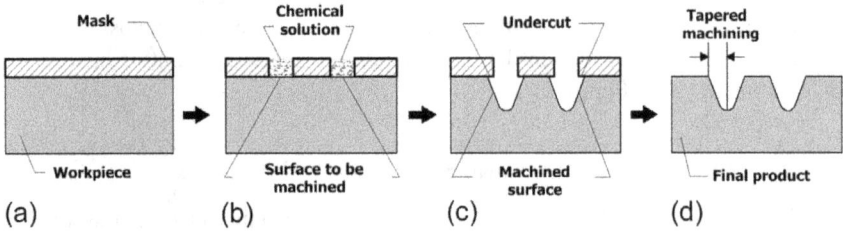

FIGURE 15.40 Material removal mechanism of chemical machining [103].

versatile [16]. Environmentally it is attractive as it is low in toxin content as well as being easier to filter, recycle, and replenish. While CM exposure to hafnium leads to inhalation, eye or skin contact, and ingestion. CM hazards depend upon the substance properties at the time of contact and concentration with alkalis and acids. The health effects include irritation, injuries due to corrosion besides burns, quicker and very often irreversible severe damaging to the eyes, and even leading to the risk of lung or larynx cancer [16].

Many of the metals very often undergo corrosion of many types which include localized attacks and uniformity in the corrosion. Conventional protection material, for instance paints, are employed to protect steel and iron from the phenomenon of rusting. Poor preparation of surfaces is the main cause for the failure of protective coatings. Improper disposal of the etchant used in CM alters the levels of the acidic and basic nature affecting fauna and flora in the water and soil. The changes in the pH value of water from 7, that is neutral, shows adverse effects on aquatic life. At a pH of 6, mollusks and crustaceans start to vanish and lead to increase mass. At a pH of 5.5, some fish like trout, whitefish, and salmon start to become extinct and also salamander eggs fail to hatch. The acidic nature of pH 4 shows adverse effects on frogs and crickets [16].

Some of the alkalis like ammonia display acute toxic effects on fish. Soil is confirmed as contaminated in the case of an acidic value of pH of 4–5, and is contaminated heavily when the pH is 2–4. In the case of soil having a value of pH of around 9–10, it is contaminated, and its alkalinity or pH is 10–12. It is then classified as heavily contaminated. Aerosols such as sulfuric acid and nitrogen or liquid substances that are corrosive in nature are the pollutants in the air, and they form

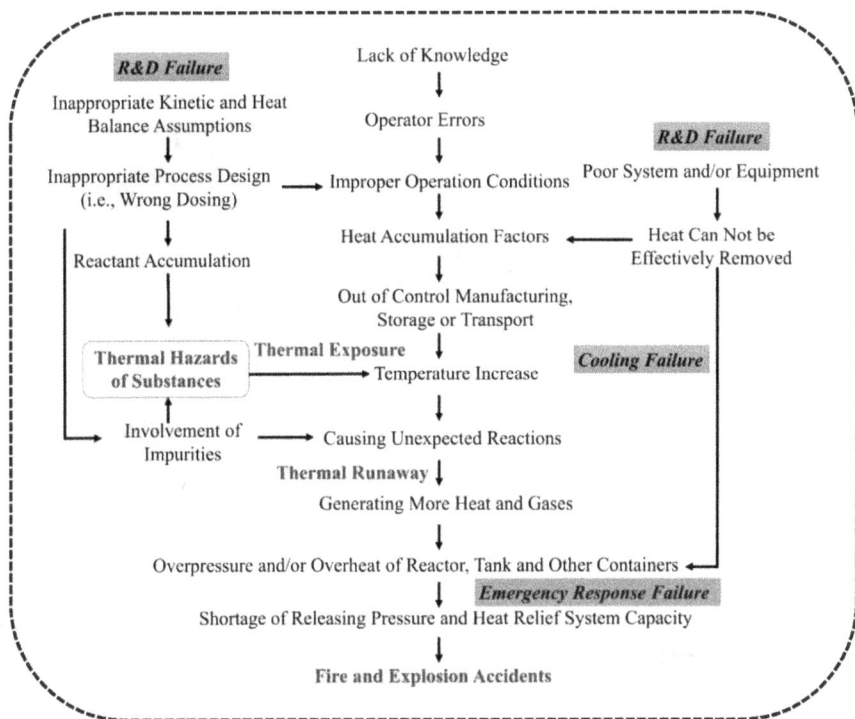

FIGURE 15.41   A schematic overview of how the thermal hazards of reactive chemicals contribute to fire and explosion accidents under a general application scenario in industry [104].

the gases which are corrosive. Gases are combined with the water to form the acids which precipitate in the rain. Acid fumes and acidic gases damage the flora [16] (Figure 15.41).

15.5   SUSTAINABILITY ISSUES IN MICROMACHINING PROCESSES

Fabrication of microparts/microcomponents is a highly expensive and energy-intensive process. However, to fabricate complex microcomponents, advanced technologies such as micromachining are necessary [105]. Micromachining processes are widely used for the fabrication of various electronic devices, especially microelectromechanical systems (MEMS). However, apart from being energy-intensive, it also generates lot of toxic waste. These issues are the real challenges and should be addressed. The energy requirement during a micromachining operation is less, but the total energy requirement including the energy consumption to maintain the desired environment in the machining center and idle time energy consumption is high in micromachining compared to conventional machining operations. Maintenance of the essential conditions in facilities, which minimizes contamination of the operational site, requires high-quality air filters, pumps, air conditioning systems, and water purifiers. The

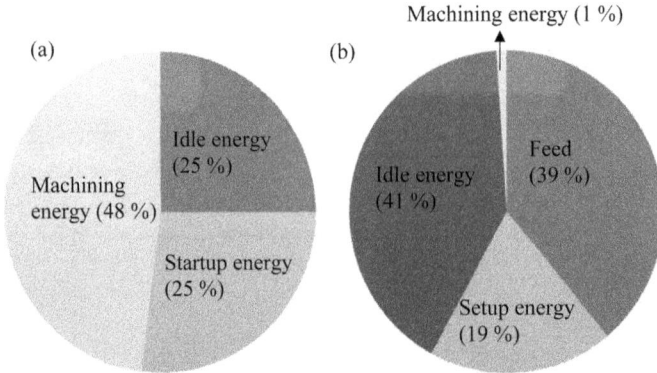

FIGURE 15.42 A comparison of energy consumption between a microdrilling system and a conventional CNC milling machine. (Adapted from [106].)

energy needed to maintain the plant in a clean room condition is about 60% of total energy consumption and it is significantly more than that for the actual workflow. Therefore, greater consideration should be given to improvising facility operations to achieve substantial energy savings. The energy spent in processing, however, is considerable, leaving space for future developments. In the event of micromachining, in comparison with traditional machining, the proportion of power consumption for productive work may be rather low. Figure 15.42 shows a comparison of energy consumption between a microdrilling system and a conventional computer numerical control (CNC) milling machine [106, 107]. From Figure 15.42, it is observed that the energy required for material removal is far lower in the case of microdrilling. However, energy consumption for the setup, stages, spindle, and idle time is too high, resulting in much lower efficiency in microdrilling operation. These issues also apply to micro-ECM, micro-EDM, and micro-ECDM as well.

15.6   CONCLUSIONS

Advanced machining processes have various sustainability issues. Based on the above discussion, all the three issues namely, social, environmental, and economic are heavily affected by the advanced machining processes. These processes are having a direct impact on operators' health and environmental conditions. Also, the efficiency of these processes is very low compared to conventional machining processes, but at the same time these techniques are able to produce the required highly complex accurate geometries and microcomponents. Industries should be aware of these sustainability-related issues and should take the necessary steps to make the process safer.

ACKNOWLEDGMENT

The authors are thankful to Elsevier for granting copyright permission to various figures to use in this manuscript with License numbers: 5116910037632 (Figure 15.1),

5116910037632 (Figure 15.4), 5116941467618 (Figure 15.5), 5116950781118 (Figure 15.6). Open access under a creative commons license: (Figure 15.8), (Figure 15.9), 5116960420636 (Figure 15.11), 5116960641123 (Figure 15.12), 5116960761520 (Figure 15.13), 5173410854118 (Figure 15.14), 5116960893204 (Figure 15.15), Open access under a creative commons license (Figure 15.16), 5116971103733 (Figure 15.17), 5116970793768 (Figure 18), 5115160215666 (Figure 15.19), 5173411457122 (Figure 15.21), 5173420185489 (Figure 15.22), 5116980473325 (Figure 15.23), 5111031397060 (Figure 15.25), 5117001472208 (Figure 15.27), 5117000232386 (Figure 15.30), 5117010323261 (Figure 15.31), 5173431252032 (Figure 15.32), 5116990868546 (Figure 15.33). Open access under a creative commons license (Figure 15.34), 5117011454002 (Figure 15.35), 5117011091903 (Figure 15.36), 5117011091903 (Figure 15.37), 5173440423140 (Figure 15.38), 5173441124484 (Figures 15.39 and 15.40), 5117021254206 (Figure 15.41), 5173441509877 (Figure 15.42).

REFERENCES

1. Cagan, Suleyman Cinar, and Berat Baris Buldum. "Variants of high speed machining." In *High Speed Machining*, pp. 283–302. Academic Press, Cambridge, Massachusetts, United States, 2020.
2. Zheng, Jun, Wang Zheng, Ankai Chen, Jinkang Yao, Yicheng Ren, Chen Zhou, Jian Wu, et al. "Sustainability of unconventional machining industry considering impact factors and reduction methods of energy consumption: A review and analysis." *Science of the Total Environment* 722 (2020): 137897.
3. Rajmohan, T., V. V. Kalyan Chakravarthy, A. Nandakumar, and S. D. Satish Kumar. "Eco friendly machining processes for sustainability-review." *IOP Conference Series*: *Materials Science and Engineering* 954, no. 1, p. 012044. IOP Publishing, 2020.
4. Dixit, Uday S., D. K. Sarma, and J. Paulo Davim. *Environmentally Friendly Machining*. Springer Science & Business Media, Verlag, London, 2012.
5. Sutherland, John W., and Kenneth L. Gunter. "Environmental attributes of manufacturing processes." In *Handbook of Environmentally Conscious Manufacturing*, pp. 293–316. Springer, Boston, MA, 2001.
6. Peng, Tao, and Xun Xu. "Energy consumption evaluation for sustainable manufacturing: A feature-based approach." In *Proceeding of the 11th World Congress on Intelligent Control and Automation*, pp. 2310–2315. IEEE, 2014.
7. Nambiar, Arun N. "Challenges in sustainable manufacturing." In *Proceedings of the 2010 International Conference on Industrial Engineering and Operations Management*, Dhaka, Bangladesh, pp. 9–10, 2010.
8. Kopac, J. "Achievements of sustainable manufacturing by machining." *Journal of Achievements in Materials and Manufacturing Engineering* 34, no. 2 (2009): 180–187.
9. Duflou, Joost R., John W. Sutherland, David Dornfeld, Christoph Herrmann, Jack Jeswiet, Sami Kara, Michael Hauschild, and Karel Kellens. "Towards energy and resource efficient manufacturing: A processes and systems approach." *CIRP Annals* 61, no. 2 (2012): 587–609.
10. Kara, Sami, and Wen Li. "Unit process energy consumption models for material removal processes." *CIRP Annals* 60, no. 1 (2011): 37–40.
11. Jain, Vijay Kumar. *Advanced Machining Processes*. Allied Publishers, New Delhi, India, 2009.

12. Dixit, Nitin, Varun Sharma, and Pradeep Kumar. "Research trends in abrasive flow machining: A systematic review." *Journal of Manufacturing Processes* 64 (2021): 1434–1461.
13. Renu, Kaviyarasi, Rituraj Chakraborty, Myakala Haritha, Koti Rajeshwari, Ademola C. Famurewa, Harishkumar Madhyastha, Vellingiri Balachandar, Alex George, and V. G. Abilash. "Molecular mechanism of heavy metals (Lead, Chromium, Arsenic, Mercury, Nickel and Cadmium) induced hepatotoxicity–A review." *Chemosphere* (2021): 129735.
14. Samiee, Fateme, Aliasghar Vahidinia, Masoumeh Taravati Javad, and Mostafa Leili. "Exposure to heavy metals released to the environment through breastfeeding: A probabilistic risk estimation." *Science of the Total Environment* 650 (2019): 3075–3083.
15. Yao, Wu, Peiyan Yang, Yuanmeng Qi, Luheng Jin, Ahui Zhao, Mingcui Ding, Di Wang, Yi Ping Li, and Changfu Hao. "Transcriptome analysis reveals a protective role of liver X receptor alpha against silica particle-induced experimental silicosis." *Science of the Total Environment* 747 (2020): 141531.
16. El-Hofyand, H., and H. Youssef. "Environmental hazards of nontraditional machining." In *Proceedings of the 4th International Conference on Energy & Environment*, pp. 474–455, 2009.
17. Di Milia, Lee, Michael H. Smolensky, Giovanni Costa, Heidi D. Howarth, Maurice M. Ohayon, and Pierre Philip. "Demographic factors, fatigue, and driving accidents: An examination of the published literature." *Accident Analysis & Prevention* 43, no. 2 (2011): 516–532.
18. Kıvrak, Elfide Gizem, Kıymet Kübra Yurt, Arife Ahsen Kaplan, Işınsu Alkan, and Gamze Altun. "Effects of electromagnetic fields exposure on the antioxidant defense system." *Journal of Microscopy and Ultrastructure* 5, no. 4 (2017): 167–176.
19. Kumar, Janender, Amrinder Pal Singh, Anurag Thakur, and Munish Mehta. "Ultrasonic machining process: A review." *Advances in Nonconventional Machining Processes* (2020): 1.
20. Swain, Santosh Kumar, Saumyadarshan Nayak, Jayprakash Russel Ravan, and Mahesh Chandra Sahu. "Tinnitus and its current treatment–Still an enigma in medicine." *Journal of the Formosan Medical Association* 115, no. 3 (2016): 139–144.
21. Mughal, Khurram Hameed, Muhammad Asif Mahmood Qureshi, and Syed Farhan Raza. "Novel ultrasonic horn design for machining advanced brittle composites: A step forward towards green and sustainable manufacturing." *Environmental Technology & Innovation* (2021): 101652.
22. Momber, Andreas W., and Radovan Kovacevic. *Principles of Abrasive Water Jet Machining.* Springer Science & Business Media, 2012.
23. Johnson, Matthew. "Sustainable design analysis of waterjet cutting through exergy/energy and LCA analysis." (2009). *Graduate Theses and Dissertations.* https://digitalcommons.usf.edu/etd/2031
24. Schweitzer, Linda, and James Noblet. "Water contamination and pollution." In *Green Chemistry*, pp. 261–290. Elsevier, Amsterdam, Netherlands, 2018.
25. Sharma, R. K., M. Yadav, and R. Gupta. "Water quality and sustainability in India: Challenges and opportunities." In *Chemistry and Water*, pp. 183–205. Elsevier, Cambridge, United States, 2017.
26. Islam, Md Saiful, Md Kawser Ahmed, Mohammad Raknuzzaman, Md Habibullah-Al-Mamun, and Muhammad Kamrul Islam. "Heavy metal pollution in surface water and sediment: A preliminary assessment of an urban river in a developing country." *Ecological Indicators* 48 (2015): 282–291.
27. Momber, Andreas W., and Radovan Kovacevic. *Principles of Abrasive Water Jet Machining.* Springer Science & Business Media, Verlag, London, 2012.

28. Selvan, Chithirai Pon, Divya Midhunchakkaravarthy, Swaroop Ramaswamy Pillai, and Sahith Reddy Madara. "Investigation on abrasive waterjet machining conditions of mild steel using artificial neural network." *Materials Today: Proceedings* 19 (2019): 233–239.

29. Radford, John Dennis. *Production Engineering Technology.* Macmillan International Higher Education, Basingstoke, London, 1980.

30. Ardon-Dryer, Karin, Caroline Mock, Jose Reyes, and Galit Lahav. "The effect of dust storm particles on single human lung cancer cells." *Environmental Research* 181 (2020): 108891.

31. Tjoe-Nij, Evelyn, Christophe Rochin, Nathalie Berne, Alessandro Sassi, and Antoine Leplay. "Chemical risk assessment screening tool of a global chemical company." *Safety and Health at Work* 9, no. 1 (2018): 84–94.

32. Sabarinathan, P., V. E. Annamalai, and K. Rajkumar. "Sustainable application of grinding wheel waste as abrasive for abrasive water jet machining process." *Journal of Cleaner Production* 261 (2020): 121225.

33. Praphawatvet, Tuangrat, Jay I. Peters, and Robert O. Williams III. "Inhaled nanoparticles—An updated review." *International Journal of Pharmaceutics* (2020): 119671.

34. Kim, Dasom, Zi Chen, Lin-Fu Zhou, and Shou-Xiong Huang. "Air pollutants and early origins of respiratory diseases." *Chronic Diseases and Translational Medicine* 4, no. 2 (2018): 75–94.

35. Niemann, Bernd, Susanne Rohrbach, Mark R. Miller, David E. Newby, Valentin Fuster, and Jason C. Kovacic. "Oxidative stress and cardiovascular risk: Obesity, diabetes, smoking, and pollution: Part 3 of a 3-part series." *Journal of the American College of Cardiology* 70, no. 2 (2017): 230–251.

36. Bhattacharyya, B., and B. Doloi. "Machining processes utilizing thermal energy." *Modern Machining Technology* (2020): 161–363.

37. Sobotová, Lýdia, and Katarína Lukáčová. "Air pollution in plasma workplace." *International Multidisciplinary Scientific Geoconference: SGEM* 2 (2016): 211–218.

38. Barro-Torres, Santiago, Tiago M. Fernández-Caramés, Héctor J. Pérez-Iglesias, and Carlos J. Escudero. "Real-time personal protective equipment monitoring system." *Computer Communications* 36, no. 1 (2012): 42–50.

39. Schaeffer, Ronald. *Fundamentals of Laser Micromachining.* CRC Press, Boca Raton, Florida, USA, 2012.

40. Dubey, Avanish Kumar, and Vinod Yadava. "Laser beam machining—A review." *International Journal of Machine Tools and Manufacture* 48, no. 6 (2008): 609–628.

41. Behera, Rasmi Ranjan, and M. Ravi Sankar. "State of the art on under liquid laser beam machining." *Materials Today: Proceedings* 2, nos. 4–5 (2015): 1731–1740.

42. Rajurkar, K. P., H. Hadidi, J. Pariti, and G. C. Reddy. "Review of sustainability issues in non-traditional machining processes." *Procedia Manufacturing* 7 (2017): 714–720.

43. Yiu, Glenn. "Retinal laser injury." In *Handbook of Pediatric Retinal OCT and the Eye-Brain Connection*, pp. 210–212. Elsevier, Philadelphia, USA, 2020.

44. Singh, Nishant K., Yashvir Singh, Abhishek Sharma, Amneesh Singla, and Prateek Negi. "An environmental-friendly electrical discharge machining using different sustainable techniques: A review." *Advances in Materials and Processing Technologies* (2020): 1–30.

45. Bhattacharyya, Bijoy. *Electrochemical Micromachining for Nanofabrication, MEMS and Nanotechnology.* William Andrew, Amsterdam, Netherlands, 2015.

46. Taylor, A. F. "Electron beam welding." *Annals of Occupational Hygiene* 7, no. 3 (1964): 241–246.

47. Moser, Harry. "Growth industries rely on EDM." *Manufacturing Engineering* 127, no. 5 (2001): 62.

48. Knight, W. A., and G. Boothroyd. *Fundamentals of Metal Machining and Machine Tools*. CRC Press, Boca Raton, Florida, USA, (2005), pp. 525–531.

49. Muthuramalingam, T. "Effect of diluted dielectric medium on spark energy in green EDM process using TGRA approach." *Journal of Cleaner Production* 238 (2019): 117894.

50. Valaki, Janak B., Pravin P. Rathod, and Ajay M. Sidpara. "Sustainability issues in electric discharge machining." In *Innovations in manufacturing for sustainability*, pp. 53–75. Springer, Cham, 2019.

51. Yadav, Avinash, Yashvir Singh, Satyendra Singh, and Prateek Negi. "Sustainability of vegetable oil based bio-diesel as dielectric fluid during EDM process–A review." *Materials Today: Proceedings* (2021).

52. Gamage, Janaka R., Anjali K. M. DeSilva, Colin Harrison, and David Harrison. "Ascertaining life cycle inventory data for electrical discharge machining." *Procedia CIRP* 41 (2016): 908–913.

53. Yeo, S. H., H. C. Tan, and A. K. New. "Assessment of waste streams in electric-discharge machining for environmental impact analysis." *Proceedings of the Institution of Mechanical Engineers, Part B: Journal of Engineering Manufacture* 212, no. 5 (1998): 393–401.

54. Moreno-Ríos, Andrea L., Lesly Tejeda-Benitez, and Ciro Bustillo-Lecompte. "Sources, characteristics, toxicity, and control of ultrafine particles: An overview." *Geoscience Frontiers* (2021): 101147.

55. Sivapirakasam, S. P., Jose Mathew, and M. Surianarayanan. "Constituent analysis of aerosol generated from die sinking electrical discharge machining process." *Process Safety and Environmental Protection* 89, no. 2 (2011): 141–150.

56. Jawahir, I. S., Keith E. Rouch, O. W. Dillon, L. Holloway, and A. Hall. "Design for sustainability (DFS): New challenges in developing and implementing a curriculum for next generation design and manufacturing engineers." *The International Journal of Engineering Education* 23, no. 6 (2007): 1053–1064.

57. Brouwer, Derk H., Jose H. J. Gijsbers, and Marc W. M. Lurvink. "Personal exposure to ultrafine particles in the workplace: Exploring sampling techniques and strategies." *Annals of Occupational Hygiene* 48, no. 5 (2004): 439–453.

58. Ross, Andrew S., Kay Teschke, Michael Brauer, and Susan M. Kennedy. "Determinants of exposure to metalworking fluid aerosol in small machine shops." *Annals of Occupational Hygiene* 48, no. 5 (2004): 383–391.

59. Jeswani, M. L. "Electrical discharge machining in distilled water." *Wear* 72, no. 1 (1981): 81–88.

60. Chen, Rui, Bin Hu, Ying Liu, Jianxun Xu, Guosheng Yang, Diandou Xu, and Chunying Chen. "Beyond PM2. 5: The role of ultrafine particles on adverse health effects of air pollution." *Biochimicaet Biophysica Acta (BBA)-General Subjects* 1860, no. 12 (2016): 2844–2855.

61. Evertz, Sven, Wolfgang Dott, and Adolf Eisentraeger. "Electrical discharge machining: Occupational hygienic characterization using emission-based monitoring." *International Journal of Hygiene and Environmental Health* 209, no. 5 (2006): 423–434.

62. Koenig, W., and L. Joerres. "Aqueous solutions of organic compounds as dielectrics for EDM sinking." *CIRP Annals* 36, no. 1 (1987): 105–109.

63. Cho, Margaret Hyunjoo. "Environmental constituents of electrical discharge machining." PhD dissertation, Massachusetts Institute of Technology, 2004.

64. Yu, Zhan Bo, Takahashi Jun, and Kunieda Masanori. "Dry electrical discharge machining of cemented carbide." *Journal of Materials Processing Technology* 149, nos. 1–3 (2004): 353–357.

65. Singh, Nishant K., Yashvir Singh, Abhishek Sharma, Amneesh Singla, and Prateek Negi. "An environmental-friendly electrical discharge machining using different sustainable techniques: A review." *Advances in Materials and Processing Technologies* (2020): 1–30.

66. Tao, Jia, Albert J. Shih, and Jun Ni. "Experimental study of the dry and near-dry electrical discharge milling processes." *Journal of Manufacturing Science and Engineering* 130, no. 1 (2008).

67. Zhang, Yanzhen, Yonghong Liu, Yang Shen, Renjie Ji, Xiaolong Wang, and Zhen Li. "Die-sinking electrical discharge machining with oxygen-mixed water-in-oil emulsion working fluid." *Proceedings of the Institution of Mechanical Engineers, Part B: Journal of Engineering Manufacture* 227, no. 1 (2013): 109–118.

68. Dong, Hang, Yonghong Liu, Ming Li, Yu Zhou, Tong Liu, Dege Li, Qiang Sun, Yanzhen Zhang, and Renjie Ji. "Sustainable electrical discharge machining using water in oil nanoemulsion." *Journal of Manufacturing Processes* 46 (2019): 118–128.

69. Kellens, Karel, Wim Dewulf, and Joost R. Duflou. "Preliminary environmental assessment of electrical discharge machining." In *Globalized Solutions for Sustainability in Manufacturing*, pp. 377–382. Springer, Berlin, Heidelberg, 2011.

70. Valaki, Janak B., Pravin P. Rathod, and C. D. Sankhavara. "Investigations on technical feasibility of Jatrophacurcas oil based bio dielectric fluid for sustainable electric discharge machining (EDM)." *Journal of Manufacturing Processes* 22 (2016): 151–160.

71. Valaki, Janak B., Pravin P. Rathod, Bharat C. Khatri, and Jignesh R. Vaghela. "Investigations on palm oil based biodielectric fluid for sustainable electric discharge machining." In *Proceedings of the International Conference on Advances in Materials and Manufacturing (ICAMM-2016)*, Bangkok, Thailand, pp. 29–30, 2016.

72. Mali, Harlal Singh, and Nitesh Kumar. "Investigating feasibility of waste vegetable oil for sustainable EDM." In *Proceedings of All India Manufacturing Technology Design and Research Conference*, pp. 405–410, 2016.

73. Ng, Pei Shan, S. A. Kong, and S. H. Yeo. "Investigation of biodiesel dielectric in sustainable electrical discharge machining." *The International Journal of Advanced Manufacturing Technology* 90, no. 9 (2017): 2549–2556.

74. Saha, Sourabh K., and S. K. Choudhury. "Experimental investigation and empirical modeling of the dry electric discharge machining process." *International Journal of Machine Tools and Manufacture* 49, nos. 3–4 (2009): 297–308.

75. Sharma, Rajeev, Binit Kumar Jha, Vipin Pahuja, and Sagar Sharma. "Role of environmental friendly machining on machinability." *Materials Today: Proceedings* (2021).

76. Singh, Nishant K., Yashvir Singh, Abhishek Sharma, Amneesh Singla, and Prateek Negi. "An environmental-friendly electrical discharge machining using different sustainable techniques: A review." *Advances in Materials and Processing Technologies* (2020): 1–30.

77. Bommeli, B. "Study of the harmful emanations resulting from the machining by electro-erosion." In *Proceedings of the Seventh International Symposium on Electro Machining (ISEM VII)*, pp. 469–478, 1983.

78. Boubekri, Nourredine, and Vasim Shaikh. "Machining using minimum quantity lubrication: A technology for sustainability." *International Journal of Applied Science and Technology* 2, no. 1 (2012).

79. Goh, C. L., and S. F. Ho. "Contact dermatitis from dielectric fluids in electrodischarge machining." *Contact Dermatitis* 28, no. 3 (1993): 134–138.

80. Jose, Mathew, S. P. Sivapirakasam, and M. Surianarayanan. "Analysis of aerosol emission and hazard evaluation of electrical discharge machining (EDM) process." *Industrial Health* 48, no. 4 (2010): 478–486.

81. Wang, Xiangzhi, Zhidong Liu, Rongyuan Xue, Zongjun Tian, and Yinhui Huang. "Research on the influence of dielectric characteristics on the EDM of titanium alloy." *The International Journal of Advanced Manufacturing Technology* 72, nos. 5–8 (2014): 979–987.

82. Das, Shirsendu, Swarup Paul, and Biswanath Doloi. "Investigation of the machining performance of neem oil as a dielectric medium of EDM: A sustainable approach." *IOP* Conference *Series: Materials Science and Engineering*, vol. 653, no. 1, p. 012017. IOP Publishing, 2019.

83. Gupta, Kapil, and Munish Kumar Gupta. "Developments in nonconventional machining for sustainable production: A state-of-the-art review." *Proceedings of the Institution of Mechanical Engineers, Part C: Journal of Mechanical Engineering Science* 233, no. 12 (2019): 4213–4232.

84. Wu, Xinlei, Yonghong Liu, Xuexin Zhang, Hang Dong, Chao Zheng, Fan Zhang, Qiang Sun, Hui Jin, and Renjie Ji. "Sustainable and high-efficiency green electrical discharge machining milling method." *Journal of Cleaner Production* 274 (2020): 123040.

85. Rajurkar, Kamlakar P., M. M. Sundaram, and A. P. Malshe. "Review of electrochemical and electrodischarge machining." *Procedia CIRP* 6 (2013): 13–26.

86. Bannard, J. "Electrochemical machining." *Journal of Applied Electrochemistry* 7, no. 1 (1977): 1–29.

87. Davydov, A. D., T. B. Kabanova, and V. M. Volgin. "Electrochemical machining of titanium: Review." *Russian Journal of Electrochemistry* 53, no. 9 (2017): 941–965.

88. Kendall, Thomas, Paulo Bartolo, David Gillen, and Carl Diver. "A review of physical experimental research in jet electrochemical machining." *The International Journal of Advanced Manufacturing Technology* 105, no. 1 (2019): 651–667.

89. Rajurkar, Kamlakar P., Di Zhu, J. A. McGeough, J. Kozak, and A. De Silva. "New developments in electro-chemical machining." *CIRP Annals* 48, no. 2 (1999): 567–579.

90. Thanigaivelan, R., R. M. Arunachalam, B. Karthikeyan, and P. Loganathan. "Electrochemical micromachining of stainless steel with acidified sodium nitrate electrolyte." *Procedia CIRP* 6 (2013): 351–355.

91. Ryu, Shi Hyoung. "Micro fabrication by electrochemical process in citric acid electrolyte." *Journal of Materials Processing Technology* 209, no. 6 (2009): 2831–2837.

92. Zhengyang, X. U., and W. A. N. G. Yudi. "Electrochemical machining of complex components of aero-engines: Developments, trends, and technological advances." *Chinese Journal of Aeronautics* 34, no. 2 (2021): 28–53.

93. Perveen, Asma, and Samet Akar. "Electrochemical discharge machining: Trends and development." In *Micro Electro-Fabrication*, pp. 317–338. Elsevier, 2021.

94. Tönshoff, H. K., R. Egger, and F. Klocke. "Environmental and safety aspects of electrophysical and electrochemical processes." *CIRP Annals* 45, no. 2 (1996): 553–568.

95. Fang, Qi, Qun Zhao, Xiaolong Chai, Yingjie Li, and Senlin Tian. "Interaction of industrial smelting soot particles with pulmonary surfactant: Pulmonary toxicity of heavy metal-rich particles." *Chemosphere* 246 (2020): 125702.

96. Petersen, Haley A., Tessa H. T. Myren, Shea J. O'Sullivan, and Oana R. Luca. "Electrochemical methods for materials recycling." *Materials Advances* 2, no. 4 (2021): 1113–1138.

97. Rajurkar, K. P., H. Hadidi, J. Pariti, and G. C. Reddy. "Review of sustainability issues in non-traditional machining processes." *Procedia Manufacturing* 7 (2017): 714–720.

98. Mortazavi, Mina, and Atanas Ivanov. "Sustainable µECM machining process: Indicators and assessment." *Journal of Cleaner Production* 235 (2019): 1580–1590.

99. Yang, Ye, Wataru Natsu, and Wansheng Zhao. "Realization of eco-friendly electro-chemical micromachining using mineral water as an electrolyte." *Precision Engineering* 35, no. 2 (2011): 204–213.

100. Chen, Xiaolei, Ningsong Qu, and Zhibao Hou. "Electrochemical micromachining of micro-dimple arrays on the surface of Ti-6Al-4V with $NaNO_3$ electrolyte." *The International Journal of Advanced Manufacturing Technology* 88, nos. 1–4 (2017): 565–574.

101. Haider, J., and M. S. J. Hashmi. "8.02—Health and environmental impacts in metal machining processes." *Comprehensive Materials Processing* 8 (2014): 7–33.

102. Rajurkar, Kamlakar P., Di Zhu, J. A. McGeough, J. Kozak, and A. De Silva. "New developments in electro-chemical machining." *CIRP Annals* 48, no. 2 (1999): 567–579.

103. Bhattacharyya, Bijoy, and Biswanath Doloi. "Chapter 5: Machining processes utilizing chemical and electrochemical energy." In *Modern Machining Technology: Advanced, Hybrid, Micro Machining and Super Finishing Technology*, pp. 365–460. Academic Press, London, United Kingdom, 2020.

104. Sun, Qi, Lin Jiang, Mi Li, and Jinhua Sun. "Assessment on thermal hazards of reactive chemicals in industry: State of the art and perspectives." *Progress in Energy and Combustion Science* 78 (2020): 100832.

105. Rajurkar, K. P., G. Levy, A. Malshe, M. M. Sundaram, J. McGeough, X. Hu, R. Resnick, and A. DeSilva. "Micro and nano machining by electro-physical and chemical processes." *CIRP Annals* 55, no. 2 (2006): 643–666.

106. Yoon, Hae-Sung, Jong-Seol Moon, Minh-Quan Pham, Gyu-Bong Lee, and Sung-Hoon Ahn. "Control of machining parameters for energy and cost savings in micro-scale drilling of PCBs." *Journal of Cleaner Production* 54 (2013): 41–48.

107. Kordonowy, David Nathaniel. "A power assessment of machining tools." PhD dissertation, Massachusetts Institute of Technology, 2002.

Index

For Product Safety Concerns and Information please contact our EU
representative GPSR@taylorandfrancis.com
Taylor & Francis Verlag GmbH, Kaufingerstraße 24, 80331 München, Germany

www.ingramcontent.com/pod-product-compliance
Lightning Source LLC
Chambersburg PA
CBHW060423220326
41598CB00021BA/2266